# 影印版说明

本书是综合性手册，分四部分：金属注射成形过程、金属注射成形质量问题、特殊金属注射成形过程、具体材料的金属注射成形。主要内容包括粉末特征、复合制造、模具设计、成形优化、脱脂和烧结；以及质量问题，如原料特性、建模与仿真、常见缺陷和碳含量的控制。

本手册包含关于金属注射成形技术、质量、市场等的最新研究成果和应用实例，对金属粉末注射成形及其应用提供了权威指南，适合从事精密零件生产与应用、粉末冶金等相关领域的科研人员、技术人员使用，也可供高等院校相关专业的师生参考。

Donald F. Heaney 宾夕法尼亚州立大学工程科学和力学系副教授、Center for Innovative Sintered Products 核心成员，美国 Advanced Powder Products 有限公司董事长和执行总裁。

**材料科学与工程图书工作室**
联系电话 0451-86412421
　　　　　0451-86414559
邮　　箱 yh_bj@aliyun.com
　　　　　xuyaying81823@gmail.com
　　　　　zhxh6414559@aliyun.com

WOODHEAD PUBLISHING IN MATERIALS

影印版

# 金属注射成形手册

# Handbook of metal injection molding

Edited by Donald F. Heaney

哈尔滨工业大学出版社
HARBIN INSTITUTE OF TECHNOLOGY PRESS

黑版贸审字08-2017-080号

Handbook of metal injection molding
Donald F. Heaney
ISBN: 978-0-85709-066-9
Copyright ©2012 by Elsevier. All rights reserved.
Authorized English language reprint edition published by the Proprietor.
Copyright © 2017 by Elsevier (Singapore) Pte Ltd. All rights reserved.

Elsevier (Singapore) Pte Ltd.
3 Killiney Road
#08-01 Winsland House I
Singapore 239519
Tel: (65) 6349-0200
Fax: (65) 6733-1817

First Published 2017

Printed in China by Harbin Institute of Technology Press under special arrangement with Elsevier (Singapore) Pte Ltd. This edition is authorized for sale in China only, excluding Hong Kong SAR, Macao SAR and Taiwan. Unauthorized export of this edition is a violation of the Copyright Act. Violation of this Law is subject to Civil and Criminal Penalties.

本书英文影印版由Elsevier (Singapore) Pte Ltd.授权哈尔滨工业大学出版社有限公司仅限在中华人民共和国境内（不包括香港特别行政区、澳门特别行政区以及台湾地区）出版及销售。未经许可之出口，视为违反著作权法，将受法律制裁。

本书封底贴有Elsevier防伪标签，无标签者不得销售。

## 图书在版编目（CIP）数据

金属注射成形手册=Handbook of metal injection molding：英文 /（美）唐纳德·F.西尼（Donald F. Heaney）主编.—影印本.—哈尔滨：哈尔滨工业大学出版社，2017.10
ISBN 978-7-5603-6394-3

Ⅰ.①金… Ⅱ.①唐… Ⅲ.①粉末注射成形-技术手册-英文 Ⅳ.①TF124.39-62

中国版本图书馆CIP数据核字（2017）第001592号

| | |
|---|---|
| 责任编辑 | 张秀华　杨　桦　许雅莹 |
| 出版发行 | 哈尔滨工业大学出版社 |
| 社　　址 | 哈尔滨市南岗区复华四道街10号　邮编150006 |
| 传　　真 | 0451-86414749 |
| 网　　址 | http://hitpress.hit.edu.cn |
| 印　　刷 | 哈尔滨市石桥印务有限公司 |
| 开　　本 | 660mm×980mm　1/16　印张 37.75　插页 2 |
| 版　　次 | 2017年10月第1版　2017年10月第1次印刷 |
| 书　　号 | ISBN 978-7-5603-6394-3 |
| 定　　价 | 360.00元 |

（如因印刷质量问题影响阅读，我社负责调换）

# Handbook of metal injection molding

Edited by
Donald F. Heaney

WP
WOODHEAD
PUBLISHING

Oxford   Cambridge   Philadelphia   New Delhi

# Contents

|  | Contributor contact details | xiii |
|---|---|---|
|  | Preface | xvii |
| **1** | **Metal powder injection molding (MIM): key trends and markets** | **1** |
|  | R. M. GERMAN, San Diego State University, USA |  |
| 1.1 | Introduction and background | 1 |
| 1.2 | History of success | 2 |
| 1.3 | Industry structure | 4 |
| 1.4 | Statistical highlights | 6 |
| 1.5 | Industry shifts | 9 |
| 1.6 | Sales situation | 10 |
| 1.7 | Market statistics | 12 |
| 1.8 | Metal powder injection molding market by region | 13 |
| 1.9 | Metal powder injection molding market by application | 14 |
| 1.10 | Market opportunities | 15 |
| 1.11 | Production sophistication | 21 |
| 1.12 | Conclusion | 23 |
| 1.13 | Sources of further information | 23 |
| **Part I** | **Processing** | **27** |
| **2** | **Designing for metal injection molding (MIM)** | **29** |
|  | D. F. HEANEY, Advanced Powder Products, Inc., USA |  |
| 2.1 | Introduction | 29 |
| 2.2 | Available materials and properties | 31 |
| 2.3 | Dimensional capability | 35 |
| 2.4 | Surface finish | 35 |
| 2.5 | Tooling artifacts | 35 |
| 2.6 | Design considerations | 40 |

Contents vii

| | | |
|---|---|---|
| 2.7 | Sources of further information | 49 |
| **3** | **Powders for metal injection molding (MIM)** | **50** |
| | D. F. HEANEY, Advanced Powder Products, Inc., USA | |
| 3.1 | Introduction | 50 |
| 3.2 | Ideal MIM powder characteristics | 51 |
| 3.3 | Characterizing MIM powders | 55 |
| 3.4 | Different MIM powder fabrication techniques | 57 |
| 3.5 | Different alloying methods | 61 |
| 3.6 | References | 62 |
| **4** | **Powder binder formulation and compound manufacture in metal injection molding (MIM)** | **64** |
| | R. K. ENNETI, Global Tungsten and Powders, USA and V. P. ONBATTUVELLI and S. V. ATRE, Oregon State University, USA | |
| 4.1 | Introduction: the role of binders | 64 |
| 4.2 | Binder chemistry and constituents | 66 |
| 4.3 | Binder properties and effects on feedstock | 70 |
| 4.4 | Mixing technologies | 84 |
| 4.5 | Case studies: lab scale and commercial formulations | 88 |
| 4.6 | References | 89 |
| **5** | **Tooling for metal injection molding (MIM)** | **93** |
| | G. SCHLIEPER, Gammatec Engineering GmbH, Germany | |
| 5.1 | Introduction | 93 |
| 5.2 | General design and function of injection molding machines | 94 |
| 5.3 | Elements of the tool set | 96 |
| 5.4 | Tool design options | 98 |
| 5.5 | Special features and instrumentation | 104 |
| 5.6 | Supporting software and economic aspects | 106 |
| 5.7 | Sources of further information | 108 |
| **6** | **Molding of components in metal injection molding (MIM)** | **109** |
| | D. F. HEANEY, Advanced Powder Products, Inc., USA and C. D. GREENE, Treemen Industries, Inc., USA | |
| 6.1 | Introduction | 109 |
| 6.2 | Injection molding equipment | 110 |
| 6.3 | Auxiliary equipment | 115 |
| 6.4 | Injection molding process | 116 |
| 6.5 | Common defects in MIM | 129 |

| | | |
|---|---|---|
| 6.6 | References | 131 |
| **7** | **Debinding and sintering of metal injection molding (MIM) components** | **133** |
| | S. BANERJEE, DSH Technologies LLC, USA and C. J. JOENS, Elnik Systems LLC, USA | |
| 7.1 | Introduction | 133 |
| 7.2 | Primary debinding | 136 |
| 7.3 | Secondary debinding | 144 |
| 7.4 | Sintering | 147 |
| 7.5 | MIM materials | 161 |
| 7.6 | Settering | 167 |
| 7.7 | MIM furnaces | 169 |
| 7.8 | Furnace profiles | 176 |
| 7.9 | Summary | 176 |
| 7.10 | Acknowledgements | 178 |
| 7.11 | References | 178 |
| **Part II** | **Quality issues** | **181** |
| **8** | **Characterization of feedstock in metal injection molding (MIM)** | **183** |
| | H. LOBO, DatapointLabs, USA | |
| 8.1 | Introduction | 183 |
| 8.2 | Rheology | 186 |
| 8.3 | Thermal analysis | 190 |
| 8.4 | Thermal conductivity | 193 |
| 8.5 | Pressure–volume–temperature (PVT) | 194 |
| 8.6 | Conclusions | 195 |
| 8.7 | Acknowledgments | 196 |
| 8.8 | References | 196 |
| **9** | **Modeling and simulation of metal injection molding (MIM)** | **197** |
| | T. G. KANG, Korea Aerospace University, Korea, S. AHN, Pusan National University, Korea, S. H. CHUNG, Hyundai Steel Co., Korea, S. T. CHUNG, CetaTech Inc., Korea, Y. S. KWON, CetaTech, Inc., Korea, S. J. PARK, POSTECH, Korea and R. M. GERMAN, San Diego State University, USA | |
| 9.1 | Modeling and simulation of the mixing process | 197 |
| 9.2 | Modeling and simulation of the injection molding process | 203 |

| 9.3 | Modeling and simulation of the thermal debinding process | 215 |
| 9.4 | Modeling and simulation of the sintering process | 224 |
| 9.5 | Conclusion | 230 |
| 9.6 | References | 231 |

| 10 | **Common defects in metal injection molding (MIM)** | **235** |

K. S. HWANG, National Taiwan University, Taiwan, R.O.C.

| 10.1 | Introduction | 235 |
| 10.2 | Feedstock | 236 |
| 10.3 | Molding | 238 |
| 10.4 | Debinding | 243 |
| 10.5 | Sintering | 250 |
| 10.6 | Conclusion | 251 |
| 10.7 | References | 252 |

| 11 | **Qualification of metal injection molding (MIM)** | **254** |

D. F. HEANEY, Advanced Powder Products, Inc., USA

| 11.1 | Introduction | 254 |
| 11.2 | The metal injection molding process | 255 |
| 11.3 | Product qualification method | 255 |
| 11.4 | MIM prototype methodology | 257 |
| 11.5 | Process control | 258 |
| 11.6 | Understanding of control parameters | 260 |
| 11.7 | Conclusion | 263 |
| 11.8 | Sources of further information | 263 |

| 12 | **Control of carbon content in metal injection molding (MIM)** | **265** |

G. HERRANZ, Universidad de Castilla-La Mancha, Spain

| 12.1 | Introduction: the importance of carbon control | 265 |
| 12.2 | Methods of controlling carbon, binder elimination and process parameters affecting carbon control | 267 |
| 12.3 | Control of carbon in particular materials | 276 |
| 12.4 | Material properties affected by carbon content | 297 |
| 12.5 | References | 297 |

| **Part III** | **Special metal injection molding processes** | **305** |

| 13 | **Micro metal injection molding (MicroMIM)** | **307** |

V. PIOTTER, Karlsruhe Institute of Technology (KIT), Germany

Contents

| | | |
|---|---|---|
| 13.1 | Introduction | 307 |
| 13.2 | Potential of powder injection molding for micro-technology | 308 |
| 13.3 | Micro-manufacturing methods for tool making | 309 |
| 13.4 | Powder injection molding of micro components | 313 |
| 13.5 | Multi-component micro powder injection molding | 325 |
| 13.6 | Simulation of MicroMIM | 328 |
| 13.7 | Conclusion and future trends | 330 |
| 13.8 | Sources of further information and advice | 331 |
| 13.9 | References | 332 |

**14  Two-material/two-color powder metal injection molding (2C-PIM)**     **338**

P. SURI, Heraeus Materials Technology LLC, USA

| | | |
|---|---|---|
| 14.1 | Introduction | 338 |
| 14.2 | Injection molding technology | 338 |
| 14.3 | Debinding and sintering | 341 |
| 14.4 | 2C-PIM products | 344 |
| 14.5 | Future trends | 346 |
| 14.6 | References | 347 |

**15  Powder space holder metal injection molding (PSH-MIM) of micro-porous metals**     **349**

K. NISHIYABU, Kinki University, Japan

| | | |
|---|---|---|
| 15.1 | Introduction | 349 |
| 15.2 | Production methods for porous metals | 351 |
| 15.3 | Formation of micro-porous structures by the PSH method | 354 |
| 15.4 | Control of porous structure with the PSH method | 360 |
| 15.5 | Liquid infiltration properties of micro-porous metals produced by the PSH method | 369 |
| 15.6 | Dimensional accuracy of micro-porous MIM parts | 374 |
| 15.7 | Functionally graded structures of micro-porous metals | 379 |
| 15.8 | Conclusion | 388 |
| 15.9 | Acknowledgements | 388 |
| 15.10 | References | 389 |

**Part IV  Metal injection molding of specific materials**     **391**

**16  Metal injection molding (MIM) of stainless steels**     **393**

J. M. TORRALBA, Institute IMDEA Materials, Universidad Carlos III de Madrid, Spain

| | | |
|---|---|---|
| 16.1 | Introduction | 393 |
| 16.2 | Stainless steels in metal injection molding (MIM) | 396 |

| 16.3 | Applications of MIM stainless steels | 403 |
| 16.4 | Acknowledgements | 409 |
| 16.5 | References | 410 |

## 17 Metal injection molding (MIM) of titanium and titanium alloys — 415

T. EBEL, Helmholtz-Zentrum Geesthacht, Germany

| 17.1 | Introduction | 415 |
| 17.2 | Challenges of MIM of titanium | 416 |
| 17.3 | Basics of processing | 422 |
| 17.4 | Mechanical properties | 425 |
| 17.5 | Cost reduction | 432 |
| 17.6 | Special applications | 435 |
| 17.7 | Conclusion and future trends | 440 |
| 17.8 | Sources of further information | 441 |
| 17.9 | References | 441 |

## 18 Metal injection molding (MIM) of thermal management materials in microelectronics — 446

J. L. JOHNSON, ATI Firth Sterling, USA

| 18.1 | Introduction | 446 |
| 18.2 | Heat dissipation in microelectronics | 447 |
| 18.3 | Copper | 451 |
| 18.4 | Tungsten–copper | 461 |
| 18.5 | Molybdenum–copper | 474 |
| 18.6 | Conclusions | 482 |
| 18.7 | References | 482 |

## 19 Metal injection molding (MIM) of soft magnetic materials — 487

H. MIURA, Kyushu University, Japan

| 19.1 | Introduction | 487 |
| 19.2 | Fe–6.5Si | 489 |
| 19.3 | Fe–9.5Si–5.5Al | 497 |
| 19.4 | Fe–50Ni | 506 |
| 19.5 | Conclusion | 513 |
| 19.6 | References | 514 |

xii    Contents

| 20 | Metal injection molding (MIM) of high-speed tool steels | 516 |

N. S. MYERS, Kennametal Inc., USA and D. F. HEANEY, Advanced Powder Products, Inc., USA

| 20.1 | Introduction | 516 |
| 20.2 | Tool steel MIM processing | 517 |
| 20.3 | Mechanical properties | 523 |
| 20.4 | References | 525 |

| 21 | Metal injection molding (MIM) of heavy alloys, refractory metals, and hardmetals | 526 |

J. L. JOHNSON, ATI Firth Sterling, USA, D. F. HEANEY, Advanced Powder Products, Inc., USA and N. S. Myers, Kennametal Inc., USA

| 21.1 | Introduction | 526 |
| 21.2 | Applications | 527 |
| 21.3 | Feedstock formulation concerns | 529 |
| 21.4 | Heavy alloys | 534 |
| 21.5 | Refractory metals | 544 |
| 21.6 | Hardmetals | 554 |
| 21.7 | References | 560 |
|  | *Index* | *569* |

# Contributor contact details

(* = main contact)

## Editor and chapters 2, 3, 6, 11, 20 and 21

Donald F. Heaney
Advanced Powder Products, Inc.
301 Enterprise Drive
Philipsburg, PA, 16866
USA
E-mail: dfheaney@4-app.com

## Chapter 1

R. M. German
College of Engineering
San Diego State University
5500 Campanile Drive
San Diego, California 92182-1326
USA
E-mail: rgerman@mail.sdsu.edu

## Chapter 4

S. V. Atre* and V. P. Onbattuvelli
Oregon State University
Corvallis, OR 97331
USA
E-mail: Sundar.Atre@oregonstate.edu

R. K. Enneti
Global Tungsten and Powders
USA

## Chapter 5

G. Schlieper
Gammatec Engineering GmbH
Mermbacher Str. 28
D-42477 Radevormwald
Germany
E-mail: info@gammatec.com

## Chapter 6

Donald F. Heaney*
E-mail: dfheaney@4-app.com

Cody D. Greene
Treemen Industries, Inc.
USA

## Chapter 7

S. Banerjee*
DSH Technologies LLC
107 Commerce Road
Cedar Grove
NJ 07009
USA
E-mail: sbanerjee@dshtech.com

C. J. Joens
Elnik Systems LLC
107 Commerce Road
Cedar Grove
NJ 07009
USA
E-mail: cjoens@elnik.com

## Chapter 8

H. Lobo
DatapointLabs
95 Brown Road
Ithaca
New York
USA
E-mail: lobo@datapointlabs.com

## Chapter 9

T. G. Kang
School of Aerospace and
  Mechanical Engineering
Korea Aerospace University
100 Hanggongdae-gil
  Hwajeon-dong
Goyang-City
Gyeonggi-do 412-791
Korea

S. Ahn, Ph.D.
School of Mechanical Engineering
Pusan National University
Busandaehak-ro 63
Geumjeong-gu
Busan 609-735
Republic of Korea (South)

S. H. Chung
Hyundai Steel Co.
167-32
Kodae-Ri
Songak-Myeon

Dangjin-Gun
Chungnam 343-711
Korea

S. T. Chung and Y. S. Kwon
CetaTech Inc.
TIC 296-3
Seonjin-Ri
Sacheon-Si
Kyongnam 664-953
Korea

S. J. Park*
Department of Mechanical
  Engineering
POSTECH
San 31
Hyoja-Dong
Pohang
Kyungbuk 790-784
Korea
E-mail: sjpark87@postech.ac.kr

R. M. German
College of Engineering
San Diego State University
5500 Campanile Drive
San Diego
CA 92182
USA

## Chapter 10

K. S. Hwang
Department of Materials Science
  and Engineering
National Taiwan University
1, Sec. 4, Roosevelt Road
106, Taipei
Taiwan
R.O.C.
E-mail: kshwang@ntu.edu.tw

## Chapter 12

G. Herranz
Universidad de Castilla-La Mancha
UCLM, Metallic Materials Group
ETSI Industriales
Avda Camilo José Cela s/n. 13071
Ciudad Real
Spain
E-mail: gemma.herranz@uclm.es

## Chapter 13

V. Piotter
Karlsruhe Institute of Technology (KIT)
Institute for Applied Materials – Materials Process Technology
Hermann-von-Helmholtz Platz 1
76344 Eggenstein-Leopoldshafen
Germany
E-mail: volker.piotter@kit.edu

## Chapter 14

Pavan Suri
Senior Development Engineer
Heraeus Materials Technology North America LLC
280 N. Roosevelt Avenue
Chandler
Arizona 85226
USA
E-mail: Pavan.Suri@heraeus.com

## Chapter 15

K. Nishiyabu
Department of Mechanical Engineering
Faculty of Science and Engineering
Kinki University
3-4-1 Kowakae
Higashi-Osaka 577-8502
Japan
E-mail: nishiyabu@mech.kindai.ac.jp

## Chapter 16

J. M. Torralba
Professor of Materials Science and Engineering
Institute IMDEA Materials
Universidad Carlos III de Madrid
Av. Universidad 30
28911 Leganés
Spain
E-mail: josemanuel.torralba@imdea.org

## Chapter 17

T. Ebel
Department of Powder Technology
Institute of Materials Research
Helmholtz-Zentrum Geesthacht
Max-Planck-Straße 1
21502 Geesthacht
Germany
E-mail: thomas.ebel@hzg.de

## Chapter 18

J. L. Johnson
ATI Firth Sterling
7300 Highway 20 West
Huntsville
AL 35806
USA
E-mail: John.Johnson@ATImetals.com

## Chapter 19

H. Miura
Department of Mechanical Engineering
Faculty of Engineering
Kyushu University

733 Motooka
Nishi-ku
Fukuoka 819-0395
Japan
E-mail: miura@
   mech.kyushu-u.ac.jp

## Chapter 20

Donald F. Heaney*
E-mail: dfheaney@4-app.com

Neal S. Myers
Kennametal Inc.
1600 Technology Way
Latrobe
PA 15650
USA
E-mail: neal.myers@
   kennametal.com

## Chapter 21

Donald F. Heaney*
E-mail: dfheaney@4-app.com

J. L. Johnson
ATI Firth Sterling
7300 Highway 20 West
Huntsville, AL 35806
USA
E-mail: John.Johnson@
   ATImetals.com

Neal S. Myers
Kennametal Inc.
1600 Technology Way
Latrobe
PA 15650
USA
E-mail: neal.myers@
   kennametal.com

# Preface

The metal injection molding (MIM) process has gained significant credibility over the last 20 years and has become prevalent in market segments previously impenetrable, including medical implants and aerospace componentry. Many variants of the technology have been developed and commercialized, resulting in over 400 commercial MIM enterprises worldwide. This book is designed to serve as an up-to-date reference book for this rapidly changing technology. The book brings together the viewpoints of leading experts from around the world to provide a clear and concise handbook that will serve both the business decision makers and technical practitioners.

Initially, a market overview provides a business view of applications, financial performance and growth potentials. This is followed by four technical sections. These sections are on general processing, quality, specialized processes, and specific materials. The general processing section provides guidelines for component design and detailed processing information on each of the MIM process steps. A quality section is provided to expand upon the limited generalized literature devoted to MIM quality. A section on specialized MIM processes provides a view of some of the unique variants achievable by modifying the MIM process. This includes topics such as micromolding, porous materials, and bonding of two MIM materials. The final section on materials is devoted to both contemporary and emerging metal systems capable of being processed using MIM technology.

I thank all the contributors for their hard work and support during the preparation of this book. By design, I hope the unique expertise of the individual contributors is clearly seen and appreciated. Finally, I thank the editorial staff of Woodhead Publishing for their unparalleled persistence and professionalism during the preparation of this book.

*Donald F. Heaney*
*Advanced Powder Products, Inc.*

# 1
# Metal powder injection molding (MIM): key trends and markets

R. M. GERMAN, San Diego State University, USA

**Abstract:** Metal powder injection molding (MIM) has been in production since the 1970s. During that time the market has expanded enormously to include a broad array of applications; the initial successes were in dental orthodontic brackets, watch cases, and firearms, but recently the technology has moved into higher performance, life-critical applications in dental implants, artificial joints, heart pacemakers, and aerospace jet engines. This chapter provides a statistical overview of MIM, its applications, growth, financial performance, and growth prospects.

**Key words:** markets, productivity, sales, applications, production statistics, capacity, financial.

## 1.1 Introduction and background

Powder injection molding (PIM) has a main subdivision, metal powder injection molding (MIM), that has penetrated many fields. This chapter captures the status of the MIM field and provides a basis for evaluating different operations, markets, and regions. Like powder metallurgy, MIM relies on shaping metal particles and subsequently sintering those particles. The final product is nearly full density, unlike press-sinter powder metallurgy. Hence MIM products are competitive with most other metal component fabrication routes, and especially are successful in delivering higher strength compared with die casting, improved tolerances compared with investment or sand casting, and more shape complexity compared with most other forming routes. Injection molding enables shape complexity, high production quantities, excellent performance, and often is lower in cost with respect to the competition. Its origin traces to first demonstrations in the 1930s. In the metallic variant, most of the growth has been after 1990, when profitable operations began to emerge following several years of incubation.

Sintered materials technologies (cemented carbides, refractory ceramics, powder metallurgy, white wares, sintered abrasives, refractory metals, and electronic ceramics) add up to a very large value, with final products reaching $100 billion per year on a global basis. About 25% of that global activity is in North America. The production of metal powders alone in North America is annually valued at $4 billion (including paint pigments, metallic inks, welding electrodes, and other uses, besides sintered bodies). Sintered carbide and metal parts production in North America is valued at near $8 billion, where metal-bonded diamond cutting tools, sintered magnets, and semi-metal products contribute significantly to industry heavily focused on automotive and consumer products.

The powder metallurgy industry consists of about 4700 production sites around the world involved in variants of powder or component production. Most popular is the press-sinter variant that relies on hard tooling, uniaxial compaction, and high-temperature sintering. Based on tonnage, about 70% of the press-sinter products are for the automotive industry. However, on a value basis the story is dramatically different; metal cutting and refractory metal industries generate the largest value. Here the products include tantalum capacitors, tungsten light bulb filaments, tungsten carbide metal cutting inserts, diamond-coated oil and gas well drilling tips, high-performance tool steels, and molybdenum diode heat sinks. Compared to the other powder technologies, the MIM variant is still relatively new and small, but it is growing at 14% per year. In 2011 MIM products were globally valued at approximately $1 billion. This sales activity is spread over about 300 actors. Thus, the average sales would be just $3 million per year for a MIM firm.

## 1.2  History of success

Powder injection molding followed behind the first developments in plastic injection molding. Early polymers were thermosetting compounds; Bakelite, the first man-made polymer, was invented about 1909. Subsequently, as thermoplastic such as polyethylene and polypropylene emerged, forming machines appeared to facilitate the shaping of these polymers a few years later. The first demonstrations of PIM were nearly coincidental with the emergence of plastic injection molding. Simultaneously in the USA and Germany during the 1930s, this was applied to the production of ceramic spark plug bodies. This was followed by the use of PIM for forming tableware in the early 1960s. Generally these were components with wide allowed dimensional variation. The MIM variant reached production in the 1970s. The time delay between early demonstration and commercialization was due to a lack of sophistication in the process equipment. The manufacturing infrastructure improved dramatically with the advent of microprocessor-controlled processing equipment, such as molders and sintering furnaces, which enabled repeatable and defect-free cycles with tighter tolerances.

About 80% of the PIM production capacity is devoted to metals, recognized as MIM, but this generally does not include other metal molding technologies such as die casting, thixomolding, and rheocasting. The first MIM patent was by Ron Rivers (Rivers), using a cellulose–water–glycerin binder that proved unsuccessful. Subsequent efforts with thermoplastic, wax-based binders did reach production at several sites.

Major attention was attracted when MIM won two design awards in 1979. One award was for a screw seal used on a Boeing jetliner. The second award was for a niobium alloy thrust-chamber and injector for a liquid-propellant rocket engine developed under an Air Force contract for Rocketdyne. Several patents emerged, and one of the most useful was issued in 1980 to Ray Wiech. From this beginning, a host of other patents, applications, and firms arose, with special activity in California. By the middle 1980s the technology landscape showed multiple actors. Many companies set up at this time without a license, simply by hiring former employees from the early firms who brought with them insight into the technology.

All of the early binder patents have expired and the wax–polymer system discovered by Ray Wiech remains the mainstay of the industry. Since the mid-1990s the use of paraffin wax has migrated to variants such as polyethylene glycol to give water solubility to part of the binder system. This has improved the concerns over solvents used to remove the binder from the molded component – simply immerse the shaped component in hot water to dissolve out most of the binder.

Thus, the MIM concept relies on plastic molding technology to shape a powder–polymer feedstock into the desired shape. The shape is oversized to accommodate shrinkage during sintering. After molding the polymer is removed and the particles densified by high-temperature sintering. The product is a shrunken version of the molded shape, with near full density, and performance attributes that rival handbook values, usually far superior to that encountered in traditional press-sinter powder metallurgy and investment casting. This success is widely employed in small, complex, and high-value components, ranging from automotive fuel injectors to watch cases.

## 1.3 Industry structure

The MIM industry structure and interactions shows generally that the firms fall into a few key focal points. Everything revolves around the custom fabricators, firms that form components to satisfy the specifications of the user community – the users are generally well-known firms such as in firearms (Glock, Colt, Remington), computers (Hewlett Packard, Dell, Apple, Seagate), cellular telephones (Motorola, Samsung, Apple), hand tools (Sears, Leatherman, Snap-on Tools), industrial components (Swagelok, Pall, LG), and automotive (Mercedes-Benz, Borg-Warner, Honda, BMW, Toyota, Chrysler). The leading conference focused on MIM started in 1990 and continues today, where participants gather to share information on technology advances. At these conferences the actors in the industry generally come from one of the following sectors:

- ingredient suppliers – polymers, powders, and ingredients for either self-mixing or commercial feedstock production; globally there are approximately 40 firms that provide most of the MIM powders, although about 400 firms supply metal powders of various chemistries, particle sizes, particle shapes, and purities; for example, in titanium about four companies out of 40 suppliers make the powders used for MIM;
- feedstock production firms – purchase raw ingredients and formulate mixtures for sales to molding firms; globally there are usually about 12 feedstock suppliers;
- molding firms – both custom and captive molders that total nearly 300 MIM operations; about one-third are captive and make parts for themselves, but many of the captive firms also perform custom fabrication; 83% of all parts production is categorized as custom manufacturing;
- thermal processing firms – own sintering furnaces and debinding equipment that provide toll services; currently only a half-dozen firms are active in this area and most are associated with furnaces fabricators; a few firms provide toll hot isostatic pressing to force 100% density when required in medical or aerospace fields;
- designers – largely systems design firms associated with large multinational firms that intersect with the MIM industry; a few independent designers are available to handle *ad hoc* projects;
- equipment suppliers – firms that design and fabricate custom furnaces, molders, mixers, debinding systems, robotic systems, and other capital devices such as testing devices; the majority of molding machine sales are from six firms, furnace sales are from eight firms, mixer sales are

from four firms, so about 20 firms constitute the key equipment suppliers;
- consumables suppliers – supply process atmospheres, chemicals, molds, polishing compounds, machining inserts, packaging materials, heating elements, and sintering substrates;
- adjuncts – including researchers, instructors, consultants, design advisors, conference organizers, trade association personnel, magazine editors, and patent attorneys.

Component production is the central activity. It is split between internal and external products, referred to as captive and custom molders. Likewise it is supported by two parallel supply routes, depending on the decision to self-mix or to purchase premixed feedstock. An example captive molder would be a firearm company that uses MIM to fabricate some of the safety, trigger, or sight components. On the other hand, custom molders also can make these same components, but just as likely may be involved in several application areas as determined by their customer base.

As outsourcing increases for multinational firms, custom fabrication grows. Accordingly, MIM from facilities owned by large firms such as Rocketdyne, IBM, AMP, and GTE as early adopters, shifted to purchasing components from captive molders focused on a variety of application areas. Some of the early captive applications included the following examples:
- dental orthodontic brackets made out of stainless steel or cobalt–chromium alloys;
- business machine components for postage meters and typewriters;
- watch components including weights, bezels, cases, bands, and clasps;
- camera components that included switches and buttons;
- firearm steel parts such as trigger guards, sights, gun bodies, and safeties;
- carbide and tool steel cutting tools such as wood router bits, end mills, and metal cutting inserts;
- electronic packages for electronic systems using glass–metal sealing alloys;
- personal care items such as hair trimmers using tool steel;
- medical hand tools for special surgical operations;
- rocket engines using specialty materials such as niobium;
- automotive air bag actuator components using hardenable stainless steels;
- special ammunition that included birdshot, armor piercing and frangible bullets;
- turbocharger rotors for trucks and automobiles formed from high-temperature stainless steels or nickel superalloys.

Since each of these MIM operations had a single field of focus, little was done to grow that portion of the industry. However, in more recent years growth in MIM has come with the shift to custom molding which services a wide variety of applications. The custom molding firms have joined together in efforts to advance the industry, via collaborative marketing efforts, promotion of material standards, publicity through annual awards, and sharing of business data. Although declining, captive molding still remains an important part of the MIM industry. Although the sales growth varies year to year, in most recent times the global sales gain has been sustained at 14% per year.

## 1.4 Statistical highlights

Measures of the MIM growth are possible through several parameters, including the following.

- Patents. Since the start of MIM the total patent generation is large, exceeding 300 by the year 2000, but in more recent years the rate of patent generation has slowed and there are today about 200 currently active patents.
- Powder sales. In 2010 more than 8000 tons of metal powder were consumed globally by MIM, with a growth rate in powder tonnage use approaching about 20% per year, but due to price reduction the value increases about 14% per year.
- Feedstock purchase. The two options of self-mixing or purchasing feedstock seem to be of equal merit. Of the top firms, 71% form their own feedstock, which is almost the same ratio for all companies independent of size, suggesting purchased feedstock is neither an advantage nor disadvantage; however, self-mixing does provide greater manufacturing flexibility.
- Mixing. For those firms mixing their own feedstock, in 2011 they generate $1.8 million in sales per mixer, but the top 20 firms that mix their own feedstock are at $7 million in sales per mixer per year.
- Sales per molder. In many countries, especially when an operation is at a high utilization, the molding machine generates at least $1 million in sales per year. Across the industry the mean sales per molder is $536 000, while the leading firms have $1.5 million in sales per molder per year.
- Sales per furnace. Furnaces come in many different sizes and designs, but across the industry sales average about $1 million per furnace per year; for the top MIM operations (with larger and continuous furnaces) the sales average $3.2 million per furnace per year.
- Continuous furnaces. In 2011 the installed capacity of high-volume continuous sintering furnaces reached 4500 tons of products per year;

these are installed with a breakdown of 38% Asia, 47% Europe, and 15% North America.
- Captive versus custom production. About a third of the firms are captive, but only 21% of the firms have more than 50% of their sales internally. The best estimate is that 17% of the production value in 2011 is for internal use.
- Sales per kg. Across the MIM industry the average is about $125 in sales per kg of powder consumed, ranging from highs of $10 000 per kg for jewelry, cutting tips, and precision wire bonding tools to $16 per kg for casting refractories. The largest ceramic application is in aerospace casting cores, and the typical is $1000 per kg. Likewise, for metals, the stainless steel orthodontic bracket contributes nearly $100 million in annual sales at an average near $650 per kg. The low tolerance tungsten cell phone eccentric weights sell for a very low price, in the $60 per kg range.
- Sales per part. Across the industry, the typical part sale price is between $1 and $2 each, but values range from 5 cent cell phone vibrator weights to $35 solenoid bodies and $400 knee implants.
- Component size. The most typical MIM part mass is in the 6–10 g range. The mass range is from below 0.02 g to over 300 g, but the mean is under 10 g. The largest MIM parts are heat dissipaters for the control systems in hybrid electric vehicles at 1.3 kg and some aerospace superalloy bodies that have similar mass and dimensions reach 200 mm. A growing aspect of MIM is the microminiature components where features are in the micrometer range and this approached $68 million per year in sales for 2010.
- Employment. Nearly 8000 people are employed in PIM globally, of which nearly 7000 people are employed in MIM, giving an average of 21 people per operation and a median of just 16 people per MIM facility. The larger firms reach upwards to 300–800 people; the largest ceramic injection molding firm once reached employment near 800 people.

Historically, about 80% of the powder injection molding field is for metallic components, or MIM. In recent times that has increased to nearly 90% metallic. Of the 366 firms that currently practice powder injection molding, the majority are located in Asia. The leading countries in terms of PIM were the USA with 106 operations, China with 69 (although expansion is rapid in China), Germany with 41, Japan with 38, Taiwan with 17, Korea with 14, and Switzerland with 12. The number of operations is not necessarily indicative of financial size, since one of the largest MIM facilities is in India, a country which only has five MIM operations, while the USA has the most firms, but they tend to be smaller. Table 1.1 provides a

*Table 1.1* Summary statistics on PIM

| | |
|---|---|
| Percentage of PIM firms in North America | 31 |
| Percentage of PIM firms in Europe | 28 |
| Percentage of PIM firms in Asia | 37 |
| Percentage of PIM firms in rest of world | 4 |
| Largest concentration of firms | USA, China, Germany, Japan |
| Percentage of firms primarily captive | 33 |
| Largest PIM firms | India, USA, Germany, Japan |

*Table 1.2* Summary of PIM global sales

| | |
|---|---|
| Total PIM sales 2010 | $1.1 billion |
| Total PIM firms | 366 |
| Total PIM employment | 8000 |
| Typical R&D staff | 2 |
| Typical profit as % of sales | 11 |
| Sales per full-time employee | $126 000 |
| Sales per molding machine | $538 000 |
| Sales per production furnace | $980 000 |
| Percent of firms self-mixing | 72 |
| Total installed number of mixers | 380 |
| Total installed number of molding machines | 1750 |
| Total installed number of furnaces | 850 |
| Percent of industry using thermal debinding | 49 |
| Percent of industry using solvent debinding | 26 |
| Percent of industry using catalytic debinding | 14 |
| Percent of industry using other debinding | 11 |
| Median part size (g) | 6 |

summary of the PIM activities. The USA and China have the most firms while the largest facility is in India.

The PIM field is approximately one-third captive and two-thirds custom production. Example captive operations include ceramic casting core production, orthodontics, surgical tools, medical implants, and firearms.

Powder injection molding includes metals, ceramics, and carbides. Together these materials amount to sales for 2010 that reached $1.1 billion, with about $1 billion in metal components. Table 1.2 provides a summary on the global sales performance. In 2011, of the 366 firms practicing PIM, some do multiple materials. Across the industry over 80% of the firms produce metallic components, 20% practice ceramic powder injection molding, 4% practice cemented carbide injection molding, and less than 1% practice composite production (mostly injection-molded silicon carbide that is infiltrated with aluminum to form Al–SiC). Obviously about 5% of the firms supply a mixture of material types.

A MIM or PIM facility basically requires mixing, molding, debinding, and sintering capabilities. Over the years a typical ratio of these production

# Metal powder injection molding: key trends and markets

*Table 1.3* Typical unit manufacturing cell in metal PIM

| | |
|---|---|
| Mixers | 1 |
| Molders | 4 |
| Debinding reactors | 2 |
| Sintering furnaces | 2 |
| Typical number of employees per cell | 20 |

*Table 1.4* Typical productivity ratios

| | |
|---|---|
| Mean sales per employee | $125 000 |
| Median employees per molding machine | 5 |
| Mean employees per molding machine | 4.2 |
| Mean employees per production furnace | 9 |
| Median employees per production furnace | 6.7 |
| Mean sales per molding machine | $536 000 |
| Median sales per molding machine | $400 000 |
| Mean sales per furnace | $976 000 |
| Median sales per furnace | $667 000 |
| Mean sales per mixer (if installed) | $1.8 million |
| Median sales per mixer (if installed) | $1.0 million |
| Median growth percentage in sales | 11 |

devices has emerged. Some of the industry unit manufacturing cell ratios are captured in Table 1.3. Note that each device is an integer (one or two or three mixers) so for the mean the value can be non-integer, but for the median only integer values are allowed.

Another metric comes from the productivity, usually measured in terms of sales or per employee. Table 1.4 gives several statistical productivity measures for the global PIM industry. In this table, the employment count is based on full-time equivalent (FTE), since several firms have significant part-time employment. Hence the typical PIM industry (metals, ceramics, carbides, and composites) seems to be as tabulated.

## 1.5 Industry shifts

The key trends in MIM on a global basis are evident by comparing year 2000 with year 2010. For many years the MIM field sustained compound annual sales growth at 22% per year with a 34% per year increase in the number of operations. In recent years the growth rates have become more modest and have stabilized near 8% per year in North America, but continue at a 30% per year pace in Asia. Globally the overall average is 14% per year for recent years.

On the basis of some important statistics, here are the changes from year 2000 to year 2010 for MIM:

- number of MIM operations decreased 34%;
- global sales increased 100%;
- employment increased 100%;
- installed molding capacity increased 79%;
- installed sintering capacity increased 86%.

These statistics indirectly indicate increased concentration of the business into the hands of fewer firms. Although a large number of companies are active in MIM, at any one time some are simply in an evaluation mode, and this was especially true in 2000. These characteristically consist of a small team (two or three people) and one or two molding machines, usually purchased feedstock, and might even rely on toll sintering. After initial exploration, many such efforts are terminated or production is transferred to an outside vendor. This is evident by the decrease in number of MIM firms from 2000 to 2010 while the overall field grew. Today, large actors consist of well over 20 molding machines.

## 1.6 Sales situation

Sales statistics were first gathered in the middle 1980s. Significant growth has produced industry maturation. The metal powder injection molding process is now accepted by several sophisticated customers, such as Bosh, Siemens, Chrysler, Honeywell, Volkswagen, Mercedes Benz, BMW, Chanel, Apple Computer, Pratt and Whitney, Samsung, Texas Instruments, General Electric, Nokia, Motorola, Rolls Royce, Continental, Stryker, LG, Sony, Philips, Seagate, Toshiba, Ford, General Motors, IBM, Hewlett-Packard, Seiko, Citizen, Swatch, and similar firms.

Most of the common engineering materials are available in MIM, but as illustrated in Fig. 1.1, based on sales, stainless steels are dominant. The global material sales (value, not tonnage) are as follows – 53% stainless steels, 27% steels, 10% tungsten alloys, 7% iron–nickel alloys (mostly magnetic alloys), 4% titanium alloys, 3% copper, 3% cobalt–chromium, 2% tool steels, 2% nickel alloys (superalloys), and 1% electronic alloys (Kovar and Invar). On a tonnage basis the stainless steel portion of powder consumption is larger, reaching upwards to 60–65% of powder consumption, and because of that large consumption the powder price is lower, further fueling the use of stainless steel. Some of the metal powders are much higher priced, such as titanium, so the sales partition based on tonnage versus dollars is skewed due to a wide range in material costs.

The metal powder injection molding sales for 2010 were in the neighborhood of $1 billion dollars. Independent reports for 2010 give estimates from $955 to $984 million. Some difficulty exists in gathering accurate information since a majority of the firms are privately held and do

Metal powder injection molding: key trends and markets 11

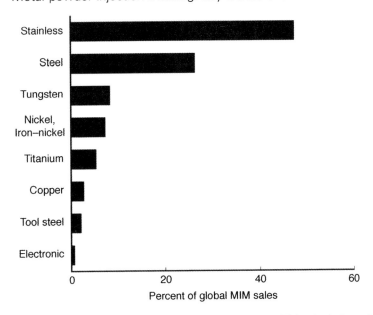

*1.1* MIM sales globally by main material category. This plot is based on percent of global sales, while other studies report based on tonnage of powder consumed.

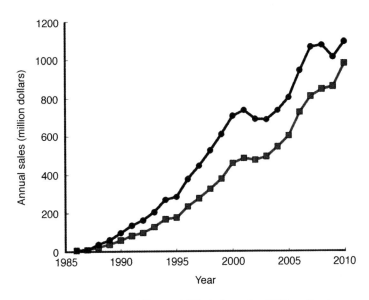

*1.2* Annual sales, in millions of US dollars, for MIM as the lower curve and all of PIM as the upper curve plotted against year.

*Table 1.5* Summary global MIM statistics

| | |
|---|---|
| 2009 total MIM sales | $860 million |
| 2010 total MIM sales | $955 million |
| Typical profit as % of sales | 9 |
| % MIM sales from self-mixing | 63 |
| Installed number of MIM mixers | 290 |
| Installed number of MIM molding machines | 1450 |

not make annual reports. Further, currency exchange rates change over the year leading to inaccurate estimates. For 2010, the best estimate on PIM sales is $1.1 billion and MIM sales is $955 million. The PIM estimate for 2009 was $920 million, so PIM grew from 2009 to 2010 by almost 20%, largely in MIM, and was ahead of the general economic recovery.

Figure 1.2 plots sales growth globally since first recorded at $9 million for 1986. This plot shows that the ceramic business contracted while MIM expanded, largely due to the aerospace slowdowns.

## 1.7 Market statistics

The largest market for MIM has historically been industrial components, including pump housings, solenoids, handles, plumbing fixtures, and fitting. This still amounts to 20% of global sales (more in Japan and less in Europe). This is based on component sales, not number of parts or firms or tons of powder, as is reported in other studies. Automotive components are the second largest market for MIM at 14% of global value (on tonnage basis it is much higher in Europe and lower in North America). Consumer products are the third largest market at 11% of the MIM sales, and these are dominated by Asia (cell phone, consumer, and computer parts). Next are the dental, medical, electronic, and firearm applications, each at 7–9% of the global MIM product value, with North America being the largest actor. Other contributions come from computers, hand tools, luggage trimmings, cosmetic cases, robots, sporting devices, and watch components.

The statistical profile for MIM firms shows that about half of the global actors are tiny, being under $1 million in annual MIM sales; sometimes these are located as pilot efforts in large companies or more commonly reflecting private ownership. Table 1.5 gives an industry summary for MIM. About 32% of the operations have captive products, but only 18% of the industry sales are primarily captive. Following the Prado Principle, in MIM the largest 20% of firms control 80% of sales. Moreover, the top 10% of MIM firms control over 60% of sales, average $25 million per year turnover and run with 20 or more molding machines and an average of 120 employees. The top 10% of the MIM industry average five employees for each molder,

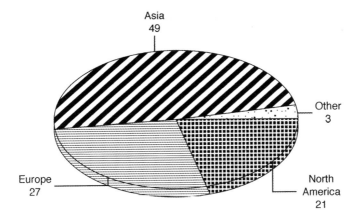

*1.3* Sales partition based on major geographic region.

*Table 1.6* Summary of regional sales for PIM

| Region | Total PIM ($ million) |
|---|---|
| North America | 316 |
| Europe | 293 |
| Asia | 446 |
| Rest of world | 27 |
| Total | 1082 |

nine people per furnace, turnover $1.5 million per molding machine each year, $3.2 million per furnace each year, and 70% self-mix.

## 1.8  Metal powder injection molding market by region

As shown in Fig. 1.3, the primary sales and growth in MIM is in Asia, but the USA remains one of the largest users and producers of ceramic PIM components. Table 1.6 summarizes the regional statistics for PIM in millions of dollars for 2009, the last year where consolidated data are reported. For reference, North American MIM was $186 million in 2009 out of the $316 million total. Of that $32 million was captive MIM. In North America, captive MIM is frequently used for orthodontic brackets and firearms, but is also used for medical applications. The valuation of those products is difficult, since the distributed cost (when the bracket is sold to the dentist) is probably $100 million, but the trade cost (bulk internal transfer cost) is more modest. For 2009 in Europe the dominant captive MIM applications were watches and similar decorative components such as specialty luggage fasteners. In 2009, the Swiss watch industry did about $12 billion in watch sales, at an average transaction of $566 per watch. Not all

*Table 1.7* Market partition by region and application in terms of percent of sales for 2007 (ROW: rest of world)

| Application | North America | Europe | Asia | ROW |
|---|---|---|---|---|
| Automotive | 30 | 28 | 18 | 0 |
| Consumer | 0 | 32 | 15 | 0 |
| Dental | 18 | 14 | 0 | 8 |
| Electronics | 6 | 0 | 41 | 0 |
| Firearms | 6 | 9 | 0 | 66 |
| Hardware | 0 | 0 | 1 | 0 |
| Industrial | 6 | 3 | 14 | 24 |
| Medical | 34 | 2 | 2 | 1 |
| Military | 0 | 1 | 2 | 1 |
| Other | 0 | 9 | 5 | 1 |

watches use MIM, but those that do have a typical MIM content in the $1–$8 range, so the 21 million watches would fit with the $32 million captive Europe sales. In Asia the captive MIM is mostly for the '3Cs' – products for computers, cellular telephones, and consumer electronics – at the assembly houses such as Foxconn.

## 1.9 Metal powder injection molding market by application

In prior reports from some of the trade associations, such as the Metal Powder Injection Molding Association in 2006, the medical component sales constituted 36% of MIM, the automotive component was 14% of sales, and hardware was 20% of sales.

In the 2007 survey, the MIM market was partitioned as given in Table 1.7 in terms of percentage of sales in each geographic region. More recently, an expanded set of application groups is used to reflect added efforts in sporting, jewelry, hand tools, aerospace, and such. Further, consumer products were separated from watches since watch production was not growing, but uses for latches, eyeglasses, and luggage clasps was growing. Accordingly, the 2010 industry partitioned by application area is given in Table 1.8. This is a combination of number of companies and their relative focus, so a firm that does only dental counts as 1, but a firm that does dental and industrial counts as 0.5 for each, and so on.

This tabulation shows that industrial components (valves, fittings, connectors) are the broadest market focus, followed by consumer (kitchen tools, toothbrush parts, scissors), and electronic components (heat sinks, hermetic packages, connectors). Automotive and medical are fast-growing areas, especially in North America.

# Metal powder injection molding: key trends and markets 15

*Table 1.8* Global market attention based on primary marketing focus for MIM firms in 2011

| Field | Percentage |
|---|---|
| Aerospace | 2 |
| Automotive | 7 |
| Casting | 2 |
| Cell phone | 2 |
| Computer | 3 |
| Consumer | 10 |
| Cutting | 2 |
| Dental | 4 |
| Electronic | 10 |
| Firearm | 6 |
| Hand tool | 2 |
| Hardware | 2 |
| Household | 0 |
| Industrial | 23 |
| Jewelry | 1 |
| Lighting | 1 |
| Medical | 8 |
| Military | 4 |
| Sporting | 2 |
| Telecom | 1 |
| Watch | 2 |
| Wear | 5 |

When considered in terms of sales value, some of the large markets with fewer actors become evident as tabulated in Table 1.9.

## 1.10 Market opportunities

Listed below are several often discussed market opportunities and some of the related information on growth aspects relative to MIM and its future.

- Consumer, cell phone, and computer uses continue to grow. A comparison of the 2007 and 2009 partition shows an increased use of MIM in hand-held devices, ranging from cell phones to portable computers. The components are small, complex, and strong, applied to switches, buttons, hinges, latches, and decorative devices. Since most of the assembly is in Asia, parts production has migrated to Asia to keep supply lines short.
- Firearms went through a rapid escalation after the November 2008 election of Obama to the Presidency in the USA, due to fear of new restrictive gun laws; although that wave passed in North America, the temporary surge offset the economic decline seen in many other fields.

*Table 1.9* Percent of global MIM sales for each market segment in 2011

| Application field | Percentage |
|---|---|
| Aerospace | 0 |
| Automotive | 14 |
| Casting | 0 |
| Cell phone | 4 |
| Computer | 4 |
| Consumer | 11 |
| Cutting | 0 |
| Dental | 9 |
| Electronic | 9 |
| Firearm | 7 |
| Hand tool | 3 |
| Hardware | 1 |
| Household | 0 |
| Industrial | 20 |
| Jewelry | 1 |
| Lighting | 0 |
| Medical | 8 |
| Military | 2 |
| Sporting | 2 |
| Telecomm | 0 |
| Watch | 2 |
| Wear | 0 |

Military procurement of firearm components has started to slow. However, smaller firearm manufacturers have started to embrace MIM.
- Industrial, hand tool, and household applications remain strong and steady, and include valve, plumbing, spraying, wrenches, multi-tools, pepper grinders, scissors, circular saws, nailing guns, and similar devices.
- Automotive applications for MIM started to escalate with use in turbochargers, fuel injectors, control components (clock mounts, entry locks, knobs, and levers), and valve lifters. This initiated in the USA for Buick and Chrysler applications, but leadership shifted to Japan with Honda and Toyota applications from integrated vendors (Nippon Piston Rings) for turbocharger and valve applications. Subsequently, European MIM shops picked up on the materials and applications opened by higher performance but smaller engines, and this wave has become global. There are many complaints over automotive parts production, but it generates large sales volumes that help to lower all costs and improve the field. All expectations are that MIM will continue to grow in the automotive sector.
- Medical applications are growing from an early base of endoscopic devices, and will become enormous as MIM becomes widely accepted.

Much of the recent growth has been in minimally invasive surgical tools and robotic devices. Early frustrations were with the time to become qualified and the relatively small production lots on many surgical tools. Now adaptations to the market show that much higher prices allow for profitable MIM production in the smaller lots. For example, with knee implants in the USA, one million replacements are made per year, so that is an attractive opportunity. However, there are left and right knees, and about 12 designs or styles. Thus the fragmentation shows on average 40 000 per year of a design, and since there are three leaders in this market, any one company might only order 12 000 of each part per year. This is a low production volume application for MIM. However, pricing allows for sales that might reach $4 million per design. So far only a few MIM firms are positioning for the production of implants, while many are seeking orders in surgical hand tools. Minimally invasive surgical tools are a prime opportunity for MIM. Micro-featured devices are frequently shown for new genetic sensors (micro-pillar, micro-texture and micro-array designs). These will be small devices, potentially used in enormous quantities for rapid blood testing and disease identification.

- Dental applications in this field long ago matured and today there are several firms involved in orthodontic bracket fabrication. However, new instrument and hand tool designs have opened up special opportunities for micro-featured designs. So MIM is moving from its strong historical position in orthodontic brackets into hand tools and special endodontic surgical devices.
- Aerospace applications for MIM have been demonstrated for 30 years. A new wave of efforts is now starting, driven by cost concerns and envisioned savings with MIM. About a dozen firms are active in this area. Like medical applications, the production volumes are often small, in the 10 000 per year range, but the unit prices are high.
- Lighting applications for MIM are limited to refractory metals and ceramics, and the developments in this area are in the hands of the big three – Sylvania, Philips, and General Electric. After much early effort the MIM viability is in serious doubt due to cost reduction and competing light-emitting diode (LED) devices. Mounts for LED devices out of copper by MIM have been displayed, but cost will probably work against MIM.
- Sporting applications have persisted for 20 years, but it appears the cost points in this field do not match well with MIM and the penetration of MIM remains small. Past successes have included metal supports for football knee braces, dart bodies, golf clubs, and running cleats.
- Jewelry applications are new to MIM and could potentially grow rapidly as alternative materials (non-gold and non-silver) become

accepted. These include titanium, high-polish stainless steel, tantalum, and even bronze.

The upside market size on a few of these is quite large, while others not listed above might grow, but the key actors are in place in Asia and it is doubtful if new entries can play a role.

Future opportunities emerging from research and development (R&D) efforts are discussed at the conferences. Some of the leading opportunities include ultra-high thermal conductivity composites (for example copper–diamond) for heat sinks. Demonstrations reaching 580 W/(m K) thermal conductivity have been shown by a Japanese MIM firm for use in supercomputers, high end servers, phased array radar systems, military electronics, hybrid vehicle control systems, gaming computers, and other applications involving high-performance computing. One such device is pictured in Fig. 1.4.

A related area is in vapor chamber designs, typically from copper, where a closed internal chamber of porous metal is used to apply heat pipe technology to a similar problem requiring heat dissipation around electronics.

Similarly, another area is LED heat sinks, where copper arrays are used to mount the semiconductor, with reports of 100 g arrays with costs as low as $0.75 per mount; these demonstrations have largely come from Asia.

Microminiature MIM for medical minimally invasive surgical tools is an area of development, involving very small components for end manipulators, such as cutters, grasps, and drug delivery. Most are made from

*1.4* A copper MIM heat transfer device used for electronic cooling (photograph courtesy of Lye King Tan).

# Metal powder injection molding: key trends and markets

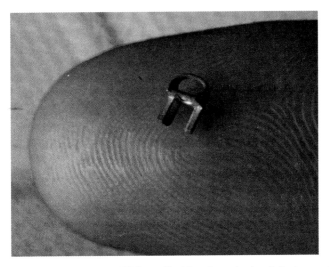

*1.5* Stainless steel MIM medical implant device (photograph courtesy of Metal Powder Industries Federation).

stainless steel and example components are being sold in the range of $2–$15 each. Figure 1.5 shows one example used in shoulder repair.

Other microminiature MIM applications involve components for cell phones, computers, hand-held electronic devices, and dental hand tools for endodontic use and dental cleaning. Implants such as dental tooth posts, components for ligament alignment, hearing canal (ear) reconstruction, drug delivery, heart valves, artificial knees, shoulders, and hips, are expected to be a billion dollar opportunity, but will require considerable dedication and resources to realize; Stryker and Medtronic have set up internal production, Zimmer and Biomed have elected to work with a few MIM shops, and Accellent has elected to be fully qualified for any applications on a custom basis.

Microarray devices with hundreds to thousands of pins, posts, or holes for disposable lab-on-a-chip devices are used in blood testing, assessment of disease, analysis of DNA to predict disease, and protein tests; the biochip market is targeted to reach $3.8 billion in sales by 2013 and considerable research is taking place to support this effort. Hewlett-Packard and Oregon State University have a small MIM facility examining options, but activity is also on-going in Germany, Singapore, and Japan.

Titanium biocompatible structures, such as for tissue affixation, implants, surgical tools, tool implants, and even sporting devices represent another development area. About 19 firms have some variant of titanium, but few have focused on medical quality. Porous titanium by MIM offers the possibility of hydroxyapatite (bone) infusion; an example MIM device for tooth implants is shown in Fig. 1.6.

20   Handbook of metal injection molding

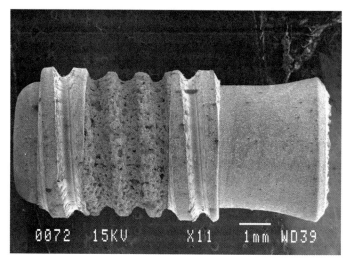

*1.6* Titanium dental implant formed by MIM with an intentional porous region for bone ingrowth (photograph courtesy of Eric Baril).

A further example is hardware tools from tool steel, such as threading devices for cast iron plumbing or water pipes, hand tools, valves and fittings, handles, forming tools, drills, dies. Also, hermetic packages for microelectronics are being developed using Kovar to enable glass to metal sealing. Figure 1.7 shows an example MIM part that is sealed with lead. This design routinely sells for $30 each.

*1.7* Hermetic Kovar microelectronic package with attached glass–metal sealed lead wires (component courtesy of Yimin Li).

# Metal powder injection molding: key trends and markets

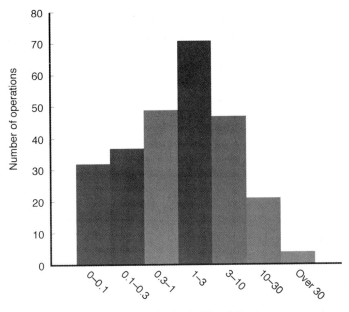

*1.8* Sales distribution chart for global MIM firms, showing the mode size is in the range from $1 million to $3 million annual sales and over half the firms are below $1 million in annual sales.

Aerospace applications exist for smaller superalloy bodies such as IN 625, 713, 718, 723, or Hastelloy X, where the high detail, good surface finish, and shape complexity offered are financially attractive for military and commercial applications. Polymer Technologies, Maetta Sciences, PCC Advanced Forming, Parmatech, Advanced Materials Technology, Advanced Powder Processing, and a few other firms are positioned for this area.

Of that total population of companies, the majority practice MIM. The distribution in annual sales for the 366 PIM firms is given in Fig. 1.8. This plot shows almost half are small, with under $1 million in annual sales. This is a partitioning that is roughly based on a factor of three step size, starting with 100 000 for the smaller firms and increasing to show five PIM firms over $30 million.

## 1.11 Production sophistication

The statistics show that MIM is at an early stage of sophistication as a net-shaping technology. The North American segment is about 90% fully in compliance with one of the ISO 9000/9001/9002 variants, about 25% in compliance with ISO 14000, but nearly 60% of the companies have no intentions along these lines. Several of the firms are into the automotive standards, but few are certified with AS 9100 B, which is required for aerospace components – Maetta Sciences, Pratt and Whitney, Polymer Technology, and PCC Advanced Forming Technology. Several of the firms are in compliance with the ISO 13485 good manufacturing practices required for medical devices, but only a few are at the standards required for implants.

Another view of the sophistication is evident by the sales distribution, where the top 10% of the firms based on annual sales control about 60% of the sales, show higher sales per full-time employee (slightly under $300 000/FTE), per molding machine ($1.5 million), per furnace ($3.2 million), and generally set the benchmark for productivity. On the other hand, 50% of the firms listed in this report have annual sales of $1 million or less, and drag down all of the industry statistics.

The key actors in MIM are small compared with several other metalworking technologies. The sophistication of the PIM industry is graphically given in Fig. 1.9. The stage 0 firms are in an evaluation mode, largely trying to determine if there is a business fit, for example possibly a plastic molder seeking material diversification. The stage 1 operation would have no serious sales, and might be setting up or shutting down or sitting idle with technology. A significant portion of the MIM industry is in this situation. A few will grow to be significant actors. Others have dabbled for many years, but have never found the right combination of customers and

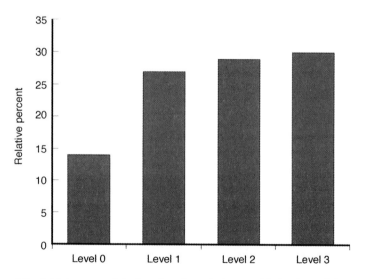

*1.9* A histogram plot showing the relative sophistication of the MIM industry based on operational characteristics, as described in the text.

technology. There are about 150 firms showing up in MIM that fit this category and they contribute about $42 million in sales (roughly $0.25 million per firm per year).

Stage 2 firms are more serious about making a business and generally are operating with several molders and are pushing to grow their business. This group generally is between one and two shifts and consists of slightly over 100 firms doing about $116 million in sales.

The stage 3 firms dominate the sales, consisting of slightly more than 100 firms doing just under $700 million in sales ($7 million per firm). These are the captive operations in dental and firearms and the major custom.

Most of the stage 2 firms are qualified by ISO audits, but only the larger firms are in compliance with aerospace, automotive, or medical quality standards.

## 1.12 Conclusion

The PIM field has exhibited enormous growth since the first sales statistics were gathered in 1986, amounting to $9 million globally then. Today, MIM is the dominant form of PIM and has sustained 14% per year growth in recent years. The number of MIM firms has not shown much change in recent times, but the size and sophistication have grown considerably. Various estimates have been offered for how far and how long MIM can sustain the growth. The informed estimates balance cost, capacity, and competitive factors, but generally agree that MIM will double yet again to reach $2 billion in annual sales by 2017. After that, significant cost reduction will be required. Unfortunately, industry-wide research and innovations seem to be lacking. The R&D personnel, publication rate, patent rate, and other leading indicators warn that MIM reached its peak of innovation in about 2005. Contemporary concepts show that patent activity is generally the best leading indicator of commercial sales and profits, so on this basis there are early warnings of MIM reaching the end of its growth. Another symptom of the end of growth comes from the fact that, rather than innovation, most industry R&D efforts have turned to cost reductions, improved quality, improved dimensional control, and improved impurity control. These shorter-term gains will not offset the longer-term needs for new materials, products, and applications as required to sustain MIM toward $2 billion in annual sales.

## 1.13 Sources of further information

Anonymous (2007), 'China reaches for a vibrant future as MIM takes off', *Metal Powder Report*, November, 12–21.

German, R. M. (2007), 'Global research and development in powder injection moulding', *Powder Injection Moulding International*, 1(2), 33–36.

German, R. M. (2009), 'Titanium powder injection moulding: A review of the current status of materials, processing, properties, and applications', *Powder Injection Moulding International*, 3(4), 21–37.

German, R. M. (2011), '*Metal Injection Molding: A Comprehensive MIM Design Guide*', Metal Powder Industries Federation, Princeton, NJ.

German R. M. and Johnson, J. L. (2007), 'Metal powder injection molding of copper and copper alloys for microelectronic heat dissipation', *International Journal of Powder Metallurgy*, 43(5), 55–63.

Itoh, Y., Uematsu, T., Sato, K., Miura, H., and Niinomi, M. (2008), 'Fabrication of high strength alpha plus beta type titanium alloy compacts by metal injection molding', *Journal of the Japan Society of Powder and Powder Metallurgy*, 55, 720–724.

Johnson, P. K. (1979), 'Award winning parts demonstrate P/M developments', *International Journal of Powder Metallurgy and Powder Technology*, 15, 323–329.

Kato, Y. (2007), 'Metal injection moulding in Asia: Current status and future prospects', *Powder Injection Moulding International*, 1(1), 22–27.

Manison, P. (2007), 'UK-based producer looks to build on current success for precision ceramic components', *Powder Injection Moulding International*, 1(1), 45–47.

Mills, B. (2007), 'Flexibility helps MIM producer meet the demands of a broad client base', *Powder Injection Moulding International*, 1(1), 20–21.

Moritz, T. and Lenk, R. (2009), 'Ceramic injection moulding: A review of developments in production technology, materials and applications', *Powder Injection Moulding International*, 32(3), 23–34.

Park, S. J., Ahn, S., Kang, T. G., Chung, S. T., Kwon, Y. S., Chung, S. H., Kim, S. G., Kim, S., Atre, S. V., Lee, S., and German, R. M. (2010), 'A review of computer simulations in powder injection molding', *International Journal of Powder Metallurgy*, 46(3), 37–46.

Piotter, V., Hanemann, T., Heldele, R., Mueller, M., Mueller, T., Plewa, K., and Ruh, A., (2010), 'Metal and ceramic parts fabricated by microminiature powder injection molding', *International Journal of Powder Metallurgy*, 46(2), 21–28.

Rivers, R. D. (1976), 'Method of injection molding powder metal parts', US Patent 4,113,480.

Sakon, S., Hamada, T., and Umesaki, N. (2007), 'Improvement in wear characteristics of electric hair clipper blade using high hardness material', *Materials Transactions*, 48, 1131–1136.

Schlieper, G. (2007), 'Leading German manufacturer works to develop the market for MIM in the automotive sector', *Powder Injection Moulding International*, 1(3), 37–41.

Schwartzwalder, K. (1949), 'Injection molding of ceramic materials', *Ceramic Bulletin*, 28, 459–461.

Venvoort, P. (2007), 'Developments in continuous debinding and sintering solutions for MIM', *Powder Injection Moulding International*, 1(2), 37–44.

Whittaker, D. (2007), 'Developments in the powder injection moulding of titanium', *Powder Injection Moulding International*, 1, 27–32.

Whittaker, D. (2007), 'Powder injection moulding looks to automotive applications for growth and stability', *Powder Injection Moulding International*, 1(2), 14–22.

Wiech, R. E. (1980), 'Manufacture of parts from particulate material', US Patent 4,197,118.

Williams, B. (2007), 'Powder injection moulding in the medical and dental sectors', *Powder Injection Moulding International*, 1, 12–19.

Williams, N. (2007), 'European MIM pioneer drives the industry forward with quality and customer satisfaction', *Powder Injection Moulding International*, 1(2), 30–34.

Ye, H., Liu, X. Y., and Hong, H. (2008), 'Fabrication of metal matrix composites by metal injection molding – A review', *Journal of Materials Processing Technology*, 200, 12–24.

# Part I

Processing

# 2
# Designing for metal injection molding (MIM)

D. F. HEANEY, Advanced Powder Products, Inc., USA

**Abstract:** In this chapter, a basic design guide for MIM components is presented. Although many of these design guidelines are similar to those used for plastic injection molding, there are subtle differences in how MIM components are designed to account for the subsequent thermal processing. For example, wall sections should not exceed a certain thickness to prevent defect formation during thermal debinding and a flat section is often designed into the component to provide uniform support for sintering. These design concepts and many others are described here as suggestions to provide guidance, since many potential components will not perfectly fall into the ideal design. As such, the opportunity exists to find clever solutions outside the advice provided here to take advantage of the MIM technology for particular applications.

**Key words:** MIM design, materials, properties, tooling features, defect reduction.

## 2.1 Introduction

Metal injection molding is a process by which powder is shaped into complex components using tooling and injection molding machines that are very similar to those used in plastic injection molding. Therefore, the component's complexity is of the same magnitude as those seen in plastic injection molding. The artifacts associated with the injection molding process (gates, ejector pins, parting line) are also similar to those seen in plastic injection molding and must be accounted for in design. However, since the MIM process requires multiple post-molding debinding and sintering steps, some design considerations such as cross-sectional thickness and geometry features require consideration.

As a general rule of thumb, components that are less than approximately 100 g and fit into the palm of your hand could be good candidates for MIM technology. A mean size of 15 g is typical for a MIM component; however,

*Table 2.1* Comparison of MIM attributes with other fabrication techniques

| Attribute | MIM | Powder metallurgy | Casting | Machining |
|---|---|---|---|---|
| Component size (g) | 0.030–300 | 0.1–10 000 | 1 + | 0.1 + |
| Wall thickness range (mm) | 0.025*–15 | 2 + | 5 + | 0.1 + |
| Percent theoretical density (%) | 95–100 | 85–90 | 94–99 | 100 |
| Percent theoretical strength (%) | 95–100 | 75–85 | 94–97 | 100 |
| Surface finish (μm) | 0.3–1 | 2 | 3 | 0.4–2 |
| Production volume | 2000 + | 2000 + | 500 + | 1 + |

*Features this small could have distortion.

*Table 2.2* Typical attributes produced by the MIM process

| Attribute | Minimum | Typical | Maximum |
|---|---|---|---|
| Component mass (g) | 0.030 | 10–15 | 300 |
| Max. dimension (mm) | 2 (0.08 in) | 25 (1 in) | 150 (6 in) |
| Min. wall thickness (mm) | 0.025 (0.001 in)* | 5 (0.2 in) | 15 (0.6 in) |
| Tolerance (%) | 0.2% | 0.5% | 1% |
| Density | 93% | 98% | 100% |
| Production quantity | 1000 | 100 000 | 100 000 000 |

*Features this small could have distortion.

components in the range around 0.030 g are possible. Table 2.1 compares the MIM process with other manufacturing processes. Notice that MIM is limited to smaller part sizes, can provide thinner wall thicknesses, has excellent surface finish and is suited for high volumes. Table 2.2 reviews the upper and lower specifications of the MIM process.

The following are some general design considerations which will be discussed in detail in this chapter.

- Avoid components over 12.5 mm (0.5 in) thick. This is a function of MIM technology and alloy, for example 4140 and alloys that use carbonyl powder can have thicker wall sections than those that use gas-atomized powders that have larger particles. Also modifications to binder systems can be made to allow thicker sections to debind.
- Avoid components over 100 g in mass; however, 300 g are possible for some technologies.
- Avoid long pieces without a draft (2°) to allow ejection.
- Avoid holes smaller than 0.1 mm (0.0039 in) in diameter.
- Avoid walls thinner than 0.1 mm (0.0039 in), although 0.030 mm walls are possible in some cases.
- Maintain uniform wall thickness; thin, slender sections attached to thick sections should be avoided to enhance flow during molding, to avoid sinks and voids, and to limit distortion during sintering.

- Core out thick areas to avoid sinks, warpage, and debinding defects.
- Avoid sharp corners. The desired radius is greater than 0.05 mm (0.002 in).
- Design with a flat surface to aid in sintering – otherwise custom ceramic setters required.
- Avoid inside closed cavities – although some technologies such as a chemically or thermally removable polymer core may be used but is not common.
- Avoid internal undercuts – although a collapsible core or extractible core mentioned above could be used but are not common.
- Design with lettering – raised or recessed.
- Design with threads – internal and external.

## 2.2 Available materials and properties

Metal injection molding is available in many of the common structural materials for medical, military, hardware, electronic, and aerospace applications. If the powder is available in the appropriate size, less than 25 microns, and the powder sinters to a sufficiently high density, without change in alloy chemistry, the material can be manufactured using the MIM process. Table 2.3 provides an overview of available materials, applications, and specific features that make these metals desirable.

Metal injection molding properties are superior to most cast products and slightly inferior to wrought products. Cast and MIM components both have microstructural pores or voids as a result of the processing methods, where the cast voids can be large and localized owing to the cooling of liquid to solid and the MIM voids are typically fine and well distributed across the microstructure after sintering. The large, localized voids of the cast material result in the inferior properties, whereas the distributed nature of the fine MIM pores provides a better microstructure for enhanced properties. Hot isostatic pressing (HIPing) can be used to attain full density. Another attribute of the MIM process is that the final product will be annealed after the sintering operation, thus materials that show work hardening strengthening in the machined state may require some form of post-sinter operation to enhance the strength after the MIM sintering operation. Table 2.4 provides typical data for many MIM structural materials.

Metal injection molding is an attractive method for the fabrication of soft magnetic materials. The MIM operation provides a net shape component that is in the annealed condition, which is a requirement for the best magnetic response. Table 2.5 provides data for soft magnetic applications. Each of these different alloys has physical attributes which makes them ideal for different applications. The 2200 alloy has magnetic properties similar to pure iron but with a greater strength. The Fe–50Ni alloy has a high

*Table 2.3* Overview of MIM materials, applications, and features

| Material family | Applications | Specific alloys | Specific feature |
|---|---|---|---|
| Stainless steel | Medical, electronic, hardware, sporting goods, aerospace, consumer products | 17-4PH | Strength, heat treatable |
| | | 316L | Corrosion resistance, ductility, non-magnetic |
| | | 420, 440C | Hardness, wear resistance, heat treatable |
| | | 310 | Corrosion and heat resistance |
| Low-alloy steel | Hardware, bearings, races, consumer goods, machine parts | 1000 series | Case hardenable |
| | | 4000 series | General purpose |
| | | 52100 | High wear resistance |
| Tool steel | Wood and metal cutting tools | M2/M4 | 61–66 HRC |
| | | T15 | 63–68 HRC |
| | | M42 | 65–70 HRC |
| | | S7 | 55–60 HRC |
| Titanium | Medical, aerospace, consumer products | Ti | Light weight |
| | | Ti–6Al–4V | Light weight, high strength |
| Copper | Electronic, thermal management | Cu | High thermal and electrical conductivity |
| | | W–Cu, Mo–Cu | High thermal conductivity, low thermal expansion |
| Magnetic | Electronic, solenoids, armatures, relays | Fe–3%Si | Low core losses and high electrical resistivity |
| | | Fe–50%Ni | High permeability and low coercive field |
| | | Fe–50%Co | Highest magnetic saturation |
| Tungsten | Military, electronic, sporting goods | W | Density |
| | | W heavy alloy | Density and toughness |
| Hardmetals | Cutting and wear applications | WC–5Co | Higher hardness |
| | | WC–10Co | Higher toughness |
| Ceramics | Wear applications, nozzles, ferules | Alumina | General purpose |
| | | Zirconia | High wear resistance |

permeability and low coercive field which makes it ideal for motors, switches, and relays. The Fe–3Si shows low core loss and high electrical resistivity in both alternating current (AC) and direct current (DC) applications. The Fe–50Co alloy has a very high magnetic saturation and is ideal for high magnetic flux density applications. Finally, if a good magnetic response is needed in conjunction with good corrosion resistance, the 430L alloy would be the alloy of choice.

Metal injection molding is a viable technique for the production of copper. The copper made with MIM exhibits good thermal and electrical conductivity, thus MIM copper is a viable option for electrical connectors and thermal management applications. Table 2.6 provides copper data in

*Table 2.4* MIM structural material properties

| Material | Density (g/cm³) | YS (MPa) | UTS (MPa) | Elongation (%) | Unnotched Charpy impact energy (J) | Macro hardness | Young's modulus (GPa) |
|---|---|---|---|---|---|---|---|
| 316L SS | 7.8 | 180 | 520 | 40 | 190 | 67 HRB | 185 |
| 17-4PH SS | 7.6 | 740 | 900 | 6 | 140 | 27 HRC | 190 |
| 17-4PH SS H900 | 7.6 | 1100 | 1200 | 4 | 140 | 33 HRC | 190 |
| 420 SS | 7.5 | 1200 | 1370 | – | 40 | 44 HRC | 190 |
| 440C SS | 7.6 | 1600 | 1250 | 1 | – | 55 HRC | 190 |
| 310 SS | 7.5 | | | | | | 185 |
| Fe | 7.6 | – | – | 20 | – | – | 190 |
| 2200 (2 Ni) | 7.6 | 125 | 280 | 35 | 135 | 45 HRB | 190 |
| 2700 (7.5 Ni) | 7.6 | 250 | 400 | 12 | 175 | 69 HRB | 190 |
| 4605 | 7.55 | 210 | 440 | 15 | 70 | 62 HRB | 200 |
| 4605 HT | 7.55 | 1480 | 1650 | 1 | 55 | 48 HRC | 210 |
| 4140 HT | 7.5 | 1200 | 1600 | 5 | 75 | 46 HRC | 200 |

*Table 2.5* MIM soft magnetic alloy properties

| Material | Density (g/cm³) | YS (MPa) | UTS (MPa) | Elongation (%) | Macro hardness (HRB) | Maximum permeability, μ max | Maximum $H_c$ (A/m) | B $_{1,990}$ |
|---|---|---|---|---|---|---|---|---|
| 2200 | 7.6 | 120 | 280 | 35 | 45 | 2300 | 120 | 1.45 |
| Fe–50Ni | 7.7 | 165 | 450 | 30 | 50 | 45 000 | 10 | 1.40 |
| Fe–3Si | 7.6 | 380 | 535 | 24 | 80 | 8000 | 56 | 1.45 |
| Fe–50Co | 7.7 | 150 | 200 | 1 | 80 | 5000 | 120 | 2.00 |
| 430L | 7.5 | 230 | 410 | 25 | 65 | 1500 | 140 | 1.15 |

*Table 2.6* Copper property comparison

| Material | Cu MIM Grade 1 | Cu MIM Grade 2 | Wrought C11000 | Cast 81100 | Cast 83400 |
|---|---|---|---|---|---|
| Density (g/cm³) | 8.5 | 8.4 | 8.9 | 8.9 | 8.7 |
| Thermal conductivity (W/m K) | 330 | 290 | 380 | 350 | 180 |
| Net shape capability | Excellent | Excellent | Difficult to machine | Difficult to cast | Easy to cast |

comparison to other forming methods. Generally speaking the electrical and thermal properties of MIM product are more affected by metal contamination, such as iron, and only modestly affected by the density, provided a sintered closed pore condition is achieved.

The purpose of the controlled-expansion alloys is to insure good mating and/or sealing with other materials as the materials change temperature. Table 2.7 provides controlled-expansion alloy data for F-15

*Table 2.7* Controlled-expansion alloys

| Material | Density (g/cm³) | YS (MPa) | UTS (MPa) | Elongation (%) | Hardness (HRB) | CTE (100°C) | CTE (200°C) | CTE (300°C) |
|---|---|---|---|---|---|---|---|---|
| F-15 | 7.8 | 300 | 450 | 24 | 65 | 6.6 | 5.8 | 5.4 |

*Table 2.8* Bioimplantable alloys

| Material | Density (g/cm³) | YS (MPa) | UTS (MPa) | Elongation (%) | Macro hardness (HRC) | Young's modulus (GPa) |
|---|---|---|---|---|---|---|
| F-75 | 7.8 | 520 | 1000 | 40 | 25 | 190 |
| MP35N | 8.3 | 400 | 900 | 10 | 8 | – |

*Table 2.9* Heavy alloys

| Material | ASTM-B-777-07 | Density (g/cm³) | YS (MPa) | UTS (MPa) | Elongation (%) | Macro hardness (HRC) |
|---|---|---|---|---|---|---|
| 90W–7Ni–3Fe | Class 1 | 17 | 607 | 860 | 14 | 25 |
| 90W–6Ni–4Cu | Class 1 | 17 | 620 | 758 | 8 | 24 |
| 95W–3.5Ni–1.5Fe | Class 3 | 18 | 620 | 860 | 12 | 27 |
| 95W–3.5Ni–1.5Cu | Class 3 | 18 | 586 | 793 | 7 | 27 |

alloy. F-15 is also known as Kovar™ and consists of 29% nickel, 17% cobalt and the balance iron. F-15 has a coefficient of thermal expansion that matches borosilicate (Pyrex) and alumina ceramics and is primarily used for hermetically sealing applications. Other MIM control expansion alloys such as Alloy 36, Alloy 42, and Alloy 48 exist and are basically iron with the percentage nickel added that matches the alloy number to adjust thermal expansion rate. Alloy 36 has a zero coefficient of thermal expansion until 100°C, Alloy 42 has low expansion until about 300°C and has thermal expansion behavior similar to many soft glasses. Alloy 48 has a thermal expansion behavior which matches soda lead and soda lime glasses.

Implantation of MIM components is a growing market where the primary alloys in use are F-75, MP35N, and titanium-based alloys. Table 2.8 provides MIM biocompatible alloy data for F-75 and MP35N. MIM titanium data exist, but are strongly dependent upon the manufacture of the product. MIM titanium and MIM titanium alloy properties are susceptible to carbon and oxygen impurities, thus, monitoring of these impurities in these alloys is paramount.

The last class of alloys discussed here is the tungsten-based heavy alloys, which are of interest because of their high density. These alloys find application in military, medical, cell phone, inertia balancing, and sporting goods applications. Some specific applications are inertia penetrators, cell phone vibration weights, golfing club weights, medical electrodes, and fishing and hunting weights. Table 2.9 provides tungsten heavy alloy data.

## 2.3 Dimensional capability

Metal injection molding is a very repeatable process with variability in the range 0.2 – 0.5%. This dimensional variability is associated with the amount of shrinkage that the component experiences from the time that it is molded to after it is sintered. Components shrink about 1% during the molding operation and an additional 15–25% after sintering. Also the ceramic fixtures that are used for component support during sintering may have variability in them, which results in variability of the components from one fixture to the next. Some extreme cases may have greater variability if the particular feature has a tendency to distort or if the feature lies along a parting line, ejector pin blemish, or a gate blemish. If a dimension of a component needs to have high precision, that feature should be embedded in one piece of steel and not have the negative effect of gates, parting lines, and ejector pins. Also, core pins that form holes may be tunneled into the far half of the tool to prevent the butt shut-off from forming flash that would cause variability in the inner diameter (ID) in that region. In general, MIM variability is superior to investment casting and inferior to high-precision machining.

## 2.4 Surface finish

Metal injection molding produces remarkable surface finish. Typically, 0.8 μm (32 μin) $R_a$ is achieved; however, a surface finish as smooth as 0.3–0.5 μm (12–20 μin) $R_a$ is possible. The surface finish is a function of the size and chemistry of powders that are used, the sintering conditions, and on any secondary operations, i.e. bead blasting or tumbling. Sandblast and beadblast have a tendency to increase surface roughness because of pitting, and tumbling has a tendency to decrease surface roughness. Component surface roughness can also be affected by the surface finish on the tooling used to manufacture the components. Electrical discharge machining (EDM) pits can be translated to the finished MIMed component.

## 2.5 Tooling artifacts

### 2.5.1 Parting line

The parting line is where the two halves of the mold intersect. Typically this can leave a witness line that is as small as 0.0003 in to as large as 0.001 in.

*2.1* Parting line blemish on a MIM component showing where the two halves of a tool come together.

This is highly dependent on the quality of the tool. An example of a parting line is shown in Fig. 2.1. In extreme cases, flashing can occur along the parting line in worn or poorly manufactured tools. These tools can be adjusted to eliminate this flash by either grinding or adjusting a tooling feature that is holding the tool 'open' during the molding operation. If the tool is worn, the parting line must be welded and the cavity must be EDMed.

The location of this parting line is a compromise between maintaining a low tool cost and ensuring that the witness line does not interfere with the functionality of the component. The location of the parting line is initially defined to make the tool as simple as possible. Ideally the parting line is placed so that all features can be handled in either side of the tool without the need for slides. Slides raise the cost of tooling considerably. However, one must also consider that greatest dimensional repeatability is obtained when the feature to be measured is in one piece of steel. Envision parting line flash variability causing variability in the dimension across the parting line. Also, the position to which the tool opens and closes is not identically the same with each cycle, owing to the presence of material on the mold face during processing. Although minor, this can result in a minor variability of ± 0.0003 in.

Another consideration about the parting line is related to draft angle.

Typically, draft begins at the parting line for long components. This enables the part to be easily removed from the tool. Parting lines are typically along one plane to minimize cost; however, the parting line can be stepped to accommodate features that could not be molded any other way or for components that require certain surfaces to be free of any witness lines for aesthetic or functional applications.

## 2.5.2 Ejector pin marks

Ejector pins are required to remove the green component from the tool. A sufficient quantity of these pins is required to ensure that the green component can be removed without distortion or cracking. As a natural consequence, witness marks where these ejector pins were located are evident. Figure 2.2 shows a typical ejector pin blemish. These witness marks become more evident as the tool ages owing to the wear between the pin and the cavity where the pin resides. The size of the pin should be selected to allow the cavity where the pin fits to be opened to accommodate larger ejector pins as the tool ages. Ejector pins are typically round, since round ejector pins are available in many standard sizes and the ejector pin housing in the cavity block is most easy to EDM. Rectangular ejector bars are sometime used in special cases; however, the radius associated with the corner of the bar causes issues with fit-up and long-term integrity of the steel in thin sections, since these corners can act as stress concentrators for crack initiation in the tool. Ejector pins are located where the greatest ejection force is required, for example near bosses, cored holes, and ribs. The

*2.2* Typical ejector pin marks shown on a MIM component.

**38** Handbook of metal injection molding

aesthetics and functionality of the finished component should also be considered when selecting an ejector pin location.

### 2.5.3 Gate locations

The gate is the location where the MIM feedstock flows into the cavity. As a consequence, a blemish will be present at this location on the finished product. A gate is typically located in the thickest section of the component and situated to ensure that uniform packing pressure is available across the component to prevent distortion during debinding and sintering. Gates are often situated so that the material flowing into the cavity impinges on a pin or another wall to prevent jetting of the molten feedstock across the cavity, which leads to surface flow defects. Other considerations for gate location are placement in non-conspicuous or secondary machining locations. Figures 2.3 to 2.6 show different gate configurations. Figure 2.3 shows a typical tab gate blemish, which is located along a parting line. Figure 2.4 shows a tab gate blemish that is recessed to prevent any gate vestige from interfering with the functionality of the device operation. Figure 2.5 shows a tunnel (also known as a subgate) blemish. Figure 2.6 shows a center gate that is produced by either a three-plate tool or a hot sprue. This type of gating is used to ensure uniform packing density along the length of the component, which is a round nose cup.

*2.3* Tab gate blemish along parting line on a MIM component.

Designing for metal injection molding 39

*2.4*  Recessed tab gate blemish along parting line.

*2.5*  Tunnel (sub-) gate blemish.

*2.6* Center gate blemish located for uniform packing pressure and concentricity.

## 2.6 Design considerations

### 2.6.1 Flats for sintering

One of the key features that should be considered when designing a MIM component is how the component will be fixtured or set during the sintering operation. The MIM material is prone to distortion during the thermal debinding and sintering operation if insufficient support is provided. To eliminate distortion, a flat is designed on the part for sintering which also allows low-cost standard flat fixturing to be utilized. The fixturing is typically ceramic for most MIM materials and may have holes in it to accommodate bosses that the component may have along the flat surface.

In cases where it is not possible to design a flat into the component, a contoured fixture which follows the shape of the component can be used. This ceramic fixture is typically designed for the as-molded green state component size. Some fixturing can have features that accommodate both the green size and the sintered size. In this design, the component shrinks from the green state support into the sintered state support. In general, the greatest amount of support is required in the green state since the softening of the polymer during the thermal debinding operation is the weakest condition of the component during the entire MIM process. Alternatives to contoured ceramic fixturing include the use of 'molded-in' supports to the actual component that can be removed using a secondary operation after sintering. Another technique is to use a cut ceramic shim that is cut to the desired height dimension of the sintered component.

### 2.6.2 Wall thickness

Wall thickness should be maintained as uniformly as possible to avoid warping and subsequent dimensional variability of the components during processing. Warpage can be the result of differences in cross-sectional thickness caused by variations in packing pressures during the molding operation, differences in binder removal time during the thermal debinding, and differences in thermal mass during the sintering operation. Other issues associated with large cross-sectional thicknesses are sinks, the potential for voids, and blister defects associated with difficulty in binder removal.

Wall thickness greater than 15 mm (0.6 in) should be avoided and wall thicknesses below 10 mm (0.4 in) are ideal. On the lower end of the spectrum, some technologies can achieve a wall thickness of 25–50 microns (0.001–0.002 in). These can be achieved over a short span, but as the span increases the likelihood of success decreases due to inability to fill or due to air entrapment.

A one-to-one ratio on section thicknesses is desired. If this is not possible because of design constraints, one should consider coring out the thick section with webs that are of the same thickness as the thin section on the component. Different wall thickness design considerations are illustrated in Fig. 2.7. Typically, the change in thicknesses should not be less than 60% of the main body of the component. If a uniform wall thickness is not allowed because of design considerations, a transition of wall thickness should follow a thickness change over a distance of three times the thickness change desired, as shown in Fig. 2.8.

### 2.6.3 Draft

Draft is a standard requirement for any type of injection mold to ensure easy release of components from the tool. Draft is a change in the tooling dimensions or an angle in a direction parallel to tool movement, as illustrated in Fig. 2.9. Draft should be specified to be as great as the design permits; however, MIM can have a minimum of draft since some binders can act as a lubricant. As a nominal recommendation, 0.5–2° of draft should be specified. As the length of the component element becomes longer or if the surface is textured, a greater draft should be used. Often the draft is selected to be within the tolerance of the dimension that the draft affects.

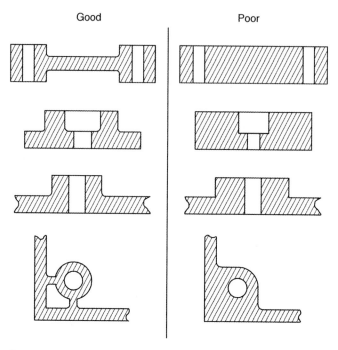

*2.7* Illustration of good and poor practice in wall thickness.

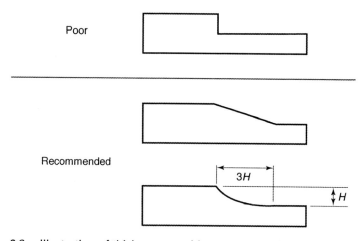

*2.8* Illustration of thickness transition recommendations.

External dimensions require none to minimal draft since the MIM material shrinks away from the wall during the cooling stage of the molding operation.

Features that the MIM material shrinks onto are most often drafted to permit easy component removal. An example would be a core pin having a

Designing for metal injection molding    43

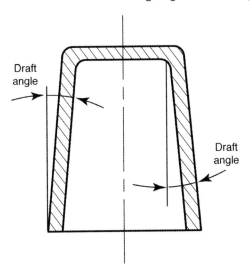

*2.9* Illustration of both an inside and outside draft angle to allow easy component removal from the tool.

draft or taper of 0.5–1° with the largest diameter dimension deepest into the tool and the small diameter dimension at the end of the core pin. This allows the part to slide easily off the pin with minimum friction, as upon initial ejection, the part is freed from all surfaces. A component may have draft on the stationary half of the tool to permit easy release during mold opening and may not have any draft on the movable half of the tool; this is to ensure that the component remains on the movable half, so that the ejectors can be used to remove the component from the tool on a consistent basis and the molding machine can be run continuously in auto cycle. The mold parting line can be located to split the draft in two different directions in order to minimize the change in dimension from one side of the part to the other. Different inside diameter drafting is illustrated in Fig. 2.10; the draft meeting in the middle requires less tolerance on the dimension than the configuration where the pin is only drafted in one direction. Reverse draft is

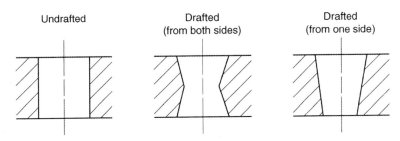

*2.10* Comparison of two drafted inside diameters with an undrafted inside diameter.

## 44  Handbook of metal injection molding

also used in some novel applications to pull the component out of mold features on the stationary half of tooling that could not otherwise be formed on the moving half of the molding machine.

### 2.6.4 Threads

Threads are possible to produce using MIM. Outside threads are commonly and economically produced using a parting line which runs the length of the thread, as shown in Fig. 2.11. This parting line can encompass each thread or it can have a 0.005 in to 0.010 in flat along its length which produces incomplete threads along the two sides next to the parting line. If the parting line is designed to produce a complete thread, flash from the parting line can interfere with thread functionality as the tool wears. When a flat is used to provide a good tool shut-off condition, as shown in Fig. 2.12, flash or vestige is prevented from interfering with the thread; however, this may lead to insufficient thread engagement strength for some applications. Alternately, a pneumatic or hydraulic actuated drive is used to rotate the threaded tool member in and out during the molding process after molding and before or during the ejection stage to form a male thread. Internal threads are produced exclusively by using a core that mimics the thread. In this type of tooling a pneumatic or hydraulic actuated drive is used to rotate the threaded core in and out as part of the molding process after molding

*2.11* Typical external thread produced by metal injection molding.

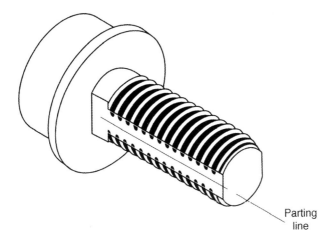

*2.12* Thread configuration with a flat to prevent flash from interfering with thread operation.

and before or during the ejection stage. This tooling method is expensive and thus is limited to high-volume applications.

Thread quality is a concern when using the MIM process. Molded threads are inferior to those threads that are machined. MIM is prone to anisotropic shrinkage which may result in shrinkage differences of a few tenths across the component; which shows up as interference on tight thread tolerances. Coarse threads are more practical than fine threads for MIM. Thread engagement length should be minimized to minimize interference of the threads with the matching component due to the potential for MIM thread variability. When exterior threads are used, e,f, and g tolerance grades should be utilized. If an internal thread is designed, a G tolerance should be specified in the design. In general, internal thread pitch diameters should be on the large side of specification and external thread pitch diameters should be on the small side of specification.

## 2.6.5 Ribs and webs

Ribs and webs are utilized to reinforce thin sections and also as replacement for thick sections. Ribs increase the bending stiffness by increasing the moment of inertia (bending stiffness = $E$ (Young's modulus) × $I$ (moment of inertia)). Consequently, ribs help to strengthen the MIM component for both processing and application. Strengthening during processing helps to enhance dimensional stability and prevent warpage. The ribs also act to enhance flow along thin sections. However, if too thick, the ribs may cause sinks on the opposite flat side where the rib mates with the main body of the component, as shown in Fig. 2.13. They may also result in warpage if the

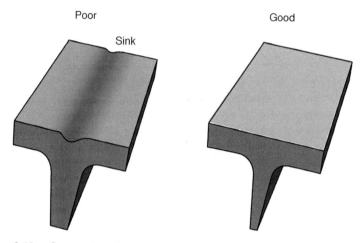

*2.13* Comparison between good and poor practice in rib design thickness for a MIM component. Notice that the oversized rib will cause a sink to form.

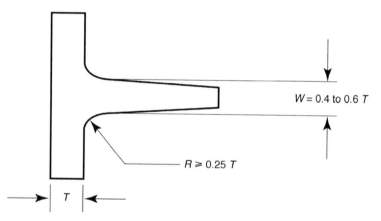

*2.14* Design rules for the design of a rib in MIM tooling.

wrong thickness of rib is utilized. Ideally the rib thickness should be 40–60% the size of the section on which it is located and the height should be no more than three times the thickness of the rib. Also, these ribs should have a good radius at their base to prevent cracking. Figure 2.14 illustrates best rib design practice. If two ribs intersect, the local thickness will be greater than an individual rib, thus coring at this intersection can be used. Gussets used for strengthening should also follow these same design rules.

Designing for metal injection molding    47

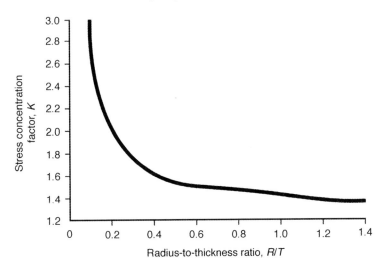

*2.15* The stress concentration factor (K) as a function of radius to thickness ratio (R/T).

### 2.6.6 Radii

Radii or fillets are an intrinsic benefit of the MIM process. They are also desirable for processability in MIM for three reasons – stress relief, component filling during molding, and simplified tooling design. Of most importance is that MIM material is generally brittle and 'notch sensitive', thus, the radii or fillets remove the stress concentrators and prevent cracking during ejection and during subsequent handling and thermal processing. As can be seen from Fig. 2.15, the stress concentration factor ($K$) gradually increases until a radius-to-thickness ($R/T$) ratio of 0.4. The actual stress is thus $K$ multiplied by calculated stress. Ideally the $R/T$ ratio should be maintained at or below 0.5, as illustrated in Fig. 2.16. As a general rule, a radius greater than 0.005 in on inside corners should be specified; however,

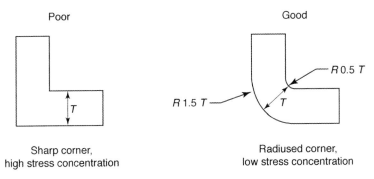

*2.16* Design considerations for radii for MIM processing.

smaller radii can be specified in some special applications. Radii also provide better flow of material in the tool during molding and reduce the potential for not completely filling out sharp corners. In tooling fabrication, the radii are in most cases easier to produce than a sharp corner. A sharp corner can be obtained if multiple pieces of steel are used, for example in a laminated tool. One condition that benefits tooling cost by not having a radius on a design is when a part geometry can be formed in only one half of the tool and the other mold half has no tooling features and is essentially flat; in this way a sharp corner is formed at the parting line and the tooling is simplified and reduced in cost.

### 2.6.7 Bosses

Bosses are often easy to utilize in a MIM component and can provide points for welding, alignment/indexing, sintering support, and for engagement with other components. Wall thickness for bosses should be designed similarly to webs, where a 0.4 to 0.6 ratio between boss thickness and the localized wall thickness is preferred to prevent the formation of sinks and distortion in the component's localized wall. Since the MIM method requires the coring out of thick sections, a loss of structural support for bolts requires that the localized area has a round boss to accommodate the added load of bolt attachment.

### 2.6.8 Undercuts

External undercuts are possible on components and do not excessively increase the cost of the tooling if they are in the direction of pull or at 90° from the parting line of the tool. They can also be added to the component by using a tooling slide if orientated along a parting line. Internal undercuts are possible by using collapsible cores or by using leachable polymeric cores. Design limitations on a collapsible core are to have a fairly large core that can have a small collapsible core feature. Internal cavities can also be manufactured by the use of sacrificial polymers, which can be chemically or thermally removed during subsequent MIM operations. The polymeric core must be selected so that it can withstand the temperature of the subsequent MIM overmolding and can be easily removed thermally or with a solvent that does not affect the polymers of the MIM material.

### 2.6.9 Decorative features

Decorative or functional features can easily be placed on the component of interest. These may be knurls, texturing, lettering, logos, part numbers, cavity identification, etc. The lowest cost tooling for lettering features are

*2.17* Lettering produced by MIM processing.

when the lettering is raised on the final component. These features are produced by simply engraving into the tool cavity or tool component. If recessed letters or features are desired, an electrode or hard turning must be used to remove the material to leave steel where the cavity features are to be located. An interesting option is to use raised features to provide color contrast on a component. For example, a component can have a black oxide finish and receive subsequent grinding of the raised features to provide a silver on black finish. Figure 2.17 shows an example of MIM manufactured lettering.

## 2.7 Sources of further information

European Powder Metallurgy Association (EPMA) (2009), *'Metal Injection Moulding – A Manufacturing Process for Precision Engineering Components'*, 2nd edition, updating 1st edition of 2004, EPMA, Shrewsbury, UK.

German, R. M. (2003), *'Powder Injection Molding – Design and Applications'*, Innovative Materials Solutions, Inc., PA, USA.

Metal Powders Industry Federation (MPIF) (2007), *'Material Standards for Metal Injection Molded Parts'*, MPIF, Princeton, NJ, USA, MPIF Standard # 35, 1993-942007 edition.

# 3
# Powders for metal injection molding (MIM)

D.F. HEANEY, Advanced Powder Products, Inc., USA

**Abstract:** This chapter is devoted to the powders used for metal injection molding and serves as an introduction to the most prevalent forms of MIM powder fabrication. The ideal powder characteristics are provided to offer guidance in selecting various grades that a user may require. The different methods of powder fabrication and resulting powder characteristics are provided. The subsequent effect of these powder characteristics on the MIM process is touched upon and illustrates the need for knowledge of powder fabrication technique in determining the proper powder for alloy development in a MIM process. Finally, the different techniques commonly used for MIM powder characterization are presented.

**Key words:** MIM powders, powder characterization, powder manufacturing techniques.

## 3.1 Introduction

Metal powders that can be manufactured to a sufficiently small size (<45 µm), loaded sufficiently into a polymeric carrier, and sintered to a sufficiently high density can be utilized for metal injection molding. Powder that is less than 22 µm average particle size is optimal. These powders can be produced using various methods; however, the different methods produce different powder characteristics that influence the final components in density, dimensions, distortion, etc. These powders are characterized using methods for small particles, thus, many characterizing methods such as screen analysis are not sufficiently accurate to monitor and predict the MIM process outcomes. This chapter is devoted to MIM powders, the different methods of powder fabrication, the characterization of MIM powders, and the subsequent effects of powder geometry or manufacturing method on the MIM process.

## 3.2 Ideal MIM powder characteristics

Many metal powders are suitable for MIM provided they are sufficiently small, sinterable, and do not exhibit strong sintering behavior at temperatures where binders are being removed. Magnesium and aluminum are not typical for MIM because of their low melting temperature and strong oxides, which interfere with sintering; however, aluminum has successfully been MIMed[1] with limited commercial success. More typical MIM alloys consist of stainless steels, low-alloy steels, tool steels, copper and its alloys, titanium and its alloys, soft magnetic alloys, refractory metals both in their pure state and as heavy alloys, and finally cemented carbides. The powders for these metals and alloys are either gas atomized, water atomized, or based on elemental mixtures of chemically or mechanically produced powders.

The ideal MIM powder characteristics are as follows:

- less than 22 μm D90 particle size for most alloys, i.e. the stainless, low-alloy steels etc; the powders for the refractory metals and cemented carbides are typically less than 5 μm D90;
- high packing density in order to load the most powder into polymer as practically possible;
- high surface purity to maintain uniform interaction with polymers and promote sintering;
- deagglomerated in the case of refractory metals and other chemically produced particles;
- spherical particle shape; however, many powders that are not spherical are used for MIM – these powders typically have lower solids loading and subsequently show greater shrinkage since they cannot be loaded at high levels into binder systems;
- sufficient interparticle friction to maintain shape during debinding; the number of particle to particle contacts per unit volume decreases as the particle size increases, thus, larger particles can experience greater distortion;
- void-free non-spongy particles to promote good sintered density and product integrity;
- minimum explosivity and toxicity; the finer the particle, the greater the surface area and the greater the possibility for explosion – this is particularly true for titanium, aluminum, and zirconium powders.

### 3.2.1 Size

Small particles are desirable to allow sinter densification of MIM components. Typically, average particle sizes less than 22 μm are used.

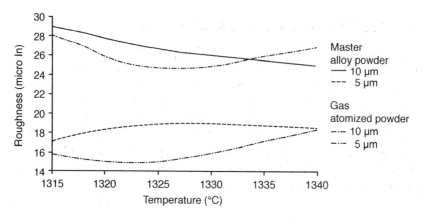

*3.1* The effect of particle size on surface finish for a 17-4PHSS alloy that was fabricated using both gas atomized and master alloy techniques.

MIM components made from refractory metals and tungsten carbide hard metal powders start with particles that are typically less than 5 μm in size. The finer particles of these metals and cermets (WC) are a result of the manufacturing method – they have too high a melting temperature to be atomized, thus, they are produced using a chemical precipitation or reduction method.

If smaller features or better surface finish are desired, a smaller particle size such as a 5 μm or 10 μm D90 should be used. The effect of particle size on surface finish is illustrated in Figure 3.1 for a 17-4PHSS material that was fabricated from both gas atomized and master alloy technique (alloying techniques are discussed later in this chapter). In this set of observations a 5 μm powder size showed a better surface finish for both the prealloyed and master alloyed technique. A slightly better surface finish was obtained for gas atomized materials as compared to master alloyed material. The alloy homogenization that takes place during the sintering of the master alloyed material may have caused the slightly greater surface roughness.

If cost is a consideration, a larger particle (> 30 μm) can be used;[2] however, the larger particle size can result in process difficulties such as screw seizure owing to the particles being trapped between the check ring and the barrel. Also, the debinding strength is reduced, which can increase distortion and the potential for defects formation. In the case of using larger particles, one should use a larger clearance check ring in the molding operation and a lower quantity of backbone polymer. The greater clearance in the screw and barrel assembly is to prevent the seizure of the check ring to the barrel as a result of the galling of particles in this tight clearance area. The lower quantity of backbone polymer is to reduce the chance for defect formation during debinding and sintering. When larger particles are packed together, they do not have as much inherent strength when compared to smaller particles, since the large particles have less particle to particle contacts per unit volume as compared to smaller particles. This concept is further supported by mineral loaded polymer which shows a greater strength with smaller particles.[3]

## 3.2.2 Size distribution

Particle size distribution is also very important. A typical particle size distribution is shown in Figure 3.2. A particle size distribution is characterized by a D10, a D50, and a D90. In Figure 3.2, the D10 ≈ 2 μm, the D50 ≈ 5 μm, and the D90 ≈ 10 μm. The D10 is a size where 10% of the particles are less than this size, a D50 is a size where 50% of the particles are less than this size, and a D90 is the size where 90% of the particles are less than this size. Essentially, the D10 and the D90 give an indication of particle sizes on the tails of the particle size distribution. Powder producers typically sell their powders as D90 sizes, for example a D90 of −22 μm has 90% of the particles less than 22 μm; however, one must be careful since some powders are sold as D80 sizes, which means that 80% of the particles are less than this size. Thus a D80 of −22 μm would have a larger particle size than the D90 of −22 μm.

If two lots of powders have the same D50, but different D10 and D90 values, the lot of material may behave very differently in compounding, molding, and sintering. A smaller D10 will have more fines and subsequently sinter better, but load poorly into a binder system. A large D90 will have more coarse particles and sinter poorly, and may show

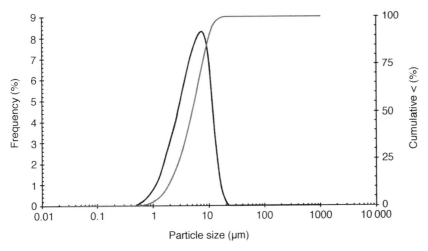

*3.2* A typical MIM powder particle size distribution where the D10 ≈ 2 μm, the D50 ≈ 5 μm, and the D90 ≈ 10 μm.

distortion or cracking since there are few particle to particle contacts per unit volume as the particle size increases. These issues are not of too much concern with gas atomized materials; however, any elemental powder, particularly those that receive an attrition step to reduce size, may have greatly different D10 and D90 values and cause process difficulties or improper component size and density from one powder lot to the next. MIM powders can be taken from dust collectors of a powder attrition operation, thus variability in the particle size may exist depending on the attrition conditions. Knowledge of the D10, D50, and D90 on each powder lot is critical to ensure that the powders being used are consistent and will behave consistently from one powder lot to the next.

### 3.2.3 Shape

Spherical powders are preferred for the high packing density and flow characteristics of MIM feedstock; however, some advantage in shape retention has been seen with slightly irregular shaped particles.[4] Enhanced shape retention for spherical particles can be obtained by decreasing the particle size to obtain greater particle to particle contacts per unit volume. Figure 3.3 shows the effect of particle shape on the packing density of monosized particles.[5,6] One can clearly see that the more spherical the powder, the higher the packing density. This also translates to the higher critical solids loading in the MIM feedstock. Consequently, the spherical powders have the potential for less shrinkage than non-spherical powders as

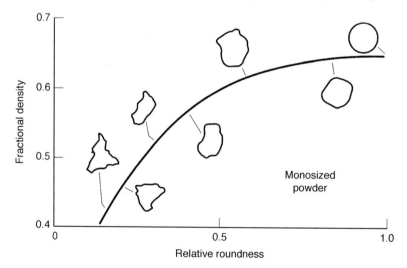

*3.3* The effects of particle shape on the packing density of monosized particles – the more spherical the powder, the higher the packing density.[5,6]

a result of their higher packing density and subsequent higher solids loading. Gas atomized powders, which are spherical, typically have the greatest critical solids loading and the potential for the least shrinkage. Elemental powders, made by chemical reduction or precipitation, typically have the greatest shrinkage since their solids loading is the lowest. Example solids loadings for gas atomized powders are 60–67 volume percent. The solids loadings for chemical precipitated or reduced powders are in the 50–62 volume percent range.

## 3.3 Characterizing MIM powders

Test methods that are typically used for conventional powder pressing such as sieve analysis (MPIF 05, ASTM B 214, ISO 4497), flow rate (MPIF 03, ASTM B 213, ISO 4490), and free flowing apparent density (MPIF 04, ASTM B 212, ISO 3923-1) are not well suited for MIM powders. These methods are appropriate for larger particles that are pressed; however, the MIM process requires powder characterization that is more suited for the smaller MIM powders. Powder characteristics such as pycnometer density, apparent density, tap density, and particle size distribution are better suited for characterizing MIM powders. Other characterization techniques such as BET (Brunauer–Emmett–Teller) surface area may be used in support of some unique applications.

### 3.3.1 Pycnometer density (MPIF 63, ASTM D 2638, ASTM D 4892)

The pycnometer density of the powder provides the theoretical density of the powder and can also provide an indication of issues with internal voids within a powder. For example, if a drop of pycnometer density is observed either within different particle size cuts or from lot to lot, a likely cause could be internal voids due to the fabrication technique. The pycnometer density also provides a practical test to evaluate the powder for any gross changes from lot to lot. Furthermore, the pycnometer density is used when determining the proper powder/polymer mixture or feedstock. Both the powders and polymers can be evaluated by pycnometer density to obtain the proper solids loading in a mixture for predictable shrinkage. For example, the knowledge of the material density allows the weight of the constituents to be used to prepare a feedstock with accurate volumetric solids loading. Typically, tooling size is determined by accurate solids loading and final sintered density; thus, pycnometer density is an essential first step for accurate feedstock determination and can be utilized to minimize the number of iterations required to obtain accurate shrinkages for tooling.

### 3.3.2 Apparent density (MPIF 28 and 48, ASTM B 417 and B 703, ISO 3923-1 and 3953)

Apparent density is the bulk density of the powder. It provides the mass per unit volume of loose packed powders. This value is a first, low-cost evaluation of a powder to determine consistency from lot to lot. A low apparent density can be an indication of fine particles and a high apparent density can be an indication of large particles. A change in apparent density can also indicate a change in the surface roughness of the powder; for example, atomization satellites may reduce apparent density. Also, if a powder is heavily agglomerated, this may appear as an increase in apparent density.

### 3.3.3 Tap density (MPIF 46, ASTM B 527, ISO 3953)

The tap density is essentially the bulk density of the powder after it has been tapped in a graduated cylinder until no appreciable change in volume is seen. Automated devices exist where the powder is mechanically tapped at a rate of 100–50 taps/min; typically 500–1000 taps is sufficient to obtain accurate densities. Hand tapping on a rubber pad can also be used if an automated device is unavailable. This density provides information on how the powders pack together and a first indication on how well the powders will load into a feedstock. Typically, the higher the powder's tapped packing density, the higher the subsequent MIM solids loading.

### 3.3.4 Particle size distribution (ASTM B 822 - 10, ISO 13320-1)

Particle size distribution (PSD) is typically measured using laser scattering or diffraction techniques for MIM powders. In this technique the 'halo' of diffracted light is measured on particles suspended in a liquid. Essentially the angle of diffraction increases as the particle size increases. The method is good for particles in the 0.1–1000 µm range. Figure 3.2 shows a particle size distribution measured using this technique for a typical MIM powder. Since each manufacturer of these devices uses a numeric algorithm to convert from 'halo' to particle size, disparity in size may exist depending on the machine that is used. However, for the particle sizes of importance to MIM, the results obtained on different manufacturing equipment are very comparable.

*Table 3.1* Manufacturing methods and attributes for MIM powders

| Manufacturing method | Relative expense | Example metals or alloys | Particle size (µm) | Particle shape |
|---|---|---|---|---|
| Gas atomized | High | Stainless steels, super alloys F75, MP35N, titanium, master alloy additives | 5–45 | Spherical |
| Water atomized | Moderate | Same as gas atomized above, except for titanium and titanium alloys | 5–45 | Semi-spherical/ irregular |
| Thermal decomposition | Moderate | Iron, nickel | 0.2–20 | Spherical, spiky |
| Chemical reduction | High/moderate | Tungsten, molybdenum | 0.1–10 | Angular, spherical |

## 3.4 Different MIM powder fabrication techniques

There are multiple powder fabrication techniques that can be used to produce MIM grade powders. These techniques primarily consist of:

- gas atomization;
- water atomization;
- thermal decomposition;
- chemical reduction.

Powders from other methods such as mechanical crushing/grounding are typically used when minor powder additions are needed for designer alloys or in an elemental powder blend to produce a particular alloy. One unique exception is the carburization of pure tungsten powder to produce tungsten carbide grade powders.[7] Table 3.1 lists powder physical attributes, their relative cost, and typical metals and alloys that are produced from these different types of powders.[8] Further information on particle fabrication techniques can be found elsewhere.[9]

Powder classification for particle size and size distribution is an important step in the production of powders for MIM since many MIM powders are taken from a larger lot of powders that have different particle sizes and must be carefully removed to ensure that the MIM powders are consistent on a lot-to-lot basis.

### 3.4.1 Gas atomization

Gas atomized powders are produced by melting the metal or alloy by induction or other method of heating and subsequently forcing the melt through a nozzle. After the liquid metal or alloy leaves the nozzle it is struck

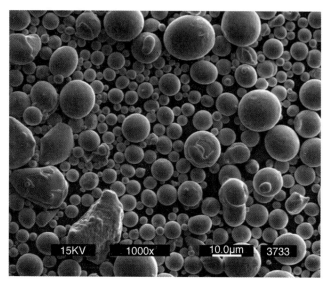

*3.4* A typical SEM image of a gas atomized powder stainless steel powder.

by a high-velocity stream of gas, which breaks the melt into fine droplets, these then solidify into spherical shaped particles during free fall. The gas is typically nitrogen, argon, or helium; however, air can be used for some powders. The air atomized particles show higher surface oxidation; thus, air atomization is not recommended for most engineering materials, particularly those that are difficult to reduce in a subsequent sintering step. Powder free fall takes place in a large container that permits the solidification of the particles prior to their contact with the container wall. If turbulence exists near the nozzle during atomization, small solidified particles will become reintroduced into the atomizing melt stream and result in satellites on the particle surface. These irregularities will interfere with the powder packing density and subsequent flow behavior in a MIM feedstock. The process can produce a wide distribution of particle sizes, which requires downstream classification such as screening or air classification. Particles that are too large are reused for future atomization. A typical scanning electron microscopy (SEM) image of a gas atomized powder is shown in Figure 3.4. These particles are known for their spherical shape, high surface purity, and high packing density.

## 3.4.2 Water atomization

Water atomization is generally similar to gas atomization in most aspects except that water and not gas is used to break up the melt stream into fine

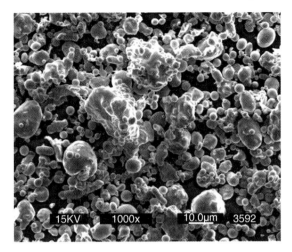

*3.5* A typical SEM image of a water atomized powder stainless steel powder.

particles. A high-pressure water jet impinges on the melt stream to rapidly break up and solidify the powders. The use of a superheated melt produces the most spherical particles and a higher water pressure results in the finest particles, thus water atomizing at superheated temperatures and at high water pressures are important for powders used for MIM. As with gas atomization, powder size classification is an important step for producing powders for MIM. A typical SEM image of a water atomized powder is shown in Figure 3.5. These particles are typically slightly irregularly shaped and show greater surface oxidation than gas atomization particles. The irregular shape does have some advantages for shape retention during debinding. The production rate using water is much higher than that with gas, thus, the cost of the water atomized material is less than gas atomized powders.

### 3.4.3 Thermal decomposition

Thermal decomposition is a chemical decomposition caused by heat and is commonly used to produce nickel and iron powders for MIM. Tungsten and cobalt are also candidates for this technique. This method produces powders that have purity $>99\%$ and particle sizes in the 0.2–20 µm range. In this process, the metal of interest is reacted with carbon monoxide at high pressure and temperature to produce a metal carbonyl. This metal carbonyl liquid is purified, cooled, and subsequently reheated in the presence of a catalyst, which results in vapor decomposition into a powder. A typical SEM image of a thermally decomposed iron carbonyl powder is shown in Figure 3.6. These powders typically have carbon impurity that must be

*3.6* A typical SEM image of a thermally decomposed iron carbonyl powder.

reduced in hydrogen prior to use, during sintering, or calculated as an alloying ingredient for low-alloy steels. If the powder is subsequently reduced prior to MIM, the powder must be deagglomerated by milling as the particles are bonded together during reduction. Furthermore, the sintering activity of these reduced powders is not as great as the unreduced powders because the finest particles become sufficiently sintered or assimilated with the larger particles during the reduction process.

### 3.4.4 Chemical reduction

Oxide reduction is one of the oldest known methods for producing powders. The process utilizes a purified oxide which is subsequently reduced using a reducing agent such as carbon to form carbon monoxide or carbon dioxide. Hydrogen can also be used to reduce the oxide powder into a metal powder. Particle size can be reduced using a lower reaction temperature; however, the reaction kinetics are slow. The process can be accelerated by using a higher temperature; however, the higher temperature results in diffusion bonds between the particles. This agglomeration must be subsequently removed by attrition or milling to a sufficiently fine particle size. If the particles are not milled, the agglomerated particles will not load properly into a binder system – resulting in high, unstable viscosity and variability in shot-to-shot mass and uniformity during the molding operation. A typical SEM image of a chemically reduced tungsten powder is shown in Figure 3.7.

*3.7* A typical SEM image of a chemically reduced tungsten powder.

## 3.5 Different alloying methods

There are three primary methods for alloying in metal injection molding: elemental, prealloy, and master alloy. Powder selection is intimately linked with alloying methods for metal injection molding because the powders are either the exact stoichiometry of interest or must be blended to acquire the appropriate stoichiometry. Therefore, to select the appropriate powder, one must know how they will alloy.

### 3.5.1 Elemental method

The elemental method requires a blend of elemental powders in the appropriate ratios to produce the desired stoichiometry. Gas atomization is used for copper or titanium. Thermal decomposition is used for carbonyl iron and carbonyl nickel. Chemical reduction is used for tungsten and molybdenum. Mechanical grinding and milling is used for milling of electrolytic chromium. Example mixtures could be carbonyl iron mixed with carbonyl nickel for nickel steels, where the carbon can be obtained directly from the carbon impurities in unreduced carbonyl iron. The mixture of iron and nickel can also be used to form magnetic 50Fe/50Ni. Others consist of electrolytic and ground chromium added to iron to form steels and stainless steels. Heavy alloys are also produced using elemental blends – for example the addition of iron and nickel carbonyl powders to chemically reduced tungsten. The practitioner must closely monitor the particle specifications from lot to lot since the methods of manufacturing these particles may produce different particle size distributions, even if they are in the producer's specification. Lot-to-lot particle size variability of minor elemental additions, <3–5% typically, have a less significant effect on the mixture's viscosity and sintering behavior. However, if the addition is a substantial component in the mixture, the lot-to-lot variability may influence the manufacturability, i.e. injection molding consistency, sintered density, and dimensions.

### 3.5.2 Prealloy method

The prealloy method is the use of a powder that is the exact stoichiometry of the alloy of interest. The powders for this alloying method are typically gas or water atomized. Example alloys are stainless steels, super alloys, and titanium alloys. These alloys are typically very consistent in particle size.

### 3.5.3 Master alloy method

The master alloy method is the use of an elemental powder with a gas or water atomized powder, where the gas or water atomized powder is enriched in certain alloying elements. An example would be one part of a gas atomized 55Cr38Ni7Mo master alloy mixed with two parts carbonyl iron powder to produce a 316L stainless steel. When the elemental powder is mixed and alloyed with the enriched atomized material, the enriched elements are diluted to the stoichiometry of interest. This method is common for most stainless steels and for some low-alloy steels. Slightly inferior properties have been observed for stainless steels;[10] however, the fine carbonyl powder used in the master alloy technique provides a means to enhance debinding strength, since a finer elemental powder provides more particle to particle contacts and a lower sintering temperature due to the smaller particle size. Care must be taken in the selection of the elemental addition; for example, if unreduced carbonyl iron powder is used for a stainless blend, the carbon may not be entirely removed in the sintering operation unless an appropriate sinter cycle is selected to allow for reduction of this powder before pore closure.

## 3.6 References

1. Tan, L. K. and Johnson, J. L. (2004), 'Metal injection molding of heat sink', *Electronics Cooling,* 1 November.
2. Baum, L. W. and Wright, M. (1999), 'Composition and process for metal injection molding', US Patent 5,993,507.
3. Bohse, J., Grellman, S., and Seidler, S. (1991), 'Micromechanical interpretation of fracture toughness of particulate filled thermoplastics', *Journal of Materials Science,* 26, 6715–6721.
4. Heaney, D., Zauner, R., Binet, C., Cowan, K., and Piemme, J. (2004), 'Variability of powder characteristics and their effect on dimensional variability of powder injection molded components', *Journal of Powder Metallurgy,* 47(2), 145–150.

5. Brown, G. G. (1950), *Unit Operations,* Wiley, New York.
6. German, R. M. (1989), *Particle Packing Characteristics,* MPIF, Princeton, NJ, USA, p. 123.
7. Lassner E. and Schubert, W. D. (1999), *Tungsten,* Kluwer Academic/Plenum Publishers, pp. 324–344.
8. German, R. M. and Bose, A. (1997), *Injection Molding of Metals and Ceramics,* MPIF, Princeton, NJ, USA, p. 67.
9. German, R. M. (2005), *Powder Metallurgy and Particulate Materials Processing,* MPIF, Princeton, NJ, USA pp. 55–90.
10. Heaney, D., Mueller, T., and Davies, P. (2004), 'Mechanical properties of metal injection molded 316L stainless steel using both prealloy and master alloy techniques', *Journal of Powder Metallurgy,* 47(4), 367–373.

# 4
# Powder binder formulation and compound manufacture in metal injection molding (MIM)

R. K. ENNETI, Global Tungsten and Powders, USA and
V. P. ONBATTUVELLI and S. V. ATRE,
Oregon State University, USA

**Abstract:** Binders play a very crucial role in processing of components by the metal injection molding (MIM) process. The present chapter discusses critical aspects regarding the binders used in the MIM process. Initial sections of the chapter focus on the function and importance of binders in the MIM process. In the next section the binder chemistries and constituents are discussed. The effect of binder properties on feedstock and the various methods used to remove the binder after injection molding are discussed in the following section. The mixing techniques used to manufacture the feedstocks and examples of typical feed stocks compositions are presented in the final sections of the chapter.

**Key words:** binders, feedstock, chemistry, binder constituents, rheology, viscosity, thermal conductivity, heat capacity, thermal debinding, defects, mixing techniques.

## 4.1 Introduction: the role of binders

Binders play a very crucial role in processing of components by metal injection molding (MIM). Binders are multi-component mixtures of several polymers. A binder typically consists of a primary component to which various additives like dispersants, stabilizers, and plasticizers are added. The basic purpose of binders is to assist in shaping of the component during injection molding and to provide strength to the shaped component. Binders act as a medium for shaping and holding the metal particles together until the onset of sintering. The binders are mixed with metal powders to make feedstocks, which are further used as starting materials for injection

*Table 4.1* Characteristics of an ideal binder system for metal injection molding process[1]

| | Desirable characteristic |
|---|---|
| Powder interaction | Low contact angle<br>Good adhesion with powder<br>Capillary attraction of particles<br>Chemically passive with respect to powder |
| Flow characteristics | Low viscosity at the molding temperature<br>Low viscosity change during molding<br>Increase in viscosity on cooling<br>Small molecule to fit between particles |
| Debinding | Degradation temperature above molding and mixing temperatures<br>Multiple components with progressive decomposition<br>Temperatures and variable properties<br>Low residual carbon content after burnout<br>Non-corrosive and non-toxic burnout products |
| Manufacturing | Easily available and inexpensive<br>Long shelf life<br>Safe and environmentally acceptable<br>Not degraded due to cyclic heating<br>High strength and stiffness<br>Low thermal expansion coefficient<br>Soluble in common solvents<br>High lubricity<br>Short chain length and no orientation |

molding. Binders are removed after molding prior to sintering of the component.

The properties of the binders have an influence on the metal particles distribution, shaping process, dimensions of the shaped component, and the final properties of the sintered component. The important characteristics required in binders are summarized in Table 4.1. The binder should have a low contact angle with metal particles. The low contact angle will result in better wetting of binder to the powder surface, which will assist the mixing and molding process. The binder and the metal particles should be inert with respect to each other. The binder should not react with metal particles and in turn the metal particles should not polymerize or degrade the binders. The binder powder mixture (feedstock) should satisfy various rheological requirements for successful molding of the components without formation of any defects. The viscosity of the feedstock should be in an ideal range for successful molding. Viscosity that is too low during the molding process will result in separation of powders and binders. On the other hand, viscosity that is too high will impair the mixing and molding process. Apart from the requirement of an ideal viscosity range during the molding process, the feedstock should also have the characteristic of a large increase in viscosity on cooling. The large increase in viscosity will assist in preserving the shape of the compact on cooling.

The binder should possess the characteristic of fast removal during debinding, without forming defects in the injection molded component. The green part is most susceptible to formation of defects during the debinding stage. The binder, which provides strength, is removed, gradually increasing the susceptibility of the green part to formation of defects. The absence of open porosity during the initial stages of thermal debinding results in the formation of defects such as cracking, blistering, etc. The stresses arising owing to trapped degradation products as a result of polymer burnout will lead to the formation of defects. To avoid this scenario, the binders are typically designed to have multiple components which undergo decomposition at different temperatures. In this case the debinding process occurs in two stages. In the first stage, one component of the binder system is removed, initially resulting in creation of open pores in the green component. During this stage the remaining components of the binder system will provide strength to the metal particles and retain the shape. The remaining components of the binder system are gradually removed during the second stage of the debinding process. The two-stage debinding process results in faster removal of binders from the green components. The binder should also possess the characteristic of complete burnout without leaving any residual carbon. The products formed due to thermal debinding should also be non-corrosive to the manufacturing equipment.

Binders used for metal injection molding must be easily available, low in cost, and should have a long shelf life. The sprues and gates are reused in the MIM process and the binder should have good recyclability and must not degrade on cyclic reheating. The binders should have high thermal conductivity and a low thermal coefficient of expansion in order to prevent defects formed as a result of thermal stresses.

It is very hard for a single binder to fulfill all the characteristics of feedstock. The binder system used in the injection molding process typically contains multiple components, each performing a specialized task. The binder system contains a major constituent and other components are mixed as additives to obtain the desired characteristics of the feedstock.

## 4.2　Binder chemistry and constituents

### 4.2.1　Binder chemistry

Thermoplastic and thermosetting are two common types of polymers. Thermoplastic polymers are formed due to repetition of small monomer groups along the chain length without cross linking. Polyethylene,

Powder binder formulation and compound manufacture in MIM    67

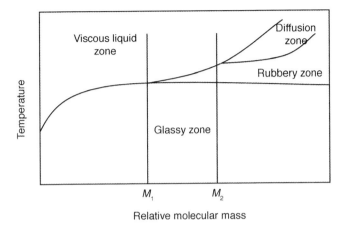

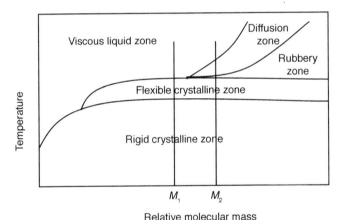

*4.1* Temperature in relation to relative molecular mass plots for amorphous (top) and crystalline (bottom) polymers.

polypropylene, polystyrene, and wax are some examples of thermoplastic polymers. The crystalline polymers have smaller chain lengths and amorphous polymers contain longer chain lengths. The amorphous polymers are more ductile in nature than crystalline polymers. The crystalline and amorphous polymers show different behaviors on exposure to high temperature (Figure 4.1). Polymers with relative mass lower than $M_1$ will display a narrow melting point, which will result in shape loss during the debinding process. Polymers with relative mass greater than $M_2$ will require a high temperature to become viscous. The high temperature will cause degradation of the polymer. Thus, ideally, the relative mass of the polymer to be used as a binder should fall between $M_1$ and $M_2$.[2]

Polymers soften and become viscous above a temperature called the glass transition temperature ($T_g$). The polymers are brittle below the glass transition temperature. Various properties like volume, thermal expansion coefficient, and enthalpy undergo major changes near the glass transition temperature. The melting point and tensile strength of the polymer depends on the molecular weight, i.e. the chain length of the polymer. The variation of this parameter with molecular weight of the polymer is shown in Figure 4.2. The viscosity ($\eta$) of the polymer also depends on the molecular weight ($M$) as per the following relation

$$\eta = \kappa M^a \qquad [4.1]$$

where $\kappa$ and $a$ are constants. The value of $a$ is very close to unity for the polymers used in the MIM process.[1]

Small chain polymers are preferred for use in MIM because of their low molecular weight. Waxes are widely used also because of their low molecular weight, thermoplastic nature, and low melting temperature. Waxes are naturally occurring ester compounds obtained from a long-chain alcohol and carboxylic acid. Some examples of waxes are paraffin wax, beeswax, and carnauba wax. The melting point of all three waxes is lower than 100°C.

Thermosetting polymers undergo cross-linking of polymer units at high temperatures. The cross-linking will result in the formation of a three-dimensional rigid structure. The rigid structure, once formed, is not reversed on reheating. The thermosetting polymer directly vaporizes on heating without melting or softening. As the thermosetting polymers do not undergo melting, the shape loss due to softening of the polymer during debinding can be avoided by using thermosetting polymers. The cross-linking of the thermosetting polymer upon heating can provide strength to the part during the debinding process. The increase in strength on debinding can be useful in

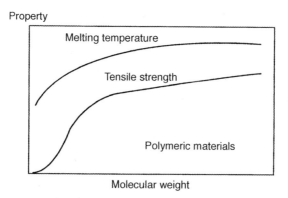

4.2 Variation of melting temperature and tensile strength with molecular weight of polymers.[1]

retaining the shape of the component. However, thermosetting polymers are rarely used in MIM. The formation of an irreversible rigid structure on heating will prevent recycling of sprues and gates from MIM. It is very difficult to burn out the thermosetting polymers completely during thermal debinding. Some residual amount of carbon is left in the components. Thus thermosetting polymers are used when there is a requirement for residual carbon during the sintering process.

### 4.2.2 Binder constituents

Dispersants, plasticizers, stabilizers and inter-molecular lubricants are some common additives added to binders in the MIM process.[2] Dispersants are added to the binder to enhance the distribution of the powder in the system. Dispersants have the unique ability to replace the powder/powder and powder/air interfaces with a powder/binder interface. The addition of dispersants enhances the solids loading.[3] The dispersants create locations for the binder to react and bond with the powder. Zinc stearate is a commonly used dispersant in the MIM process.

The plasticizers are added to the binder system to enhance the flow behavior in MIM. Camphor, dimethyl phthalate, and dibutyl phthalate are some of examples of plasticizers used in MIM. The molecules of the plasticizers contain ring-like atom groups. These ring-like atom groups will reduce the inter-molecular friction between the binder molecules and increase the flowability of the binder system.[4] Stabilizers are added to the binder system with the main aim of preventing agglomeration of the particles. The stabilizers must strongly bond to the powder particles and they should have sufficient extension in the binder blend to prevent agglomeration due to approaching particles. The metal powder particles' surface should also be completely covered by the stabilizer. In addition, the stabilizers should mix completely in the binder media used for MIM.[2]

The inter-molecular lubricants enhance the flow of the feedstock. The inter-molecular lubricants have a much lower molecular weight than the polymer and thus possess very much lower viscosity than the base polymer at the processing temperature. The inter-molecular lubricants decrease the friction between the adsorbed layer on the particles and the binder molecules. Stearic acid, wax E, wax OP, etc. are some common examples of inter-molecular lubricants.[2]

## 4.3 Binder properties and effects on feedstock

### 4.3.1 Flow: rheology as a function of shear rate, temperature, and particle attributes

The rheological characteristics, especially the viscosity of the feedstock, play a critical role while filling the mold during the injection molding process. The viscosity of the feedstock increases with addition of metal particles. The feedstocks require shear stress to exceed yield stress to initiate a viscous flow (Bingham systems). Capillary and torque rheometers are used to measure the rheological characteristics of the feedstocks. In capillary rheometers, the feedstock is forced through a small gap and the pressure drop and flow rate are measured. The torque rheometer measures the torque required for mixing the feedstock as a function of time. The capillary rheometer is preferred and widely used for characterizing the feedstocks as the test conditions (shear rates, viscosity) closely match the conditions experienced by feedstocks in the MIM process. The flow of the feedstock through the capillary rheometer is similar to the feedstock filling the mold cavity during the MIM process.

The relative viscosity, which is defined as the viscosity of the feedstock in relation to the viscosity of the binder, increases with the addition of metal particles. The relative viscosity becomes infinity when the addition of metal particles reaches a critical limit. At the limit, the feedstock becomes stiff and does not flow. The amount of metal particles in the feedstock at this limit is called the critical solids loading. In case of mono-sized spheres the limiting loading occurs at 63.7% solids. The effect of solids loading in the form of randomly distributed spherical mono-sized spheres on the viscosity of liquid was estimated by Einstein according to the following equation

$$\eta_r = 1 + 2.5\phi \qquad [4.2]$$

where $\eta_r$ is the relative viscosity and $\phi$ is the solids loading. The equation developed by Einstein can be used for solids loading of less than approximately 15%.[1] Various modeling studies have been further carried out to estimate the variation of relative viscosity with solids loading.[1, 5-8] Some of the equations developed from these studies are shown below

$$\eta_r = A(1 - \phi_r)^{-n} \qquad [4.3]$$

$$\eta_r = A\phi/(1 - B\phi) \qquad [4.4]$$

$$\eta_r = (1 - \varepsilon\phi)^{-2} \qquad [4.5]$$

$$\eta_r = 1 + A\phi_r + B\phi_r^2 \tag{4.6}$$

$$\eta_r = \left[A\bigg/\left(1 - \frac{\phi}{\phi_{\max}}\right)\right]^2 \tag{4.7}$$

$$\eta_r = A\phi^2/[1 - \phi_r^n] \tag{4.8}$$

$$\eta_r = \left(1 - \frac{\phi}{\phi_{\max}}\right)^{-2} \tag{4.9}$$

$$\eta_r = (1 + A\phi_r)/(1 - \phi_r) \tag{4.10}$$

$$\eta_r = \left(1 - \phi - A\phi^2\right)^{-n} \tag{4.11}$$

$$\eta_r = C\left\{(\phi/\phi_{\max})^{1/3}\bigg/[1 - (\phi/\phi_{\max})]^{1/3}\right\} \tag{4.12}$$

where $\eta$ is mixture viscosity, $\eta_r$ is relative viscosity, $\phi$ is solid volume fraction, $\phi^{\max}$ is maximum solid volume fraction, $\phi_r$ is relative solids loading ($\phi/\phi^{\max}$), and $A$, $B$, and $n$ are constants.

The models developed for estimating the relative viscosity of feedstocks must satisfy certain conditions. The relative viscosity of the only binder should be equal to 1. The first derivative of viscosity must be equal to 2.5 for zero volume fraction and the relative viscosity should be infinity at the maximum solids loading.

$$\lim_{\phi \to 0}[\eta_r] = 1.0 \tag{4.13}$$

$$\lim_{\phi \to 0}\left[\frac{d\eta_r}{d\phi}\right] = 2.5 \tag{4.14}$$

$$\lim_{\phi \to \phi_{\max}}[\eta_r] = \infty \tag{4.15}$$

The most suitable model based on experimental studies to estimate the change in relative viscosity with relative solids loading is listed in equation [4.16]

$$\eta_r = \eta/\eta_b = A(1 - \phi_r)^{-n} \tag{4.16}$$

where $\eta_r$ is the relative viscosity and $\phi_r$ is the relative solids loading. The exponent $n$ is found to be equal to 2.0 and the coefficient $A$ is typically near

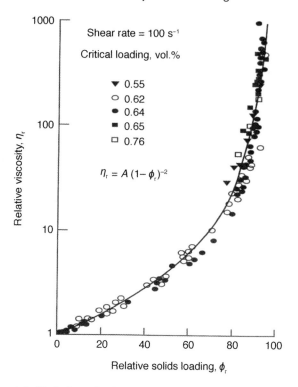

*4.3* Variation of relative viscosity with solids loading estimated from equation [4.16].[1]

unity.[1] The relative viscosity variations for various systems as shown by equation [4.16] are shown in Figure 4.3.

*Effect of shear rate*

The viscosity of the MIM feedstock is very sensitive to the shear rate. At lower shear rates the feedstocks show a yield point which corresponds to a very high viscosity. As the stresses exceed the yield point, the viscosity of the feedstocks decreases with further increase in shear rate. At higher shear rates the particles increase in volume (dilate) as they slide past one another, resulting in an increase in viscosity. The shear dilation is absent at low solids loading due to the high amount of liquid and little ordering of particles. The polydisperse systems exhibit a gradual change in viscosity with shear rate when compared to mono-sized systems. The dilation process is also absent in case of polydisperse systems. The surface active agents lower the viscosity of the feedstock but have no effect on the shape of the viscosity versus shear rate curves.

## Powder binder formulation and compound manufacture in MIM

*Effect of temperature*

The viscosity of the feedstock decreases with the increase in temperature. The decrease in viscosity with increase in temperature will be greater for the feedstock than in the case of pure binder. The larger decrease in viscosity in the feedstocks is primarily due to the combined effect of the decrease in viscosity of binder and change in volume fraction of the solids as a result of difference in thermal expansion coefficients. The lower viscosity of the feedstock at higher temperature can be utilized to increase the solids loading during the injection molding process. The mixing of the feedstocks is also easy because of the lower viscosities at high temperature. However, molding the feedstock at higher temperatures will result in an increased processing time due to longer times required for cooling and also causes defects in the molded samples due to larger contraction of binder on cooling from the higher processing temperature.

*Effect of particle characteristics*

The viscosity of the feedstocks increases with decrease in particle size of the metal particles. The smaller sized metal particles have high surface area and interparticle friction and these characteristics result in higher viscosity of the feedstocks. The powders having high packing density require a lower amount of binder to attain the viscosity required for molding. The solids loading can be increased significantly with a good packing powder. Powders with a larger difference in particle size can be tailored to improve the packing density and decrease the viscosity of the feedstock for any given fixed solids loading. The change in relative viscosity with the composition (% of small particles) for a bimodal mixture of spheres at a solids loading of 55% and a particle size rate of 21 is shown in Figure 4.4. The relative viscosity decreases initially reaching a minimum point prior to increasing again with continued addition of smaller size particles.

The shape of the metal particles has a significant effect on the relative viscosity of the feedstock. Irregular shaped powders due to high interparticle friction and low packing density result in feedstocks with high viscosities. The small sized spherical powders are ideal for injection molding because of their better flowability and low interparticle friction. The critical solids loading decreases as the particle size becomes more non-spherical. The spherical powder has the largest critical solids loading. The agglomeration of the powders also results in increase in the viscosity of the feedstock. The increase in viscosity becomes larger with the agglomerate size and with the number of particles in the agglomerate. The variation of relative viscosity with particle shape is shown in Figure 4.5.

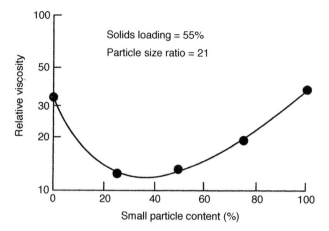

*4.4* Variation of relative viscosity with % small particle content at solids loading of 55% and particle size ratio of 21.[1]

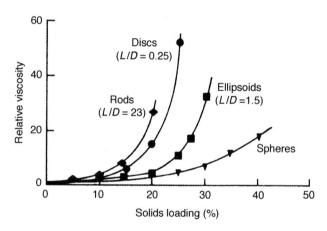

*4.5* The effect of particle shape on the relative viscosity with increase in solids loading.[1]

## 4.3.2 Solidification: thermal conductivity, heat capacity

During injection molding, the component after shaping is cooled in the die under external pressure. The pressure is removed once the compact has cooled completely. The difference in thermal properties of the binder and the metal powders results in development of internal stresses in the compact during the cooling stage. These stresses result in distortion of the compact once it is ejected from the mold. The feedstock should have the characteristic of large increase in viscosity on cooling to avoid distortion of the compacts during cooling and subsequent handling. The rule of mixtures provides a very good estimate of the thermal properties of the

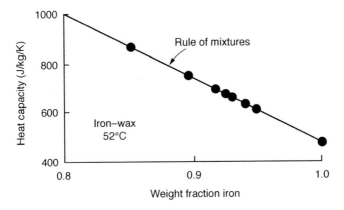

4.6 Variation of heat capacity of iron–wax feedstock with increase in iron content.[1]

feedstock. The thermal properties of the feedstock estimated according to the rule of mixtures are given below

$$A_f = A_b + \phi(A_m - A_b) \qquad [4.17]$$

where $A_f$ is the property of the feedstock, $A_b$ is the property of the pure binder, $A_m$ is the property of the metal, and $\phi$ is the solids loading.

The feedstock should possess high thermal conductivity to minimize the formation of shrinkage cracks. The thermal conductivity values of the feedstocks lie in between pure binder and metal particles. The thermal conductivity of the feedstock ($k_f$) can be estimated as

$$k_f = k_b(1 + A\phi)/(1 - B\phi) \qquad [4.18]$$

where $k_b$ is the thermal conductivity of binder, $A$ and $B$ are constants, $\phi$ is solids loading.[1] The heat capacity of the feedstock can also be estimated from the rule of mixtures. The variation of heat capacity of the feedstock (iron powder and wax) with the addition of iron particles is shown in Figure 4.6. The estimation based on rule of mixtures shows a very good agreement with the measured values of the heat capacity of the feedstock.

## 4.3.3 Removal: solubility, thermal degradation

After injection molding the next step involves removing the binders prior to sintering. The classic and widely used method of removing the binders from the green component is by providing thermal energy and burning out the binder. Polymer burnout is the widely applied method of binder removal and is commonly termed thermal debinding. The microstructure of the injection molded component in the green state is shown in Figure 4.7. All

76  Handbook of metal injection molding

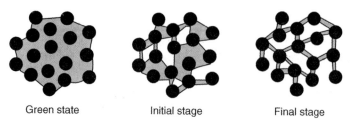

Green state          Initial stage          Final stage

*4.7* Microstructural changes of the compact during thermal debinding.[9]

the pores in the component are completely filled by binders (saturated state). The absence of open pores and the use of a large amount of binders make the burnout process very critical. During initial stages, defects like cracking, bloating, etc. are formed due to the stresses from the trapped gas formed by the decomposition of binders. In order to prevent the formation of defects, the binders are removed in two stages. Various processes like wicking and solvent extraction were developed for the initial stage of binder removal, with the main intention to form some open pores for the rapid removal of binders in the second stage by burnout. During the initial stage, lower molecular weight binders are removed and the high molecular weight binder (backbone polymer) provides handling strength to the component. Subsequently, the high molecular weight binder is removed by burnout.

*Wicking*

The wicking process relies on the capillary extraction of the liquid polymer to extract the polymer from the green part. During wicking, the green part is placed in a fine powder and heated to softening temperature of the polymer. At softening temperature the liquid polymer is extracted by the surrounding fine powder by capillary flow. Fine alumina is usually used as the wicking powder. The wicking technique has advantages in reducing the polymer removal time and also helps in preventing shape loss or distortion during polymer removal. However, the use of fine powder may result in contamination of the green part, which may require further cleaning.

*Solvent extraction*

This process relies on selective dissolution of polymer in a solvent to form an open porous network. For example, heptane is a widely used solvent to extract the lower molecular weight polymers (waxes). The polymer removal time increases with increase in thickness of the sample and with decreasing particle size. Temperature control at which the solvent extraction occurs is very critical. Too low a temperature will result in faster diffusion of solvent in the green component, resulting in swelling and cracking of the green part. On the other hand, too high a temperature will result in slumping of the green part due to softening of the background polymer. Solvent extraction has the advantage of reducing the polymer removal time and improved shape retention when compared to thermal debinding. However, the disposal and recycling of the used solvent is a major drawback of this process. Owing to the stringent environment regulations, the use of solvent extraction may be limited in future applications.

*Thermal debinding*

Thermal debinding is the widely applied method for removing the binders from the green component. The simplicity of the method, devices used, and its applicability for mass production to combine with the sintering operation has made thermal debinding the foremost method for polymer removal in industry. Even though the equipment used and the method of thermal debinding is fairly simple, the process itself is very complex. Polymer removal by burnout involves chemical and physical mechanisms. The chemical mechanisms occur because of the thermal degradation of polymer into volatile species – pyrolysis. The physical mechanisms involve diffusion of the volatile species to the surface of the compact, as well as changes in the polymer distribution within the green body. Depending on the material, its thermal conductivity, porosity, pore size, and related features, it is possible for mass diffusion or thermal diffusion to be controlling steps. As a further complication, component heating depends on heat transfer and the reaction enthalpies associated with pyrolysis events, leading to thermal–kinetic effects that couple to the chemical–physical aspects. During thermal debinding, the strength of the component decreases first due to thermal softening of the polymer and subsequently due to loss of the polymer. Simultaneously stresses (thermal, gravitational, and residual) act on the component which can generate cracks or distortion as the polymer degrades. Apart from macroscopic defects, any microscopic defects caused during thermal debinding are exaggerated during the subsequent sintering process. Contrary to general impressions, sintering does not heal debinding defects, but instead tends to amplify those defects.

Improper thermal debinding can result in carbonaceous residues that usually degrade the mechanical, optical, thermal, magnetic, or electronic properties of the sintered component. To prevent the various defects, it is common practice to employ extremely slow heating cycles, with negative implications on equipment requirements, capital expenditures, and productivity; in several situations thermal debinding is a production bottleneck. On average, the cost is about $2.5 per kg (component mass, not binder mass) to remove the polymer from small injection molded components in a typical 8 h

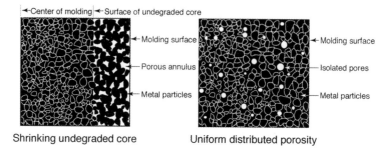

*4.8* Porosity structures used for modeling the polymer burnout process.[9]

cycle.[10] Curiously, the binder content in the component might be just 5–10 wt%, so the cost of its removal is roughly ten-fold more than the cost of the binder.

Various modeling studies developed for thermal debinding of injection molded samples are based on two hypothetical porosity developments. In the first case, the polymer–vapor interface is modeled to recede in a linear way in the compact as the polymer is removed, resulting in a porous outer layer and a core with liquid polymer. This type of porosity structure is termed 'shrinking undegraded core or series model'.[11–20] In the second case, porosity is considered to form uniformly in the compact because of distribution of liquid polymer under the influence of capillary pressure. This type of configuration is termed 'uniform distributed or parallel model'. The microstructure of the two porosity models is shown in Figure 4.8. 'Shrinking undegraded core structure' is considered appropriate for the thermal debinding in an oxygen atmosphere. In this case, the thermal debinding starts from the surface of the compact. The rate of thermal debinding depends on the exposed surface area and diffusion of oxygen in the compact. In the case of thermal debinding in neutral atmospheres, the absence of a well-defined liquid–vapor interface was experimentally observed.[21–23] The microstructure change during the thermal debinding process using scanning electron microscope (SEM) technique is shown in Figure 4.9. The formation of pendular bonds during softening of binder can be clearly seen in Figure 4.9.

*Strength model*

A model based on the *in situ* strength of the component can be used to predict the defect formation during thermal debinding. According to the model, defects in the sample occur when the stresses or pressure due to various activities in the burnout process exceed the *in situ* strength of the compact. The *in situ* strength of the sample with localized bonding due to

# Powder binder formulation and compound manufacture in MIM

Softening stage

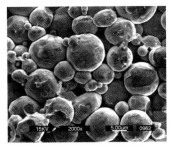
Burnout stage

*4.9* Microstructure changes during thermal debinding process.[9]

the polymer can be estimated by the Rumpf equation[24,25]

$$\sigma_c = 1.1 \frac{1-\varepsilon}{\varepsilon} \frac{H}{d^2} \qquad [4.19]$$

where $\varepsilon$ is the porosity in the agglomerate, $H$ is the bonding force, and $d$ is the diameter of the particle. The bonding force $H$ is estimated from

$$H = \sigma_0 \left(\frac{\pi x^2}{4}\right) \qquad [4.20]$$

where $x$ is the average of the contact diameter between the particles.

Taking the stress concentration factor into consideration, the strength of the agglomerate is given as

$$\sigma_c = 1.1 \frac{1-\varepsilon}{K\varepsilon} \frac{H}{d^2} \qquad [4.21]$$

where $K$ is the stress concentration factor and is obtained from

$$\ln(K) = 0.457 + 0.175 \ln\left(\frac{x}{8d}\right) + 0.095 \left[\ln\left(\frac{x}{8d}\right)\right]^2 \qquad [4.22]$$

Suri et al.[25] observed the morphology of agglomerates for powder injection molding feedstock of tungsten alloy powder mixed with a paraffin wax–polypropylene binder. The observations suggested that the average diameter of the contact between the particles ($x$) is 0.3 times the diameter of the particles ($x = 3d$). Using the same approximation, the value of green strength is estimated for carbonyl iron powder which was used as the candidate material by Ying et al.[19,20] to estimate the stresses developed in the compact during the thermal debinding process. The *in situ* strength of the carbonyl iron compact was estimated as 10.6 MPa for $\varepsilon$ value of 0.4. The *in situ* strength decreased to 0.79 MPa with increase in $\varepsilon$ to 0.9. The stresses in the compact simulated by Ying et al.[19,20] were in the order of 10 MPa due to polymer content change and 0.1 MPa due to gas pressure and temperature change. The strength of the agglomerates as predicted by the modified Rumpf model is also in the same range as the stresses estimated by Ying et al.[19, 20] The analysis shows that modeling with the approach of *in situ* strength variation during thermal debinding is a promising way to develop heating cycles to minimize the formation of defects. The *in situ* strength of the compact is also expected to be dependent on the softening (viscosity) and burnout characteristics of the polymer. Thus the final equation for the *in situ* strength of the compact during the thermal debinding process should also include parameters based on the softening and degradation behavior of the polymer. Further research studies are required to finalize the model based on the *in situ* strength principle. These studies would require measuring *in situ* strength during the thermal debinding process and correlating measured values to the softening and burnout characteristics of the polymer.

*Defect formation*

Understanding the phenomena of defect evolution during thermal debinding and predicting critical heating cycles to avoid them has been the focus of research for many studies.[26–32] A summary of the defects observed in various studies is listed in Table 4.2. These defects are more commonly observed during initial stages of thermal debinding for large sized ceramic (small particle size) injection molding components. Several theories have been reported in the literature explaining the high probability of cracking during thermal debinding from ceramic injection molded components. These theories attribute the small particle size and brittle nature of the ceramics to the cracking tendency. However, none of the theories has been experimentally proven. The probable explanation of formation of defects in ceramics can be attributed to the angular shape of the particles. The angular shape restricts re-arrangement of the particles under the influence of increase in internal pressure during thermal debinding. This results in further increase in internal pressures and cracking of the component. On the other hand, metallic particles are usually more spherical and re-arrange (interparticle movement) under the influence of internal pressures and thus prevent the accumulation of high pressures and cracking of the component. Ceramic particles were found to serve as nuclei for heterogeneous bubble formation during thermal debinding.[28] The molding pressure during injection molding was found to have an effect on bloating defect formation during burnout of ethylene vinyl acetate (EVA) from SiC particles. The bloating was proved to be due to acetic acid formed during degradation of EVA. The bloating was found to be absent in the case of samples shaped at high injection molding pressure.[29, 30] Several studies[30, 31] showed the beneficial effects of wicking prior to thermal debinding in reducing crack

*Table 4.2* Defects observed during the thermal debinding process

| Ref. no. | Materials | | Sample shape and dimensions | Heating rate (°C/min) | Defects |
|---|---|---|---|---|---|
| | Powder | Polymer | | | |
| 26 | $Al_2O_3$ (0.53 μm) | 40.2 vol% polymer mix of 4.60 wt% EVA (Elvax 250), 2.30 wt% EVA (Elvax 260), 4.60 wt% paraffin wax, 1.20 wt% stearic acid, and 0.85 wt, % oleic acid | Cylinder of 5.8 mm diameter and 25 mm length | 0.05, 0.2 | Bloating and cracking |
| 27 | $Al_2O_3$ (0.6 μm) | 2 wt% polystyrene and 12 wt% paraffin oil | Square: 4 x 4 mm$^2$ | 2, 4.5 | Blistering, bloating and cracking |
| 3 | $Al_2O_3$ | 50 vol% poly (α-methylstyrene) | Plates of 40 mm x 40 mm x 3.4–2.9 mm | 0.05–0.048 | Bloating and cracking |
| 28 | $SiO_2$ | 35 vol% proprietary wax | 31 mm diameter cylinder | 0.033 | Cracking |
| 29 | SiC (0.75 μm) and minor additions of $Al_2O_3$ and $Y_2O_3$ | 49 vol% ethylene vinyl acetate (EVA) | Discs of 25.4 mm diameter and thickness 0.5–8 mm | 0.5, 2, 8 | Bloating |
| 30 | $ZrO_2$ (0.25 μm) | 50 vol% polymer mixture of paraffin wax, vinyl acetate and stearic acid (volumetric ratio of wax : polymer is 60:40) | Parallelepiped moldings 4 x 5 x 60 mm | 0.5 | Cracking |
| 31 | $Si_3N_4$ 6 wt% $Y_2O_3$, 2 wt% $Al_2O_3$ | 40 vol% polymer mix of 90 wt% paraffin wax, 5 wt% epoxy thermo setting material and 5 wt% oleic acid | Turbine rotors | 0.17 | Cracking |
| 32 | Al | 35 vol% polymer mix of isotactic polypropylene, microcrystalline wax, and stearic acid | Bars 12.7 mm x 6.35 x 3, 6 mm | 0.02–2.1 | Bloating and cracking |

formation in samples. However, the beneficial effects of wicking were absent in case of large sized components like turbine rotors, where the formation of cracking was found to be inevitable even at slow heating rates.

The requirement of extremely slow thermal debinding cycles (extending from hours to days) to successfully remove the polymer without forming defects can be acknowledged after noticing the heating rates at which defects were observed in various experimental studies (Table 4.2).

*Carbon contamination*

The residual carbon left in the component due to incomplete debinding was found to have an impact on the magnetic, electrical, mechanical, and sintered properties. The carbon content after thermal debinding was reported to be dependent on the following parameters:

- the degradation behavior of the polymer;
- interaction of polymer with the powder particles;
- chemistry of the powder surface;
- thermal debinding atmosphere;
- oxidation temperature of the powder.

Low molecular weight polymers which degrade by evaporation and chain secession showed no carbon contamination. The carbon residue increases linearly with increase in molecular weight of the binders.[33,34] The effect of the gaseous atmosphere (nitrogen, hydrogen, and a mixture of nitrogen and hydrogen) on the carbon residue and carbon content of Fe–2Ni steel has been reported.[35,36] Thermal debinding of samples in a nitrogen and pure hydrogen atmosphere resulted in complete removal of polymer. However, addition of hydrogen to nitrogen resulted in incomplete burnout of polymer and carbon residue in the samples. Recombination of the burnout products in the presence of hydrogen was explained as one of the reasons for the increase in carbon content of the powder.

Some research studies have shown that the carbon residue in the burnout sample is dependent on the oxidation temperature of the powders.[37–40] The carbon residue was reported to be a minimum in case of powders like silicon and alumina, which do not oxidize during the thermal debinding process. The high oxidation temperature of these metals has been reported to aid in effective conversion of carbon to volatile gases.[37,39] On the other hand, iron-based powders like low-alloy Fe–Ni, which undergo severe oxidation at low temperatures, have revealed residual carbon after the thermal debinding process. Excessive formation of oxides at the surface of the powder has been explained as one of the reasons for the carbon residue in these parts. It was hypothesized that the formation of oxides on the surface of the powders prevented easy out-gassing of the degraded polymer, resulting in carbon residue in the burnt out sample.[36, 38, 40]

In general it can be concluded that the residual carbon content in the burnout samples is primarily dependent on the degradation behavior of the polymer in the chosen burnout atmosphere. The low molecular weight polymers which degrade by evaporation or chain secession have the lowest probability of leaving carbon residue after the burnout process.

*Distortion*

Very few studies have been published in the literature regarding distortion or dimensional changes during thermal debinding. The *in situ* observation of shape loss during thermal debinding was recently published.[41] The shape loss experiments were carried out using die compacted, gas atomized 316L stainless steel powders mixed with 1 wt% polyethylene co-vinyl acetate polymer and on injection molding samples made from gas atomized 316L powders, polypropylene, and polyethylene. The study showed that shape loss occurs primarily during the softening of the binder. The shape loss was also found to be dependent the degradation behavior of the binder. A recovery of shape loss was observed on burnout in the case of binders which degrade by formation of a double bond product. A recovery of shape loss was observed for gas atomized 316L powder and 1 wt% polyethylene co-vinyl acetate binder, as shown in Figure 4.10. No recovery of shape loss was observed in the case of injection molding samples (Figure 4.11). In this case the samples broke after thermal debinding due to poor strength.

The total dimensional change during thermal debinding of injection molded SiC and EVA feedstock was modeled as the summation of extension due to thermal expansion and shrinkage due to loss of binder. The dimension change due to thermal expansion was estimated by the rule of mixtures.[42] Use of high solids loading, small particle size, and irregular shaped powders was suggested as an ideal combination to minimize distortion during thermal debinding. The study also suggested faster removal of polymer by applying a vacuum to minimize distortion. Increasing the heating rate was not suggested for faster removal of binder. Kipphut and German[43] stated that most of the distortion during thermal debinding occurs due to flow by viscous creep during the softening of the polymer. Experiments were carried out with iron powders of different particle sizes and their binary mixtures. Interparticle friction between the particles was identified as an important factor for shape retention. The angle of repose at tap density (compacted angle) was correlated to the distortion behavior during thermal debinding. Based on their results, they recommended powders with a compacted angle over 55° to minimize the distortion.

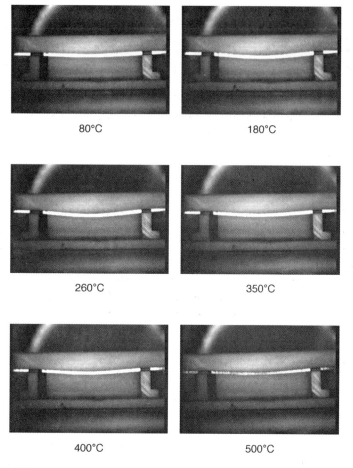

*4.10* *In situ* images during polymer burnout of samples made from admixed powders of gas atomized 316L stainless steel and 1 wt% EVA.[41]

## 4.4 Mixing technologies

The feedstock for injection molding is prepared by mixing metal powder and the binder system. The main objectives of mixing are to obtain a uniform coating of binder on the metal particle surface, to mix all of the components of the binder system (polymer, wetting agents, surface active materials) uniformly, to break down powder agglomerates, and to yield a uniform feedstock with no powder or binder segregation. Various factors like particle size, shape, size distribution, and binder properties affect the mixing behavior of the feedstocks.

During the mixing process the large clusters of particles are initially broken down under the influence of shear action. As mixing is continued the

# Powder binder formulation and compound manufacture in MIM

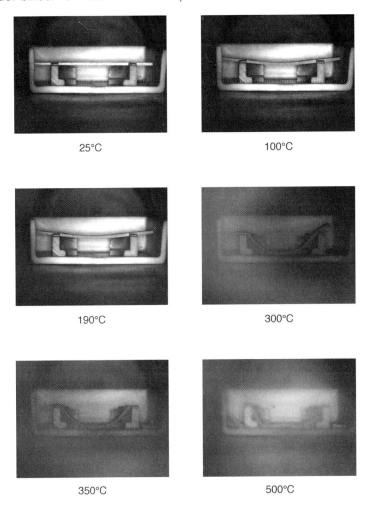

4.11 *In situ* observation of shape loss during polymer burnout for injection molded samples made from gas atomized 316L stainless steel and polymer mix of polypropylene and polyethylene.[41]

clusters size decreases and the binder becomes dispersed in the interparticle pores. The homogeneity ($M$) of the feedstock is estimated based on the following equation

$$M = M_O + \exp(kt + C) \quad [4.23]$$

where $M_O$ is the initial mixture homogeneity, $t$ is the time of mixing, and $C$ and $k$ are constants which depend on the powder and binder characteristics, agglomeration, and surface condition of the metal powder.

Typically the mixing process starts with heating the binder with a high melting point. Remaining constituents are added to the mix based on their respective melting points. The constituents are added according to progressively decreasing melting point. The metal powders are added once the binder constituents are mixed. The temperature of the feedstock decreases sharply with the addition of metal particles. The decrease in temperature is due to the high heat capacity of the binder. In some systems the metal particles are added during the melting of the high melting point binders and before adding the low melting point components. This practice results in achieving a uniform coating of binder on the metal particles. The final mixing of the feedstock is carried out in a vacuum to degas the feedstock. The presence of air in the feedstock can result in the formation of defects during injection molding. The feedstock after mixing is discharged from the equipment and precautions should be taken to prevent segregation of the feedstock during this process. It is preferable to solidify the feedstock in the homogeneous condition. Continuous mixing of the feedstock during cooling can also result in obtaining an homogeneous feedstock. The temperature of mixing has to be chosen appropriately in case of thermoplastic binders. The temperature of mixing is carried out at intermediate temperatures. Mixing at low temperatures, at which the mixtures still possess high yield strength, will cause cavitation defects in the injection molded parts. Mixing at too high a temperature can result in binder degradation, resulting in a lowering of viscosity and separation of the powder from the binder.

The inhomogeneities in the MIM feedstock occur as a result either of the binder separating away from the metal particles or segregation of the metal particles in the binder. The size segregation of the particles is dominant in the case of metal powders with wide size distribution. During agitation, the smaller particles fill the interstitial pores between large particles resulting in segregation of powders. The segregation will be more in the case of larger particle size difference in the metal powders. Smaller particles size and irregular shaped metal particles due to higher interparticle friction will show low segregation. A binder with high viscosity can also lower segregation in the metal powders.

Smaller and irregular particles are more prone to agglomeration and require longer mixing times to obtain homogeneous feedstocks. The agglomeration of the particles also has a negative effect on the solids loading in the feedstock. The solids loading decreases from 0.67 to 0.37 for mono-sized spheres due to agglomeration. Addition of surface active agents to the binder system will prevent agglomeration of the feedstocks. A powder with wide particle size distribution will result in separation from the binder. The separation will be predominant in the case of a binder with low viscosity.

The variation of torque with mixing time of the feedstock is shown in

Powder binder formulation and compound manufacture in MIM 87

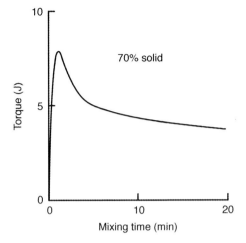

*4.12* Variation of torque with time of mixing.[1]

Figure 4.12. The initial torque is for mixing the pure binder. The torque continues to increase with the addition of metal particles. The increase in torque is due to a decrease in temperature of the feedstock because of the high thermal conductivity of the metal particles and an increase in homogeneities. The torque required for mixing decreases as the agglomerates are broken and liquid resulting from melting of the binder is released into the feedstock. The torque continues to decrease as a greater amount of liquid is released with continuous mixing. The torque reaches a steady state where the rate of mixing equals the rate of demixing. An homogeneous mixture will show a steady torque with mixing time.

The viscosities of the feedstocks vary with the shear rate. The distance of the shear region in mixing will influence the homogeneity of the feedstock. Several high-shear mixer designs are used in MIM to obtain uniform distribution of shear rate and homogeneous feedstock composition. Some of the mixtures are single screw extruder, twin screw extruder, twin cam, double planetary, Z-blade mixtures, etc. Out of all the available mixtures, the twin screw extruder is most successful as it combines high shear rate and short dwell time at high temperature. The equipment consists of twin screws that counter rotate and move the feedstock through the heated extruder barrel. The discharge from the equipment is in the form of a uniform cylindrical product. The cost of the twin screw extruder is high and the double planetary mixer is mostly preferred as it provides a good balance of cost, quality, and productivity.

## 4.5 Case studies: lab scale and commercial formulations

The examples of lab scale binder systems used for metal injection molding are summarized in Table 4.3.

*Table 4.3* Examples of binder composition used in injection molding

| Binder composition | Metal | Reference number |
|---|---|---|
| 41.3 wt% starch, 23.3 wt% glycerol, 28.5 wt% linear low-density polyethylene, 1.9 wt.% citric acid, 5 wt% stearic acid | 316L stainless steel | 44 |
| 45 wt% low-density polyethylene, 55 wt% paraffin wax, 5 wt% stearic acid | 316L stainless steel | 45 |
| 45 wt% low-density polyethylene, 45 wt% paraffin wax, 10 wt% stearic acid | 316L stainless steel | 45 |
| 30% paraffin wax, 10% carnauba wax, 10% bees wax, 45% ethylene vinyl acetate, 5% stearic acid | 316L stainless steel | 46 |
| 30% paraffin wax, 10% carnauba wax, 10% bees wax, 45% polypropylene, 5% stearic acid | 316L stainless steel | 46 |
| 25% paraffin wax, 20% carnauba wax, 20% bees wax, 25% ethylene vinyl acetate, 5% polypropylene, 5% stearic acid | 316L stainless steel | 46 |
| 64% paraffin wax, 16% microcrystalline paraffin wax, 15% ethylene vinyl acetate, 5% high-density polyethylene | 17-4 PH stainless steel | 47 |
| 63% paraffin wax, 16% microcrystalline paraffin wax, 15% ethylene vinyl acetate, 5% high-density polyethylene, 1% stearic acid | 17-4 PH stainless steel | 47 |
| 59% paraffin wax, 16% microcrystalline paraffin wax, 15% ethylene vinyl acetate, 5% high-density polyethylene, 5% stearic acid | 17-4 PH stainless steel | 47 |
| 55% paraffin wax, 16% microcrystalline paraffin wax, 15% ethylene vinyl acetate, 5% high density polyethylene, 9% stearic acid | 17-4 PH stainless steel | 47 |
| 50% high-density polyethylene, 50% paraffin wax | HS12-1-5-5 high-speed steel | 48 |

*Table 4.3 (cont.)*

| Binder composition | Metal | Reference number |
|---|---|---|
| 65% paraffin wax, 30% polyethylene, 5% stearic acid | Copper | 49 |
| 79% paraffin wax, 20% ethylene vinyl acetate, 1% stearic acid | Iron–nickel | 50 |
| 79% paraffin wax, 20% high-density polyethylene, 1% stearic acid | Iron–nickel | 50 |
| 79% paraffin wax, 10% high-density polyethylene, 10% ethylene vinyl acetate, 1% stearic acid | Iron–nickel | 50 |
| 85% paraffin wax, 15% ethylene vinyl acetate | 316L stainless steel | 51 |
| 65% paraffin wax, 35% ethylene vinyl acetate | 316L stainless steel | 51 |
| 70% paraffin wax, 5% stearic acid, 25% polyethylene | 316L stainless steel | 51 |
| 75% paraffin wax, 20% polyethylene, 5% ethylene vinyl acetate | 316L stainless steel | 51 |
| 75% paraffin wax, 15% polyethylene, 10% ethylene vinyl acetate | 316L stainless steel | 51 |
| 55% paraffin wax, 25% polypropylene, 5% stearic acid, 15% carnauba | Iron–nickel | 52 |
| 35% polypropylene, 60% paraffin wax, 5% stearic acid | W–Cu | 53 |
| 40% polypropylene, 55% paraffin wax, 5% stearic acid | W–Cu | 54 |

## 4.6 References

1. German, R. M. (1990), *'Powder Injection Molding'*, Metal Powder Industries Federation, Princeton, NJ.
2. Liang, S., Tang, Y., Huang, B., and Li, S. (2005), 'Chemistry principles of thermoplastic polymer binder formula selection for powder injection molding', *Transactions of the Nonferrous Metals Society of China*, 14(4), 763–768.
3. Mutsuddy, B. C., (1994), 'Rheology and mixing of ceramics mixtures used in plastic molding', in *Chemical Processing of Ceramics*, eds Lee, B. I. and Pope, E. J. A. Marcel Dekker Inc, New York, pp. 239–261.
4. Liang, S. Q. and Huang, B. Y. (2000), *'Rheology for Powder Injection Molding'*, Central South University Press, Changsha, pp. 90–103.
5. Huang, B., Liang, S., and Qu, X., (2003), 'The rheology of metal injection molding', *Journal of Materials Processing Technology*, 137, 132–137.
6. Kitano, T. *et al.*, (1981), 'An empirical equation of the relative viscosity of

polymer melts filled with various inorganic fillers', *Rheologica Acta,* 20, 207–209.
7. Wildemuth, C. R. and Williams, M. C. (1984), 'Viscosity of suspensions modeled with a shear dependent maximum packing fraction,' *Rheologica Acta,* 23, 627–635.
8. Dabak, T. and Yucel, O. (1987), 'Modeling of the concentration and particle size distribution effects on the rheology of highly concentrated suspensions', *Powder Technology,* 52, 193–206.
9. Enneti, R. K. (2005), 'Thermal analysis and evolution of shape loss phenomena during polymer burnout in powder metal processing', PhD thesis, Pennsylvania State University, USA.
10. German, R. M. (2003), 'The impact of economic batch size on cost of powder injection molded (PIM) products', in *Advances in Powder Metallurgy and Particulate Materials,* Metal Powder Industries Federation, Princeton, NJ, USA, vol. 8, pp. 146–159.
11. Matar, S. A., Edrrisinghe, M. J., Evans, J. R. G., Twilzell, E. H., and Song, J. H. (1995), 'Modeling the removal of organic vehicle from ceramic or metal moldings: the effect of gas permeation on the incidence of defects', *Journal of Materials Science,* 30, 3805–3810.
12. Matar, S. A., Edirsinghe, M. J., Evans, J. R. G., and Twilzell, E. H. (1995), 'The influence of monomer and polymer properties on the removal of organic vehicle from ceramic and metal moldings', *Journal of Material Research,* 10(8), 2060–2072.
13. Matar, S. A. Edirisinghe, M.J. Evans, J. R. G., and Twizell, E. H. (1996), 'Diffusion of degradation products in ceramic moldings during pyrolysis: Effect of geometry', *Journal of American Ceramic Society,* 79(3), 749–755.
14. Song, J. H., Edirisingle, M. J., and Evans, J. R. G. (1996), 'Modeling the effect of gas transport on the formation of defects during thermolysis of powder moldings', *Journal of Materials Research,* 11(4), 830–840.
15. Calvert, P. and Cima, M. (1990), 'Theoretical models for binder burnout', *Journal of American Ceramic Society,* 73(3), 575–579.
16. Lam, Y. C., Ying, S., Yu, S. C. M., and Tam, K. C. (2000), 'Simulation of polymer removal from a powder injection molding compact by thermal debinding', *Metallurgical and Materials Transactions A,* 31A, 2597–2606.
17. Ying, S., Lam, Y. C., Yu, S. C. M., and Tam, K. C. (2001), 'Two-dimensional simulation of mass transport in polymer removal from a powder injection molding compact by thermal debinding', *Journal of Material Research,* 16(8), 2436–2451.
18. Ying, S., Lam, Y. C., Yu, S. C. M., and Tam, K. C. (2002),'Simulation of polymer removal from a powder injection molding compact by thermal debinding', *Key Engineering Materials,* 227, 1–6.
19. Ying, S., Lam, Y. C., Yu, S. C. M., and Tam, K. C. (2002), 'Thermal debinding modeling of mass transport and deformation in powder-injection molding compact', *Metallurgical and Materials Transactions A,* June, 33B, 477–488.
20. Ying, S., Lam, Y. C., Yu, C. M., and Tam, K. C. (2002), 'Thermo-mechanical simulation of PIM thermal debinding', *International Journal of Powder Metallurgy,* 38(8), 41–55.
21. Lewis, J. A. Cima, M. J., and Rhine, W. E. (1994), 'Direct observation of

preceramic and organic binder decomposition in 2-D model microstructures', *Journal of American Ceramic Society*, 77(7), 1839–1845.
22. Hwang, K. S. and Tsou, T. H. (1992), 'Thermal debinding of powder injection molded parts: observations and mechanisms', *Metallurgical and Materials Transactions A*, 23A(10), 2775–2782.
23. Cima, M. J., Dudziak, M., and Lewis, J. A. (1989), 'Observation of poly(vinyl butyral)-dibutyl phthalate binder capillary migration', *Journal of American Ceramic Society*, 72(6), 1087–1090.
24. Shubert, H. (1975), 'Tensile strength of agglomerate', *Powder Technology*, 11,107–119.
25. Suri, P., Atre, S. V., German, R. M., and de Souza, J. P. (2003), 'Effect of mixing on the rheology and particle characteristics of tungsten-based powder injection molding feedstock', *Materials Science and Engineering*, A356, 337–344.
26. Trunec, M. and Cihlar, J. (1997), 'Thermal debinding of injection molded ceramics', *Journal of European Ceramic Society*, 17, 203–209.
27. Chartier, T. Ferrato, M., and Baumard, J. F. (1995), 'Influence of the debinding method on the mechanical properties of plastic formed ceramics', *Journal of European Ceramic Society*, 15, 899–903.
28. Zhang, J. G., Edirisinghe, M. J., and Evans, J. R. G. (1989), 'A catalogue of ceramic injection molding defects and their causes', *Industrial Ceramics*, 9, 72–82.
29. Hrdina, K. E., Halloran, J. W., Kaviany, M., and Oliveira, A. (1999), 'Defect formation during binder removal in ethylene vinyl acetate filled system', *Journal of Materials Science*, 34, 3281–3290.
30. Tseng, W. J. and Hsu, C. K. (1999), 'Cracking defect and porosity evolution during thermal debinding in ceramic injection molding', *Ceramics International*, 25, 461–466.
31. Bandyopadhyay, G. and French, K. W. (1993), 'Injection-molded ceramics: Critical aspects of the binder removal process and component fabrication', *Journal of European Ceramic Society*, 11, 23–34.
32. Pinwill, I., Edirisinghe, E., and Bevis, M. J. (1992), 'Development of temperature-heating rate diagrams for the pyrolytic removal of binder used for powder injection molding', *Journal of Materials Science*, 27, 4381–4388.
33. Lee, S. H., Choi, J. W., Jeung, W. Y., and Moon, T. J. (1992), 'Effects of binder and thermal debinding parameters on residual carbon in injection moulding of Nd(Fe,Co)B powder', *Powder Metallurgy* 42(1), 41–44.
34. Howard, K. E., Lakema, C. D. E., and Payne, D. A.(1990), 'Surface chemistry of various poly(vinyl butyral) polymers adsorbed onto aluminum', *Journal of American Ceramic Society*, 73(8.1), 2543–2546.
35. Streicher, E., Renowden, M., and German, R. M.(1991), 'Atmosphere role in thermal processing of injection molded steel', *Advances in Powder Metallurgy, Powder Injection Molding*, 2, 141–158.
36. Phillips, M. A., Streicher, E. C., Renowden, M., German, R. M., and Friedt, J. M. (1992), 'Atmosphere process for the control of the carbon and oxygen contents of injection molded steel parts during debinding', *Proceedings of the Powder Injection Molding Symposium*, pp. 371–384.
37. Edirisinghe, M. J. (1991), 'Binder removal from moulded ceramic bodies in different atmospheres', *Journal of Materials Science Letters*, 10, 1338.

38. Howard, K. E., Lakeman, C. D. E., and Payne, D. A. (1990), 'Surface chemistry of various poly(vinyl butyral) polymers adsorbed onto aluminum', *Journal of American Ceramic Society*, 73(8), 2543.
39. Gillissen, R. and Smolders, A. (1986) 'Binder removal from injection molded ceramic bodies', *Proceedings of the 6th World Congress on High Tech Ceramics*, Milan, Italy.
40. Kraemer, O. and Hsuum, P. (2003), 'Injection molding of 316L stainless steel powder', in *Advances in Powder Metallurgy and Particulate Materials*, Metal Powder Industries Federation, Princeton, NJ, USA, vol.8, p. 1141.
41. Enneti, R. K., Park, S. J., German, R. M., and Atre, S. V. (2011), 'In situ observation of shape loss during polymer burnout in powder metal processing', *International Journal of Powder Metallurgy*, 47(3), 45–54.
42. Hrdina, K. E. and Halloran, J. W. (1998), 'Dimensional changes during binder removal in a moldable ceramic system', *Journal of Materials Science*, 33, 2805–2815.
43. Kipphut, C. M. and German, R. M. (1991), 'Powder selection for shape retention in powder injection molding', *International Journal of Powder Metallurgy*, 27(2), 117–124.
44. Abolhasani, H. and Muhamad, N. (2010), 'A new starch-based binder for metal injection molding', *Journal of Materials Processing Technology*, 210, 961–968.
45. Setasuwon, P., Bunchavimonchet, A., and Danchaivijit, S. (2008), 'The effects of binder components in wax/oil systems for metal injection molding', *Journal of Materials Processing Technology*, 196, 94–100.
46. Supriadi, S., Baek, E. R., Choi, C. J., and Lee, B.T. (2007), 'Binder system for STS 316 nanopowder feedstocks in micro-metal injection molding', *Journal of Materials Processing Technology* 187–188, 270–273.
47. Li, Y., Liu, X., Luo, F., and We, J. (2007), 'Effects of surfactant on properties of MIM feedstock', *Transaction of the Nonferrous Metals Society of China*, 17, 1–8.
48. Dobrzánski, L. A., Matula, G., Herranz, G., Várez, A., Levenfeld, B., and Torralba, J. M. (2006), 'Metal injection moulding of HS12-1-5-5 high-speed steel using a PW-HDPE based binder,' *Journal of Materials Processing Technology*, 175, 173–178.
49. Moballegh, L., Morshedian, J., and Esfandeh, M. (2005), 'Copper injection molding using a thermoplastic binder based on paraffin wax', *Materials Letters* 59, 2832–2837.
50. Huang, B. Liang, S., and Qu, X. (2003), 'The rheology of metal injection molding', *Journal of Materials Processing Technology* 137, 132–137.
51. Shimizu, T., Kitazima, A., Nose, M., Fuchizawa, S., and Sano, T. (2001), 'Production of large size parts by MIM process', *Journal of Materials Processing Technology*, 119, 199–202.
52. Lin, K. H. (2011), 'Wear behavior and mechanical performance of metal injection molded Fe–2Ni sintered components', *Materials and Design*, 32, 1273–1282.
53. Yang, B. and German, R. M. (1997), 'Powder injection molding and infiltration sintering of superfine grain W–Cu', *International Journal of Powder Metallurgy*. 33(4), 55–63.
54. Yang, B. and German, R. M. (1994), 'Study on powder injection molding ball milled W–Cu powders', *Tungsten Refractory Metals*, 2, 237–244.

# 5
# Tooling for metal injection molding (MIM)

G. SCHLIEPER, Gammatec Engineering GmbH, Germany

**Abstract:** Starting with a general description of the design and function of injection molding machines, a basic understanding of the elements of the tool set and tool design is conveyed. Particular attention is given to the various options for the design and position of gates and their effect on mold filling. Molding parts with undercuts or threads requires special tool elements such as lifters, collapsible cores, and threaded cores. Hot runner molds increase the output and avoid recycling of sprues and runners. Computer software is available for mold fill studies and cost calculation.

**Key words:** mold materials, multiple gating, undercut design, hot runner molds, mold fill studies by computer simulation.

## 5.1 Introduction

This chapter deals with the tooling for MIM parts. The intention is to provide the MIM process engineer with a basic understanding of the design options for MIM tooling. This should enable him to communicate with the tool designer on a deeper level of understanding. For more detailed information about MIM tooling he should refer to the vast choice of specialist literature devoted to plastic injection molding tooling, which for the most part is also valid for metal injection molding.

First of all, the process engineer should be familiar with the basic design and the main functions of an injection molding machine. These features are introduced in Section 5.2. The next step is the knowledge of the elements of a tool set, as outlined in Section 5.3. On the basis of this knowledge, the various general mold design options are described in Section 5.4. This includes mold materials, oversize design, gating options, and venting. The design for undercuts involving the use of lifters, threaded and collapsible cores is also part of this section. Special features like hot runner molds and instrumentation are included in Section 5.5 and finally Section 5.6 gives some information about available supporting software and economic aspects.

94    Handbook of metal injection molding

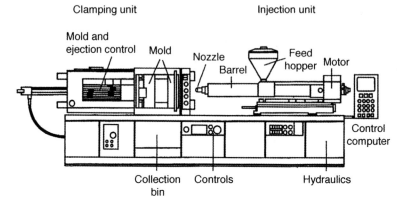

5.1 General design of an injection molding machine (from *ASM Handbook*, Vol. 7 (ASM, 2008), p. 358). Reprinted with permission from ASM International. All rights reserved.

## 5.2 General design and function of injection molding machines

Corresponding to the great importance of injection molding technology in modern manufacturing industries, the choice of injection molding machines worldwide is immense. In spite of the substantial differences in the design of injection molding machines from various suppliers all of them have certain general design features in common which are necessary to carry out and control the injection molding process. The essential components are the injection unit and the clamping unit, see Fig. 5.1. The tooling is attached to the clamping unit.

The injection unit and clamping unit are usually arranged horizontally, as shown in the schematic picture. This is generally the most convenient way to produce MIM parts in a fully automatic way. Many machines allow the injection unit and the clamping unit to swivel to a vertical position. Then it is possible to load inserts into the mold and parts will not fall freely after ejection. This may be advantageous for very delicate parts.

The whole machine is mounted on a rigid frame. A hydraulic pressure supply provides the necessary power to close the clamping unit and keeps it closed against the pressure of the feedstock in the mold. In fact, the maximum clamping force is the main characteristic by which the power and size of an injection molding machine is defined. The other drives of the machine are usually electric. There are also machines on the market which are fully electric.

Control instruments for pressure and temperature serve to monitor the process and a computer controls the operation of the machine. The collection bin under the mold can either be used to collect the products falling out of the mold, or a basket or conveyor belt or pick-and-place handling system may be installed.

The injection unit is essentially a heated barrel with a screw or plunger transporting the molten feedstock to the injection nozzle at the end pointing towards the clamping unit. The screw or plunger is driven by the motor at its rear end.

Feedstock granules are fed into the heated barrel through the hopper. Inside the barrel the feedstock is then heated, compressed, homogenized, and finally injected into the mold cavity. The barrel is usually equipped with a screw along its axis, which exerts a forward movement on the feedstock when rotating. A metering valve at the front end separates the exact amount of feedstock for the next shot from the reservoir inside the barrel. The molten feedstock is injected into the mold through a nozzle at the hot end of the barrel.

The control computer of the machine allows the barrel temperature to be adjusted in several independent zones. The first heated zone serves to melt the feedstock quickly, as the abrasiveness of the molten feedstock is much less than that of the solid granules. In the following zones the temperature is gradually increased and homogenized inside the barrel.

An elevated barrel temperature is required for melting the feedstock and the action of the screw exerts pressure and torque on the feedstock that brings it to the optimum viscosity. If the temperature is too low, the feedstock may freeze before the mold cavity is completely filled. An excessively high temperature leads to a very low viscosity that will cause problems, such as molten feedstock dripping out of the nozzle opening, increasing formation of flash due to binder squeezing out through gaps between mold parts, prolonged cooling times, and so on. Therefore the barrel temperature will be kept as low as possible, just to have enough heat to securely fill the mold cavity. It should also be considered that some heat is generated in the barrel by the frictional forces between screw and feedstock, so not all of the heat that is required must be introduced from outside.

The screw is usually designed with a reduction of the feedstock volume while it is transported forward. A schematic diagram of the screw profile is shown in Fig. 5.2. The feedstock enters the feed zone and is heated and plasticized there. The turning screw transports the feedstock towards the compression zone where the reducing cross-section causes the compression. The third section is the metering zone where a further homogenization is achieved. The feedstock is intensively sheared, homogenized, and compressed by the turning screw. The rotation of the screw determines the pressure inside the barrel, and a forward and backward movement of the screw serves to inject a well-defined quantity of feedstock into the mold with each molding cycle. A non-return valve at the front end of the screw prevents molten feedstock from being pressed back into the barrel.

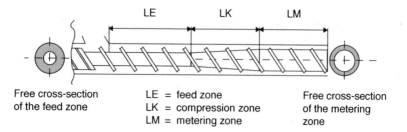

5.2 Schematic diagram showing profile and cross-sections of a feed screw (Arburg).

## 5.3 Elements of the tool set

In its simplest form, a tool set for powder injection molding looks as shown in Fig. 5.3. It is composed of a set of steel plates and has clamping plates at both ends. The front clamping plate is firmly attached at the side facing the injection unit; it is sometimes referred to as the stationary plate. This has a central hole surrounded by a locating ring where the nozzle of the injection unit is centered when the feedstock is injected into the mold through the sprue bushing. Two cavity plates behind the front clamping plate contain one or more mold cavities which are machined into the space between them. The front cavity plate is stationary together with the front clamping plate and the rear cavity plate is attached to the rear clamping plate. The stationary side of the mold is known as the A side, and the moving side is known as the B side.

The temperature of the mold cavities is controlled by means of hot water or oil, which is circulated through cooling channels inside the cavity plates. The proper design of cooling channels and the optimum mold temperature depend largely on the type of feedstock and require some experience. It is impossible to give any general guidelines. The mold temperature must be low enough to freeze the feedstock quickly and high enough to allow all mold cavities to be completely filled before the feedstock starts to freeze. The BASF Catamold typically uses oil heating because of its high temperature requirement, whereas wax/polymer and water-soluble systems typically perform well with water temperature control.

The cavity plates are closed during molding, then separated from one another after the molded parts have solidified so that the parts can be ejected. Mold cavities should be so designed that the adhesive forces are higher at the rear cavity plate than at the front plate. In this way, the molded parts will stick to the rear plate when the plates are separated and can be ejected by the ejector pins; otherwise the parts will not fall off the mold. Some tools have spring-loaded ejector pins in the A side to ensure that the parts remain on the B side when the mold opens.

While the front cavity plate is firmly attached to the front clamping plate,

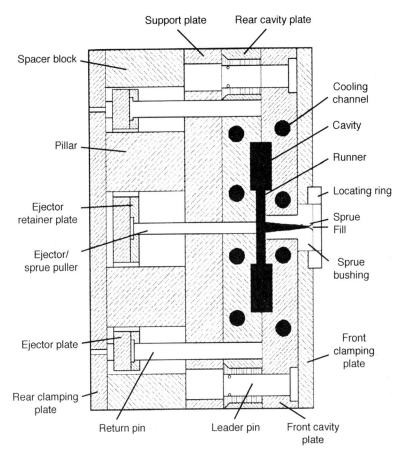

*5.3* Basic elements of a tool set for metal injection molding (from *ASM Handbook*, Vol. 7 (ASM, 2008), p. 357). Reprinted with permission from ASM International. All rights reserved.

the rear cavity plate is held in position by the support plate. It is connected to the rear clamping plate by the spacer block. The exact position of the two cavity plates when the mold is closed is guaranteed by four leader pins at the corners of the tool set. Holes in the spacer plate leave room for moving elements such as ejector pins and return pins. Their ends are attached to the ejector plate, which moves inside the spacer plate, and they are held in place by the ejector retainer plate. Depending on the size and quantity of the molded components, several ejector pins may be required. The central ejector pin serves to pull the sprue out of the sprue bushing when the parts are ejected.

The decision about the number of ejector pins and their position is made by the mold designer on the basis of an analysis of the adhesive forces between mold and part. This includes consideration of the shrinkage during cooling. Owing to this shrinkage, parts will generally be released easily from their outer contours, but may stick on inner contours. Areas with great adhesion can be flat surfaces or complex geometries.

Mold release may be supported by draft angles. These are angles on surfaces which are parallel to the direction in which the mold opens. An angle of 1° is usually sufficient.

The basic elements that form part of each injection molding tool set have been introduced. Most of these components are standardized and can be purchased at moderate cost from specialist suppliers. MIM parts manufacturers rely on these suppliers to save costs and maintain a high quality standard of their molds. Only the mold cavity plates must be custom manufactured.

For more complex part geometries the mold cavities may include cores, retractable slides, threads, etc. An additional plate between the two cavity plates is sometimes introduced that provides more flexibility to position the gate or gates. These are the so-called three-plate molds; for example, a gate position in the center of the part requires a three-plate mold.

## 5.4 Tool design options

Although injection molding of MIM feedstocks is generally very similar to injection molding of thermoplastics, a few differences should be kept in mind when talking about mold design. MIM feedstock is brittle and often sticks to the parting line. Therefore MIM tools need to be tighter than tools for plastic injection molding to prevent flashing.

### 5.4.1 Mold materials

MIM powders are not compressible and therefore MIM feedstocks are often more abrasive in the injection molding process as compared to regular thermoplastic materials. The feedstock may further damage the mold in the parting line with repetitive mold opening and closing. Consequently there is an increased risk of erosion in the mold cavities and also in the barrel of the molding machine. Preventive measures are wear resistant coatings inside the barrel and on the screw surface and the use of highly wear resistant materials like tool steels and cemented carbide for all tool parts that are in direct contact with the feedstock.

### 5.4.2 Oversize design

The significant shrinkage of MIM components in the processing steps following injection molding, i.e. during binder removal and particularly during sintering, must be anticipated in the mold dimensions. The shrinkage is empirically determined as the linear shrinkage of a rectangular bar 26–30 mm long and 4–6 mm wide, with dimensions given for the mold cavity. When the length of the mold cavity $L_0$ and that of the sintered bar $L$ have been measured, the shrinkage $\delta$ can be calculated as

$$\delta = \frac{L_0 - L}{L_0} \quad [5.1]$$

The value of $\delta$ is usually multiplied by 100 to give the shrinkage in percent. While the shrinkage is a characteristic of the material, tool designers prefer to use the tool cavity expansion factor $Z$, which is given as the ratio of the mold dimension $L_0$ and the final part dimension $L$.

$$Z = \frac{L_0}{L} \quad [5.2]$$

Since these two parameters are very similar mathematically, care should be taken not to confuse them. It should be noted that the *shrinkage* $\delta$ is related to the dimension of the mold cavity $L_0$ and the *cavity expansion factor* $Z$ is related to the final part dimension $L$. As the part dimension is always smaller, the cavity expansion factor suggests a larger shrinkage than there really is. For example, a cavity expansion factor of $Z = 1.18$ would correspond to a shrinkage of $\delta = 15.25\%$.

The tool cavity expansion factor $Z$ can be calculated from the shrinkage according to

$$Z = \frac{1}{1 - \delta} \quad [5.3]$$

By using the expansion factor, the mold designer will expand all linear dimensions and radii according to $L_0 = Z * L$. Angles are not generally changed. Outer dimensions are initially set to the lower end of the tolerance band and inner dimensions to the upper end. Then it is possible to correct the mold cavity when the first samples have been sintered and to allow for some tool wear. This is also known as 'steel-safe' condition. Fine tuning of the mold dimensions is often necessary when close part tolerances are required.

## 5.4.3 Gating options and venting

The sprue is a conical channel of typically around 6 mm diameter with a taper of 5° through which the feedstock enters the mold. The feedstock speed is reduced by the growing diameter. At the end of the sprue there is a small recess where the front of the feedstock stream, which has already

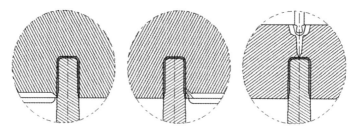

5.4 Tab gate (left), sub-gate (center), and three-plate tool (right) (Dynamic).

cooled off, is deposited. From here the runners branch off to the sides. A continuous flow is achieved if the total runner cross-section is similar to the maximum sprue cross-section. The runner design and the position of the mold cavity, or cavities, should be such that the forces exerted on the mold by the feedstock are symmetrical with respect to the central axis of the mold in order to minimize the formation of flash.

The point where feedstock enters the mold cavity is called the gate. This is the point with the smallest cross-section in the entire tooling system and the highest flow speed and pressure. One or more gates may be designed on each mold cavity.

The gate position is the most important measure to control the flow of molten feedstock inside the mold cavity. Since the runner is usually in the mold parting plane, the easiest option is to place the gate in the parting plane as well. This is the tab gate, see Fig. 5.4. The flow of material should be directed against an opposing wall so that the feedstock fills the cavity continuously without jetting. Also it should be designed at the thickest section of the part so that the cross-section of the feedstock stream always remains constant or diminishes. If the cross-section increases, the pressure in the mold and flow speed will drop dramatically so that it becomes difficult to fill the cavity completely without defects.

The jetting phenomenon occurs where a liquid leaves containment at high speed under high pressure into a free space; in everyday life, for example, it is known as the garden hose effect. The liquid is dispersed into fine droplets. If this occurs in the mold, the molded parts will most likely contain air bubbles which lead to internal defects after sintering, weld lines, and surface defects. This must be avoided by all means.

The gate can be displaced from the parting plane up to several millimeters by designing a short channel through the cavity plate. This is the sub-gate, also known as the tunnel gate. Sub-gate channels cannot be longer than a few millimeters because the feedstock freezes very quickly in such a small cross-section. If the gate is required at a great distance from the parting plane, three-plate tooling is necessary. Runner and gate are then led to the mold cavity through the third plate. The three-plate tooling has the additional advantage that the sprue and runner are automatically sheared off and separated from the parts when the parts are ejected.

When several gates are designed on a part, the molten feedstock enters the mold cavity from all gates simultaneously. This reduces the time required to fill the mold cavity, but also imposes risks on the part quality. Where two streams of feedstock unite, the frozen front areas may not completely weld together. The result is a so-called weld line at the part surface. Weld lines are often only an optical deficiency as long as the feedstock welds together perfectly, but they may also be indicative of internal defects.

Multiple gates are particularly recommended if there is a risk that the feedstock could freeze before the cavity is entirely filled, i.e. if the flow path is long and/or the wall thickness is small. The gate positions must ensure that the various streams of feedstock completely unite without weld lines and without trapping air.

The sprue and runner material can be easily broken off at the gate and separated from the compacts. Nevertheless the parts will exhibit some irregularity at the fracture surface. If possible the part surface at the gate should be designed a little lower than the neighboring surface so that it does not impede the function of the part. If this is impossible, the gate area must be mechanically flattened on the green part after injection molding.

The mold parting line is the trace left on the molded part where the two cavity plates are separated. The mold designer will try to place the parting line so that the function of the component is not negatively affected. In some cases it is essential to remove the parting line mechanically for functional or aesthetic reasons.

The mold cavity is filled with air when the feedstock is pressed into the mold. Proper venting is therefore very important; i.e. the mold designer has to foresee ways for the air to leave the mold. Air can often escape through the clearance between ejector pins and the cavity plate or in the parting plane and elsewhere, but if necessary venting channels 0.005–0.01 mm deep are ground into the parting plane.

A thorough understanding of the feedstock flow in the mold cavity is necessary for proper design of gates and vents. Vents should preferably be at the ends of the cavity which are filled last. If air is trapped inside the mold cavity, it cannot be completely filled with feedstock.

Gaps and vents do not only perform the useful function of letting air out of the mold, they also provide opportunities for the feedstock to squeeze out. The result is flash formation on the green parts, which is undesirable and necessitates additional work such as subsequent deburring. Wherever possible, all necessary secondary operations are usually done on the green part rather than on the sintered component because the green parts are much softer and easier to process.

### 5.4.4 Undercut design

Undercuts in MIM parts require mold elements that move in a direction lateral to the opening movement during the ejection procedure. This can be achieved by angle pins which are attached to the front clamping plate. An example is shown in Fig. 5.5. When the mold is opened, which here is in the vertical direction, the block with the angle pin is forced to move sideways and the part with an undercut is released and can be ejected.

Another option to produce parts with an undercut is to attach the angle lifter to the ejector plate and let it travel forward with the ejection movement, as shown in Fig. 5.6. The end of the lifter forms the undercut and moves sideways with the ejection movement so that the part is released.

The principle of lateral motion of tool parts under a slight angle can be applied in many ways. Figure 5.7 shows an example of a lifter composed of two halves which are attached to the ejector plate. When the lifter moves upward during ejection, the two halves approach each other and set the part free.

Even more complex internal geometric features, including undercuts, are possible with collapsible cores. These are relatively complex tooling elements which collapse when their inner support elements are withdrawn. The core shown in Fig. 5.8 is composed of an inner supporting element and six outer elements. When the mold is opened, the inner support is withdrawn and the outer elements move inwards and release the part. Undercuts up to 12% are generally possible with these cores, in some cases even 17%.

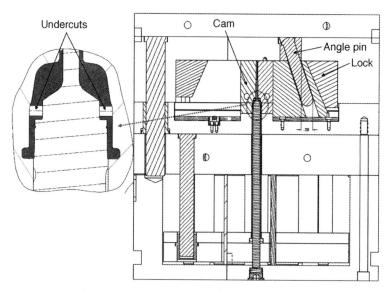

*5.5* Undercuts can be shaped by laterally moving elements (Dynamic).

Tooling for metal injection molding 103

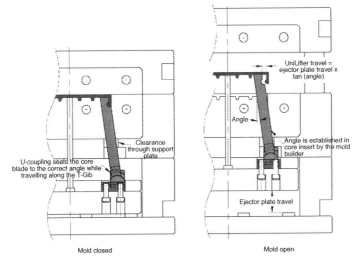

*5.6* Angle lifters for shaping undercuts (Dynamic).

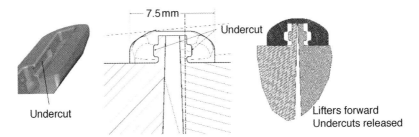

*5.7* Part with undercut (left) and mold design (Dynamic).

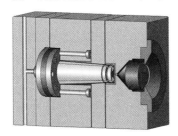

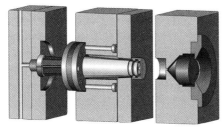

*5.8* Collapsible core in molding (top) and ejection position (bottom) (Wiedemann).

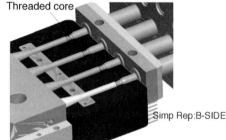

*5.9* Part with inner thread (left) and threaded cores in a mold with four cavities (Dynamic).

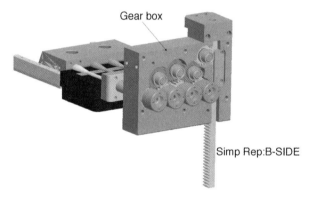

*5.10* Gearbox for synchronized unscrewing (Dynamic).

Internal threads in MIM parts are shaped by means of threaded core rods. These are unscrewed before the part is ejected from the mold. An example is shown in Fig. 5.9 for a multiple cavity tooling. The set of threaded cores can be unscrewed simultaneously with a gearbox, as shown in Fig. 5.10.

## 5.5 Special features and instrumentation

The stream of feedstock entering the mold can be imagined like a stream of lava flowing down a volcano. The inside is liquid and the surface, which is in contact with the environment, is solid. When the feedstock comes into contact with the cool walls of sprue and runner, a thin skin of frozen feedstock is immediately formed. The hot, molten feedstock then flows in the core of the runner channel. The frozen skin eventually grows thicker, the cross-section of the molten core is reduced, and the resistance against a further supply of feedstock increases. When the cross-section of the runner

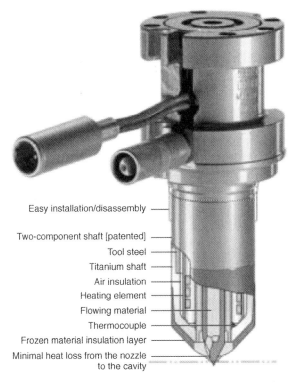

*5.11* Hot runner injection nozzle (Günther Heißkanaltechnik).

is entirely frozen, it is impossible to further sustain the injection pressure inside the mold.

A more advanced technology uses electrically heated hot runner nozzles to keep the feedstock temperature in the entire feeding system permanently above the freezing point. Hot runner nozzles are pre-manufactured components which are integrated into the mold. They are equipped with a power supply and a thermocouple for temperature control. Besides single nozzles, as shown in Fig. 5.11, manifolds are also available, for multiple cavity tooling.

Hot runner technology allows the injection pressure to be transmitted over a longer distance. The gate towards the mold cavity can be sealed with a needle after the cavity has been filled. The temperature distribution inside a hot runner mold must be carefully analyzed to make sure that the mold cavities are cool enough for the product to freeze quickly and still keep the runners liquid.

Hot runner molds have a special design at the injection side of the tooling, see Fig. 5.12, where the injection is from the top. The sprue is replaced by the hot runner nozzle, whose tip forms part of the mold cavity.

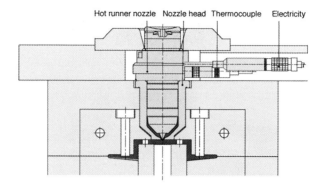

*5.12* Design of a single hot runner mold (Günther Heißkanaltechnik).

In cold runner technology the entire shot, that is the molded parts including runners and sprue, is ejected from the mold after it has frozen and the sprue and runner material is recycled (i.e. it is crushed and added to fresh feedstock). Hot runner molds are more expensive to manufacture, but allow savings by reducing the feedstock waste and sometimes reduce the molding cycle times because the mold cavity often freezes faster than the runners. The significantly higher cost is only justified, however, for very high volume production.

The quality of MIM parts depends very much on the consistency of the injection molding process. It is therefore desirable to reduce variations in temperature and pressure in the mold as far as possible. In the case of multiple cavities, a symmetrical arrangement is preferred where the path of the feedstock to each mold cavity is the same. Sensors are available for the direct measurement of temperature and pressure inside the mold cavities. With feedback of this information to the control computer, the barrel temperature and injection pressure can be adjusted in order to keep the mold temperature and pressure constant within narrow limits.

## 5.6 Supporting software and economic aspects

In the mold making sector, the MIM industry can rely on the vast plastics injection molding industry and their suppliers. The resources of mold designers, mold makers, suppliers of standardized mold components, sensor elements and instrumentation, mold design software, simulation, and even calculation software are immense.

Although the parameters of the injection molding machine can have a significant influence on the part quality, the most important factor in the injection molding process for the quality of MIM products is certainly the quality of the mold, with the greatest consideration given to its design and manufacture.

For economic reasons, independent mold designers often work for plastic injection moulders and powder injection moulders at the same time, but they are not involved in the subsequent manufacturing processes. Therefore they usually do not receive direct feedback about the performance of their molds and cannot immediately react to any drawbacks of their tooling. In the long run this may be a disadvantage for MIM. There is even a growing tendency to import dies and molds from countries with low labor cost. A recent survey of the Center for Promotion of Imports from Developing Countries (CBI) based in Rotterdam in the Netherlands stated that the demand for tooling, dies, and molds in Europe amounted to €11.6 billion in 2007 and 8.5% of all tooling was imported from developing countries in that year (CBI, 2009).

Mold fill studies based on computer simulation can be a valuable tool in mold design. This software must be specially adapted to MIM feedstocks because the high density of metal powder particles in comparison to the thermoplastic binder can lead to unexpected effects during injection molding. An experiment using a two-phase computer model to predict the stages of mold filling on a wheel of 50 mm diameter found that the density was not uniform in the green compact. Plate I (see color section between pages 222 and 223) shows the part at 25, 50, 75, and 100% mold fill. The different colors represent the volume fractions of powder in the feedstock. Higher powder volume fractions mean higher green densities.

This effect was interpreted as follows. The flow speed of the feedstock during injection is so high that a change in the direction of flow causes considerable forces on powder particles, owing to their high moment of inertia, and this leads to a powder/binder segregation resulting in density variations in the green compact. The consequence of density variations is non-uniform shrinkage during sintering and therefore less dimensional accuracy on the sintered parts.

In many cases economic aspects have a great influence on mold design as well. The number of mold cavities, use of hot or cold runner mold, and other variables depend not only on technical considerations, but also on matters of economy and efficiency. The use of computer calculation software allows the tool manufacturing costs to be calculated depending on the number of cavities, instrumentation, and so on. The user can quickly assess the most economic combination of many parameters and even determine the required injection pressure, clamping force, and cycle time. Advanced mold calculation software delivers not only the costs of the mold, but also the most important molding parameters and the manufacturing costs of the injection molding step.

## 5.7 Sources of further information

ASM (2008), *ASM Handbook, Vol. 7, Powder Metal Technologies and Applications,* ASM International, Materials Park, Ohio, USA.

Bryce, D. M. (1996), *Plastic Injection Molding: Manufacturing Process Fundamentals,* Society of Manufacturing Engineers, Dearborn, MI, USA.

CBI (2009), *The EU Market for Tooling, Dies and Moulds,* Centrum tot Bevordering van de Import uit ontwikkelingslanden, Rotterdam, The Netherlands.

*European Tool and Mould Making* (2007), IX(9), Sellers Media, Wiesbaden, Germany.

German, R. M. (2003), *Powder Injection Molding – Design and Applications,* Innovative Material Solutions, Inc., State College, PA, USA.

Kazmer, D. O. (2007), *Injection Mold Design Engineering,* Hanser Publishers, Munich, Germany.

McGillivray, P. (2007), 'Mold making for PIM', Dynamic Engineering, Minneapolis, MN, USA, *presented at the PIM 2007 MPIF Conference on Powder Injection Molding,* Orlando, USA.

Osswald, T. A., Turng, L. S., and Gramann, P. J. (2007), *Injection Molding Handbook,* 2nd edition, Hanser Publishers, Munich, Germany.

Rees, H. (1996), *Understanding Product Design for Injection Molding,* Hanser Publishers, Munich, Germany.

Rosato, D., Rosato, M., and Rosato, D. (2000), *Injection Molding Handbook,* 3rd edition, Kluwer Academic Publishers, Norwell, Massachusetts, USA.

Song, J., Barrière, T., Liu, T., and Gelin, J. C. (2007), 'Experimental and numerical analysis on the densification behaviours of the metal injection moulded components', *Proceedings of EUROPM 2007,* Toulouse, France, European Powder Metallurgy Federation, Shrewsbury, UK.

# 6
# Molding of components in metal injection molding (MIM)

D. F. HEANEY, Advanced Powder Products, Inc., USA and
C. D. GREENE, Treemen Industries, Inc., USA

**Abstract:** Metal injection molding (MIM) utilizes the same shape-making technology as thermoplastic injection molding; however, there are subtle differences in the equipment and molding techniques. In this chapter a review of the equipment, molding process parameters, shrinkage behavior, and common MIM molding defects and their solutions is presented. Where applicable, the similarities and contrasts with molding of thermoplastics will be discussed.

**Key words:** MIM molding equipment, micromolding, anisotropic shrinkage, molding defects.

## 6.1 Introduction

Metal injection molding (MIM) is a process that utilizes the vast shape making capability of injection molding technology to manufacture structural metallic components. Unlike molding of thermoplastics where the molded component is often the final product, these components must subsequently be thermal processed to remove the polymers and sinter densified into structural components. Both thermoplastic and MIM processes utilize temperature and pressure to form components, thus, similarities in the science of shape forming exist between the two technologies. In this chapter a review of the technology of injection molding is presented from the context of MIM. Where applicable, the similarities and contrasts with molding of thermoplastics will be discussed.

The following Sections 6.2 to 6.4 give an overview of the injection molding machines and tooling used and needed for the injection molding process. Emphasis on the reciprocating screw is given since this configuration dominates the industry.

## 6.2 Injection molding equipment

The first injection molding machine was patented in 1872 by two brothers, John and Isaiah Hyatt.[1] This machine was simple and utilized a plunger injection method to fill a cavity. This technology is still used today and is often referred to as low-pressure injection molding since this configuration can only produce low pressures during the injection step, as compared to a reciprocating screw configuration. This low-pressure method is utilized for powder injection molding (PIM) of ceramics and carbides. Low-pressure PIM technology uses a wax-based binder system that has a low viscosity. Low-pressure plunger machines are also used to form the wax patterns for modern investment (lost wax) casting. The initial capital cost and operating cost of low-pressure molding machines is lower; however, the productivity and dimensional repeatability are inferior to that of reciprocating screw technology. The first reciprocating screw was invented in 1946, by American inventor James Watson Hendry for thermoplastics. A combination of the screw and plunger technology exists in modern thermoplastic and MIM for micromolding technology. Although injection molding of thermoplastics is over 100 years old, the injection molding process was not recognized as a metal processing technology until 1979, when two design awards were given to MIM components; since then MIM has become one of the leading net-shape metal forming processes.[2]

### 6.2.1 Conventional injection molding machines

Metal and plastic injection molding use essentially the same type of machine to produce near net-shape components. The injection molding machine consists of an injection unit and a clamping unit. The injection and the clamping unit are actuated by hydraulic or electromechanical force depending upon make and model of the machine. Hybrid machines that combine hydraulic and electromechanical features also exist. Plate II (see color section between pages 222 and 223) shows a photograph of a typical hydraulic injection molding machine used for MIM.

The injection unit is composed of the barrel, heater bands, hopper, nozzle, screw, and screw tip. Figure 6.1 shows the typical injection unit component configuration. The screw tip consists of a retainer, a check ring and rear seat, as shown in Fig. 6.2. The screw tip experiences significant wear and should be closely monitored for dimensional change, particularly on the check ring and the rear seat, which provides the seal to allow for injection at high pressures. These components are essential in providing a homogeneous feedstock to the melt delivery system of the mold. The injection unit should be sized so that the shot size is 20–80 % of the barrel capacity, but ultimately depends on the sensitivity of the binder system in the feedstock and the

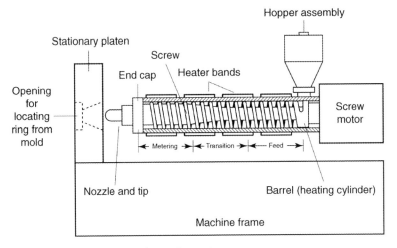

6.1 Injection unit configuration.

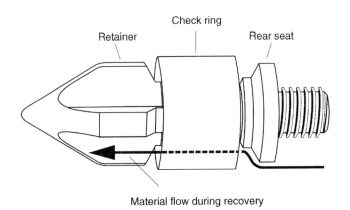

Material flow during recovery

6.2 Screw tip configuration.

required volume to fill the part completely. This ensures that the feedstock has adequate time to reach equilibrium temperature and not degrade the binder system.

The clamping unit consists of the tie bars, platens, and ejection system. Some clamping units are designed without tie bars, which allows for larger tools to be utilized since the tie bars are absent from taking space on the platens. The clamping unit is where the mold is mounted to the injection molding machine. The tie bar spacing and platen size are the limiting factors in the envelope size of tool that can be utilized in a machine. Clamp tonnage is the limiting factor in how large a component's surface area can be to successfully run in the machine. If the machine clamping tonnage is too low, the component will flash. Note that flash can also be caused by a worn or an improperly fitted tool or some other process issue, such as too high an injection speed. Typical tonnage requirements are about 8–10 tons per square inch of projected area of runner and part. MIM components are typically made on presses in the 100 ton or lower range; however, larger presses can be used in special applications.

The most significant differences between the plastic injection molding machine and the metal injection molding machine are in the injection unit and control system. The screw, screw tip, and barrel of the machines used for MIM are typically hardened to resist the abrasive nature of injection molding metals. Hardened and abrasion resistant components are also used on machines specified for glass-loaded nylon or other compounds that have abrasive fillers. Wear on these components will most often be seen on the feed zone of the screw directly below the hopper; this is logical since the material at this point has not yet melted, so it behaves like a solid. The other location for wear will be the check ring and seat as the check ring experiences a significant amount of shear due to the material flowing through the small spacing orifice during injection and the seat must close on MIM feedstock during each cycle; thus, they must be replaced periodically. MIM molding machines should be equipped with precise screw position regulation controls to ensure better shot-to-shot consistency, particularly with smaller components; however, a standard control system is adequate for many MIM components. Precise screw position regulation controls are also utilized for high-precision plastic injection molding; however, the effect of poor molding processes is exacerbated in the MIM process due to the significant amount of shrinkage that MIM experiences during the sintering operation. This shrinkage is significantly affected by the weight of the molded component and the weight is more variable with poor screw position regulation.

Another consideration with MIM is the screw configuration. MIM materials are typically shear sensitive and less compressible than pure polymers, thus the requirement for the screw is different. In general, the screw has three different regions – the feed zone near the hopper, the transition zone, and the metering zone near the screw tip, as shown in Fig. 6.1. The size and depth of these different regions affect how well the material feeds and how much shear is put into the material. In the simplest sense, the compression ratio is the ratio between the flight depth of the feed zone and the metering zone. However, this is true only if the flight width and spacing of the screw geometry and the length of the feed and metering zone are identical. The true compression ratio is the volume ratio of the volume within flights between the feed zone and the metering zone.[3] The higher the compression ratio, the more shear heating and compression the feedstock will experience. The compression ratios used for MIM screws are typically lower than for plastic screws. A typical general purpose thermoplastic compression ratio is 2.3, process sensitive materials such as polyvinyl chloride (PVC) and liquid silicone have compression ratios less than 2.3, and non-sensitive materials may have compression ratios up to 3.0. MIM material compression ratios are typically in the 1.7–2.2 range; however, larger compression ratios have been used and in some cases might be desirable, for example, where materials may have a significant amount of air bubbles as a result of poor mixing, they could be better processed with a higher compression ratio screw and for materials that have less shear sensitivity.

### 6.2.2 Micro injection molding machines

Micro-sized components can be fabricated using both conventional molding machines and those specifically designed for components of this size scale. When using conventional molding machines, the runner and component must make a significant enough shot size to register as a change on the injection unit controller. This can be accomplished by using a larger runner system as compared to the component of interest or by using a small-sized screw. The down-side to the method of using a larger runner is the waste material associated with the larger runner, although this material can be reused. Also, the smaller the screw, the greater the shear rate at the screw tip in order to obtain the same volumetric fill rate as compared to a larger screw. This higher shear rate results in powder/binder separation and subsequent galling at the screw tip.

Also available on the market are machines specifically designed for micro-molding. These machines possess a two-stage injection unit. The first stage consists of a screw that feeds material very similarly to a reciprocating screw. In this stage the material is melted and metered to eliminate any air voids into a plunger or screw configured barrel. This second-stage plunger or screw is used to provide small, very accurately controlled dosages of material into the cavity. The pre-metered material has a more uniform consistency with little air void space, which provides for a more accurate shot size and pressure control during the second-stage injection. These two-stage machines are typically more costly than conventional molding machines. A sketch of a micro-molding injection unit configuration is shown in Fig. 6.3. The fabrication of micro components is covered in Chapter 13.

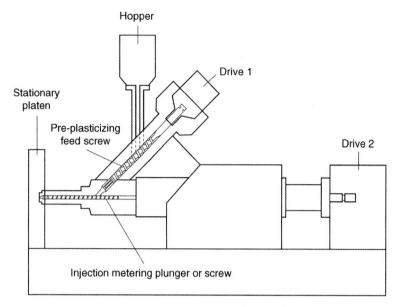

*6.3* Two-stage micro injection molding injection unit configuration.

### 6.2.3 The mold

The mold is a significant factor in successful injection molding of both plastics and MIM feedstock. The mold performs several key functions. It contains the scaled-up component geometry cavity and melt delivery system, acts as a heat exchanger to remove or add heat to ensure defect-free molding, mates with the injection unit, withstands the clamping and injection forces, and ejects the molded component.

Special care must be taken when designing the mold. The component must be easily ejected to avoid damaging the green components. A significant difference between thermoplastic injection molding and MIM is that the MIM material is very brittle as compared to most thermoplastics, thus, tooling that allows easy ejection of components is essential. The mold must be equipped with a temperature control system that will provide uniform and sufficient temperature to all regions of the part, to ensure even cooling rates, which influences dimensional stability and defect production. Blisters are a common defect when the mold is unable to sufficiently remove the amount of heat that the molten MIM compound brings to the tool. Mold design is addressed in Chapter 5.

## 6.3 Auxiliary equipment

This section gives an overview of the auxiliary equipment used and needed for the injection molding process.

### 6.3.1 Material drying

Feedstock with hygroscopic binder systems will require the feedstock to be dried before processing – particularly if the molding is performed in a high-humidity, non-air-conditioned atmosphere. Typical systems that contain polyethylene glycol (PEG) or polyvinyl alcohol (PVA) would be candidates for drying. The drying can be accomplished using desiccant bed or compressed air dryers. Drying the material helps to eliminate moisture-related defects such as blisters and powder binder separation. Machine metering problems can also be a moisture-related symptom. These problems often show up in the humid time of summer and disappear during the dry winter months.

### 6.3.2 Mold temperature controllers

The mold temperature controller's main function is to maintain a desired mold temperature setting appropriate for the component and feedstock combination being molded. The mold temperature controller circulates cooling fluid through the mold's cooling circuits to provide equilibrium to the heat flux experienced by the mold for every cycle. It is common to run separate mold temperature controllers for each mold half. This set-up allows for greater process control and processability. If a hot sprue is used on the stationary half of the tool to bring the molten MIM feedstock to the parting line, this half of the tool requires a greater amount of cooling, thus having a separate controller here will allow the temperatures of the two mold halves to be more accurately controlled.

### 6.3.3 Granulators

Granulators are essentially rotary grinders that are used to grind scrap parts and melt delivery systems (sprues and runners) into feedstock sized granules for reprocessing. This allows the molder to reduce waste and produce components more cost effectively. Since these granulators will be used to cut polymers that are loaded with metal powders, the wear of the cutting blades is great. As such, the blades should be manufactured from carbide or a tool steel with high wear resistance. A design that allows resharpening of blades is desirable. Many granulators are designed so that the blades must be completely replaced when they are worn. The last consideration with respect to granulators is to have a granulator that is easy to clean thoroughly to avoid cross contamination between MIM materials. Having a granulator to serve each material may be desirable.

### 6.3.4 Part removal

Once the part has cured or solidified in the tool, it can be dropped into a container, dropped onto a conveyor, or picked out of the molding machine by a robot or by an operator when the machine uses a semi-automatic cycle. Conveyors can be either continuous and/or indexing. For parts with high green strength this can be an acceptable form of part removal; however, fragile components or components that require no cosmetic defects such as 'dings' or 'scuffs' must avoid this technique. Robots are employed when a high level of automation is needed for cost considerations and/or when green strength is low and the parts are susceptible to contact damage. Robots allow for precise placement of the green components into green machining stations and/or sintering fixtures.

## 6.4 Injection molding process

### 6.4.1 Overview

The molding process begins once the mold and barrel have reached the molding equilibrium temperature. The reciprocating screw in the heated barrel augurs the required volume of material, or the shot size, to completely fill the gate, runner, sprue, and cavity, yet still maintain a small amount of material in front of the screw for a cushion. The cushion is critical for providing ample material for complete part filling and it accommodates for any feedstock variation. If insufficient cushion is available, the screw will bottom in the barrel resulting in low-mass components, voids, and incomplete cavity fill. Once the shot size is achieved, the mold closes and the screw tip now acts as a plunger via a non-return valve or check ring that allows the feedstock to be injected through the nozzle tip of the injection unit, into the sprue, runner, gate, and part geometry of the mold. Other technologies, such as a smear tip (which is a conical device used in place of a non-return valve at the discharge end of an injection screw) are available if the metal being molded galls on the check ring and causes binding of the check ring on the barrel, or if the feedstock is particularly susceptible to shear degradation. These are best suited for higher viscosity materials and for small parts where holding pressure is not as critical to fully pack out the mold cavity. The use of a smear tip reduces the available injection pressure and hold pressure since it allows material to flow past, as compared to a conventional check ring and seat configuration.

The time required to fill the mold cavity to 92–98% full is called the fill or injection forward time. The fill time is controlled by the injection velocity profile set on the machine controller and is dependent upon the viscosity of the melt and the runner and component geometry. This velocity is referred to as the fill rate or fill velocity. The fill rate can be profiled in many different ways and is a bit of an art, as it is strongly influenced by material behavior and tooling configurations. For example, the profile might start slowly during runner filling to prevent gate defects and jetting into the part, followed by a fast injection, and finally slowed down prior to switchover to hold pressure. In other cases, the fill rate is profiled to first inject at a fast velocity to prevent flow lines or wrinkling in the part at the beginning of fill and then the velocity is reduced at the end of fill. Reduction of fill rate prior to completely filling the component is essential to reduce the chance of over-packing the part or causing flash. Flash is the material that escapes through the parting line of the two mold halves or along two pieces of steel in the mold, such as along a slide face or along the interface between ejector pin and its cavity.

Once the cavity is 92–98% full, the process transfers from a velocity-controlled filling stage to a pressure-controlled stage that finishes filling the last 2–8% of the cavity and compensates for the volumetric shrinkage that occurs due to the pressure–volume–temperature (PVT) behavior of the feedstock as it cools in the mold cavity. The switchover does not occur at 100% full cavity in order to avoid over-packing the cavity, which often causes flash. The pressure applied during the final filling and compensation stage is commonly referred to as the hold or pack pressure and the duration is referred to as hold or pack time. The hold time is the time in which hold pressure is applied until the gate has frozen and feedstock can no longer be pushed into the part geometry. A gate freeze study (part weight versus hold time) is used to determine the appropriate hold time. This ensures that the gate is frozen, the cavity is completely packed, and no material can escape the mold cavity once the hold pressure is released. Figure 6.4 shows a typical gate freeze study. In the example shown, the hold pressure needs to be held for 3 s to ensure complete filling of the component and to avoid sink defects.

Upon completion of part filling and packing the part is subject to a cooling time to ensure that the component green strength is adequate to withstand the ejection forces and removal. Concurrent with the cooling time is shot size generation, which is called plastification. During plastification the screw reciprocates at a set r/min to augur material in front of the check ring. During screw rotation the material must overcome the backpressure set on the machine controller to push the screw back in the barrel and generate the required shot size. The backpressure generates additional shear heating and ensures a fully compressed feedstock. An increase in back-pressure can be used to remove any voids in the material. If the material is

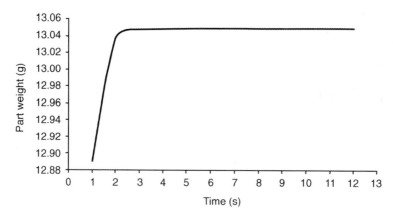

*6.4* Graphical representation of a typical gate freeze study used to ascertain the optimum pack time. A 3 s hold time would be optimal in this case.

difficult to plastify or feed in the screw, a decrease in backpressure is used. It is very important that the r/min and backpressure are set so plastification is completed 2 s prior to completion of the cooling time to ensure consistent cycle times.

Once the cooling time and plastification are complete the clamp decompresses, the mold opens and the green component is ejected and removed, at which point the process can be repeated. The complete molding cycle is depicted in Fig. 6.5.[4]

### 6.4.2 Molding parameters

Although there are many parameters in molding and all find importance in particular molding conditions, the most critical molding parameters are mold and melt temperature, injection speed, switchover point and method, hold pressure and time, and cooling time. Note that the following discussion presents generalizations since all these parameters are interrelated, which can convolute troubleshooting and process optimization. For example, an increase in mold temperature can increase the ability to fill a component completely; however, this can be overshadowed by injection speed, switchover pressure, and hold pressure.

*Mold and melt temperature*

The mold and melt temperature should be set at a temperature that will produce defect-free parts. In other words, these are often changed within a

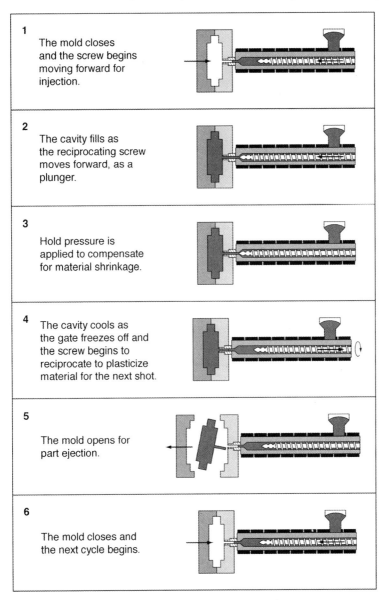

| | |
|---|---|
| 1 | The mold closes and the screw begins moving forward for injection. |
| 2 | The cavity fills as the reciprocating screw moves forward, as a plunger. |
| 3 | Hold pressure is applied to compensate for material shrinkage. |
| 4 | The cavity cools as the gate freezes off and the screw begins to reciprocate to plasticize material for the next shot. |
| 5 | The mold opens for part ejection. |
| 6 | The mold closes and the next cycle begins. |

*6.5* Injection molding process overview.[4]

process window to reduce temperature-related defects such as blisters, gate blemishes, short shots, weld lines, etc. The melt temperature is typically 10–20°C above the melting temperature of the backbone polymer, which is the highest temperature polymer in the compound. If the temperature is set too low, poor flow characteristics like flow lines could be evident. Also, at too low a temperature, the backbone polymer may show as separate strands in the component and leave voids in the final sintered product. This is also evident when the material is compounded at too low a temperature or for too short a time. Low temperature also shows as variability in green part mass and variability in switchover pressure or position depending on the method of switchover. When the temperature is too high, the material will experience more shrinkage in the tool and this may result in cracking. Also, too high a temperature can result in blister defects, flash of the lower melting temperature polymer along the parting line and into the interface between the ejector pin and the cavity block, or problems in the injection unit where the powder/polymer separation can cause the metal to gall onto the check ring of the screw. Furthermore, too high a melt temperature results in degradation of the polymer carrier and can result in short material life.

Mold temperatures influence how well the component can fill and the production of defect-free components. Too low a temperature can cause the component not to fill completely and to have flow defects. Too high a temperature can result in blisters and flashing.

Typical melt temperatures for common wax–polymers systems are 150–190°C and the mold temperature is 25–55°C. Typical melt temperatures for catalytic systems are 200–260°C and the mold temperature is 100–150°C.

*Injection speed*

MIM injection speeds are typically higher than those for polymers because of the higher thermal conductivity of the feedstock as compared to pure polymers. The injection speed is typically set at the minimum injection speed required to completely fill the component without any defects. Too low an injection speed will result in surface imperfections such as flow lines, poor weld lines, and incomplete fill. Too high an injection speed will result in flash and gate blemishes due to powder/binder separation and potentially voids, if the injection is too fast to allow the air inside the cavity to escape at the mold vents.

*Switchover point and method*

The ideal injection molding process is one where the component is 98% filled under velocity control and then switched to a pack control for the remaining 2% of component volume. Many studies have been conducted to evaluate different methods of switching from velocity control to pack control during the injection molding operation.[5] The four switchover methods of most importance are position, hydraulic pressure, time, and cavity pressure. Of the four, cavity pressure is the most repeatable. In this technique, a pressure transducer located in the melt stream is used to detect a pressure, which is used to signal the machine to stop velocity control and start pack control. This transducer can be located at the nozzle tip, in the runner, or in the component cavity. The component cavity or runner is the most accurate; however, it is the most expensive since it must be incorporated into each tool, which is an added tooling expense. In general, the closer you are to your actual component cavity, the more accurate your switchover.

Switchover using position is a practical method to obtain consistent production. In this technique, a linear transducer is used to signal that the screw has reached a pre-selected position. This technique ensures that the screw reaches the same position for each cycle, even if the screw begins to foul. Because of this potential fouling, other safety controls such as screw torque limiting or statistical process control (SPC) of switchover pressure is used to ensure that the screw or screw tip does not become damaged. This type of safety control is also used with the pressure transducer switchover technique.

Time and hydraulic pressure switchover methods should be avoided. Time switchover uses an injection time to switch from speed to pressure control. This method can be variable since the material may have different pressures at switchover due to variability in the feeding behavior of the material. Hydraulic pressure switchover uses the pressure of the hydraulics in the injection unit to switch into pack control fill. This technique can be accurate; however, over long run MIM production, if the check ring becomes fouled, the screw does not reach the same position for the same pressure, and thus the components will begin to lose mass or show defects such as sinks or short shots. Thus, this technique is not recommended for MIM.

*Hold pressure and time*

Hold pressure and time are important to ensure that the component is completely packed out. Typically, the hold pressure is selected to be approximately the same pressure at switchover, depending on material behavior and tooling configuration. An example to the contrary is a component geometry which may have a feature with a thin section that is difficult to fill, but the thickness of the component is relatively low; in this case a high switchover pressure may be required to obtain fill, but a low hold pressure may be desirable as the part does not need the high pressure to compensate for sink formation because it is a relatively thin component. A gate freeze study is used to determine the hold time. Signs of insufficient hold pressure or time are voids and sinks. Signs of excessive hold pressure or hold time are warpage, gate defects, and flashing. The hold pressure can also be profiled to reduce molded-in stresses that may show up as warpage or cracking in subsequent debinding and sintering.

*Cool time*

Cool time is important to ensure that components are completely solidified prior to ejection. If the component is cooled for too long or too short a time, the component can be damaged during ejection. Cooling that is too short results in damage such as ejection pin damage on the component or ejection warpage. Cooling that is too long results in cracking of the component during ejection. Short cooling times can be used to remove damage-prone components from the tool when they still have some elasticity as a result of being under-cured. A mold closing delay can also be used for this process step to ensure that the tool cools for the next cycle.

### 6.4.3 Shrinkage

The manipulation of the injection molding process conditions for the MIM process can modify the associated volumetric shrinkage, which subsequently affect the final sintered dimensions.[6,7] For instance, as melt temperature, mold temperature, and part thickness increase, so does the amount of shrinkage. It is well known that polymers have positive coefficients of thermal expansion and are highly compressible in the molten state. As a result, the volume that a given mass of material occupies will change with both temperature and pressure.[8] Multiple studies have determined that hold pressure has the largest effect on shrinkage in plastics.[9–11] All studies agreed that increasing hold pressure decreased the amount of shrinkage. Some of the process variable/shrinkage relationships are generalized in Fig. 6.6.[4] These shrinkage characteristics also prove to be true for MIM components.

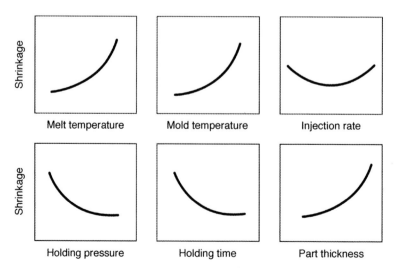

*6.6* Typical shrinkage relationships with various process parameters.[4]

## 6.4.4 PVT effect

Accurate linear-shrinkage prediction of MIM components for tool scale-up is commonly achieved by a combination of estimating sintering shrinkage and prototyping or through historical shrinkage data gathering. When a new material is being developed, sintering shrinkage calculations for mold scale-up can underestimate the final sintered shrinkage by up to 2%. Historically, linear shrinkage ($Y_S$), as shown in equation [6.1], was calculated as a function of the initial solids loading ($\phi$) of powder in the feedstock, the theoretical density of the powder ($\rho_t$), and the final sintered density ($\rho$), as described in the following equations.[2]

$$Y_S = 1 - \left(\frac{\phi}{\rho/\rho_t}\right)^{1/3} \qquad [6.1]$$

However, when the sintering shrinkage ($Y_S$) is used for mold scale-up ($Z_s$), as shown in equation [6.2], components are found to be undersized. Therefore, the engineer must manipulate the tool dimensions, feedstock formulation, and the sintering process to achieve the target dimensions specified by the design engineer.

$$Z_C = \frac{1}{1 - Y_C} \qquad [6.2]$$

Inaccuracy of component shrinkage can be corrected by considering the shrinkage associated with the injection molding process as defined by the MIM feedstock's PVT behavior during injection molding of a MIM feedstock. The molding shrinkage can be estimated from the PVT behavior of the molten powder/binder feedstock and combined with sintering shrinkage for accurate shrinkage. The PVT relationships are commonly represented by a two-dimensional diagram as shown in Fig. 6.7. The PVT graph for polymer-behaving systems, like MIM feedstock, normally depicts the specific volume plotted against temperature at various pressure isobars. Essentially, by using these data the injection molding process can be traced on the PVT diagram, if the processing conditions and several key material properties are known. Once the cycle is traced on the PVT diagram it is then possible to predict the volumetric shrinkage due to injection molding ($S_v$), as shown in equation [6.3],[12] of the component by subtracting the specific volume of the component at equilibrium ($v_e$), point 5, from the specific volume of the component at the time of gate freeze ($v_{gf}$), point 4. Although, specific volume does give an accurate depiction of the volumetric shrinkage

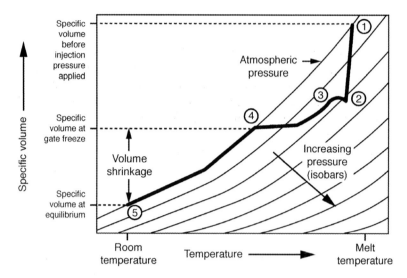

*6.7* PVT diagram with molding cycle trace. 1. Polymer in barrel at melt temperature, 1–2. Pressure increases as cavity fills, 2–3. Switches to holding phase, 3–4. Pressure decreases as melt solidifies, 4. Gate freezes off, 4–5. Part shrinks until equilibrium is reached (after R. Malloy[8]).

in a part, linear shrinkage ($S_L$) is much more useful to the mold engineer.

$$S_V = 1 - \left(\frac{v_e}{v_{gf}}\right) \qquad [6.3]$$

Assuming that the part undergoes isotropic volumetric shrinkage, the linear shrinkage equation is then given by equation [6.4][8]

$$S_L = 1 - (1 - S_V)^{1/3} \qquad [6.4]$$

A PVT diagram for a 63 vol% carbonyl iron wax–polymer MIM feedstock is presented in Fig. 6.8. From this diagram, it is possible to see the different hold pressures and subsequent specific volume associated with these hold pressures. Upon comparison of these data, as obtained from the modified two-domain Tait PVT model, with actual measured volumetric behavior, one can clearly see the correlation of volumetric shrinkage with pack pressure. This is shown in Fig. 6.9. The accuracy of this predicted green shrinkage from the modified two-domain Tait PVT model was within 6.5% error of the experimental results in this particular analysis. Both the green shrinkage and the PVT predicted shrinkage have a slope of $-0.0002$ (mm$^3$/mm$^3$)/MPa. Thus, the holding pressure during injection molding can significantly affect final molding component dimensions and mass of components, which subsequently influence the final component dimensions.

A combination of both sintering shrinkage (equations [6.1] and equation

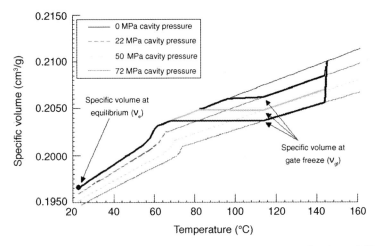

*6.8* PVT diagram with the injection molding cycle traces for the 30 MPa, 60 MPa, and 90 MPa cycles for a 63 vol% carbonyl iron wax–polymer feedstock.

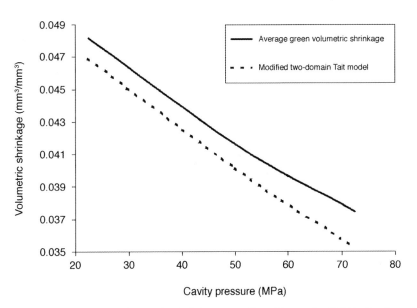

*6.9* MIM volumetric shrinkage predicted from the PVT behavior of a MIM feedstock compared to the average measured green volumetric shrinkage.

[6.2]) and molding shrinkage (equations [6.3] and [6.4]) is required to obtain accurate shrinkage predictions. The combined linear shrinkage ($Y_C$) can then be used to more accurately calculate the combined mold scale-up ($Z_C$) required to account for molding and sintering shrinkage, equations [6.5] and [6.6].[7]

$$Y_C = 2 - \left(\frac{\phi}{\rho/\rho_t}\right)^{1/3} - \left(\frac{v_e}{v_{gf}}\right)^{1/3} \qquad [6.5]$$

$$Z_C = \frac{1}{1 - Y_C} \qquad [6.6]$$

In Fig. 6.10, the theoretical linear sintering shrinkage, as derived from equation [6.1], which considers only the shrinkage from sintering, is compared with average measured sintered linear shrinkage and the shrinkage calculated using equations [6.5] and [6.6]. A significant disparity exists between equation [6.1] predicted results and measured sintering shrinkage. This is only corrected when the PVT effect is incorporated into the shrinkage equation. Therefore, shrinkage at molding contributes to the total shrinkage of sintered components. When no PVT effect is considered, the sintered shrinkage as defined by equation [6.1] underpredicts dimensions by an average of 0.015 mm/mm (1.5%) and does not account for the effect of cavity pressure on the final shrinkage of the sintered components.

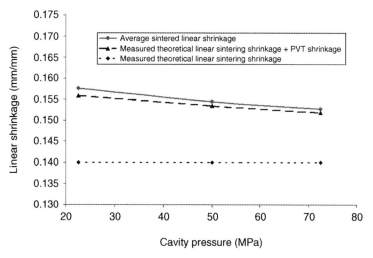

*6.10* Comparison of measured sintering shrinkage with the shrinkage predicted from sintering and sintering plus the PVT shrinkage.

## 6.4.5 Anisotropic shrinkage

The melt history in injection molding of both MIM and thermoplastics can be quite complex and can result in anisotropic shrinkage. The material is also affected by orientation during injection, multiple pressures during the injection process, part geometry, flow patterns, time, and cooling rates. All of these factors may lead to anisotropic shrinkage. Figure 6.11 shows a MIM component that was characterized for anisotropic behavior. The dimensions characterized were the thickness at gate (denoted 'Thickness gate'), thickness at end of fill ('Thickness EOF'), width at gate ('Width gate'), width at end of fill ('Width EOF'), and length ('Length'). After molding, these component dimensions were measured and compared to the tooling to characterize shrinkage, as shown in Fig. 6.12. Interestingly, the thickness of the component experiences the greatest shrinkage. Logically, of the two thickness measurements, the gate area experiences the least amount of shrinkage since the holding pressure would be greatest in this region. The reason for the greater shrinkage in the thickness direction is attributed to internal stresses created within the component as the polymer component of the feedstock freezes off in layers.[13, 14] The layers create a mechanical coupling within the solidifying and shrinking component. Figure 6.13 illustrates the cross-flow direction, in-flow direction, and through-thickness direction of flow. The layers exhibit a strong interlocking force in the cross-flow and in-flow directions, Fig. 6.14. Therefore, the largest amount of volumetric shrinkage is achieved through the thickness of the component.[4, 11, 14, 15] Lee and Dubin[16] explain that the outer layers crystallize at higher pressures. However, once the gate freezes off, the solidifying core of the

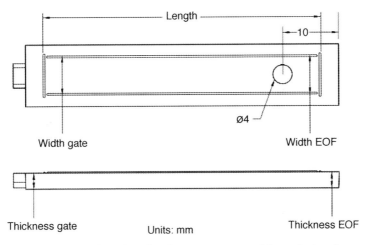

*6.11* Labels used to describe features measured for anisotropic shrinkage analysis.

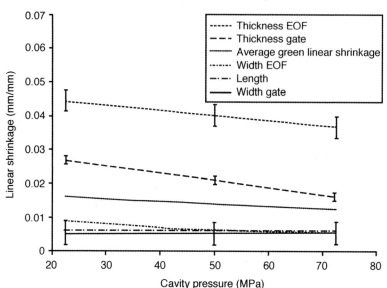

*6.12* Measured experimental green anisotropic shrinkage.

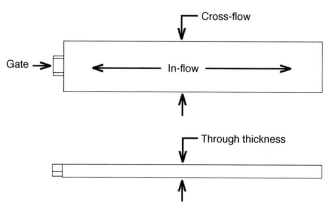

*6.13* Description of flow direction terminology.

component is crystallizing at lower pressures, therefore, more volumetric shrinkage occurs.

Anisotropic shrinkage in filled MIM and unfilled injection molded polymer materials is a very common phenomenon.[17–20] However, there are many factors that influence the shrinkage anisotropy. For semi-crystalline materials, cooling rate, polymer and/or filler orientation in the direction of flow during crystallization, mold restraints, internal stresses, and pressure gradients also affect the magnitude and direction of the shrinkage.

Anisotropic shrinkage makes tool scale-up extremely difficult. The most

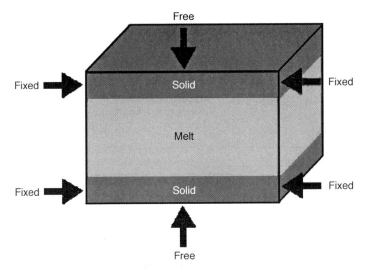

*6.14* Description of mechanical coupling. The through-thickness direction is free to shrink. The in-flow and cross-flow directions are constrained due to internal stresses and mechanical coupling.

common practices for avoiding non-uniform stress distributions caused by anisotropic shrinkage include uniform cooling of all cavity and core components, utilizing a decreasing hold pressure profile to ensure even packing, and the use of mold filling and packing simulation software to optimize the process.

## 6.5 Common defects in MIM

Injection molding defects in the MIM process can be apparent directly after molding or they may not manifest themselves until after subsequent process steps. Molding is a multi-variable process where the variables are heavily interactive, thus, there are multiple ways to solve a problem. Also, the solution to one issue can cause defects of a different form. Table 6.1 is a guide to common defects and potential solutions.

*Table 6.1* Common defects and solutions for MIM molding

| Defect | Potential causes | Potential solution |
|---|---|---|
| Blistering | Improperly dried material | Dry hygroscopic material<br>Reduce humidity in plant |
| Blistering | Poor heat removal from tool | Increase cycle time<br>Reduce mold temperature<br>Reduce melt temperature<br>More cooling channels in tool<br>Higher thermal conductivity mold components |
| Blistering | Localized shear heating | Reduce injection speed<br>Larger gates to reduce shear heating |
| Blistering | Polymer degradation | Use higher MW polymers<br>Use antioxidant in binder |
| Cracking | Ejection issue | Increase mold temperature<br>Reduce pack pressure<br>Decrease melt temperature<br>Polish tool |
| Cracking after solvent | Molded in stress | Increase venting<br>Increase mold temperature<br>Post mold anneal |
| Flash | Tooling not closed | Adjust tooling for better clamping<br>Worn tooling, dress parting line<br>Clean mold faces of dirt |
| Flash | Too low clamp tonnage | Higher tonnage machine<br>Reduce injection pressure by slowing injection<br>Reduce pack pressure |
| Flash | Material viscosity | Increase viscosity by increasing solids loading (will cause shrinkage change)<br>Increase viscosity by high molecular weight polymers |
| Gate blemish | Powder/binder separation at gate | Decrease injection speed<br>Decrease pack pressure |
| Incomplete fill | Insufficient material in cavity prior to gate freeze | Increase shot size<br>Increase pack pressure<br>Increase injection speed<br>Increase venting<br>Increase melt temperature<br>Increase mold temperature |
| Incomplete fill | Material feeding issue | Decrease backpressure on recovery<br>Decrease size of feedstock and regrind<br>Dry material for hygroscopic material<br>Decrease fines in feedstock |

*Table 6.1 (cont.)*

| Defect | Potential causes | Potential solution |
|---|---|---|
| Sinks | Poorly filled part | Increase pack pressure or pack time<br>Increase gate size<br>Increase backpressure for screw recovery<br>Decrease melt temperature to decrease material shrinkage<br>Increase melt temperature to allow greater fill<br>Increase venting |
| Variable component mass | Shot size variation | Clean check ring<br>Increase recovery backpressure<br>Increase cushion |
| Voids | Air entrapment | Decrease injection speed<br>Increase backpressure during recovery<br>Increase pack pressure and time |
| Warpage | Component distorts during ejection | Increase cool time<br>Decrease mold temperature<br>Better cooling in tool |
| Warpage | Pressure gradient in component | Decrease hold pressure |
| Weld lines | Premature material solidification | Increase injection speed<br>Increase melt temperature<br>Increase pressure for switchover to pack control |
| Weld lines | Gas entrapment in mold | Increase venting on part and runner<br>Decrease injection speed |
| Wrinkles | Premature material solidification | Increase injection speed<br>Increase melt temperature |

## 6.6 References

1. Hyatt, J. and Hyatt, I. (1872), US Patent 133329, 19 November 1872.
2. German, R. and Bose, A. (1997), *Injection Molding of Metals and Ceramics*, Metal Powder Industries Federation (MPIF), Princeton, NJ, USA, p. 13.
3. Womer, T. W. (2000), 'Basic screw geometry', *Proceedings of ANTEC 2000, Annual Technical Conference.*
4. Moldflow Pty Ltd (1991), *Warpage Design Principles: Making Accurate Plastic Parts*, Moldflow Pty Ltd, Kilsyth, Victoria, Australia.
5. Piemme, J. C. (2003), *Effects of Injection Molding Conditions on Dimensional Precision in Powder Injection Molding*, MS thesis, Pennsylvania State University, USA.
6. Greene, C. and Heaney, D. (2004), 'The PVT effect on final sintered MIM components', *Proceedings of ANTEC 2004, Annual Technical Conference*, vol 1: Processing, pp. 713–717.
7. Greene, C. D. and Heaney, D. F. (2007), The PVT effect on the final sintered

dimensions of powder injection molded components, *Materials and Design*, 28, 95–100.
8. Malloy, R. (1994), *Plastic Part Design for Injection Molding*, Hanser/Gardner, NY, USA.
9. Wang T. and Yoon C. (1999), 'Effects of process conditions on shrinkage and warpage in the injection molding process', *SPE Technical Papers*, pp. 584–588.
10. Pontes, A., Oliveira, M., and Pouzada, A. (2002), 'Studies on the influence of the holding pressure on the orientation and shrinkage of injection molded parts', *SPE Technical Papers*, pp. 516–520.
11. Jansen, K., Van Dijk, D., and Husselman, M. (1998), 'Effect of processing conditions on shrinkage in injection molding', *Polymer Engineering and Science*, 38(5), 838–846.
12. Shay, R. M., Poslinski, A. J., and Fakhreddine, Y., (1998), 'Estimating linear shrinkage of semicrystalline resins from pressure–volume–temperature (PVT) data', *SPE Technical Papers*.
13. Pötsch, G. and Michaeli, W. (1990), 'The prediction of linear shrinkage and warpage for thermoplastic injection moldings', *SPE Annual Technical Papers*, pp. 355–358.
14. Zöllner, O. (2001), *The Fundamentals of Shrinkage of Thermoplastics*, Bayer Corporation.
15. Bushko, W. C., and Stokes, V. K. (1996), 'Dimensional stability of thermoplastic parts: modeling issues', *SPE Technical Papers*, pp. 482–485.
16. Lee, C. S. and Dubin, A. (1990), 'Shrinkage and warpage behavior of injection molded Nylon 6 and PET bars and plates', *SPE Annual Technical Papers*, pp. 375–381.
17. Greene, C. D. and Heaney, D. F. (2004), 'The PVT effect on final sintered MIM components', *SPE Technical Papers*.
18. Mamat, A. Trochu, F. and Sanschagrin, B. (1995), 'Shrinkage analysis of injection molded polypropylene parts', *Polymer Engineering Science*, 35(19), 1511–1520.
19. Binet, C., Heaney, D. F., Piemme, J. C., and Burke, P. (2002), 'An investigation of orientation and how it relates to anisotropic shrinkage', *Proceedings of the PM2TEC 2002 World Congress*, MPIF, 16–21 June, Orlando, Florida.
20. White, G. R. and German, R. M. (1993), 'Dimensional control of powder injection molded 316L stainless steel using in-situ molding correction', In *Advances in Powder Metallurgy and Particulate Materials*, Metal Powder Industries Federation, Princeton, NJ, no. 5, pp. 121–132.

# 7
# Debinding and sintering of metal injection molding (MIM) components:

S. BANERJEE, DSH Technologies LLC, USA and
C. J. JOENS, Elnik Systems LLC, USA

**Abstract:** The chapter deals with processing MIM parts after they have been molded. Primary debinding depends on the feedstock type being used. The reason for primary debinding and the process for major feedstock types are discussed. The equipment and process for the commercially used feedstocks are described. Secondary debinding and sintering have been combined into one step from the old three-step process used in the original MIM process. Thermal debinding of the secondary binder is described and the consequences of incomplete debinding discussed. Sintering, mass transport during sintering and the practical aspects of sintering are explained. Guidelines are provided for obtaining optimum properties by MIM for most MIM materials including the effect of different atmospheres. Settering of parts with both flat surfaces and complex shapes and setter materials are described. Both continuous and batch equipment may be used.

**Key words:** MIM, debinding, primary binder, solvent debinding, catalytic debinding, water-soluble binder, secondary binder, thermal debinding, TGA, diffusion bonding, sintering, surface area of powder, MIM materials, furnace atmospheres, oxide reduction, setters, ceramics, furnace profiles, MIM furnaces, continuous furnace, batch furnace, graphite furnace, refractory metal furnace.

## 7.1 Introduction

This chapter describes how the part is processed to obtain the solid metal part after it has been molded. This section would also apply to parts made by other consolidation methods using a feedstock or a metal/binder mixture where the binder initially holds the powder together.

In order to understand debinding it is necessary to understand the thinking behind the formulation of the MIM feedstock. The feedstock is a mixture of binders and metal powders that makes the mixture moldable in an injection molding machine and later permits the easy removal of the binders such that the molded shape is retained after binder removal and sintering, resulting in a solid part in the molded shape. A detailed explanation is given in the book by German and Bose.[1]

From the debinding viewpoint, the feedstock is made basically of two binder components. The purpose of the first component, which is easily removable at a low temperature, is to open up a network of pores through which the second binder may escape at a later stage, while the second binder holds the powders in place until the temperature is high enough for diffusion bonds between the metal powders to take place so that the molded shape is retained when all the binders are gone. Other components are added to enable the binders to wet the powder surfaces and to create a bond between the two major binder components. These components are either removed with the primary binder or in some cases may need to be removed in the same manner as the backbone binder.

The two major binders require two debinding steps, usually called primary debinding and secondary debinding. All binders must be removed before moving on to the higher temperatures where sintering takes place. Sintering temperatures are much higher than the volatilization temperatures of the binders. Hence any residual binder escaping rapidly from the part could cause the part to crack and lose its shape integrity or form soot which may affect the composition of the material.

The original MIM patents are based on the wax-based feedstock, but many different binder systems have been developed since. Each system needs to be treated differently for purposes of debinding because of the different polymers used in the binder systems.

### 7.1.1 Binder systems

Different binder systems have come up based on the principles above. Here we discuss the ones being used most from a commercial point of view in order to understand how the polymers are removed from them. These binder systems are listed below.

1. The first systems based on the Strivens[2] and Weich[3–5] patents used waxes and oils for the primary binder and a polyolefin as the secondary binder with an addition of a rubber-like material to bond these two materials. Binder systems based on waxes are still the most widely used feedstock type in the USA.
2. Polyacetal is broken down first in the presence of an acid catalyst.[6,7] This is the basis of the BASF Catamold binder system[8] using polyacetal

and a polyolefin. The Catamold feedstock is the most used ready-made feedstock.
3. The primary binder used is water soluble, for example, as described by Rivers[9] or Hens and Grohowski[10] or sold by PolyMIM.[11] Feedstocks based on these systems are becoming more popular as halogenated and other organic solvents are becoming more difficult to use.
4. The primary binder easily evaporates or sublimes at room temperature or at a relatively low temperature; for example, water used as the primary binder, when it is mixed with agar to form a gel in the process developed by Allied Signal[12] or naphthalene, the main ingredient of traditional moth balls, is used as the primary binder in a system developed at the Pacific Northwest National Laboratories.[13, 14] These feedstocks show promise because almost no equipment is needed for primary debinding.

### 7.1.2 The first MIM systems

Although the water-soluble system by Rivers[9] and the wax-based system by Weich[3–5] were developed almost in the same time frame, the wax-based system proliferated faster as Weich and his partners went their own way and the initial model was to license the process to grow the industry.

In the wax-based system, the primary binders were thermally debound very slowly in an electrically heated air oven. This low-temperature burnout (LTB) oven required a very slow ramp of about 0.5°C/min and long hold times at the peak temperature of about 200°C, followed by oven cooling to about 70°C before the carts were removed from the oven. Processing times through the LTB oven for orthodontic parts weighing between 0.1 g and 0.05 g were close to 24 h. Hence part thickness was limited to obtain a reasonable processing time. Also, the long exposure to low temperatures degrades the backbone binders, making the part extra fragile.

The parts were then transferred to a high-temperature debind (HTB) oven, which ran under a reducing or protective atmosphere, where the secondary binder was removed and the parts pre-sintered at temperatures up to about 1050°C to reach sufficient strength to be handled and moved to a third furnace where sintering to the full density would take place. The third furnace was usually a vacuum furnace. This choice stemmed from the fact that the furnaces with protective atmospheres that could handle the large amount of binder released did not go to a high enough temperature to reach the full density desired and vacuum furnaces which did reach the high temperatures desired could not handle the binders released.

Today the MIM process has been improved and requires less time, although there are still a few companies who are manufacturing parts using the old three-step method described above. The process used today uses one piece of equipment to perform the primary debinding depending on the feedstock type used in the process. The parts are then moved to a 'debind and sinter' furnace where the parts go through the secondary debinding operation, followed by ramping on for the sintering operation. Most of these furnaces for the standard MIM materials reach 1450°C and the furnace processing time is usually less than a day.

The three-step process is considered to be obsolete, although there are still a few manufacturers using this process. Thermal debinding of the primary binder and secondary debinding with pre-sintering are not discussed in this chapter.

## 7.2 Primary debinding

Most feedstocks have at least two binders. When the primary binder, sometimes called the soluble binder, is removed a network of interconnected pores is formed. The polymer evaporating at the higher temperature uses this network of interconnected pores to escape without cracking or degrading or causing sudden stresses that degrade or distort the molded part.

During thermal debinding of the primary binder, as the temperature rises, increasing amounts of binder evaporate from the entire binder mass. Sudden evaporation of a binder fraction with no clear exit path causes the part to rupture and lose integrity. Hence the evaporation rate of the binder must be extremely slow to prevent part rupture. This makes thermal debinding of the primary binder a difficult and time-consuming process that has been discontinued by most MIM processors. The desired method is to remove the primary binder starting from the outside of the part, moving in to the core, in such a manner that the secondary binder is affected as little as possible. Since this is achieved by solvents, organic or water, as well as the catalytic method, these three types of feedstocks have found the widest applications. Debinding of these binder systems have been discussed in detail in this chapter.

The Allied Signal/Honeywell feedstock[12] lost momentum when Honeywell bought Allied Signal. This material needs only an air drying oven to dry the moisture for primary debinding. Although this process requires little capital expenditure for primary debinding, there seem to be no users left today. The binder system from Pacific Northwest Laboratories,[13,14] which uses naphthalene that sublimes at room temperature as the primary binder, is relatively new and is not fully commercialized as yet. These systems have not been discussed.

## 7.2.1 Solvent debinding of wax-based systems

In North America the most prevalent binder system is the wax-based system because most of these were derived from a Weich type feedstock system.[3-5] Waxes and oils dissolve in a large number of solvents. The process is relatively quick if the solvent used is at a temperature where the waxes and oils are in the liquid stage.[3] The parts are immersed in the solvent for a duration long enough to dissolve all the primary binder. The solvent is then removed and the now porous parts are dried to a solvent-free condition before being put into the sintering furnace. The method most widely used to determine if all the primary binder is removed is to weigh and catalog 30 samples and remove one part every 30 min, dry it and check the weight loss. When the weight loss stops then the parts are fully debound. If the density of the feedstock is known and the theoretical density of the brown (debound) part is available, measurement of the brown density of the part using a helium pycnometer[15, 16] is the most accurate method to determine whether the part has been properly debound or not. Also note that the time required to debind is a function of the debinding method, the size of the part, as well as the particle size of the powders used to make the part. For example, the same part molded from 20 µm gas atomized copper powder, 10 µm carbonyl iron powder, and 1 µm alumina powder using the same feedstock required 3 h, 6 h, and 22 h of solvent debinding time respectively.

Initially solvent debinding units were probably heated open tanks. Environmental regulations have forced most of these open tanks out of use in most advanced countries. Solvent debinding units permitted by the regulations were modified parts-washing units or dry cleaning machines. These are already fitted with proper environmental controls and have closed-loop circulation and distillation systems which prevent the release of the solvent to the atmosphere and the exposure of the workers to the solvent. 99% plus of the solvent vapors are extracted by drying the parts under vacuum and using a carbon canister from which the captured solvent may be reused. Only a very small fraction of the solvent used is released to the atmosphere per run. The required process steps are programmed into the programmable logic controller (PLC) which controls the solvent debinding unit. A simplified schematic diagram for the solvent debinding system is shown in Fig. 7.1. The parts are immersed in clean solvent for the set time to dissolve all the primary binders. When the binders are removed, the solvent is moved to a 'dirty tank'. While the parts are dried in the chamber under vacuum, the dirty solvent is distilled and the clean solvent is stored in the 'clean tank' for reuse in the next cycle. Figure 7.2 shows such a typical solvent debinding system marketed by Elnik Systems,[17] where all the tanks, distillation unit, pumps, and filters are hidden behind the panels.

Some of the older solvents used are 1,1,1 trichloroethylene, perchlor-

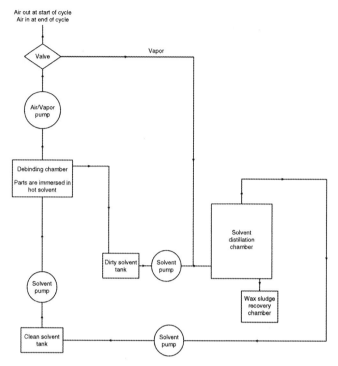

7.1 Schematic diagram of a solvent debinding system.

oethylene, and *n*-propyl bromide. Since these halogenated chemicals either deplete the ozone layer or are considered to be carcinogens, in most countries their vapors must be totally captured. Many countries have either banned or are banning the use of ozone depleting chemicals and require the use of non-ozone depleting organic solvents, such as hexane, alcohols or acetone, which are considered to be green solvents. Hence, the modern solvent debind units may also be set up as explosion proof units to work with such organic solvents.

### 7.2.2 Supercritical solvent debinding

In the supercritical solvent debinding[18] process the waxes are removed using liquid carbon dioxide as the solvent at about 50°–70°C. This process is very attractive because the solvent is carbon dioxide, which is also considered to be a green solvent, and the dissolution of the wax requires a very short time. After the soaking process, the liquid carbon dioxide is removed from the chamber and the pressure is reduced, which allows carbon dioxide to evaporate. The waxes are left behind in the solid form and can easily be cleanly removed. However, to keep the carbon dioxide in the liquid form at

Debinding and sintering of metal injection molding components 139

*7.2* Typical solvent debinding system marketed by Elnik Systems.

these high temperatures, pressures as high as 350 bars are required. Special high-pressure vessels are required, which has limited the diameter of the chambers, limiting the use for very small parts only. The costs of making conventional size chambers are prohibitive and have slowed commercialization of this process.

### 7.2.3 Catalytic debinding of Catamold feedstock

The catalytic debinding (CD) oven works with the Catamold binder system only.[8] The Catamold binder system comprises two major ingredients: (a) polyoxymethylene (POM), which is also known as polyacetal or polyformaldehyde and (b) polyethylene. POM, whose formula is $(-H_2-C-O-)n$, breaks down into formaldehyde, $(H_2-C=O)$, in the presence of nitric acid between 100°C and 140°C and leaves behind the metal powder skeleton held together by the remaining polypropylene. The formaldehyde formed is explosive in the presence of air, hence a nitrogen flow is maintained to remove all oxygen from inside the oven. The ratio of nitric acid vapor to nitrogen must be below 4% for proper catalytic reaction. Although formaldehyde itself is toxic, it burns cleanly and easily. Hence all CD ovens have a burner to burn the formaldehyde coming out of the exhaust of the oven.

There are many manufacturers that make CD ovens and all of them work based on the above principle. Most of these are batch ovens, although at least one manufacturer makes a continuous CD oven in tandem with their continuous sintering furnace. Figure 7.3 shows a CD oven manufactured by Elnik Systems. The Elnik CD ovens are the only ones controlled by a

*7.3* A catalytic debinding (CD) oven.

# Debinding and sintering of metal injection molding components 141

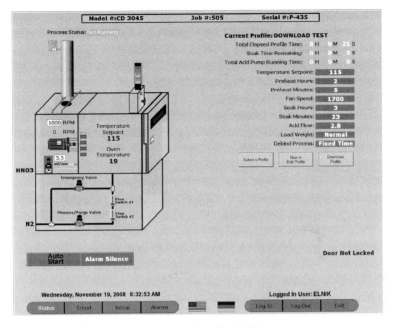

7.4 The main computer screen of an Elnik CD oven.

computer (PC). The main computer screen of this oven is shown in Fig. 7.4. The main screen shows all the combinations possible. The computer uses an Excel spreadsheet to develop the profile program which is downloaded to the PLC. This permits the user to make programs using all the variables in the process and allows the recording of all the programs and the trends for all the debinding runs performed by the oven.

BASF provides a minimum weight loss percentage that needs to be achieved for complete debinding in their product data sheets.

## 7.2.4 Debinding of water-soluble systems

In case of the water-soluble systems, the primary binder is water soluble while the secondary binder is not. Parts made out of the water-soluble binders are immersed in a tank until all the primary binders are removed. After all the binders are removed they are put in an air oven to dry. After complete drying the parts are ready to move on to the next stage. Most of these units are home-made ones where readily available water baths have been modified to dissolve the primary binder and separate air ovens are used for drying.

Elnik Systems took this concept and has produced the first specialized water debinding system where the parts are dried in the same chamber by hot air or vacuum without handling. The unit allows the water temperature

142   Handbook of metal injection molding

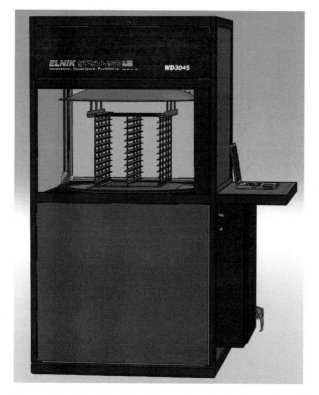

*7.5* Water debind unit with a 4.5 cu. ft (127 liters) work volume.

to be controlled at temperatures up to 70°C. The unit permits fully automatic cycling via the built-in PLC control and computer. Figure 7.5 shows a WD 3045 for a 4.5 cu. ft (127 liters) work volume.

### 7.2.5 Primary debinding guidelines

The basis for debinding and the equipment used for the three major types of feedstocks commercially available or used by houses that make their own feedstock have been discussed. In the case of solvent debinding, water being the solvent for water-soluble materials, the dissolution of the solute into a solvent depends on the solubility of the solute, the temperature of the solvent, and the solute concentration gradient. Hence, the debinding parameters depend on the specific binder and solvent combinations. Some of the following issues require attention.

- The debinding temperature or the solvent temperature when the parts are immersed; when the primary binder is liquefied, dissolution rates are much faster than in the solid state.

*Table 7.1* Typical solvents and their boiling points

| Solvent | Boiling point (°C) | Flash point (°C) |
|---|---|---|
| Organic | | |
| Acetone | 56 | −18 |
| Hexane | 69 | −7 |
| Heptane | 98.4 | −4 |
| Isopropyl alcohol | 82 | 12 |
| Halogenated | | |
| n-Propyl bromide (NPB) | 70 | None |
| Trichloroethylene | 87 | None |
| Perchloroethylene | 121 | None |
| n-Methyl pyrrolidone (NMP) | 204.3 | None |

- Reaction or interaction of the solvent with the secondary or other tertiary binders at the debinding temperature should not cause part distortion from the melting of a binder constituent or by adsorption of the solvent by the secondary binder.
- The solvent should not have a high vapor pressure at debinding temperatures and become a fire hazard. This is especially true for modern organic solvents with low flash points. Often such equipment must be housed in separate explosion-proof rooms. Table 7.1 shows some of these solvents with their boiling points and flash points. Most organic solvents have very low flash points. Also included in this table are some of the popular halogenated solvents and their boiling points. The halogenated solvents do not have flash points.
- Care must also be taken regarding local waste disposal issues when releasing solvents or distillation wastes which are contaminated by the primary binder. Also, just because a binder is water soluble does not mean the solution may be dumped in the sewer system.

In the case of all the binder systems discussed, the reaction is taking place from the outside of the part and proceeds towards the center of the part. In case of the solvent systems the binder has to dissolve into the solvent and in case of the catalytic reaction, the catalyst has to be present at the surface for the reaction to proceed. Hence in all cases the debinding processing time is a function of the thickness of the part in conjunction with the other individual process parameters.

Table 7.2 shows some binder systems with the debinding method, temperatures used for debinding, and approximate debinding rates achieved for regular MIM parts. As the part thickness increases over 10 mm significant decrease in the debind rates may be observed when extra time will be needed to completely debind the parts.

*Table 7.2* Debinding temperature and rates for different binder systems

| Primary binder | Secondary binder | Primary debind method | Debind temperature (°C) | Debind rate (mm/h) |
|---|---|---|---|---|
| Wax-based | | Solvent | | |
| Paraffin wax | Polypropylene | Heptane | 50 | 1.5 |
| Synthetic wax | Polypropylene | Perchloroethylene | 70 | 2 |
| Water-soluble Polyethylglycol | | | | |
| (PEG) 200 | Polypropylene | Water | 40 | 0.3 |
| PEG | Polyacetal | Water | 60 | 0.5 |
| Catamold | | | | |
| Polyacetal | Polyethylene | Nitric acid catalyst | 120 | 1.5 |

## 7.3 Secondary debinding

At the low temperatures where the primary binders are removed no diffusion bonding between powder particles takes place. It is the secondary binders, also known as the backbone binders, and interparticle friction that hold the powder particles together and maintain the shape provided by the injection molding after the primary binders are removed. Removal of the primary binders generates a network of interconnected pores within the brown molded part which enhances the ability of the secondary binder to be removed without causing defects in the components.

The secondary binders are removed thermally. This is achieved by heating the parts slowly to the temperature where the secondary binder evaporates and holding the parts at this temperature until all the binders are removed. More than one temperature hold may be necessary if there is more than one secondary binder present. Often, a small amount of the primary binder is also left behind inside and on the surface of the part after the primary debinding by the solvent or water debinding process. A short hold for this remnant primary binder is also customary.

A thermo-gravimetric analysis (TGA) of the feedstock, before and after primary debinding, is the best method to determine the hold temperatures for the secondary binders. The TGA is usually ramped at the ramp rate the parts would see during thermal debinding in the sintering furnace. During thermal debinding in the sintering furnace the ramp rates are slow. Holds are placed at each of the temperatures where the weight loss for each of the components is completed. Table 7.3 gives some debinding temperatures used for some typical secondary binders. Figure 7.6 shows a typical curve from a TGA run of a 4140 wax-based feedstock. Based on this analysis holds were placed at 380°C, 440°C, and 540°C, which resulted in the carbon content of the alloy coming out at the middle of the specification. Hold times are a

# Debinding and sintering of metal injection molding components

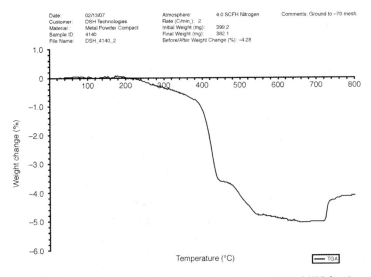

7.6 TGA curve for a solvent debound wax-based 4140 MIM feedstock.

Table 7.3 Thermal debinding temperatures of common secondary binders

| Secondary binder | Thermal debind temperature (°C) |
|---|---|
| Polypropylene | 450–500 |
| Polyethylene | 500–600 |
| Polyacetal | 300–450 |

function of the maximum wall thickness of the part and will depend on the individual binder system.

By the end of the thermal debinding cycle, tiny diffusion bonds are formed between the powder metal particles which augment the interparticle friction and help retain the shape given to the part by injection molding. In case of ceramic molded parts, the particle sizes are of the order of one magnitude finer than that of metal powders. The finer particles result in narrower pores, when longer debind times are necessary for removing the binders. For ceramics, no sintering takes place at the thermal debinding temperatures. In case of these very fine particle sizes, inter particle friction is the dominant cause that helps maintain the shape after removal of all the binders.

### 7.3.1 Incomplete binder removal

There are several reasons for incomplete binder removal but they all result in two factors, wrong debinding temperature and/or insufficient debinding time at temperature. Gas flow rate is also an important factor in debinding. Insufficient gas flow prevents the evaporated binders from being removed and allows the binders to be retained in the part. The result is the same effect as the binder being removed at the wrong temperature. Gas flow is a processing matter which depends on the binder system and content, as well as the powder particle size and distribution.

The correct debinding temperature is best determined using a TGA, as explained earlier. The hold duration is a function of the binder system, the amount of binder, the pore size and length (which are a function of the particle size and distribution), and the wall thickness of the part. However, in spite of the heat transfer and thermodynamics classes, many engineers forget that just because a thermocouple reads 600°C (or whatever temperature the hold is supposed to be at), it does not mean that the entire load inside a furnace is at 600°C. It takes time for the furnace temperature to equalize at all regions inside the furnace load and the hold time start point must be counted from the time only after the furnace temperature has equalized.

If the parts inside the furnace do not dwell at the correct temperature for the correct time there is binder inside the parts when the furnace moves on to the next ramp to reach the sintering temperature. The remaining binder comes out at temperatures higher than where it should have been removed and at this point the heating rate is also faster. Depending on the amount of binder left the following may happen.

- Sudden evaporation of binder can occur leading to cracking and/or blistering of the parts.
- Rapid evaporation of the binder can occur, causing the binder to be deposited on the insulation of the heating elements and the furnace, followed by conversion of the binders into soot and graphite with rapid rise in temperature. Multiple deposits in this manner reduce the life of the hot zone of the furnace by coating the heating element insulation, thus shorting them to ground. The life of the heating elements could also be reduced. These would require rework of the furnace at a reduced interval requiring additional capital expenses.
- Rapid temperature rise inside the parts can also occur, causing the polymer to crack and leave soot deposits inside the parts, resulting in a high carbon content and a change in the desired chemical composition; this in turn leads to property changes in the parts produced, such as

partial melting or slumping, distortion, low ductility and brittleness, high hardness, poor corrosion resistance, etc.

It is thus extremely important to remove the secondary binder properly to obtain the desired properties in the part and to obtain the optimum life for the furnace.

## 7.4 Sintering

The purpose of the exercise in metal injection molding is to remove the binder from the parts and convert the powder mass into a strong metal part without losing the molded shape. This final process of converting the part into a strong mass is achieved by sintering. The parts need to be subjected to a higher temperature in a suitable atmosphere for this to happen.

### 7.4.1 Sintering definitions

Sintering is a term used in many industries. The current author's (S.B.) first contact with the term was an extractive metallurgy class where students learnt about iron ore fines and limestone fines being mixed together and fired to make a solid but porous mass that could be easily charged into a blast furnace. The iron pillar of Delhi in India, made in the late 4th century AD and brought to its current location in the 11th century, is said to be powder forged from magnetite ores. The last time I saw it, it had no rust and you could touch it. Today there is a fence around it to prevent erosion from the touches of the daily tourist. Sintering is used to make ceramics, refractory metals, spark plugs, cemented carbides, self-lubricating bearings, electrical contacts, insulators, oxide nuclear fuels, structural, magnetic, aerospace, and medical parts to name a few, with many more possible applications in the future.

There are thus many different definitions depending on the industry that is served. German[19] provides a working definition of sintering as follows: 'Sintering is a thermal treatment for bonding particles into a coherent, predominantly solid structure via mass transport events that often occur on the atomic state. The bonding leads to improved strength and a lower system energy'. The glossary in ASTM B 243-09a[20] says in Section 3.3.1 Sinter, v. 'to increase the bonding in a mass of powder or a compact by heating below the melting point of the main constituent.'

### 7.4.2 Sintering theories

There are many different theories on the mechanisms of sintering. Although an older work, an excellent review is provided by Thümmler and Thomma.[21]

148  Handbook of metal injection molding

The theories are far behind the sophistication and degree of the practice of sintering. German's[19] book *Sintering Theory and Practice* does an excellent job in trying to fill this gap and explain the various mechanisms of sintering. For those who like explanations without mathematics, Pease and West[22] provide a comprehensive explanation of sintering.

Sintering lowers the surface energy of the powder mass by the formation of interparticle bonds which reduce the surface area. With the growth of the interparticle bonding significant changes occur in the pore structure resulting in the improvement in many properties, such as, strength, ductility, corrosion resistance, conductivity, and magnetic permeability. These changes are of primary interest in industrial applications including MIM. Also, in MIM the dramatic elimination of pores results in large shrinkages and the variables in the entire process have to be within control to obtain the process repeatability that keeps the sintered dimensions within the desired tolerance limits.

### 7.4.3 Mass transport mechanisms

Atoms move and vibrate all the time, even in the solid state. This movement combined with the need to remove free surfaces in order to reduce surface energy results in the formation of solid bonds between particles. Heat enhances this effect. As the free surfaces are removed pore migration occurs by the movement of vacancies using grain boundaries as conduits. Further energy reduction occurs by the reduction of grain boundaries by grain growth, after which further homogenization may occur by bulk diffusion. If the process continues for infinite time the two spheres in the model should end up as one sphere, the state of the least energy.

Sintering theories assume that sintering is taking place between two spheres of the same size, which are at a point contact, under isothermal conditions. Mass flow in response to the driving force for sintering is determined. Under these conditions mass transport may occur by surface transport mechanisms and by bulk transport mechanisms.

Figure 7.7 depicts the two-sphere sintering model. The spheres touch, then the neck forms and begins to grow. During this time surface transport mechanisms predominate by moving atoms from surface sources, namely, evaporation and condensation (E-C), surface diffusion (SD) and volume diffusion (VD). In all these cases the atoms travel from the surface of the particles and end up at the point of contact of the particles, increasing the bond between particles without any change to the distance between the two particles but causing the formation and some growth of the neck between the two particles.

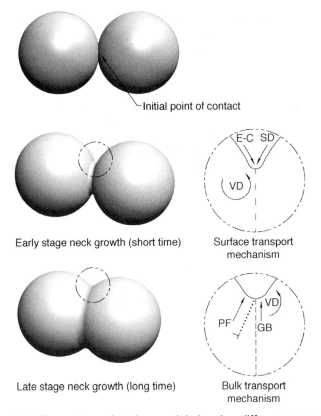

7.7 Two-sphere sintering model showing different stages of progression of sintering and mass transport mechanisms.

With later stage neck growth, the centers of the two spheres begin to get closer to one another and shrinkage is observed. Bulk transport mechanisms predominate at this time and provide for neck growth and elimination of pores using atoms from inside the two particles by plastic flow (PF), viscous flow, grain boundary diffusion (GB), and volume diffusion (VD). Only bulk transport mechanisms can cause shrinkage.

Here is a brief overview of the individual transport mechanisms.

- Evaporation and condensation: materials with high vapor pressures or those reacting with the sintering atmospheres to form a volatile species are potential candidates for sintering controlled by the evaporation and condensation process. NaCl, $TiO_2$, $H_2O$, and $Si_3N_4$ are typical examples showing this behavior. For most commercial metallic systems the vapor phase transport contributions are considered to be negligibly small.
- Viscous flow: this is the usual sintering mechanism for amorphous materials. The viscosity of most amorphous materials declines with

increasing temperature and any external pressure aids the process. Since there are no grain boundaries and the amorphous structure is like that of a liquid, full of defects, bonding progresses until curvature gradients are eliminated.

- Surface diffusion: surfaces of crystalline solids are full of defects and are never smooth, even after the best polishing techniques. The atoms are moving from one defect site to another and end up at the point of contact between the two particles to form contact and neck formation thereby reducing the surface energy between the two particles. Smaller particles result in increased curvature and lower activation energy for the process. The activation energy is also lower than that for the bulk diffusion processes. The practical application is that it causes the formation of bonds between particles at low temperatures which help retain the shape of loosely bound powder aggregates, as in cases like slurry casting of powders and powder injection molding.
- Volume diffusion: this occurs by the migration of atoms through interstitial defects such as dislocations and vacancies from inside the grains in a crystalline material. Volume diffusion can cause both neck formation and densification.
- Grain boundary diffusion: grain boundaries are made up of huge quantities of defects that stem from the mismatch between the orientations of adjacent grains. These boundaries become the conduit for atoms to move to close pores, or conversely vacancies from the pores to move out to the surface. Grain boundary diffusion has a lower activation energy than volume diffusion and hence is the mode of mass transfer that occurs before volume diffusion.
- Plastic flow: this occurs by the annihilation of vacancies by dislocation climb and its occurrence is debated when there is no external force. Under isothermal sintering conditions it is considered to be a transient process. Plastic flow is the major transport mechanism in the presence of an external pressure.

### 7.4.4 Stages of sintering

The sintering stages represent the changes in the powder mass after consolidation into a shape to the final densified object. Not all applications are the same, for example for pressed and sintered powdered metal structural parts, the high green densities are achieved by pressing the particles together at high pressure and large contact surfaces are present before sintering. On the other hand, for filter or self-lubrication bearing applications a specific pore size is desired and the parts are not sintered to be fully dense. MIM represents a condition similar to loose powder sintering, where all the stages of sintering apply.

Debinding and sintering of metal injection molding components 151

The first stage is adhesion, rearrangement, and repacking which begin with weak van der Waals' forces. Small amounts of rotation and twisting occur at high temperature to obtain lower energy states with respect to grain orientations.

The next stage is the growth of a sinter bond between contact points formed by the above process and leads to the initial stage neck growth. This is the initial stage of sintering where the early portion of neck growth occurs but there is no or only minor densification taking place.

This stage is followed by the intermediate stage, where neck growth resulting in densification takes place. The necks lose their identity and pores become rounded but are still interconnected.

During the final stage of sintering the interconnected open pores close and become isolated closed pores. Grain growth also occurs as the pores become closed, which causes the diffusion process to slow down. While the previous stages are relatively rapid, the final stage of sintering slows down asymptotically as the density passes the 95% mark and nears 99% depending on the system and the sintering temperature.

Figure 7.8 shows a scanning electron micrograph of a sintered iron–cobalt–vanadium Permandur pre-alloyed material, with powder particles on the surface showing neck growth and interparticle grain boundaries. These particles show some interconnected pores with the fully sintered area in the background. The fully sintered structure in the background shows that the

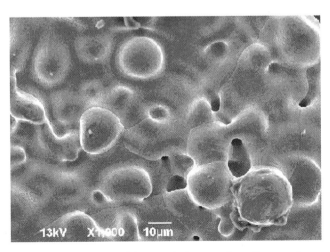

*7.8* SEM of a sintered iron–cobalt–vanadium Permandur pre-alloyed material, with powder particles on the surface showing neck growth and interparticle grain boundaries. In the background is the sintered structure where particle identities are lost and larger intra-particle grain boundaries are seen.

particle identities have been lost and larger intra-particle grain boundaries caused by grain growth have formed with the progress in sintering.

Figure 7.9(a) shows the microstructure of a pre-alloyed cobalt–chrome (F 75) alloy MIM part as sintered at 1200°C. The grains are small and there is a considerable amount of porosity, some of which is inside the grains. Figure 7.9(b) shows the structure of the same alloy as sintered at 1300°C. The grains are considerably larger than in the parts sintered at 1200°C, the total porosity is considerably less, and the majority of the pores are inside the grain boundaries. The sizes of the individual pores are also larger. Sintering has progressed between the two temperatures causing grain growth, pore migration along the grain boundaries, and pore coalescence and isolation inside the grain boundaries.

### 7.4.5 Sintering practices

While sintering models have also been made with other configurations, such as sphere–plate, wire–wire, cone–plate, cone–cone, etc., sintering practice involves many particles, not all of the same size but with different particle sizes, and not all spherical but with the shapes of the particles varying, depending on how the powders are made. Heating conditions are never isothermal. This is further complicated by the presence of other elements, which come from the powder manufacturing system or which are added on purpose for specific reasons. Diffusion of one element into another plays a major role as do the phase diagrams involving the different elements. Ternary or quaternary systems often show low melting eutectics at temperatures lower than that for similar binary systems which would bring in the effect of the presence of liquid phases.[23, 24]

When powders are consolidated into shapes the contacts formed between particles depend on the consolidation process, as well as the particle shape, particle size, and the particle size distribution. The term consolidation process is a catch-all phrase for all powder systems, for powders that could be pressed in a die or isostatically; hot, warm or cold; slip or slurry cast; injection or compression molded with a binder; loose powder in dies, and so on. In case of consolidation with pressure, the pressure causes plastic deformation in the cold, warm or hot conditions and contacts between particles are enhanced. In the other processes without external pressure they are packed to a near ideal random packing approximating the tap density of the powder. Conditions in MIM correspond to this group of consolidation without external pressure.

The powder mix used to make parts is not necessarily a homogeneous pre-alloyed material and mixtures of powders are used very often. How these elements affect one another also has important effects on the sintering process, depending on how the elements react with one another. Adding

# Debinding and sintering of metal injection molding components 153

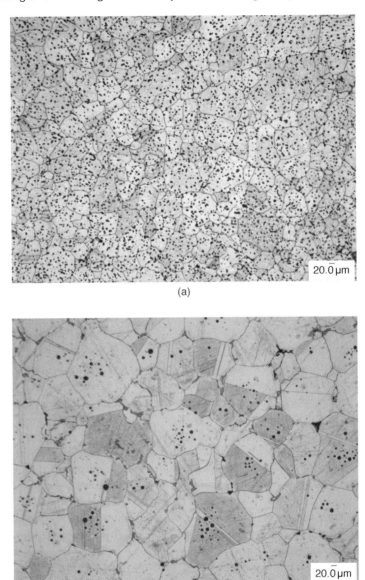

7.9 (a) Cobalt–chrome alloy (F 75) sintered at 1200°C (courtesy of J. Alan Sago, Accellent). (b) The same cobalt–chrome alloy (F 75) sintered at 1300°C (courtesy of J. Alan Sago, Accellent).

elemental powders may often be the least expensive method to make the sintered alloy. However, in some cases the reaction between some elements may leave no other recourse than blending a master alloy with an elemental powder to make the particular alloy. There are a number of possibilities when elemental powders are added to one another.

- The two components are mutually intersoluble, when homogenization takes place. The formation of stainless steels from the elemental powder blends or master alloy elemental powder blend is one such example.
- The base (matrix) is soluble in the alloy addition but not the other way around, which leads to enhanced sintering. The addition of a small percent of Ni to W results in what is referred to as activated sintering, which allows W to be sintered at temperatures as low as 1400°C.
- The alloy addition is soluble in the matrix (base) but not the other way around. This results in the addition to be dissolved in the matrix but since the matrix is not soluble in the additive, the additive is left full of voids and porosity which results in swelling. This is a situation you would want to avoid during sintering.
- Both phases are mutually insoluble; this is the situation in case of composite materials when the properties of the two components are necessary for the application. One such example is oxide dispersion strengthened alloys when $Al_2O_3$ is dispersed in a metallic matrix, such as certain Incoloy products.
- The formation of a liquid phase which needs to be discussed separately.

### 7.4.6 Sintering in the presence of a liquid phase

Many sintering systems generate a liquid. Often an element or compound is added which has no reaction or solubility with the matrix, even when it is a liquid, when they do not even wet the matrix. Some examples are lead in bronze or MnS in stainless steels. Such liquids do not aid or affect the sintering process but remain as droplets inside the matrix, which sinters by the solid state diffusion process.

On the other hand when a liquid with even a limited reaction with the matrix is present and when the liquid wets the matrix, the liquid provides a diffusion path with mass transport rates much faster than those by solid state diffusion. Hence, the formation of too much liquid can cause the rest of the skeleton to collapse; the part slumps in such cases. However, a controlled amount of liquid phase can be really beneficial to the sintering process because the faster mass transport rates result in rapid densification.

Liquid phases may be present in two forms; when the liquid phase is present during the entire sintering hold period it is known as a persistent

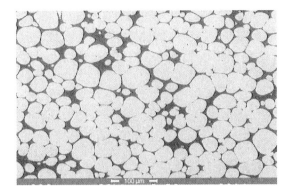

*7.10* Photomicrograph of a 90W–7Ni–3Fe alloy showing rounded grains of tungsten in a Ni–Fe–W alloy matrix, an example of solution–re-precipitation during liquid phase sintering (courtesy Animesh Bose, Materials Processing, Inc.).

liquid phase. The other form is when the liquid phase solidifies during the sinter hold period, and this is called a transient liquid phase.

Persistent liquid phase sintering is classified in two types: the first is when mixed powders are used and one is heated up to form a liquid. Typical examples are heavy alloys, such as W–Fe–Ni where the alloy is heated to form liquid Fe–Ni in which the W has limited solubility, or WC–Co where Co dissolves some WC and forms an eutectic but WC dissolves a very minimal amount of Co. Figure 7.10 shows a photomicrograph of a 90W–7Ni–3Fe alloy showing rounded grains of tungsten in a Ni–Fe–W alloy matrix. During sintering the Fe–Ni melts and dissolves tungsten in the liquid phase resulting in the rounding of the tungsten particles. The excess tungsten beyond the solubility limit precipitates inside the liquid. This is a typical example of solution – re-precipitation during liquid phase sintering.

The second type is supersolidus liquid phase sintering (SLPS) and occurs when a pre-alloyed powder is heated above the solidus to result in a small amount of liquid that causes melting of the surface and the grain boundaries within the particles, causing solution re-precipitation and densification. A typical example of where SLPS is used is in the case of M2 type tool steels. Figure 7.11 shows a typical structure of an as-sintered M2 tool steel, with small globules of the carbide phase in the matrix and some larger amounts along some of the grain boundaries.

There are two types of transient liquid phase sintering. The first is reactive sintering, when an element A forms a compound with another B with an exothermic reaction resulting in heat and the compound AB. One such example is NiAl. The second is when the other transient liquid phase disappears because of diffusion of an element to form a solid solution, such

*7.11* Typical structure of an as-sintered M2 tool steel, with small globules of the carbide phase inside the grains and some larger amounts along some of the grain boundaries (courtesy Clint Torgerson, Britt Manufacturing).

as carbon which had formed an eutectic with iron and chromium, which solidifies when the carbon diffuses into the matrix.[23, 24]

### 7.4.7 Sintering in MIM

While everything on sintering discussed above applies to MIM, the above models are based on conventional powder technologies. MIM does have some special requirements. In MIM the binder is used to carry the powder into the molds and make the desired shapes. Hence MIM powders are much finer than powders used in conventional powder metallurgy. Powders used in conventional powder technologies are usually smaller than 150 μm with about 25% fines where the fine powders are less than 45 μm. Typical MIM powders that are based on carbonyl iron powders have a $D_{90}$ of about 15 μm while typical pre-alloyed powders have a $D_{90}$ of about 22 μm. Some components have also been made with a $D_{90}$ of about 45 μm, but this is on the high side and requires special check rings in the molding machines with extra large clearances. In case of micro MIM parts, where the parts or certain features of the parts are in the micrometer range, particle sizes used are even finer and powders having $D_{90}$ of 2 μm and 3 μm are commercially available for such applications. These fine particles represent a high amount of surface area and hence a large amount of surface energy, which influence the sintering and densification mechanisms for MIM.

## 7.4.8 Effect of powder size and surface area

The powder particle size has a direct relationship with the surface area of the powder. In order to understand the effect of surface area in MIM let us calculate the surface area of 1 g of powder, assumed to be made up of 100% of 10 µm particles, to simplify the calculations. Then,

Volume of one particle = $(4/3)\pi(d/2)^3$ where $d$ is the diameter
$= (\pi/6) \times 10^{-15}$ m$^3$.
Weight of one particle = density × particle volume
$= 7.8 \times (\pi/6) \times 10^{-15}$ kg
where the density of carbonyl iron powder is 7.8 kg/m$^3$.

Hence,
Number of particles in 1 g of 10 µm powder
$= 1 \times 10^{-3}$/weight of one particle
$= [6/(7.8\ \pi)] \times 10^{12}$ particles
Surface area of one particle = $\pi d^2$
$= \pi \times 10^{-10}$ m$^2$

Thus, the surface area of 1 g of the 10 µm powder is the number of particles in the 1 g of powder times the surface area of one powder particle, which works out to be about 77.5 m$^2$.

To give a perspective, 1 g of 10 µm powder, which is a little over a pinch of powder, has a surface area of approximately 80 m$^2$ (850 ft$^2$), the size of a small apartment. In spite of what the calculations show, this fact is difficult to either visualize or comprehend, compared to 1 g of iron as one spherical particle, which has a diameter of 6.25 mm and a surface area of 127.8 mm$^2$, or a square with sides measuring 11.3 mm only. In a typical $D_{90}$ 10 µm powder used in MIM the particles are finer and the surface area is even larger and this entire area is a lake of exposed metal atoms with this humongous surface energy. While this surface energy aids the sintering process, this huge surface area also leaves the material prone to oxidation with exposure to the ambient atmosphere, so that a can of open powder would show increasing oxygen content with increasing time of exposure. For this reason, sintering atmospheres and hence the material of furnace construction play very important roles in the properties of the parts processed.

## 7.4.9 Sintering atmospheres

The extremely large surface area of MIM powders results in oxidation of the powder surface and the extent depends on the reactivity of the metal powder used with the ambient atmosphere. For certain extremely reactive metals like titanium, which reacts with oxygen, nitrogen, hydrogen, and carbon, it is necessary to prevent the powder from coming in contact with the atmosphere and keep the surface protected all the way through until the part is produced. In the other extreme case of noble metals, because the metals do not tarnish, no protective atmosphere is needed during storing the powders.

However, the reactivity of the powders increases in most cases with temperature. Hence during MIM debinding and sintering the powders need to be protected from further oxygen pick-up and also the existing oxygen picked up during storage should be reduced. Some carbon-containing materials use their carbon content to reduce the surface oxygen present in the powders. The Ellingham diagrams[25] provide excellent guidance regarding which atmosphere to use for which metal. Accordingly, a reducing or neutral atmosphere is necessary to protect the different materials being sintered. The individual systems are discussed in Section 7.5 on MIM materials.

### 7.4.10 Sintering results

To make a part consistently and repeatably by MIM, every step in the MIM process must be repeatable, from the powders used to the sintering step. In each step there are a number of variables and these must be controlled within a window which yields the desired results. In case of debinding every bit of the binder must be removed, which means there is a minimum processing time and any additional time beyond the minimum has almost no negative effect. In case of sintering, which is controlled by diffusion rates for the different materials, time and temperature are the major variables. The end result has to be the correct dimensions, which means the correct density and shrinkage resulting in the correct physical properties. Continuation of the sintering time beyond a certain level has little effect on the density but a profound effect on grain growth. Extremely large grains could have a detrimental effect on some physical properties.

Figure 7.12 shows the effect of grain growth. The same M2 material as in Figure 7.11 is seen here, except it was sintered for a longer time. While the grains have grown considerably, most of the liquid phase has migrated to the grain boundaries resulting in heavy carbide deposits. This would reduce the toughness and the ductility of the sintered material.

Figure 7.13 shows a plot of density with time. The starting material has a density corresponding to the green density. After sintering the part shrinks and densifies to the upper ninety-something percent range with time. Since the progress in densification is diffusion dependent, it is asymptotic and slows with time as the density nears 100%. Increasing the temperature moves the curve up and to the left. In a production furnace the temperature

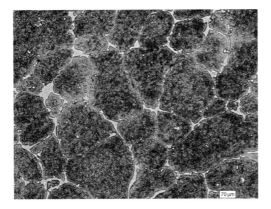

*7.12* Structure of an as-sintered M2 tool steel, over sintered, showing grain growth and the majority of the carbide phase along the grain boundaries (courtesy Clint Torgerson, Britt Manufacturing).

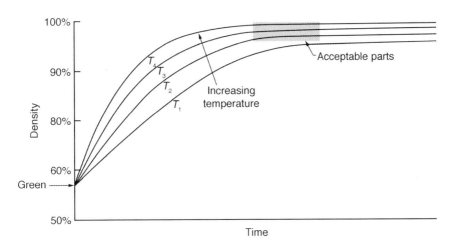

*7.13* Plot of sintered density in relation to time of sintering.

is not the same at every place and there are internal variations because of the distance of the part from the heating elements, as well as for other reasons, such as gas flow, shielding effects, and so on. Under these real conditions, a box, bounded by a minimum time and temperature and a maximum time and temperature, which results in densities (or critical dimensions) that fall within the part specifications defines the results of acceptable sintering. Such a hypothetical box is also marked in this figure.

Table 7.4 List of MIM materials

| Material | Debind temperature (°C) | Preferred debind atmosphere | Sintering temperature range (°C) | Preferred sintering atmosphere | Note code |
|---|---|---|---|---|---|
| **Iron** | | | | | |
| FN02 | 250–650 | $N_2$ | 1180–1290 | $N_2$ | A |
| FN08 | 250–650 | $N_2$ | 1180–1290 | $N_2$ | A |
| FN50 | 250–650 | $N_2$ | 1180–1280 | $N_2$ | A |
| Silicon iron | 250–650 | $H_2$ | 1180–1300 | $H_2$ | B |
| **Tool steel** | | | | | |
| H11 | 250–650 | $N_2$ | 1200–1275 | $N_2$ | A |
| M2 | 250–650 | $N_2$ | 1180–1250 | $N_2$ | A |
| M4 | 250–650 | $N_2$ | 1180–1250 | $N_2$ | A |
| M42 | 250–650 | $N_2$ | 1180–1250 | $N_2$ | A |
| T15 | 250–650 | $N_2$ | 1200–1270 | $N_2$ | A |
| **Steel** | | | | | |
| 1040 | 250–650 | $N_2$ | 1100–1270 | $N_2$ | A |
| 4340 | 250–650 | $N_2$ | 1100–1270 | $N_2$ | A |
| 4140 | 250–650 | $N_2$ | 1100–1270 | $N_2$ | A |
| 8620 | 250–650 | $N_2$ | 1100–1290 | $N_2$ | A |
| 42CrMo4 | 250–650 | $N_2$ | 1100–1290 | $N_2$ | A |
| 100Cr6 | 250–650 | $N_2$ | 1190–1290 | $N_2$ | A |
| **Stainless steel** | | | | | |
| 17-4 PH | 250–650 | $H_2$ | 1200–1360 | $H_2$ | B |
| 316L | 250–650 | $H_2$ | 1250–1380 | $H_2$ | B |
| 410 | 250–650 | $H_2$ | 1250–1375 | $H_2$ | B |
| 420 | 250–650 | $N_2$ | 1200–1340 | $N_2$ | A |
| 440C | 250–650 | $N_2$ | 1200–1280 | $N_2$ | A |
| 17-7 PH | 250–650 | $H_2$ | 1200–1340 | $H_2$ | B |
| 18/8 | 250–650 | $H_2$ | 1200–1340 | $H_2$ | B |
| 304 | 250–650 | $H_2$ | 1250–1375 | $H_2$ | B |
| Panacea | 250–650 | $H_2$ | 1180–1275 | $H_2/N_2$ | B |
| **Tungsten-based** | | | | | |
| WC | 250–650 | $N_2/H_2$ | 1250–1390 | $N_2/H_2$ | A |
| W–Cu | 250–650 | $H_2$ | 1150–1400 | $H_2$ | B |
| **Copper-based** | | | | | |
| Pure Cu | 250–650 | $H_2$ | 950–1050 | $H_2$ | C |
| Bronze | 250–650 | $H_2$ | 850–1000 | $H_2$ | B |
| Spinodal | 250–650 | $H_2$ | 850–1000 | $H_2$ | B |
| **Precious metals** | | | | | |
| 14–22 kt gold | 250–650 | $N_2/H_2$ | 850–1000 | $N_2/H_2$ | A |
| Sterling silver | 250–650 | $N_2/H_2$ | 850–1000 | $N_2/H_2$ | A |
| **Titanium** | | | | | |
| Ti | 250–650 | Ar/vacuum | 1130–1220 | Ar/vacuum | C |
| Ti 6/4 | 250–650 | Ar/vacuum | 1140–1250 | Ar/vacuum | C |

*Table 7.4 (cont.)*

| Material | Debind temperature (°C) | Preferred debind atmosphere | Sintering temperature range (°C) | Preferred sintering atmosphere | Note code |
|---|---|---|---|---|---|
| High-temperature alloys | | | | | |
| Inconel 718 | 250–650 | Ar | 1200–1280 | Vacuum | C |
| Hastelloy X | 250–650 | $H_2$ | 1200–1270 | $H_2$ | C |
| HK–30 | 250–650 | Ar/$H_2$ | 1200–1280 | Ar/$H_2$ | C |
| GMR–235 | 250–650 | Ar | 1200–1280 | Ar/$H_2$ | C |
| Haynes 230 | 250–650 | Ar | 1200–1260 | 96Ar/4$H_2$ | C |

Note: A: Can be processed with no difficulties in either graphite or in refractory metal furnaces; B: Can be processed with reduced properties in graphite furnaces under $N_2$ or with optimum properties in refractory metal furnaces under $H_2$; C: Must be processed in refractory metal furnaces.

## 7.5 MIM materials

MIM may be used to make parts from a wide variety of materials. Table 7.4 shows a list of materials which have been used to make parts by MIM. This list is not all inclusive. Most often, powder availability for new materials is the big question regarding whether parts may be made cost effectively from this material or not.

### 7.5.1 Effect of reactivity of materials

As mentioned above the large surface area could be the cause for a large amount of oxygen to be picked up by the powder. The oxygen removal from the part depends on the reactivity of the material. The type of furnace used would depend on the reactivity of the metal and the atmosphere desired in the furnace.

*Precious metals*

In case of gold and some other precious metals that do not tarnish, no oxides are formed. In case of silver, the oxides break up to release oxygen below sintering temperatures. In such cases the oxides are no problem. But in most of the other materials processed a reducing agent is needed to remove the oxygen. The two reducing agents used for most materials are carbon and hydrogen.

*Metals whose oxides are easily reducible*

Copper, tin, iron, nickel, cobalt, manganese, molybdenum, and tungsten are examples where the oxides are easily reduced by both carbon and hydrogen. However, carbon must be added in the solid form or be present in the form of carbon monoxide. Since carbon monoxide is the form of oxide that is more stable at higher temperatures, using gases containing carbon monoxide, such as endo-gas, does not provide a satisfactory reducing atmosphere because two carbon monoxide molecules break down into carbon dioxide and carbon, which is deposited as soot. Copper and bronze parts are the exception where endothermic gases may be used. Adding carbon in the solid form introduces the question: what happens when you add too much carbon? The answer depends on the alloy system and the phase diagrams involved.

*Carbon steels*

Sintering carbon-containing steels to obtain the same desired carbon content from batch to batch requires a critical control of both the oxygen and carbon contents of the mixture. The oxygen content of the powder will cause the carbon content to be depleted because of the reduction of the oxides by carbon. To achieve the desired carbon content, for example, 0.4% in a 4140, the powder must have an extra amount of carbon that reduces the oxygen content of the powder. These parts are usually sintered in nitrogen to provide an inert atmosphere. Argon or even a vacuum may be used if nitrogen pick-up is not desired. Hydrogen will reduce the oxides but it will also remove all the carbon in the alloy by the formation of methane and hence is not used. In a graphite furnace hydrogen may not be used above 800°C because it will attack all the graphite inside the furnace from the hot zone to the heating elements and the graphite retort or load carrier used to carry the parts.

To prevent carbonyl iron powder from oxidizing further between its packaging and use and to prevent agglomeration of the particles, a number of manufacturers put a thin layer of silica over each iron particle. This works well with carbon steels because further oxidation of the iron powder is minimized, which allows for easier carbon control. The carbon reduces the silica to silicon during the sintering process and the silicon dissolves in the matrix. Since the layer of silica is thin, the amount of silicon added is well within the standard specifications for the alloys.

*High alloyed steels containing carbon*

Heavily alloyed carbon-containing steels, such as tool steels and martensitic stainless steels, have larger carbon contents than low alloyed steels. These are mostly pre-alloyed powders and often have layers of chromium oxides, silicon dioxide, and oxides from the other elements, which are all reduced by the carbon in the alloy. These are usually sintered in neutral atmospheres, such as nitrogen or argon or a vacuum.

*Corrosion resistant low carbon steels*

Oxides of chromium and silicon require a high temperature and a very low moisture (product of the reduction process) content to be reduced by hydrogen. If carbon is added for sintering in a neutral atmosphere, any excess carbon will form chromium carbide and the purpose of having a low-carbon stainless steel will be defeated. Hence stainless steels and other high-chromium-containing alloys should be sintered in a hydrogen atmosphere. This requires the use of a refractory metal batch furnace or a high-temperature continuous furnace, such as a pusher or a walking beam with 100% hydrogen atmosphere.

MIM stainless steels have been made in three different ways: by using elemental powder blends, by using a master alloy blended with carbonyl iron powder, and by using the pre-alloyed powder. Figure 7.14 shows the

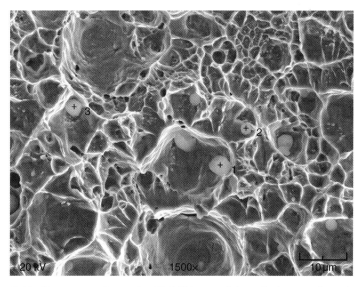

*7.14* Fracture surface of a 17-4 PH material made from carbonyl powder blended with master alloy powder sintered in nitrogen showing many spherical inclusions.

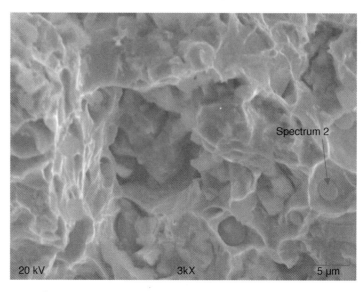

*7.15* Fracture surface of gas atomized 17-4 PH powder sintered in vacuum shows a smaller number of inclusions compared to that seen in Fig. 7.14.

fracture surface of a 17-4 PH alloy made from a carbonyl iron powder and master alloy blend that was sintered in a nitrogen atmosphere. An energy dispersion X-ray analysis (EDXA) shows that the large number of spherical inclusions seen are silica, which probably come from the silica coatings used on the carbonyl iron powder. The same mixture sintered in hydrogen showed no such inclusions. A pre-alloyed powder sintered in vacuum also showed spherical inclusions, but to a smaller amount, as seen in Fig. 7.15. These inclusions were chromium oxides as determined by EDXA and arise from the surface oxide layer over each 17-4 PH powder particle. Again, the same material sintered in hydrogen showed no inclusions. Corrosion properties were the best for hydrogen sintering and worst for the master alloy based material sintered in nitrogen, while vacuum sintering resulted in mediocre corrosion properties.[26] MIM stainless steels made in the three-step process where the final sintering was done in a vacuum also showed chromium oxide spheroids because the pre-sintering temperature was not high enough to reduce the chromium oxides by hydrogen. Many manufacturers world-wide produce stainless steel components in vacuum or nitrogen in graphite furnaces, which result in sub-optimum properties because of the oxide inclusions. The presence of these oxide inclusions also results in a poorer polish on finishing compared to parts made in hydrogen that have no oxide inclusions. Of course, optimum properties are not needed

*Debinding and sintering of metal injection molding components* 165

for all applications, for example cell phones and laptop computer hinges have to last the life of the appliance, which is taken to be a couple of years.

*Tungsten alloys*

Although tungsten is listed above where it can be reduced by both carbon and hydrogen, tungsten is a carbide former and any excess carbon will result in carbides. Hence heavy alloys, where carbides are not desired, are usually sintered in hydrogen. Since the oxides of tungsten are reducible at relatively low temperatures wet hydrogen is often used to remove any excess carbon and prevent the formation of blisters. Most tungsten alloys are sintered by either activation sintering with a sintering aid or by a liquid phase that involves a solution–reprecipitation process as discussed above. Tungsten carbide and other hard metals are usually bonded with either cobalt or iron–nickel type binders. Here too the method of sintering is solution–reprecipitation. Usually a nitrogen atmosphere with a small amount of hydrogen is used for sintering carbides in order not to lose carbon from the carbides.

*Titanium and its alloys*

Titanium is considered to be the most difficult to sinter with respect to controlling the oxygen content because titanium dioxide cannot be reduced by either carbon or hydrogen. Hence any oxides formed on the powder particle surface remain in or on the part. Titanium will also react with

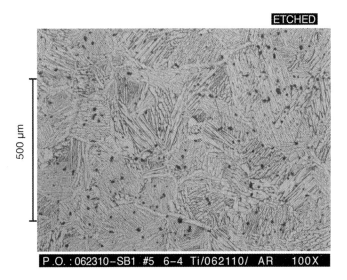

*7.16* Structure of MIM Ti–6Al–4V sintered under argon.

carbon, hydrogen, and nitrogen. Graphite furnaces should be avoided because of the carbon atmosphere present. Traditionally, titanium has been sintered under high vacuum between $10^{-5}$ and $10^{-6}$ mbar, but at DSH Technologies, experience has shown that a low partial pressure sweep of pure argon between 50 and 70 mbar may also be used to obtain similar, if not better, results. Figure 7.16 shows the structure of MIM Ti–6Al–4V sintered under argon.

*High-temperature alloys*

High-temperature alloys or superalloys are another challenge. They contain chromium for oxidation resistance and elements like aluminum, vanadium, tantalum, niobium, and titanium added as gamma prime formers to harden the matrix of the materials. Hydrogen would form hydrides in case of tantalum, niobium and titanium, and cause hydrogen embrittlement in superalloys. Carbon is also not desired in these materials as the carbides would reduce the amount of gamma prime precipitates. Oxides of aluminum and titanium cannot be reduced by either carbon or hydrogen. Vacuum or argon is the atmosphere of choice in most cases.

*Other materials*

The discussion so far has been limited to conventional materials. MIM has been attempted for refractory type materials too, such as rhenium, niobium, molybdenum, platinum, etc. One limitation for such materials is the furnaces needed, namely a furnace with tungsten shields, furniture, and heating elements. This makes the furnace very expensive because tungsten cannot be cut or punched the way most metals can. Powder availability is another factor which depends on the applications.

## 7.5.2 Powder availability

Many materials that have been studied in the laboratory have not translated into large volumes of MIM parts. One such example is titanium and its alloys. Papers have been written on MIM titanium for years but the actual volume of MIM titanium parts being made is still a very small quantity with just a few manufacturers world-wide. The quality of powder needed is extremely tight and hence the powders are expensive, especially because the quantities are not yet available. Applications are still waiting for the powder prices to come down.

## 7.6 Settering

Sintering temperatures used in MIM are usually close to the melting range of the alloys and in some cases also involve a liquid phase. To prevent any reaction between the parts and the charge carriers a layer of inert ceramic is used as a barrier between the two. While continuous furnaces use a ceramic charge carrier, that is usually not the case for batch furnaces. The charge-carrying shelves in a graphite furnace are usually also graphite. Direct contact of graphite in a number of alloying systems would result in the formation of a low melting eutectic. In case of refractory metal furnaces, molybdenum or a molybdenum alloy is used as shelving material. Molybdenum reacts with ferrous materials to form a liquid phase. Hence in both these cases a non-reactive barrier, such as a ceramic, is a must.

Alumina-based ceramics are the most widely used for most materials except for titanium alloys which react with alumina. Alumina goes through a phase transition around 900°C. Rapid cooling through this region causes the alumina to crack from the stresses due to this phase transformation. Hence the 96% alumina grade is the least expensive and most common material used. Zirconia-toughened alumina (ZTA) is also gaining popularity because of its higher strength and the absence of the stresses generated during rapid cooling through this temperature range.

Titanium and its alloys are placed on stabilized zirconia. Since titanium has a greater affinity towards oxygen than most of the other elements, the purity of the raw materials is very important to minimize the adsorption of oxygen from the setter materials. Zirconia stabilized with yttria (YSZ) is the best among the usual materials. Calcia-stabilized zirconia has been used successfully, but the chances of the calcia having impurities is greater than that of pure yttria. High-purity yttria plates have also been used but these are even more expensive and difficult to find than the YSZ plates. Traces of typical impurities found in ceramic materials, such as $Na_2O$, $SiO_2$, FeO, $Fe_2O_3$, etc., all give up the oxygen content to the titanium during sintering and reoxidize when the plates are in the atmosphere. Hence titanium parts do pick up a small amount of oxygen from most zirconia plates during sintering.

Since MIM components are sintered at high temperatures, during sintering the material has no strength. The forces exerted by gravity are enough to distort the parts unless they are properly supported. Hence most designers try to build in a flat surface on which the part may be staged.

When there are no flat surfaces available there are two options.

1. Add appendages to the part to create a flat surface. These additions would have to be machined to bring the shape to the original design, adding a recurring cost. Figure 7.17 shows a part where an appendage

*7.17* Appendage added to the part in front to obtain a flat surface. The added section is cut from about the mark on the part (courtesy Karl Lewis, Lewis Machine Tools).

has been added to create the flat staging surface desired. The mark on the part shows an approximate position from where the machine cut will be made.

2. Create a part-specific surface with ceramics. The cost for such ceramics would depend on the complexity and how the shaped ceramic is created.

When the part has a complex shape and machining is not an option then a contoured ceramic is chosen. Figure 7.18 shows the setter created for a high end spoon concept to be made by MIM. A concept sketch is shown for a setter for a specially shaped part in Fig. 7.19. These special setters are part specific and cannot be used for any other part. Other simpler setters, such as

*7.18* Setter for a high end spoon concept.

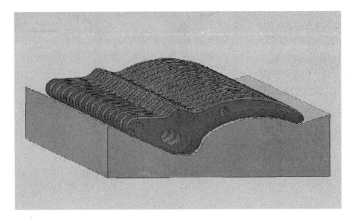

*7.19* Setter concept for a special part.

a rectangular groove for an overhang or a V-groove for cylindrical parts, may be used for simpler parts.

The special setters may be made in three different ways.

1. They may be machined or ground to have the special contour opposite to that of the part. For a few samples a machinable ceramic may be used, but if the need is to manufacture thousands of parts then hard ceramics are essential and the setters become very expensive.
2. Another method is to take a machinable ceramic, make the contour, and then coat the contoured setter with fine ceramic particles and sinter the setter to obtain a harder material. This is less expensive than machining hard ceramics.
3. The third method is to make a reverse mold and cast a ceramic slurry into the mold. Then the ceramic is dried and sintered to form the final shape. The base of the setter would be ground flat. This is the least expensive of the three methods.

In general if a design change of the part to create a flat surface allows it to perform its function then that is the most economical way to make the part.

## 7.7  MIM furnaces

Today MIM furnaces carry out two functions: removal of the secondary binder followed by ramping up and holding at the desired high temperature for sintering. These furnaces have evolved from the three-step process used when MIM was first invented.

The first of the three steps was removal of the primary binders. During the second step, the secondary binder was removed. Since this required a higher temperature, usually between 400°C and 600°C, a protective atmosphere was necessary and the equipment had to deal with the large amount of binder by products being released during the process. In order to give the parts handling strength, the parts needed to be pre-sintered at temperatures between 600°C and around 1100°C. In the laboratory it is not unusual to use lower temperatures of 600–900°C, while in production it is more the norm to go close to the maximum temperature capability of the furnace, usually up to 1150°C, in order to carry out the reduction of the oxides present to the maximum ability. In the third step the parts were heated up to between 1250°C and 1400°C to obtain the final desired density. These high-temperature furnaces were vacuum furnaces which could not deal with the large amounts of binders released during secondary debinding, because of which the intermediary second step was necessary.

The major drawbacks with the three-step process are the double handling of the parts and the time needed to process the materials. The parts had to be cooled down to be handled and moved to another furnace. This required restaging the parts because the two furnaces were not designed to work with one another.

### 7.7.1 Evolution of MIM furnaces

As the owners of the Wiech process[3–5] separated and set up their own shops, variations in the process began to show. Multi Metal Molding in San Diego, CA used a continuous mesh furnace for secondary debinding and pre-sintering and a continuous pusher furnace for the sintering to final density. Form Physics in San Diego, CA also used a similar system. Although Multi Metal Molding was dissolved somewhere in the late 1980s or in the early 1990s, the continuous furnaces evolved to become one unit which combined secondary debinding with sintering.

Brunswick in Deland, FL wanted to eliminate the time taken to remove the parts from one furnace to another and decided to make their own furnace. They came up with a bell jar furnace where an Inconel retort was heated by an exterior removable furnace core. The Inconel retort limited the maximum temperature to 1250°C, pushing the material beyond its usable limits, but it allowed processing of stainless steels in pure hydrogen. The major drawbacks were the time per cycle with 6–10 h at final temperature, the inability to process carbon-bearing steels, and frequent reworking of the Inconel retort to maintain shape. This process too was licensed to a number of companies. FloMet, which bought Brunswick, is still manufacturing its furnaces in house.

A couple of manufacturers started out making carbon-bearing steel parts. They used graphite vacuum furnaces run under a partial pressure of nitrogen. The vacuum furnaces were fitted with claw type dry pumps that were not affected by the secondary binder. These were the first one-step debinding and high-temperature sintering furnaces. The drawbacks of these furnaces are that processing under hydrogen was not possible and the pumps required frequent cleaning because the binder traps were dysfunctional. Again, low-carbon materials, such as stainless steels, cannot be processed to obtain optimum corrosion properties in the absence of a hydrogen atmosphere as by this method.

A number of manufacturers tried to convert sintering furnaces with molybdenum hot zones into MIM furnaces. These early attempts led to limited success until Randall German, then at Pennsylvania State University, asked Claus Joens of Elnik Systems to develop a MIM furnace based on his vacuum furnaces. Elnik Systems developed the first truly MIM batch furnace that could sinter under vacuum, hydrogen, nitrogen, and argon or a mixture of two of these gases at a desired partial pressure from 15 mbar to 900 mbar or even under high vacuum, if the furnace had the high vacuum option.

### 7.7.2 Continuous furnaces

The continuous furnaces used in MIM are based on the high-temperature sintering furnaces used in conventional sintering but adapted for the large amounts of binders emitted by the parts. The first section is a low-temperature section for debinding and requires a special gas sweep to make sure the binder breakdown products do not reach the sintering section of the furnace. The travel speed of the parts and hence the duration of stay in this section must allow the parts to be debound completely. The debind duration is set for the thickest part cross-section the furnace will see. The parts then move on to the high-temperature sintering section of the furnace. This travel rate of the plates also fixes the sinter time at temperature. Typical sintering

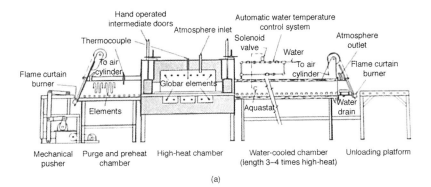

7.20 (a) Schematic diagram of a pusher furnace; (b) a CM pusher furnace (continues on next page).

172  Handbook of metal injection molding

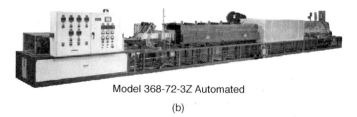

Model 368-72-3Z Automated

(b)

*7.20* (continued)

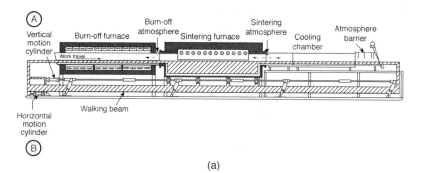

(a)

(b)

*7.21* (a) Schematic diagram of a walking beam furnace; (b) a Cremer MIM Master walking beam furnace (courtesy Ingo Cremer and Harald Cremer).

temperatures used are in the 1300–1400°C range. After the sintering zone the parts go through a cool down zone before they come out of the furnace.

CM Furnaces in the USA made the first pusher furnace for MIM while Cremer Furnaces based in Germany used their walking beam design for MIM. Some other furnace makers have been making continuous furnaces for MIM relatively recently.

Figure 7.20(a) shows the schematic diagram of a pusher furnace and Fig. 7.20(b) shows a picture of a CM pusher furnace. Figure 7.21(a) is a schematic diagram of a walking beam furnace and Fig. 7.21(b) shows a Cremer walking beam furnace.

Continuous furnaces are built with refractory bricks inside a gas-tight metal shell. Hence it takes a while for the temperature inside the furnace to stabilize, because of which start-up times after a shut down are long. If you are making millions of the same part, the continuous furnace is the best option. But this is not the best if you need to make sintering temperature changes for different materials or make debinding hold time changes for different part thicknesses. Also, whether you are running parts or not, you will need to run gas to protect the heating elements, which makes idling of these furnaces very expensive.

*7.22* A typical graphite furnace with a graphite hot zone and graphite elements (courtesy David Smith, Advanced Forming Technology – PCC).

*7.23* A refractory metal MIM furnace with a molybdenum hot zone and retort made by Elnik Systems.

### 7.7.3 Batch furnaces

Batch furnaces are popular with those working with a number of different materials and part mixes that require production flexibility. Figure 7.22 shows a typical graphite batch furnace. In this case a graphite box shelving with parts loaded on shelves is loaded into the hearth. Three zones are used to control the temperature and you use your experience to judge when the entire load mass reaches constant temperature. This type of a furnace works well for carbon-containing steels because the furnace is usually equipped to use a nitrogen atmosphere only. Hydrogen would react with the graphite to form methane and hence may not be used. This limits the materials that may be processed to obtain optimum properties. Since these furnaces run under nitrogen or vacuum only, safety requirements are less than those furnaces running under hydrogen. Because graphite costs less than refractory metals and is easier to machine, and as the safety requirements are considerably less, graphite furnaces are relatively inexpensive.

Centorr in the USA was the first company to offer refractory metal sintering furnaces for MIM with the two-step process but their debinding system was not optimal. Thermal Technologies in Santa Rosa, CA was also one of the first to be in this field, but Thermal Technologies also wanted to make MIM parts and was subsequently sold and had to leave the MIM business. Elnik Systems, which formerly manufactured vacuum furnaces for the defense industry, is the only furnace manufacturer specializing in equipment for the MIM industry. Figure 7.23 shows a refractory metal

MIM furnace[17] with a molybdenum hot zone and a molybdenum retort. The furnace can process most materials to their optimum properties using hydrogen or argon or nitrogen atmospheres or vacuum. The furnace is completely computer controlled and uses the Microsoft Excel program to create and download the furnace recipes. Even trap cleaning is automated and is accomplished by the simple push of a virtual button on a computer screen. The ability to run hydrogen mandates certain safety measures in the furnace, the refractory metal market is volatile, and the raw materials are costly. These requirements make these furnaces more expensive than graphite furnaces.

Elnik Systems also make a MIM furnace similar to their refractory metal furnace with a graphite hot zone and retort with characteristics and controls similar to the all-metal furnace for those wanting to process carbon-containing materials only.

### 7.7.4 Comparing batch and continuous furnaces

Whether a batch furnace or a continuous furnace should be used depends on the application. Continuous furnaces are capable of providing large throughput of parts and are chosen when throughput is the major requirement. All parts in the same line of travel see the same temperature, although it is difficult to introduce survey thermocouples on parts.

Gas consumption in a continuous furnace is high, because of the large openings on both sides of the furnace, but power consumption is lower because of the smaller cross-section of the furnace compared to a batch furnace. The major drawbacks of the furnace are their inflexibility regarding materials and part size, the high cost of idling, and long times needed to start up or cool down these furnaces for repair and maintenance. These furnaces also require a large amount of floor space.

Batch furnaces are very flexible regarding the size of parts and the materials to be run. When there are no parts to run there is no power or gas consumption. During normal running gas consumptions are lower and power consumptions are higher than in continuous furnaces but there is no power consumption during cooling or idling. Temperatures inside the batch furnace can be directly measured using survey thermocouples on the parts.

Graphite furnaces are limited with respect of the processing atmospheres that may be used in them. Hence the flexibility of graphite batch furnaces regarding the ability to process different materials is less than those with refractory metal hot zones and heating elements. Refractory metal MIM furnaces are the most flexible among all the different furnaces available because they permit the use of hydrogen atmospheres, argon or nitrogen, as well as high vacuum if this option is included. This permits these furnaces to process high-carbon materials, stainless steels, and cobalt–chrome alloys, as well as super alloys, titanium and its alloys, and intermetallic materials with the proper choice of atmospheres.

## 7.8 Furnace profiles

Whether a batch furnace or a continuous furnace is used, the basic profile remains the same. The principle is to heat up a brown part slowly to the temperature where each of the individual secondary binders evaporates and hold the temperature long enough for all the binders to be removed completely. In case of a batch furnace a slow ramp is used because a fast ramp causes the temperature at the center of the furnace to lag even more and this lag depends on the thermal load inside the furnace. In case of a continuous furnace the ramp rate is a function of the travel speed of the parts inside the furnace. Here the distance of the parts from the heat source is relatively small, so the large lag between the outer edges and the center of the furnace in case of the batch furnaces is not observed. Once all the binders are eliminated the temperature is ramped to the sintering temperature, where the parts are held for a time period between 1 and 4 h depending on the desired density. The exact temperatures would depend on the boiling point of the different binders and the sintering temperature of the material being sintered.

Both types of furnaces have been used to make parts with excellent properties in commercial parts provided proper process controls and the correct atmosphere are maintained.

## 7.9 Summary

MIM feedstocks are made with two major components, the primary one that comes out easily, starting from the outside of the part and moving into the part, resulting in the opening up of pore channels. The secondary binder component holds the powder particles together as the temperature rises until small interparticle diffusion bonds begin to form within the part. These bonds, together with interparticle friction, result in sufficient strength of the part to retain its shape at the higher temperature when the secondary binder begins to vaporize and the vapors may escape through the open pore channels created by the removal of the primary binder. Three types of feedstock are in commercial use today:

1. a wax-based feedstock where the wax is removed with an organic solvent;
2. a polyacetal-based feedstock where the polyacetal is removed by the catalytic reaction in presence of nitric acid vapor;
3. one where the primary component is water soluble.

Sophisticated PLC controlled primary debinding equipment is available for all of these feedstock types. These equipments result in automated and repeatable processing with full documentation capabilities for quality systems for the aerospace, automotive, and medical device industries.

Secondary debinding is a thermal process which used to be carried out in a separate second step. Sintering of the parts was done in a third step, typically in a vacuum furnace. Today most of the MIM parts are debound and sintered in the same equipment. Hold temperatures for thermal debinding of the secondary binder are best determined by running a TGA on a brown part run at the heat-up rate used for debinding. Hold times depend on the binder material(s), part thickness, and the particle size of the powders. Incomplete debinding at the hold temperatures causes the binders to come out rapidly at higher temperatures, which could affect the parts in respect of the shape integrity and carbon content, as well as the life of the heaters and hot zone of the sintering furnace.

Sintering is the process by which bonds between the particles occur and the MIM part shrinks to the desired size. The theories, which are based on simple two-particle or other simple models, are behind the practice of sintering. Mass transport during sintering may occur by surface transport mechanisms or bulk transport mechanisms. Sintering occurs in the following stages for loose packed powders: first, adhesion, rearrangement, and repacking; then a sinter bond grows between particles and forms the neck between particles; then neck growth continues with densification until finally pores become closed, diffusion slows, and grain growth begins.

In practice, sintering happens between many particles that are not all of the same size and not spherical. There are minor compositional variations within the same material and added alloying elements may be present. Most systems are not simple binary ones. When sintering and homogenization are happening together, the equivalent phase diagram helps in understanding the process. Solubility and diffusion rates between the constituents play major roles. Liquid phase formation could also aid sintering.

MIM powders are smaller than powders used for conventional powder metal applications. A 1 g of 10 μm spherical powder has a surface area of 77.5 $m^2$, that of a small apartment. This is a huge amount of exposed area which easily absorbs oxygen from the atmosphere. The amount of oxide formed depends on the reactivity of the metal. This surface needs protection from oxidation and those oxides already formed need to be removed, if possible. Hence, sintering atmosphere plays an important role in MIM. Hydrogen, nitrogen, argon or vacuum may be used to protect the material depending on the metal being processed.

Most materials can be processed by MIM. Powder availability drives the economics. The type of furnace and protective atmosphere used is a function of the reactivity of the materials being sintered. Carbon-containing steels and stainless steels may be processed in graphite furnaces, but low-carbon stainless steels and chromium-containing alloys should be processed in refractory metal or continuous furnaces under a hydrogen atmosphere at high temperatures to obtain oxide-free structures and optimum properties. Titanium and superalloys require high vacuum or argon partial pressures in refractory metal furnaces.

MIM parts must be placed on a non-reactive ceramic material on top of the charge carrier to prevent interaction with the parts and the charge carrier material. MIM materials have no strength at sintering temperatures. Designing in a flat surface on which the part may be placed is the most economical option. If a shaped surface contour on the part is the only option, special setters must be constructed to nest this shape.

Both continuous and batch furnaces are available for the debinding and sintering of MIM parts. Continuous furnaces are preferred when very high throughput is the most important criterion for the operation, whereas when the operation needs to be more flexible to accommodate different materials and part sizes, batch furnaces are chosen. Graphite and all-refractory-metal-type batch furnaces are both used to make MIM parts although the materials processed optimally in a graphite furnace are limited. Refractory metal furnaces are, however, more flexible than graphite furnaces. Furnace profiles are similar for both continuous and batch furnaces. Both furnace types can result in parts with excellent properties if proper process controls are maintained throughout the process.

## 7.10 Acknowledgements

The authors would like to thank Animesh Bose, Materials Processing Inc., Ingo Cremer and Harald Cremer, Cremer Furnaces, Karl Lewis, Lewis Machine Tools, James Neill, CM Furnaces, Maria Vesna Nicolic, Institute for Multidisciplinary Research, Beograd, Serbia, J. Alan Sago, Accellent, David Smith, Advanced Forming Technology-PCC and Clint Torgerson, Britt Manufacturing for the permission to use pictures or micrographs, Robert Sanford of TCK, S. A. and S. K. Tam of Ormco (Sybron Dental) for discussions, and James Casillo of Elnik Systems for help with drawings.

## 7.11 References

1. German, R. M. and Bose, A. (1997), *Injection Molding of Metals and Ceramics*, Metal Powder Industries Federation, Princeton, NJ, USA.
2. Strivens, M. A. (1960), 'Formation of ceramic moldings', US Patent No. 2,939,199, 7 June.
3. Weich, Jr, R. E. (1980), 'Manufacture of parts from particulate material', US Patent No. 4,197,118, 8 April.

4. Weich, Jr, R. E. (1981), 'Method and means for removing binder from a green body', US Patent No. 4,305,756, 15 December.
5. Weich, Jr, R. E. (1986), 'Particulate material feedstock, use of said feedstock and product', US Patent No. 4,602,953, 29 July.
6. Farrow, G. and Conciatori, A. B. (1986), 'Injection moldable ceramic composition containing a polyacetal binder and process of molding', US Patent No. 4,624,812, 25 November.
7. Wingefeld, G. and Hassinger, W. (1991), 'Process for removing polyacetal binder from molded ceramic, green bodies with acid gases', US Patent No. 5,043,121, 27 August.
8. Krueger, D. C. (1996), 'Process for improving the debinding rate of ceramic and metal injection molded products', US Patent No. 5,531,958, 2 July.
9. Rivers, R. D. (1978), 'Method of injection molding powder metal parts', US Patent No. 4,113,480, 12 September.
10. Hens, K. F. and Grohowski, Jr, J. A. (1997), 'Powder and binder systems for use in powder molding', US Patent No. 5,641,920, 24 June.
11. www.polimer-chemie.de
12. Fanelli, A. J. and Silvers, R. D. (1988), 'Process for injection molding ceramic composition employing an agaroid gel-forming material to add green strength to a preform', US Patent No. 4,734,237, 29 March.
13. Nyberg, E. A., Weil, K. S., and Simmons, K. L. (2009), 'Method of using a feedstock composition for powder metallurgy forming of reactive metals', US Patent No. 7,585,458 B2, 8 September.
14. Nyberg, E. A., Weil, K. S., and Simmons, K. L. (2010), 'Feedstock composition and method of using same for powder metallurgy forming of reactive metals', US Patent No. 7,691,174 B2, 6 April.
15. Sanford, R. and Banerjee, S. (2009), 'The importance of a helium pycnometer in powder metal injection molding,' In *Advances in Powder Metallurgy and Particulate Materials, Part 4 — Powder Injection Molding (Metals and Ceramics)*, MPIF, Princeton, NJ, USA.
16. Sanford, R. and Banerjee, S. (2009), 'Using a helium pycnometer as a quality tool in powder metal injection molding', *Euro PM2009 Conference Proceedings*, Vol. 2, EPMA, Shrewsbury, UK.
17. www.elnik.com
18. www.appliedseparations.com
19. German, R. M. (1996), *Sintering Theory and Practice*, John Wiley and Sons, New York, USA.
20. ASTM (2010), ASTM B 243-09a: 'Standard terminology of powder metallurgy', Metallic and Inorganic Coatings; Metal Powders and Metal Powder Products, *Annual Book of ASTM Standards*, Section 2, Vol. 02.05, p. 41, ASTM, West Conshohocken, PA, USA.
21. Thümmler, F. and Thomma, W. (1967), 'The sintering process', *Metals Review*, 12, 69–108.
22. Pease, III L. F. and West, W. G. (2002), *Fundamentals of Powder Metallurgy*, Metal Powder Industries Federation, Princeton, NJ, USA.
23. Banerjee, S., Gemenetzis V., and Thümmler, F. (1980), 'Liquid phase formation during sintering of low alloyed steels with carbide based master alloy additions', *Powder Metallurgy International*, 23, 126–129.

24. Banerjee, S., Schlieper, G., Thümmler, F., and Zapf, G. (1980), 'New results in the master alloy concept for high strength steels', in *Modern Developments in Powder Metallurgy*, Vol. 12, *Principles and Processes*, eds H. Hausner, H. Antes, and G. Smith, Metal Powder Industries Federation, Princeton, NJ, USA, pp. 143–157.
25. Gaskell, D.R. (1981), *An Introduction to Metallurgical Thermodynamics*, McGraw Hill, New York, USA, p. 287.
26. Banerjee, S. and Joens, C. J. (2008), 'A comparison of techniques for processing powder metal injection molded 17-4 PH materials', in *Advances in Powder Metallurgy and Particulate Materials, Part 4 – Powder Injection Molding (Metals and Ceramics)*, Metal Powder Industries Federation, Princeton, NJ, USA.
27 Bradbury, S. (1996), *Powder Metallurgy Equipment Manual*, third edition, Metal Powder Industries Federation, Princeton, NJ, USA.
28. http://www.cmfurnaces.com/pdfs/cat_300.pdf
29. http://www.cremer-furnace.com/F/3/3_f.html

# Part II

Quality issues

# 8
# Characterization of feedstock in metal injection molding (MIM)

H. LOBO, DatapointLabs, USA

**Abstract:** This chapter covers the measurement of the physical properties of MIM feedstock. Characterization of the rheological, thermal behaviors will be covered, accompanied by real measured properties that exhibit the behavioral trends seen with these materials. The effect of binders and fillers will be noted. Rheological characterization will primarily revolve around the shear and temperature sensitivity of the materials using capillary rheometers. Comments will be made about general trends and expectations for such materials. Thermal characterizations will include differential scanning calorimetry (DSC) work for melting and crystallization behavior, characteristics of multi-component binder systems, and analysis of key transitions. Thermal conductivity will be covered because of the high conductivity exhibited by MIM feedstocks and the consequent effect on processability. The use of such data in computer-aided engineering (Moldflow) will be noted along with comments and cautionary notes.

**Key words:** MIM feedstock, properties, rheology, thermal, flow, slip, PVT, thermal analysis, thermal conductivity, processing.

## 8.1 Introduction

Metal injection molding feedstocks are composed fundamentally of a fine metal powder and complex binder system. The binder is designed to provide flow characteristics during the metal injection molding process and structural rigidity to the resulting solid part during the subsequent debinding and sintering processes. A number of binder systems have been designed, including wax–polymer systems, water-soluble systems, catalytic debindable systems, and water–gel-based systems. With the exception of the latter, the binder systems are solid at room temperature and go through a few melting transitions, one for each component of the binder until the

Table 8.1 Comparison table showing typical thermal properties of some materials

| Material | Density (kg/m$^3$) | Mass heat capacity (J/kg K) | Volumetric heat capacity (J/m$^3$ K) | Thermal conductivity (W/m K) |
|---|---|---|---|---|
| Tungsten | 19 300 | 134 | 2 586 200 | 163 |
| Iron | 7900 | 440 | 3 476 000 | 76 |
| MIM | 5100 | 700 | 3 570 000 | 3 |
| POM acetal | 1400 | 1480 | 2 072 000 | 0.28 |
| Wax | 766 | 2500 | 1 915 000 | 0.25 |

entire system is molten. Processing occurs at temperatures above the final melt temperature of the binder system but well below the decomposition temperature.

The molding of MIM feedstocks is similar to plastics injection molding processes with a few notable exceptions. The thermal characteristics of these materials differ greatly from plastics. The density is much higher, being rather closer to metal than the binder. This is because, while volumetrically the amount of binder is close to 40%, this translates to a weight-based composition exceeding 90% metal. Table 8.1 shows the density of some feedstocks, metals and plastics.

The measured mass heat capacity $C_{pm}$ is also similarly affected, being apparently quite low, again closer to metal than the binder system. This number is only apparently small thanks to conventions of the measurement rather than the actual heat capacity of the material. In reality, heat capacity is a volumetric term defined as

$$C_{pv} = \rho * C_{pm} \quad [8.1]$$

where $C_{pv}$ is the volumetric heat capacity (J/m$^3$ K), $\rho$ is the density (kg/m$^3$), and $C_{pm}$ is the mass heat capacity (J/kg K).

Table 8.1 illustrates this difference for a few common materials in comparison to a typical MIM feedstock. It can be observed that the volumetric heat capacity is not dramatically different for these materials even though the measured mass specific heat is.

Thermal conductivity in MIM feedstocks is significantly higher than conventional plastics with numbers ten times higher than plastics being commonplace. Obviously, this is due to the high content and thermal conductivity of the metal powder. Highly conductive metals will contribute to an increase in the thermal conductivity of the MIM feedstock. Surprisingly, however, the contribution is not proportional, as conventional wisdom would suppose. Instead, the conductivity remains closer to that of the binder system with a relatively smaller dependence on the conductivity of the metal. This is because the heat flow is controlled by the binder system, which constitutes the continuous matrix through which all the heat must flow. Therefore, while heat will flow quickly through each powder particle, the particles are not directly connected and the heat must flow through the intervening binder layer before it reaches the next particle. This layer thus constitutes the greatest resistance to heat flow and provides the correspondingly largest contribution to the thermal conductivity. In previous work with glass-fiber-filled plastics, Lobo and Cohen[1] showed that the following equation reasonably correlated thermal conductivity with filler content

$$\frac{1}{k} = \frac{\phi_1}{k_1} + \frac{\phi_2}{k_2}$$
[8.2]

where $\phi$ is the volume fraction and $k$ is the thermal conductivity. The numbers 1 and 2 denote the continuous and dispersed phases respectively. Table 8.1 gives thermal conductivity of some common materials in comparison to a typical MIM feedstock.

The most important characteristic of MIM feedstocks is the melt rheology, the flow properties of these materials. The rheological behavior of these materials is complex, showing dependencies on temperature and shear rate. The viscosity is observed to decrease with increasing temperature in a manner similar to what is seen in plastics. The temperature dependency unfortunately can be much more complicated than plastics. Plastics are by and large homopolymeric systems, being composed of a single polymer, but binder systems usually are not. When a plastic melts, the entire material moves from solid to melt. The resulting melt viscosity then proceeds to fall predictably with increasing temperature. With binder systems, as each component of the binder system melts, there is a dramatic change in the viscosity, until all components are molten. This process obviously reverses during cooling. This complex behavior is not necessarily detrimental to the processability of MIM feedstocks; it is just more difficult to understand and predict so that computer modeling using commonly available software is not so easy.

Another effect not commonly encountered with polymers is slip.[2] Newtonian fluids such as water have a parabolic velocity profile with zero velocity at the wall to a maximum at the center of the flow cavity. Polymers tend to have a uniform non-parabolic shear profile. With highly filled systems such as MIM feedstocks, the velocity at the wall is not zero and a plug flow type profile is observed. This phenomenon also causes some difficulty in modeling because most computer-aided engineering (CAE) codes do not account for slip phenomena, assuming only a Rabinowitsch correction, which captures the unique non-parabolic flow profiles typical to polymeric flows.

*Table 8.2* Characteristics of MIM feedstocks used in this study

| Material | A | B | C | D |
|---|---|---|---|---|
| Binder | Polyacetal | Water soluble | Wax polymer | Wax polymer |
| Powder | 17-4PH | 17-4PH | 17-4PH | 17-4PH |
| Content | 90% | 91.2% | | |
| Particle size | | | 22 µm | 12 µm |
| Process temperature | 190°C | 180°C | 160°C | 170°C |
| Mold temperature | 99°C | 38°C | 40°C | 45°C |

In the rest of this chapter, experimental methods commonly used to characterize MIM feedstock will be described. The resulting data are explained, particularly unique characteristics that differentiate them from conventional polymers. For the purpose of explaining these properties, four MIM feedstocks were chosen, each with the compositional characteristics described in the Table 8.2.

## 8.2 Rheology

Rheological measurements typically mean the measurement of viscosity versus shear rate. Data are taken at several temperatures. Two instruments are commonplace for rheology: the capillary rheometer and the cone and plate rheometer.

The capillary rheometer consists of a barrel containing the fluid being tested. One end has a precision capillary die; the other end is fitted with a piston capable of forcing the fluid through the die at a precise flow rate. Pressure drop across the die is measured by means of a pressure transducer located a short distance above the capillary die. The extrudate discharges to the air, so that the pressure measured by the pressure transducer is the pressure drop itself. Knowing the speed of the piston and the diameter of the barrel, the flow rate through the capillary is also known. With this, the viscosity can be calculated as follows

$$\dot{\gamma}_{ap} = \frac{4Q}{\pi R^3} \qquad [8.3]$$

where $\dot{\gamma}_{ap}$ is the apparent shear rate (/s), $Q$ is the flow rate (mm³/s), and $R$ is the capillary diameter (mm).

$$\sigma_W = \frac{\Delta P \cdot R}{2L} \qquad [8.4]$$

where $\sigma_W$ is the wall shear stress (Pa), $\Delta P$ is the pressure drop across the capillary die (Pa), and $L$ is the length of capillary die (mm).

Also

$$\eta_{ap} = \frac{\sigma_w}{\dot{\gamma}_{ap}} \quad [8.5]$$

where $\eta_{ap}$ is the apparent shear viscosity (Pa s).

The rheometer barrel can be heated so that measurements at different temperatures can be achieved. The instrument is well described by Nelson.[3] Capillary rheometers measure what is termed apparent shear viscosity versus apparent shear rate. This is because the data contain known errors that can be corrected for. Bagley[4] developed a correction for the additional pressure drop related to the contraction flow at the entrance to the die and the corresponding expansion at the die exit. The pressure drop related to these extensional effects can be removed by repeating the test with dies of the same diameter but different length-to-diameter ($L/D$) ratios.

Rabinowitsch[5] developed a correction for the fact that the axial velocity profile in the die is not parabolic owing to the shear thinning nature of polymers. While this mathematical correction is routinely performed for polymers, some caution may be advisable in the case of MIM feedstocks because of the possibility of slip at the die wall. Wall slip is a phenomenon that is not seen in polymers; here, the velocity of the fluid in contact with the wall is not zero as is the case with Newtonian fluids and polymers. This results in a plug flow-like behavior. Hatzikiriakos and Dealy[6] describe a method to characterize slip by performing measurements with capillary dies of different diameter but the same $L/D$ ratio. In this case, any difference in viscosity would be due to slip. It is then possible to calculate the slip velocity versus shear rate, thereby developing a characterization of this phenomenon. Commercial simulation codes do not typically take slip into consideration.

Looking at some rheological data, it is useful to define some terms. In this context, Newtonian behavior refers to fluids whose viscosity is independent of shear rate. A plot of viscosity against shear rate is a horizontal line. Shear thinning fluids are those whose viscosity decreases with increasing shear rate. On a log viscosity versus log shear rate plot, such data appear as a straight line with negative slope. Sample A shows shear thinning behavior with no tendency toward a Newtonian region over the shear rate range of interest to processing, as seen in Fig. 8.1. In the figure, the lines represent a Cross-WLF equation, which is commonly used to describe the non-Newtonian, non-isothermal change in viscosity with shear rate. The data are presented at three temperatures where it can be seen that the viscosity declines with increasing temperature. The rate of change of viscosity with temperature is of interest to processing. A large change suggests a higher sensitivity of viscosity to temperature. In such cases, a small change in temperature will cause a large change in viscosity, significantly altering the flow characteristics of the material.

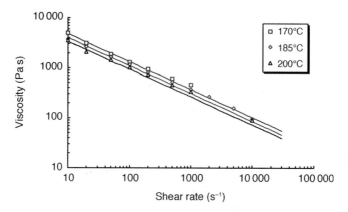

*8.1* Viscosity of sample A showing shear thinning behavior.

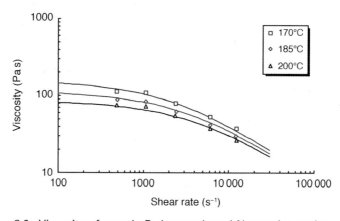

*8.2* Viscosity of sample B shows a broad Newtonian region.

Figure 8.2 shows the viscosity of sample B, a feedstock with Newtonian behavior at low shear rates. As the shear rate increases the feedstock is observed to transition to shear thinning behavior. Again, the Cross-WLF equation is observed to be capable of handling this behavior. Sample C (Fig. 8.3) exhibits similar 'shear-thinning only' behavior to sample A, but its viscosity is significantly lower than sample A because of the different binder system. The viscosity of sample C is lower than sample D (Fig. 8.4), possibly due to compositional or particle size differences.

Mooney slip velocity measurements were performed on samples C and D to evaluate the effect, if any, on particle size (Fig. 8.5). It is observed that the wall slip is greater for sample D, which has a smaller particle size. The figure also indicates that the wall slip will increase greatly with increasing shear stress, transitioning more and more toward plug flow at very high apparent shear rates.

Characterization of feedstock in metal injection molding 189

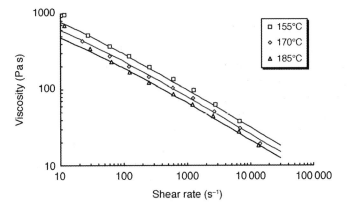

*8.3* Viscosity of sample C.

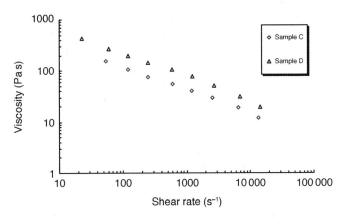

*8.4* Comparing the viscosity of samples C and D.

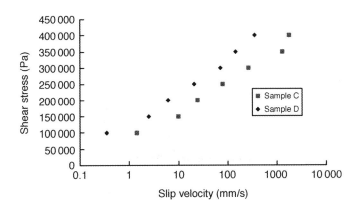

*8.5* Mooney slip velocity data for samples C and D.

## 8.3 Thermal analysis

Differential scanning calorimeters are typically the instrument of choice for the characterization of the melting and solidification of MIM feedstocks. The measurement of specific heat is also accomplished using ASTM E1269.[7] The instrument comprises two identical furnaces, one containing a specimen of the material and the other acting as an empty reference. By heating the two furnaces identically at a programmed rate of temperature rise, it is possible to determine the difference in heat needed to keep both furnaces heating at an identical rate. An alternative approach measures the temperature differential as both furnaces are subjected to identical quantities of heat. Additional details are provided by Salamon and Fielder.[8]

During the actual experiment, milligram quantities of the sample are used. The specimen is encapsulated, typically in an aluminum pan, which serves to contain it and prevent leakage and damage to the apparatus. The scanning data are taken by computer and plots of heat flow versus temperature permit the observation of melting and solidification phenomena (Fig. 8.6).

Thermal analysis provides an insight into the behavior of the binder system of the MIM feedstock. Binder systems often contain waxes and semi-crystalline polymers such as polyethylene, polypropylene, and acetal (POM). In polymeric systems the change in state, from melt to solid and vice versa, is attributable to either a glass transition or a semi-crystalline transition. Polymers in their solid states can either be solidified glasses or contain a partially crystallized phase that solidifies the polymer well above the glass transition temperature. When the solidification is due to a glass transition, the polymer becomes brittle and glassy; the temperature at which this occurs is termed the glass transition temperature. Polymers above their glass transition tend toward a rubbery or leathery state, eventually

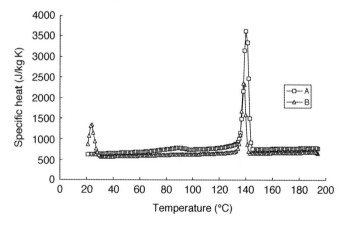

*8.6* Cooling mode specific heat and transition plots for samples A and B.

becoming fully molten when heated to a sufficiently high temperature. The glass transition is reversible, implying that solidification occurs by simply reversing the change; cooling below the glass transition temperature returns the polymer to a glassy state.

In the case of semi-crystalline polymers, the glass transition is less dramatic as the polymer continues to be quite solid above the glass transition. With increasing temperature, a state is reached where the crystallized fraction begins to melt. This is a kinetic process affected by both time and temperature. Nucleating agents can accelerate the onset of crystallization by providing sites for the initiation of crystallization. The semi-crystalline polymer becomes molten when this transition is complete. Because the process is kinetic, the temperature at which the semi-crystalline polymer melts depends on the heating rate. A faster heating rate corresponds to a higher melt temperature. The solidification of a molten semi-crystalline polymer requires that the melt be super-cooled below the melting point. This super-cooling is needed to create sufficient driving force for the initiation of the crystallization process. Once crystallization is initiated, the conversion is usually fast and the polymer is solid when complete. Note that 100% crystallization does not occur. In Fig. 8.6, the higher temperature peaks observed for samples A and B are typical of the crystallization of the polymeric components of binder systems.

The understanding of these phenomena is vital to comprehending the manner in which polymers transition from melt to solid. The situation with MIM feedstocks is even more complex, with multiple such transitions as each component of the binder system solidifies at its own temperature. The challenge then lies in determining which of these transitions is the crucial transition that takes the whole system from melt to solid. This is not often easily determined from thermal analysis because the size of a given peak is not always an indicator of a major transition. In such cases, the no-flow temperature can provide useful indication of the major transition. The no-flow temperature is an approximate indication of the solidification transition of a material; it measures the temperature at which the material can no longer flow through a capillary die as it is gradually cooled from its melt processing temperature in a capillary rheometer. Table 8.3 shows a comparison of first cooling transition onset temperatures and no-flow temperature for various feedstocks.

While it may be argued that the no-flow temperature lacks precision, in the particular case of materials with multiple transitions, it clearly indicates which differential scanning calorimetry (DSC) transition is the cause of solidification of the feedstock. In the case of the data presented, the higher temperature transitions are the ones of consequence.

Wax, the other common component of binders, can be viewed to have similar characteristics to a simple semi-crystalline polymer. Waxes transition

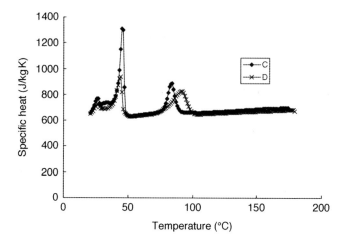

*8.7* Specific heat and transitions of samples C and D, which have the same binder system.

*Table 8.3* Comparing DSC cooling transitions with no-flow temperature

| Material | A | B | C | D |
|---|---|---|---|---|
| No-flow temperature (°C) | 157 | 151 | 108 | 121 |
| DSC cooling transition | 144 | 141 | 88 | 99 |

at low temperatures, below 100°C, and show large transition peaks; observe the lower temperature peak of sample B in Fig. 8.6. However, while waxes may aid feedstock flow by acting as flow enhancers, they do not typically control the melt–solid transition; this is more often than not controlled by the solidification of the polymer.

Specific heat is an additional property that can be obtained from thermal analysis. The measurement requires an additional baseline experiment carried out with empty pans in the DSC which serves as a subtractive reference for the data. With these additional data, the specific heat is conveniently measured as a function of temperature. Data are taken either in heating or cooling mode, with the cooling mode being in common use for the purpose of MIM feedstock modeling.

Figure 8.7 shows specific heat and transition data for samples C and D, which contain the same binder systems. It can be observed that there is close correspondence between the two curves. The highest transition is shifted lower for sample C, suggesting some kind of kinetic effect on the initiation of crystallization.

## 8.4 Thermal conductivity

Measurements of thermal conductivity can be carried out using a variety of instruments. Most are designed for the testing of materials in the solid state. Since our interest is with molten MIM feedstocks, the line-source method is preferred. This is a transient method permitting rapid measurement of thermal conductivity. The test method is covered in ASTM D5930.[9] The test instrument consists of a heated barrel that maintains the material at a constant temperature. The measurement is performed by a line-source needle probe that is inserted into the fluid. The line-source probe contains a heater that runs the entire length of the probe and a temperature sensor located half way along the length of the probe. At equilibrium, the line-source heater sends a heat wave into the fluid. The temperature rise in the probe is recorded with time for 30–45 s. A plot of temperature against log time theoretically yields a straight line, the slope of which can be directly related to thermal conductivity by the following equation

$$k = \frac{CQ'}{4\pi \, \text{Slope}} \quad [8.6]$$

where $k$ is the thermal conductivity (W/m K), $Q'$ is the heat input per unit length of line source (W/m), $C$ is the probe constant and

$$\text{Slope} = \frac{T_2 - T_1}{\ln(t_2/t_1)} \quad (K)$$

where $T_2$ is the temperature recorded at time $t_2$ and $T_1$ is the temperature recorded at time $t_1$.

Data at multiple temperatures are obtained by equilibrating the instrument at each test temperature before applying the temperature transient described above. The fluid can be solidified *in situ*, permitting solid state measurements to be made. Measurements are typically taken in the cooling mode, although it is possible to reheat the solidified material and thereby obtain data in the heating mode. Lobo[10] provides additional details in his book.

Thermal conductivity of MIM feedstocks tends to be rather high relative to plastics. This can result in some scatter as the upper limit of capability of the test instrument is approached with an expected variability of 10%. Typical data are shown in Fig. 8.8. It can be observed that melt state data are typically fairly constant, with a slight increase in melt thermal conductivity being observed as temperature decreases. As the feedstock cools and solidifies, various phases of the binder system begin to crystallize. Because the thermal conductivity values of the crystalline phases are higher than the surrounding amorphous base, the overall thermal conductivity is observed to rise as the material cools.

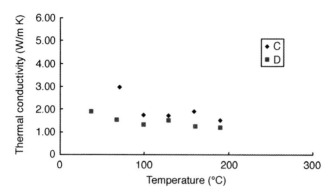

*8.8* Thermal conductivity of samples C and D.

Consequently, the solid thermal conductivity is typically higher than the melt thermal conductivity. If the binder system contains no waxes or polymers of a semi-crystalline nature, then this behavior is not observed and the solid thermal conductivity is close to or lower than the melt thermal conductivity.

The implications for processing are now quite easily understood. The overall high thermal conductivity of these materials renders them able to conduct heat at a faster pace than typical polymers, usually by about ten times. Molten MIM feedstock is able to solidify much faster as it readily gives heat to the cold mold. The processing becomes more sensitive to the choice of melt and mold temperature. As the melt solidifies, it becomes even more conductive, potentially aggravating the process. Simulating this process also becomes tricky because injection-molding codes are tuned to low thermal conductivity plastics and the predicted onset of the transition becomes extremely important in determining the point of solidification.

## 8.5  Pressure–volume–temperature (PVT)

One of the most complex pieces of data related to understanding the processing behavior of MIM feedstocks is the equation of state data commonly referred to as the pressure–volume–temperature (PVT) relationship. A high-pressure dilatometer is widely used for these measurements. The preferred test specimen is a molded component as the actual feedstock may have trapped air and change the effective compressibility of the material. The test involves encapsulating the specimen in a confining fluid and then compressing the fluid, thereby generating a hydrostatic pressure field surrounding the specimen. The change in volume is measured.

Data are typically taken by performing pressure cycles at a given temperature (isothermal compression) with the process being repeated after heating and equilibrating at the next test temperature. Other measurement

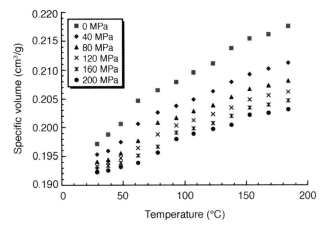

*8.9* PVT relationship of sample C.

*Table 8.4* Melt density of feedstocks

| Material | A | B | C* | D |
| --- | --- | --- | --- | --- |
| Pressure (MPa) | 2 | 4 | 0 | 5 |
| Temperature (°C) | 185 | 210 | 168 | 170 |
| Density (kg/m³) | 4610 | 4980 | 4630 | 4730 |

*Taken from PVT data.

configurations can be used. Each will give different data owing to the various kinetic effects at work as the material melts or solidifies. Figure 8.9 shows PVT data for a feedstock (sample C). It is seen that at a particular temperature, the specific volume, which is the inverse of the density, decreases with increasing pressure. The specific volume increases with temperature and is described by the volumetric thermal expansion. Multiple transitions can be observed in the data. Because of the cost and complexity of such data, it is sometimes convenient simply to replace these data by a single point value, the melt density. Table 8.4 lists melt densities of some feedstocks.

Knowing this and the solid density, one has a crude approximation of the volumetric thermal expansion of the material and thereby the shrinkage.

## 8.6 Conclusions

The processing of MIM feedstocks is similar to plastic injection molding. The physical properties of these materials differ significantly from plastics, as described above. These differences impact the molding of MIM feedstocks. By understanding and accounting for this it is possible to process MIM with considerably higher levels of accuracy. Simulation can be of assistance to MIM processing, particularly if the simulation is able to account for the significant differences that exist in the properties of MIM feedstocks as compared to conventional plastics. Typical challenges that often arise come from the inability of the simulations to handle very high thermal conductivity. In such cases, the simulation cools the material much faster than in real life, leading to underprediction of flow length. The inability to account for wall slip is another issue of some consequence as it can be seen that this phenomenon is prevalent in MIM feedstocks but is less common in plastics, for which the CAE codes were originally designed. Lastly, high density of these materials may make it necessary to consider inertial effects, something that is less important with plastics which have very high viscosities and low densities. Material data for the feedstocks used in this study are available for download to commercially available injection molding CAE at www.matereality.com.

## 8.7 Acknowledgments

We would like to acknowledge the contributions of Professor Donald Heaney for arranging for the MIM feedstocks, Smith Metal Products, Netshape and Advanced Powder Products for kindly providing materials, and Mr Jamie Antosh, Mr Will Liguori and Mr Brian Croop of DatapointLabs, who were responsible for the generation of the material data reported in this chapter.

## 8.8 References

1. Lobo, H. and Cohen, C. (1990), *Polymer Engineering Science,* 30, 65 (1990).
2. Shaw, M. (1986), *1986 SPE ANTEC Proceedings,* p. 707.
3. Nelson, B. (2003), 'Capillary rheometry', in *Handbook of Plastics Analysis,* eds Lobo, H. and Bonilla, J. V., Ch. 2, pp. 43–78, Marcel Dekker.
4. Bagley, E. B. (1957), *Journal of Applied Physics,* 28, 624.
5. Rabinowitsch, B. Z. (1929), *Physical Chemistry,* 145, 1.
6. Hatzikiriakos, S. and Dealy, J. M. (1992), 'Wall slip of HD polyethylene', *Journal of Rheology,* 36(4).
7. ASTM E1269–11 (2011), 'Standard test method for determining specific heat capacity by differential scanning calorimetry'.
8. Salamon, A. and Fielder, K. (2003), 'Practical uses of differential scanning calorimetry for plastics', in *Handbook of Plastics Analysis,* eds Lobo, H. and Bonilla, J. V., Ch. 3, pp. 79–110, Marcel Dekker.
9. ASTM D5930–09 (2009), 'Standard test method for thermal conductivity of plastics by means of a transient line-source technique'.
10. Lobo, H. (2003), 'Thermal conductivity and diffusivity of polymers', in *Handbook of Plastics Analysis,* eds Lobo, H. and Bonilla, J. V., Ch. 5, pp. 129–154, Marcel Dekker.

# 9
# Modeling and simulation of metal injection molding (MIM)

T. G. KANG, Korea Aerospace University, Korea,
S. AHN, Pusan National University, Korea, S. H. CHUNG,
Hyundai Steel Co., Korea, S. T. CHUNG, CetaTech Inc.,
Korea, Y. S. KWON, CetaTech, Inc., Korea, S. J. PARK,
POSTECH, Korea and R. M. GERMAN, San Diego State
University, USA

**Abstract:** Powder injection molding (PIM) is a means of mass-producing various metal or ceramic three-dimensional components in large production quantities at low costs. Computer simulations used for plastic injection molding and powder metallurgy have been applied to PIM, but the high solid content in the powder–binder feedstock makes for differences in density, conductivity, shear rate sensitivity, and other factors. Thus, simulations for the PIM process build on the success demonstrated in plastics and powder metallurgy, but rely on modifications to properly handle mixing, molding, debinding, and sintering. In this chapter the basic models and numerical simulations are introduced in sequence to treat mixing, molding, debinding, and sintering to make predictions relevant to defect-free, high-quality production. The simulations are applied to optimization for a few applications.

**Key words:** mixing, debinding, sintering, simulation, particle tracking, master decomposition curve.

## 9.1 Modeling and simulation of the mixing process

In analysis of the mixing process as relevant to PIM, we are mostly concerned with distributive mixing without molecular diffusion, governed by the flow kinematics of specific mixer geometry for given material properties and operating conditions. This is the case for flows with an infinite Péclet number. The flow problem in this mixer can be solved using the finite element method (FEM). As for mixing analysis with the kinematics of fluid–particle interactions like the PIM process, a particle tracking method can be employed.[1] A distribution of particles at a down-channel location is used to characterize the progress of mixing both qualitatively and quantitatively. A measure of mixing, called the information entropy, based on the particle distribution, can be introduced to quantify a degree of mixing.

### 9.1.1 Modeling

It is assumed that materials to be mixed are homogeneous generalized Newtonian fluids and the flow is governed by viscous force only, neglecting inertia force, which is a reasonable assumption in the mixing analysis of highly viscous powder injection molding feedstock. In the creeping flow regime, continuity and momentum balance equations are given as follows

$$\nabla \cdot \mathbf{u} = 0 \quad \text{and} \quad -\nabla p + \nabla \cdot (2\eta \mathbf{D}) = 0 \qquad [9.1]$$

where $\eta$ is the viscosity, $\mathbf{D}$ is the rate-of-deformation tensor, $p$ is the pressure, and $\mathbf{u}$ is the velocity. The viscosity of a feedstock is generally represented by a function of the shear rate $\dot{\gamma}$, temperature $T$, and pressure $p$. The Cross-WLF (Williams–Ladel–Ferry) model is one of viscosity models for the feedstock in PIM, defined as follows

$$\eta(\dot{\gamma}, T, p) = \frac{\eta_0}{1 + (\eta_0 \dot{\gamma}/\tau^*)^{1-n}} \qquad [9.2]$$

where $\eta_0$ is the zero-shear-rate viscosity described by the WLF model, given by

$$\eta_0 = D_1 \exp\left[-\frac{A_1(T - T^*)}{A_2 + T - T^*}\right] \qquad [9.3]$$

where $A_2 = \tilde{A}_2 + D_3 p$ and $T^* = D_2 + D_3 p$ and the materials parameters, $A_1$, $\tilde{A}_2$, $D_1$, $D_2$, and $D_3$, are determined by curve-fitting using the experimental viscosity data. Although PIM feedstock has a more involved rheological response, the corrections are not significant.

### 9.1.2 Numerical methods

*Finite element formulation*

To obtain the velocity field and pressure distribution, the set of governing equations, equation [9.1], together with proper boundary conditions, are solved using the Galerkin FEM. The weak formulation for the incompres-

Modeling and simulation of metal injection molding    199

sible Stokes equations is given by

$$\int_\Omega (\nabla \cdot \mathbf{u})q d\Omega = 0 \text{ and } \int_\Omega 2\eta \mathbf{D}(\mathbf{u}) : \mathbf{D}(\mathbf{w}) d\Omega - \int_\Omega p(\nabla \cdot \mathbf{w}) d\Omega$$
$$= \int_\Gamma \mathbf{t} \cdot \mathbf{w} d\Gamma \qquad [9.4]$$

where $\Omega$ is the computational domain, $\Gamma$ is the boundary, $\mathbf{w}$ is the weighting function for the velocity $\mathbf{u}$, $q$ is the weighting function for the pressure $p$, and $\mathbf{t}$ is the traction force defined by

$$\mathbf{t} = -p\mathbf{n} + 2\mu \mathbf{D}(\mathbf{u}) \cdot \mathbf{n} \qquad [9.5]$$

In three-dimensional flow simulations, the resulting matrix equation after numerical integration is a huge sparse symmetric matrix. Thus, one needs an efficient method to solve the resulting matrix equation. A parallel direct solver, such as PARDISO,[2] can be used to solve the resulting sparse matrix on a parallel computer with multiple CPUs.

*Particle tracking method*

Mixing analysis based on a particle tracking scheme consists of three steps:

1. flow analysis to obtain a velocity field of a mixing device;
2. particle tracking step to obtain a distribution of particles at the end of the mixer;
3. quantification of mixing from the obtained particle distribution.

To track the position of particles, the problem to be solved is an ordinary differential equation

$$d\mathbf{x}/dt = \mathbf{u} \qquad [9.6]$$

where $\mathbf{x}$ is the particle position vector, $\mathbf{u}$ is the particle velocity, and $t$ is the time. A 4th-order Runge–Kutta method is widely used to integrate the set of ordinary differential equations, equation [9.6], with respect to time. However, in some cases the problem can be simplified such that a two-dimensional problem may be solved instead of a three-dimensional one. By changing the original equation, one can reduce the dimensions of the system and integration can be performed using a fixed spatial increment along the down-channel direction. This approach has several advantages such as faster solution and ease of representing the dynamics at any cross-sectional area in terms of the down-channel direction. This scheme is restricted to problems where all the axial components of the velocity field are positive in the whole domain, i.e. no backflow. The resulting two-dimensional problem is represented as follows

$$\frac{dx}{dz} = \frac{u}{w} = \tilde{u} \text{ and } \frac{dy}{dz} = \frac{v}{w} = \tilde{v} \qquad [9.7]$$

where $u$, $v$ and $w$ denote the velocity components along the $x$, $y$ and $z$ coordinate, respectively. Thus, the particles are tracked along the axial position rather than time.

### 9.1.3 Applications

*Mixer and mixing materials*

A static mixer, called the Kenics mixer as a mixing device for PIM feedstock preparation, operates in the Stokes flow regime where inertia is negligible. Figure 9.1 depicts a representative geometry of a six-element static mixer (Fig. 9.1(a)), with mixing elements twisted by 180° in alternating directions (Fig. 9.1(b)), that is used in simulations.

The viscosity of the PIM feedstock of carbonyl iron at 63 vol% solids loading is measured by means of a capillary rheometer, following ASTM 3835-96 for three temperatures versus a range of shear rates. From these data we determined the seven coefficients of the Cross-WLF model by curve-fitting: $n = 0.4999$, $\tau^* = 0.0005734$ Pa, $D_1 = 1.50 \times 1011$ Pa s, $D_2 = 373.15$ K, $D_3 = 0$ K/Pa, $A_1 = 6.30$ K, and $\tilde{A}_2 = 51.6$ K.

*Working principle of the Kenics mixer based on flow characteristics*

First the flow problems are solved using the FEM. The mixer geometry is divided into 344 510 ten-node tetrahedral elements with 519 155 nodes. A quadratic interpolation for the velocity and a linear interpolation for the

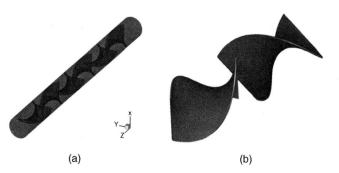

*9.1* Kenics static mixer. (a) Shaded image of a mixer with six helical elements; (b) two mixing elements twisted by 180° in alternating directions, called LR-180 elements.

Modeling and simulation of metal injection molding 201

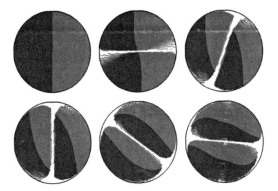

*9.2* Working principle of the Kenics mixer by the two grey tints representing fluid–particles mixtures.

pressure are used, which satisfy the compatibility condition of the interpolation functions for the velocity and pressure. The mixing principle of the Kenics mixer is also based on the baker's transformation consisting of repeated stretching, cutting, stacking, which is realized in a pipe flow with blades twisted by a fixed angle in alternating directions. Figure 9.2 illustrates the progress of mixing in the first half period, with the blade rotating 180° in a counterclockwise direction. The two fluids are split horizontally by the mixing element (blade) and the split materials are deformed by the helical motion leading to an increase in the number of striations at the end of the operation. At the beginning of the next half-period, the materials are split vertically and similar operations (stretching and stacking) are repeated. The same operations are repeated to the number of mixing elements inserted in the circular channel.

*Mixing analysis*

A particle tracking method is used to analyze mixing. Given velocity field, the particles initially located at the interface between the two fluids with different power-law index $n$ are tracked to the end of the mixer geometry. The more uniform the distribution at a cross-sectional area, the better the mixing. The final result of the particle tracking is evident in the distribution of tracer particles. Evaluation of the mixing uniformity is made from the information on the distribution of the particles. A proper measure of mixing is the uniformity of the particle distribution, for example the information entropy of the particles at any cross-section.[1,3] Along these lines, characterization of the mixing progress requires first the division of the mixer cross-section into a number of cells. Then, for a certain particle configuration, the mixing entropy $S$ is defined as a sum of the information entropy of individual cells constituting the cross-sectional area, defined by

$$S = -\sum_{i=1}^{N} n_i \log n_i \qquad [9.8]$$

where $N$ is the number of cells and $n_i$ the particle number density at the $i$th cell. Instead of directly using the information entropy in the mixing analysis, we employ a normalized entropy increase $S^*$ as a measure of mixing, given by

$$S^* = \frac{S - S_0}{S_{max} - S_0} \qquad [9.9]$$

where $S_0$ is the entropy at the inlet and $S_{max}$ is the maximum possible entropy, defined by $\log N$, which is the ideal case with a uniform distribution of the particles, i.e. $n_i = 1/N$. Figure 9.3 shows the progress of mixing characterized by the normalized entropy increase defined in equation [9.9]. In this plot, the influence of the index $n$ on the progress of mixing is not significant, but this is an example of the parameter study and design process.

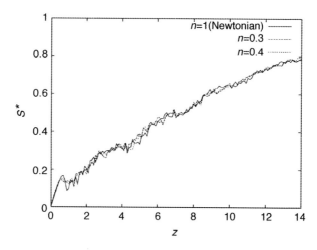

*9.3* The progress of mixing characterized by the normalized entropy increase $S^*$ along the axial direction $z$ depending on the index $n$ in the viscosity model.

## 9.2 Modeling and simulation of the injection molding process

Models and simulations used for plastics have been applied to PIM, but the high solid content often makes for differences that are ignored in the plastic simulations. Several situations demonstrate the problems, such as powder–binder separation at weldlines, high inertial effects such as in molding tungsten alloys, and rapid heat loss such as in molding copper and aluminum nitride. Also, powder–binder mixtures are very sensitive to shear rate. Thus, the computer simulations, to support molding build from the success demonstrated in plastics, and adapt those concepts in new customized PIM simulations for filling, packing, and cooling.

### 9.2.1 Theoretical background and governing equations

A typical injection molded component has a thickness much smaller than the overall largest dimension.[4] In molding such components, the molten powder–binder feedstock mixture is highly viscous. As a result, the Reynolds number (a dimensionless number characterizing the ratio of inertia force to viscous force) is much smaller than 1 and the flow is modeled as a creeping flow with lubrication, as treated with the Hele–Shaw formulation. With the Hele–Shaw model, the continuity and momentum equations for the melt flow in the injection molding cavity are merged into a single Poisson equation in terms of the pressure and fluidity. Computer simulation is usually based on a 2.5-dimensional approach because of the thin wall and axial symmetry. But the Hele–Shaw model has its limitations and cannot accurately describe three-dimensional (3D) flow behavior in the melt front, which is called fountain flow, and special problems arise with thick parts with sudden thickness changes, which cause race-track flow. Nowadays, several 3D computer-aided engineering simulations exist that successfully predict conventional plastic advancement and pressure variation with changes in component design and forming parameters.[5] In this section, we will focus on the 2.5-dimensional approach rather than a full 3D approach, because the 2.5-dimensional approach is more robust and commonly accepted by industry.

*Filling stage*

Powder injection molding involves a cycle that repeats every few seconds. At the start of the cycle, the molding machine screw rotates in the barrel and moves backward to prepare molten feedstock for the next injection cycle while the mold closes. The mold cavity fills as the reciprocating screw moves forward, acting as a plunger, which is called the filling stage. During the filling stage, a continuum approach is used to establish the system of governing equations as follows.

- Mass and momentum conservation: with the assumption of incompressible flow, the mass conservation, also called continuity equation, is expressed as

$$\frac{\partial u}{\partial x} + \frac{\partial v}{\partial y} + \frac{\partial w}{\partial z} = 0 \qquad [9.10]$$

where $x$, $y$, and $z$ are Cartesian coordinates and $u$, $v$, and $w$ are corresponding orthogonal velocity components. As for the momentum conservation, with lubrication and the Hele–Shaw approximation, the Navier–Stokes equation is modified for molten feedstock during filling stage as follows[6, 7]

$$\frac{\partial P}{\partial x} = \frac{\partial}{\partial z}\left(\eta \frac{\partial u}{\partial z}\right)$$

$$\frac{\partial P}{\partial y} = \frac{\partial}{\partial z}\left(\eta \frac{\partial v}{\partial z}\right)$$

$$\frac{\partial P}{\partial z} = 0 \qquad [9.11]$$

where $P$ is the pressure, $z$ is the thickness, and $\eta$ is the viscosity of PIM feedstock. Combining equations [9.10] and [9.11] with integration in the $z$-direction (thickness direction) gives

$$\frac{\partial}{\partial x}\left(S\frac{\partial P}{\partial x}\right) + \frac{\partial}{\partial y}\left(S\frac{\partial P}{\partial y}\right) = 0 \qquad [9.12]$$

where

$$S \equiv \int_{-b}^{b} \frac{z^2}{\eta} dz \qquad [9.13]$$

Equation [9.12] is the flow governing equation for the filling stage. This is exactly the same form as the steady-state heat conduction equation obtained by substituting temperature $T$ into $P$ and thermal conductivity $k$ into $S$. In this analogy, $S$ is the flow conductivity or fluidity. As a simple interpretation of this flow governing equation, molten PIM feedstock flows from the high-pressure region to the low-pressure region, and the speed of flow depends on the fluidity $S$.

- Energy equation: in accordance with the lubrication and Hele–Shaw approximations during the filling stage, the energy equation is simplified as follows

$$\rho C_p \left( \frac{\partial T}{\partial t} + u \frac{\partial T}{\partial x} + v \frac{\partial T}{\partial y} \right) = k \frac{\partial^2 T}{\partial z^2} + \eta \dot{\gamma}^2 \qquad [9.14]$$

where $\rho$ is the molten PIM feedstock's density, $C_p$ is the molten PIM feedstock's specific heat, $\dot{\gamma} = \sqrt{(\partial u/\partial z)^2 + (\partial v/\partial z)^2}$ the generalized shear rate, and $k$ is the thermal conductivity of the feedstock.

In addition, we need a constitutive relation to describe the molten PIM feedstock's response to its flow environment during cavity filling, which requires a viscosity model. Several viscosity models for polymers containing high concentrations of particles are available. Generally they include temperature, pressure, solids loading, and shear rate and selected models will be introduced later in this chapter. The selection of a viscosity model depends on the desired simulation accuracy over the range of processing conditions, such as temperature and shear rate, as well as access to the experimental procedures used to obtain the material parameters.

Once we have the system of differential equations from continuum-based conservation laws and the constitutive relations for analysis of the filling stage, then we need boundary conditions. Typical boundary conditions during the filling stage are as follows:

- boundary conditions for flow equation: flow rate at injection point, free surface at melt-front, no slip condition at cavity wall;
- boundary conditions for energy equation: injection temperature at injection point, free surface at melt-front, mold-wall temperature condition at cavity wall.

Note that the only required initial condition is the flow rate and injection temperature at the injection node, which is one of the required boundary conditions.

*Packing stage*

When mold filling is nearly completed, the packing stage starts. This precipitates a change in the ram control strategy for the injection molding machine, from velocity control to pressure control, which is called the switchover point. As the cavity nears filling, the pressure control ensures full filling and pressurization of the filled cavity prior to freezing of the gate. It is important to realize the packing pressure is used to compensate for the anticipated shrinkage in the following cooling stage. Feedstock volume shrinkage results from the high thermal expansion coefficient of the binder, so on cooling there is a measurable contraction. Too low a packing pressure before the gate freezes results in sink marks due to the component shrinkage, while too high a packing pressure results in difficulty with ejection. Therefore appropriate pressurization prior to cooling is critical for component quality. For the analysis of the packing stage, it is essential to include the effect of melt compressibility. Consideration is given to the melt compressibility using a dependency of the specific volume on pressure and temperature, leading to a feedstock-specific pressure–volume–temperature ($pvT$) relationship, or the equation of state. Several models are available to describe the $pvT$ relation of PIM feedstock, such as the two-domain modified Tait model and IKV model. These models predict an abrupt volumetric change for both semi-crystalline polymers used in the binder and the less abrupt volume change for amorphous polymers used in the binder.

With the proper viscosity and $pvT$ models, the system of governing equation for the packing stage based on the continuum approach is as follows.

- Mass conservation: the continuity equation of compressible PIM feedstock is expressed as

$$\frac{\partial \rho}{\partial t} + \frac{\partial(\rho u)}{\partial x} + \frac{\partial(\rho v)}{\partial y} + \frac{\partial(\rho w)}{\partial z} = 0 \qquad [9.15]$$

with the assumption that pressure convection terms may be ignored in the packing stage this becomes

$$\kappa \frac{\partial p}{\partial t} - \beta \left( \frac{\partial T}{\partial t} + u \frac{\partial T}{\partial x} + v \frac{\partial T}{\partial y} \right) + \left( \frac{\partial u}{\partial x} + \frac{\partial v}{\partial y} \right) = 0 \qquad [9.16]$$

where $\kappa$ is the isothermal compressibility coefficient of the material ($\partial \rho / \rho \partial p$) and $\beta$ measures the volumetric expansivity of the material ($\partial \rho / \rho \partial T$). Those are easily calculated from the equation of state. Note that the same momentum conservation is used as equation [9.12] regardless of the material to be considered as compressible or not.

- Energy equation: the energy equation is derived as

$$\rho C_p \left( \frac{\partial T}{\partial t} + u \frac{\partial T}{\partial x} + v \frac{\partial T}{\partial y} \right) = k \frac{\partial^2 T}{\partial z^2} + \eta \dot{\gamma}^2 + \beta T \frac{\partial p}{\partial t} \qquad [9.17]$$

That is, the shear rate for the compressible case in packing is for practical purposes the same as for the filling phase.

Typical initial and boundary conditions during packing stage are as follows:

- initial conditions: pressure, velocity, temperature, and density from the results of the filling stage analysis;
- boundary conditions for equations of mass and momentum conservations: prescribed pressure at injection point, free surface at melt-front, no slip condition at cavity wall;
- boundary conditions for energy equation: injection temperature at injection point, free surface at melt-front, and mold-wall temperature condition at the cavity wall, which is interfaced with the cooling stage analysis.

*Cooling stage*

Of the three stages in the injection molding process, the cooling stage significantly affects the productivity and the quality of the final component. Cooling starts immediately upon the injection of the feedstock melt, but formally the cooling time is referred to as the time after the gate freezes and no more feedstock melt enters the cavity. It lasts up to the point of component ejection, when the temperature is low enough to withstand the ejection stress. In the cooling stage, the feedstock volumetric shrinkage is counteracted by the pressure decay until the local pressure drops to atmospheric pressure. Thereafter, the material shrinks with any further cooling, possibly resulting in residual stresses due to non-uniform shrinkage or mold constraints (which might not be detected until sintering). In this stage, the convection and dissipation terms in the energy equation are neglected since the velocity of a feedstock melt in the cooling stage is almost zero.[8–10] Therefore, the objective of the mold-cooling analysis is to solve only the temperature profile at the cavity surface to be used as boundary conditions of feedstock melt during the filling and packing analysis. When the injection molding process is in steady state, the mold temperature will fluctuate periodically over time during the process due to the interaction between the hot melt and the cold mold and circulating coolant. To reduce the computation time for this transient process, a 3D cycle-average approach is adopted for the thermal analysis to determine the cycle-averaged temperature field and its effects on the PIM component. Although the mold temperature is assumed invariant over time there is still a transient for the PIM feedstock,[11] leading to the following features.

- Mold cooling analysis: under this cycle-average concept, the governing equation of the heat transfer for injection mold cooling system is written as

$$\nabla^2 \bar{T} = 0 \qquad [9.18]$$

where $\bar{T}$ is the cycle-average temperature of the mold.

- PIM component cooling analysis: without invoking a flow field, the energy equation is simplified as

$$\rho C_p \frac{\partial \bar{T}}{\partial t} = k \frac{\partial^2 \bar{T}}{\partial z^2} \qquad [9.19]$$

Typical initial and boundary conditions applied during the cooling stage are as follows:

- initial conditions: temperature as calculated from the packing stage analysis;
- boundary conditions for the mold: interface input from the PIM feedstock cooling analysis, convection heat transfer associated with the coolant, natural convection heat transfer with air, and thermal resistance condition from the mold platen;
- boundary conditions for the component: interface input from the mold cooling analysis.

Note that the boundary conditions for the mold and PIM feedstock cooling are coupled to each other. More details on this are given in Section 9.2.2 on numerical simulation below.

### 9.2.2 Numerical simulation

As far as the numerical analysis of injection molding is concerned, several numerical packages are already available for conventional thermoplastics, and one may try to apply the same numerical analysis techniques to PIM. However, the rheological behavior of a powder–binder feedstock mixture is significantly different from that of a thermoplastic. Hence, the direct application of methods developed for thermoplastics to PIM requires caution.[12, 13] Commercial software packages, including Moldflow (Autodesk, San Rafael, CA, USA), Moldex3D (CoreTech System Co., Ltd, Chupei City, Taiwan), PIMsolver (CetaTech, Sacheon, Korea), and SIMUFLOW (C-Solutions, Inc., Boulder, CO) are available for PIM simulation. These commercial software packages have been developed and are developing based on their own technical and historical backgrounds, which results in their own pros and cons. Further, several research groups have written customized codes, but generally these are not released for public use.

*Filling and packing analysis*

For numerical analysis of the filling and packing stages of PIM, both the pressure and energy equations must be solved during the entire filling and packing cycle. This is achieved using the FEM for equation [9.11] while a finite difference method (FDM) is used in the $z$-direction (thickness), making use of the same finite elements in the $x$–$y$ plane for solving equation [9.12].

The FDM is a relatively efficient and simple numerical method for solving differential equations. In this method, the physical domain is discretized in the form of finite-difference grids. A set of algebraic equations is generated as the derivatives of the partial differential equations and these are expressed by finite differences of the variable values at the grid points. The resulting algebraic equation array, which usually forms a banned matrix, is solved numerically. Generally, the solution accuracy is improved by reducing the grid spacing. However, since the FDM is difficult to apply to a highly irregular boundary or a complicated domain typical of injection molding, the use of this method has to be restricted to regular and simple domains, or used with the FEM as a FDM–FEM hybrid scheme.[5]

For the numerical analysis of the filling process of PIM, one has to solve both the pressure equation and energy equation during the entire filling cycle until the injection mold cavity is filled. A FEM method is employed to solve equations [9.11] and [9.12] while FDM is used in the thickness-wise or $z$-direction, making use of the same finite elements in the $x$–$y$ plane.[7]

*Cooling analysis*

For the numerical analysis of the cooling process in PIM, the boundary element method (BEM) is widely used due to its advantage in reduction of the dimensionality of the solution. The BEM discretizes the domain boundary rather than the interior of the physical domain. As a result, the volume integrals become surface integrals, then the number of unknowns, computation effort, and mesh generation are significantly reduced.[5] A standard BEM formulation for equation [9.18] based on Green's second identity leads to the following

$$\alpha T(\mathbf{x}) = \int_S \left[ \frac{1}{r} \left( \frac{\partial T(\zeta)}{\partial n} \right) - T(\zeta) \left( \frac{1}{r} \right) \right] dS(\zeta) \qquad [9.20]$$

Here $\mathbf{x}$ and $\mathbf{z}$ relate to the positional vector in the mold, $r = |\mathbf{z} - \mathbf{x}|$, and $\alpha$ denotes a solid angle formed by the boundary surface. Equation [9.18] for two closed surfaces, such as defined by the component shape, leads to a redundancy in the final system of linear algebraic equations, so a modified procedure is used.[14] For a circular hole, a special formulation is created

210    Handbook of metal injection molding

based on the line-sink approximation. This approach avoids discretization of the circular channels along the circumference and significantly saves computer memory and time. For the thermal analysis of a PIM component, the FDM is used with the Crack–Nicholson algorithm for time advancement. The mold and PIM component analyses are coupled with each other in boundary conditions so iteration is required until the solution converges.

*Coupled analysis between filling, packing, and cooling stages*

The filling, packing, and cooling analyses are coupled to each other. When we analyze the filling and packing stages, we have the cavity wall temperature as a boundary condition for the energy equation. This cavity wall temperature is obtained from the cooling analysis. On the other hand, when we analyze the cooling stage, we have the temperature distribution of the powder–binder feedstock mixture in the thickness direction at the end of filling and packing as an initial condition for the heat transfer of powder–binder feedstock. This initial temperature distribution is obtained from the filling and packing analysis. Therefore, the coupled analysis among the filling, packing, and cooling stages might be made for accurate numerical simulation results.[11]

Typical procedure for computer simulation of a powder injection molding process consists of three components: input data, analysis, and output data. The quality of the input data is essential to success. The pre-processor is a software tool used to prepare a geometric model and mesh for the component and mold; it includes material data for feedstock, mold, and coolant, and processing conditions for filling, packing, and cooling processes. Figure 9.4 shows one example of geometry modeling and mesh

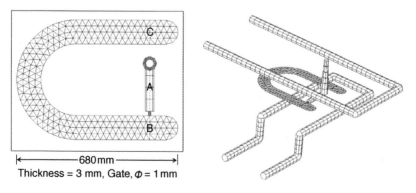

9.4 Geometry modeling and mesh generation for U-shape part including delivery system and cooling channels (pressure measurements at A, B, and C).[11]

generation for a U-shape component, including the delivery system and cooling channels.

### 9.2.3 Experimental – material properties and verification

*Material properties for the filling stage*

Successful simulation of the filling stage during PIM depends on measuring the material properties, including density, viscosity, and thermal behavior. Among these, the viscosity of the PIM feedstock and its variation with temperature, shear rate, and solid volume fraction are special concerns.[15–17] The following procedure is an example of the method used to obtain these material properties. For this illustration, assume a spherical stainless steel powder in a wax-polymer binder.

- First, melt densities, heat capacities, and thermal conductivities of the feedstock are required. These are obtained using a helium pycnometer, differential scanning calorimeter, and laser flash thermal conductivity device.
- Second, the transition temperature for the feedstock is measured, again using differential scanning calorimetry.
- Third, the feedstock viscosity is measured in relation to key parameters. The rheological behavior of the feedstock is measured by capillary rheometry. By using high length-to-diameter ratio capillaries, the pressure loss correction, called Bagley's correction, is avoided. Rabinowitch's correction is extracted to obtain the true shear rate from the apparent shear rate for a non-Newtonian fluid, characteristic of the feedstock.[13] The variation of viscosity with temperature is determined by testing the feedstock at different temperature above the transition temperature. Then, the feedstock viscosity data are modeled using standard loaded polymer concepts based on equations [9.2] and [9.3].

*Material properties for the packing stage*

To simulate the packing stage, the Hele–Shaw flow of a compressible viscous melt of PIM feedstock under non-isothermal conditions is assumed. For this, the two-domain modified Tait model is adopted to describe the phase behavior of the feedstock.[6] A dilatometer is used to measure dimensional changes as a function of temperature and other variables and the results are extracted by curve fitting.

Figure 9.5 gives one example of *pvT* material properties based on the following two-domain modified Tait model for the packing stage simulation

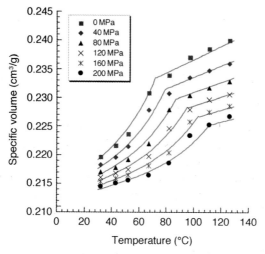

9.5 Pressure–volume–temperature (pvT) data of the feedstock.[11]

for a stainless steel feedstock. For the solid–liquid phase

$$v(p, T) = v_0(T)[1 - 0.0894 \ln(1 + p/B(T))] + v_t(p, T) \qquad [9.21]$$

with $v_0(T) = b_1 + b_2\bar{T}$, $B(T) = b_3 \exp(-b_4\bar{T})$, $\bar{T} = T - b_5$, $v_t(p, T) = b_7 \exp(b_8\bar{T} - b_9 P)$, and the transition temperature, $T_g$, which is calculated as $T_g(p) = b_5 + b_6 p$.

*Material properties for the cooling stage*

For the cooling stage simulation, material properties of mold material and coolant need to be measured.

*Verification*

Verification of the simulation is a critical step prior to any effort to optimize a design based on simulations. This verification usually includes the validation of the model used in developing the software. To demonstrate the verification of the simulation tool through experiment, we selected the U-shaped test mold shown in Fig. 9.4 with the stainless steel PIM feedstock reported above and an H13 mold. Three pressure transducers were used to compare the simulation results with the experimental data. The cavity thickness is 3 mm and the gate diameter is 1 mm. The coolant inlet temperature is 20°C, the inlet flow rate is 50 cm³/s, and the cooling time is 10 s.

Figure 9.6 describes some of the simulation results obtained using PIMsolver. Plate III (a) (see color section between pages 222 and 223) shows

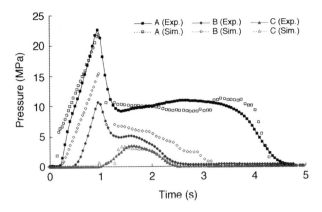

*9.6* Pressure–time plot at three points indicated in Fig. 9.4.[11]

the filling pattern, indicating how the mold cavity fills as a function of time. The filling time is 1.28 s. Plate III (b) shows the average mold-temperature distributions in K on the upper and lower surfaces of the cavity from the cooling analysis results. The highest average temperature is 51°C and occurs at the base of the U and the lowest temperature is 34°C at the runner inlet. The mold wall temperature is not uniform and the difference between maximum and minimum values is 18°C. This variation is large enough to cause a significant difference in the solidification layer development during the packing stage. Therefore, one might expect the simulation to have a significant error in pressure prediction during the packing stage without consideration of the cooling effect.

To examine the validation, the pressure traces are compared between simulation and experiment, as shown in Fig. 9.6 using the three positions indicated in Fig. 9.4. Figure 9.6 gives the pressure–time plot obtained from the experiment and simulation. The simulation results are obtained from the filling and packing analyses at a constant cavity wall temperature of 30°C. The results with the distributed mold wall temperature interfaced with the results from cooling analysis, as shown in Fig. 9.6, explain this deviation. If we consider the cavity wall temperature distributions from the cooling analysis (coupled analysis), then that temperature enables the best agreement to the experimental results.

### 9.2.4 Applications

This section presents simulation results from some of the 2.5D examples to demonstrate the usefulness of the CAE analysis and optimization capability of the PIM process. We will demonstrate how to use this basic information from the simulation tool to predict injection molding related defects and will present the systematic way of using the CAE tool to develop an optimal injection molding process.

*Basic capability – short shot, flash, weld line, air vent, and other features*

This section will demonstrate use of the simulation results to predict typical molding defects. There are many kinds of molding defects and we can identify them mainly as basic defects, dimensional defects, and other defects. The basic defects are traced to the molding parameters.

For the basic defects, simulations use the pressure field analysis to predict short shots and flashing, as well as the filling pattern to identify trapped air and weld line location. A short shot occurs when the molded part is filled incompletely because insufficient material is injected into the mold. Flash is a defect where excessive material is found at locations where the mold separates, notably the parting surface, moveable core, vents, or venting ejector pins. The causes of flash are low clamping force, gap within the mold, molding conditions, and improper venting.

Weld line and the resulting mark or knit line is another flaw that is also a potential weakness in a molded plastic part. Weld lines are formed by the union of two or more streams of feedstock flowing together, such as when flow passes around a hole, insert, or in the case of multiple gates or variable thickness in the component. Consequently, a weld line reduces the strength of the green component and leaves an undesirable surface appearance, and should be avoided when possible. The results from the computer analysis are used to predict the weld line location. An air trap or air vent is a defect by air that is caught inside the mold cavity. The air trap locations are usually in areas that fill last. The air trap is predicted from the filling pattern analysis and can be avoided by reducing the injection speed, and enlarging or properly placing vents. Other defects, such as burn marks, flow marks, meld lines, jetting, surface ripples, sink marks, and suchlike are also accessible using computer simulation tools in the design process stage. Especially important in production is the control of factors related to dimensional uniformity. These are analyzed by checking all three main stages of the injection molding. Plate IV (see color section between pages 222 and 223) shows some predicted defects from injection molded PIM components.

*Optimization of filling time*

Filling time is an important variable that is optimized to reduce the required injection pressure. From the CAE filling analysis, by varying the filling time the injection pressure is calculated. When plotting the required injection pressure versus various filling times, the optimal filling time ranges are

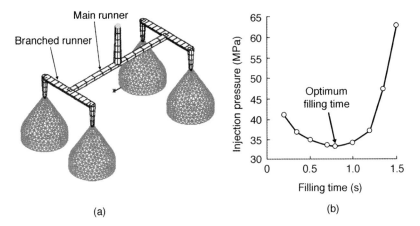

*9.7* Filling time optimization by using a CAE tool for PIM: (a) mesh used for optimization study; (b) optimum filling time finding which minimizes injection pressure.[18]

determined for the lowest injection pressure. The curve is U-shaped because, on the one hand, a short fill time involves a high melt velocity and thus requires a higher injection pressure to fill the mold. On the other hand, the injected feedstock cools more with a prolonged fill time. This results in a higher melt viscosity and thus requires a higher injection pressure to fill the mold. The shape of the curve of injection pressure versus fill time depends very much on the material used, as well as on the cavity geometry and mold design. If the required injection pressure exceeds the maximum machine capacity, the process conditions or runner system must be modified. Figure 9.7 shows the optimum filling time selection for the selected multi-cavity mold.

## 9.3 Modeling and simulation of the thermal debinding process

As PIM is a binder-assisted forming technique, removal of the binder without loss of product integrity is a crucial point. The process of binder removal is termed debinding. A multi-component binder can overcome distortion problems by leaving one component in the compact to hold the particles in place while a lower stability component is removed. Often, the lower molecular weight binder component is dissolved into a fluid in a process termed solvent debinding. Thermal debinding, a conventional heat treatment in a gas atmosphere, is used to remove higher molecular weight components of the binder systems. Previously, debinding cycles were based on 'trial and error' until an adequate time–temperature cycle was achieved.

With the use of a multi-component binder system, achieving an economical and effective debinding cycle while maintaining the compact's shape can be quite difficult. To counter this, thermogravimetric analysis (TGA) and differential thermal analysis (DTA) are used to determine the weight percent of the binder component left as a function of time and temperature. This protocol is to show heating at points of rapid weight loss to avoid compact damage. This expedites the debinding process and lowers the overall cost. But still, a large number of TGA and DTA experiments are needed to optimize the percentages of each binder component in a multi-component binder system. To minimize TGA experiments, a master decomposition curve (MDC) is proposed.[19]

### 9.3.1 Theoretical background and governing equations

*Master decomposition curve for a single reaction step*

Depolymerization of polymeric binders can be described by first-order reaction kinetics. The remaining weight fraction of a polymer, $\alpha$, is expressed as[20]

$$\frac{d\alpha}{dt} = -K\alpha \qquad [9.22]$$

where $t$ is the time and $K$ is the rate constant for thermal degradation and follows the Arrhenius equation

$$K = k_0 \exp\left[-\frac{Q}{RT}\right] \qquad [9.23]$$

in which $k_0$ is the specific rate constant, $Q$ is the apparent activation energy for thermal degradation, $R$ is the gas constant, and $T$ is the absolute temperature. Combining equations [9.22] and [9.23], and integrating

$$-\int_1^\alpha \frac{d\alpha}{\alpha} = -\ln\alpha = \int_0^t k_0 \exp\left[-\frac{Q}{RT}\right]dt = k_0\Theta \qquad [9.24]$$

where $\Theta$ is termed the work of decomposition, defined as follows

$$\Theta(t,T;Q) \equiv \int_0^t \exp\left[-\frac{Q}{RT}\right]dt \qquad [9.25]$$

The remaining weight fraction of a polymer, $\alpha$, is related to the work of decomposition $\Theta$ from equation [9.24] as follows

$$\alpha(\Theta;k_0) = \exp[-k_0\Theta] \qquad [9.26]$$

Modeling and simulation of metal injection molding 217

The above equation defines the MDC, wherein we can merge the decomposition curves for a given polymer with any decomposition cycles using the concept of work of decomposition, $\Theta$.

*Calculation of apparent activation energy*

To calculate the work of decomposition, $\Theta$ from equation [9.25], apparent activation energy, $Q$ needs to be determined. The activation energy for polymer decomposition can be determined from TGA data in a method developed by Kissinger.[21] The Kissinger method utilizes the temperature $T_{max}$ at which maximum rate of weight loss occurs at various heating rates as follows

$$\frac{d}{dt}\left(-\frac{d\alpha}{dt}\right) = 0 \text{ and } T = T_{max} \text{ at the maximum rate of weight loss}$$

[9.27]

Under the condition of constant heating rate $r$, that is $dT/dt = r$, equation [9.27] can be expressed as

$$\frac{rQ}{RT_{max}^2} = k_0 \exp(-Q/RT_{max}) \text{ at } T = T_{max}$$

[9.28]

or

$$\ln\left[\frac{r}{T_{max}^2}\right] = Q\left(-\frac{1}{RT_{max}}\right) - \ln\left[\frac{Q}{k_0 R}\right] \text{ at } T = T_{max}$$

[9.29]

Thus from equation [9.29], we can plot a graph between $\ln[r/(T_{max})^2]$ and $-1/RT_{max}$ from the TGA experiments with several constant heating rates. The slope of this plot gives the apparent activation energy, $Q$ of the reaction.

*MDC for multi-reaction steps*

Typically, the TGA curves of polymer degradation follow a single sigmoid path. But multi-component binder systems may have two or more sigmoids due to the different molecular weights, bonding groups, and degradation paths of various polymer components. Each sigmoid represents a rate-controlling step whose activation energy differs from those of other controlling steps. Powder–binder systems tailored for PIM often consist of several components. These components may be categorized mainly into two groups according to their molecular weight. Low molecular weight ingredients, such as solvents and plasticizers, either evaporate or are decomposed at rather low temperatures. High molecular weight polymers, possessing higher thermal stability, decompose at relatively high temperatures. Consequently the TGA curve often has more than two sigmoids. One sigmoid of this curve could be a superposition of several sigmoids that have similar activation energies. Two sigmoids of this curve could mean two groups of reaction steps with sufficiently separate activation energies. Each sigmoid may be described by three kinetic parameters in an Arrhenius-type equation. These parameters are reaction order, activation energy, and frequency factor. Some powders may have catalytic effects on the pyrolysis rate; however, the shape of the pyrolysis curve with powders is similar to that without powders. Therefore the mathematical form of polymer pyrolysis can still be applied. Two polymers of the binder system used in this study are decomposed during the thermal debinding process.

$$\alpha = \omega\alpha_1 + (1 - \omega)\alpha_2 \quad\quad [9.30]$$

where $\alpha$ is the mass ratio of the initial masses of two polymers, $\alpha_1$ is the mass ratio to initial mass of low molecular weight polymer, $\alpha_2$ is the mass ratio to initial mass of high molecular weight polymer, and $\omega$ is the ratio of initial mass of low molecular weight polymer to initial mass of the two polymers.

The mathematical form that is generally applied to the TGA curve of polymer is modified to describe the TGA curve of organic pyrolysis with powders. In the present analysis we assume that the kinetics are solely controlled by the temperature.

$$-\frac{1}{k_0\beta}d\alpha = \exp\left[-\frac{Q}{R}\left(\frac{1}{T} - \frac{1}{T_t}\right)\right]dt \quad\quad [9.31]$$

if $T \geq T_t$ (or $\alpha \geq \omega$), $\alpha = \omega\alpha_1 + (1 - \omega), \beta = \dfrac{\alpha + \omega - 1}{\omega}, k_0 = k_{01}, Q = Q_1$

if $T < T_t$ (or $\alpha < \omega$), $\alpha = (1 - \omega)\alpha_2, \beta = \dfrac{\alpha}{1 - \omega}, k_0 = k_{02}, Q = Q_2$

where $T_t$ is the transition temperature between the first and second sigmoids, subscripts 1 and 2 on $\alpha$, $Q$, and $k_0$ denote low- and high-temperature degradation, respectively. A decomposition kinetic curve with more than two sigmoids can be expressed in a similar manner. The Kissinger method is applied to estimate the activation energy, as described in the previous section.

The left-hand side of equation [9.31] is a function of only the mass ratio $\alpha$

*Table 9.1* Characteristics of binder components used in the binder system[22]

| Binder | Paraffin wax | Polypropylene | Polyethylene | Stearic acid |
|---|---|---|---|---|
| Density (g/cm$^3$) | 0.90 | 0.90 | 0.92 | 0.94 |
| Melting (°C) | 42–62 (peak 58) | 110–150 (peak 144) | 60–130 (peak 122) | 74–83 (peak 78) |
| Decomposition (°C) | 180–320 | 350–470 | 420–480 | 263–306 |
| Activation energy, $Q$ (kJ/mol) | 77 | 241 | 299 | 100 |
| Specific rate constant $k_0$ (s$^{-1}$) | $3.330 \times 10^4$ | $2.19 \times 10^{15}$ | $8.12 \times 10^{18}$ | $2.25 \times 10^7$ |

and material properties except $Q$, which then becomes

$$\Phi(\alpha; k_0) \equiv \int_1^\alpha -\frac{1}{k_0 \beta} d\alpha \qquad [9.32]$$

The right side of equation [9.31] is denoted by

$$\Theta(t, T; Q, T_t) \equiv \int_0^t \exp\left[-\frac{Q}{R}\left(\frac{1}{T} - \frac{1}{T_t}\right)\right] dt \qquad [9.33]$$

which depends only on $Q$, $T_t$, and the time–temperature profile. Note $\Phi(\alpha)$ is a characteristic that quantifies the effects of binder system components on the decomposition kinetics. The relationship between $\alpha$ and $\Phi(\alpha)$ is defined as the MDC, which is unique for a given powder and binder system and is independent of the decomposition path, given the assumptions described above.

### 9.3.2 Applications

*Calculation of activation energy*

A typical example of wax-polymer binder used in PIM is listed in Table 9.1 with characterization of each binder component. TGA is a mass change method in which the mass of the sample is measured as a function of temperature while the sample is subjected to a controlled temperature program. This can be achieved as a function of the increasing temperature or isothermally as a function of time with a controlled atmosphere. The TGA is performed with results for each binder component at three different heating rates to calculate the activation energy by the Kissinger method. This activation energy is then used to develop the MDC.

Figure 9.8(a) shows the TGA of polypropylene at three different heating rates. It is evident that the decomposition range is 350–490°C. Figure 9.8(b) gives the peak temperatures at maximum weight loss of polypropylene.

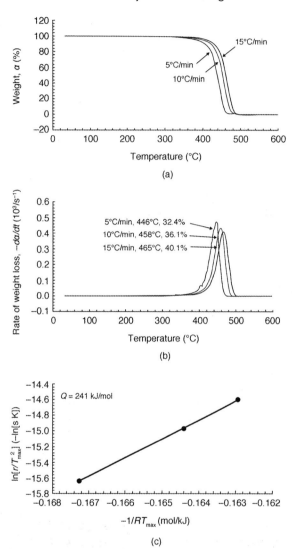

9.8 TGA results and Kissinger method for calculating activation energy for polypropylene.[22] (a) TGA result; (b) temperatures at maximum rate of weight loss; (c) temperature dependency.

These are plugged into equation [9.29] to plot a graph between $\ln[r/(T_{max})^2]$ and $-1/RT_{max}$ (Figure 9.8(c)). The slope of this graph gives the apparent activation energy for polypropylene. Table 9.1 gives the material parameters of each binder component for plotting the MDC.

Modeling and simulation of metal injection molding 221

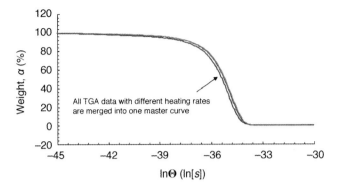

*9.9* MDC for polypropylene, showing all TGA.

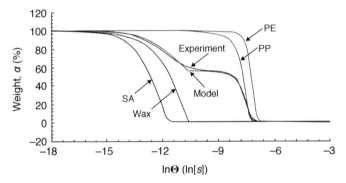

*9.10* Synthesis of overall decomposition behavior for the binder system.[22]

*MDC for single reaction-step decomposition*

Equations [9.31], [9.32], and [9.33] are then used to plot the MDC for polypropylene (Fig. 9.9). Polypropylene exhibits a similar decomposition behavior at all heating rates. Similarly MDCs for other binder components (paraffin wax, polyethylene, and stearic acid) were generated from their respective TGAs (not shown here) based on calculated activation energies by the Kissinger method. The overall MDC for the binder system is plotted from the individual MDCs of its components. Individual decomposition behavior of binder components is important in basic understanding but decomposition behavior of the whole binder system is of more interest in PIM. This behavior is shown in Fig. 9.10 for the binder system. It can also be seen that the model is in close agreement with the experimental results. In addition, the remaining weight of each binder component can be predicted and monitored at any time–temperature combination based on the constructed MSCs.

*MDC for multi-reaction step decomposition*

The decomposition behavior of the feedstock during thermal debinding is imperative in deciding the debinding cycle for the injection molded parts. This makes MDC for the feedstock an attractive tool in both cost and effort optimization. To demonstrate this concept and find out the processing parameters for debinding, niobium feedstock (57% solids loading) is used.[23, 24] Figure 9.11 shows the niobium feedstock MDC. A two-step decomposition behavior is observed in which during the first step, the lower molecular weight binder components (wax and stearic acid) decompose and later the higher molecular weight binder components (polypropylene and polyethylene) go away. It is also observed that the decomposition is rate dependent, i.e. at a particular temperature more binder remains at a higher heating rate. This is not the case for the pure binder. There is an effect of powder on the decomposition behavior.

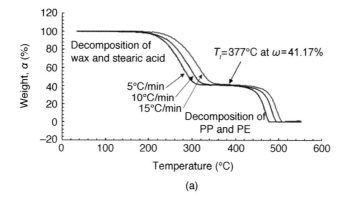

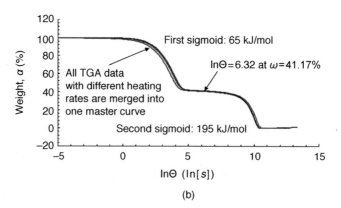

*9.11* Multi-reaction step MDC of Nb feedstock.[22] (a) TGA for feedstock 1; (b) MDC for feedstock 1.

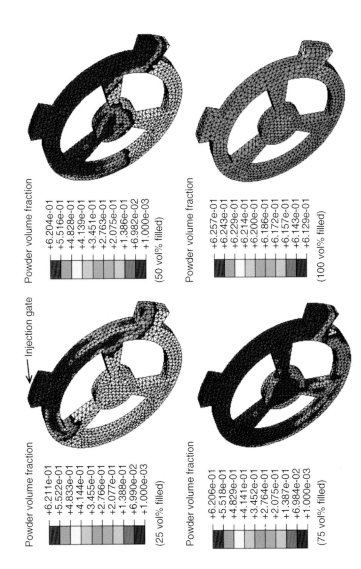

*Plate 1* Study of mold filling by computer simulation (European Powder Metallurgy Association).

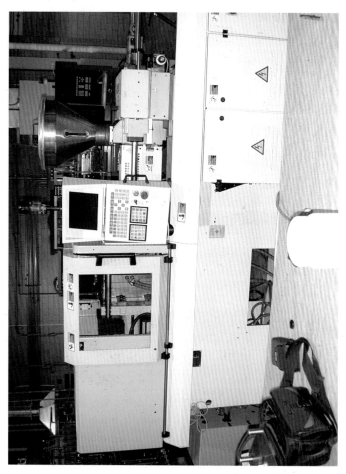

*Plate II* Typical hydraulic reciprocating screw injection molding machine used for metal injection molding.

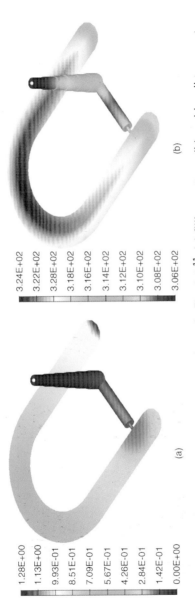

*Plate III* Simulation results with geometry input in Fig. 9.4.[11]: (a) filling pattern; (b) mold wall temperature.

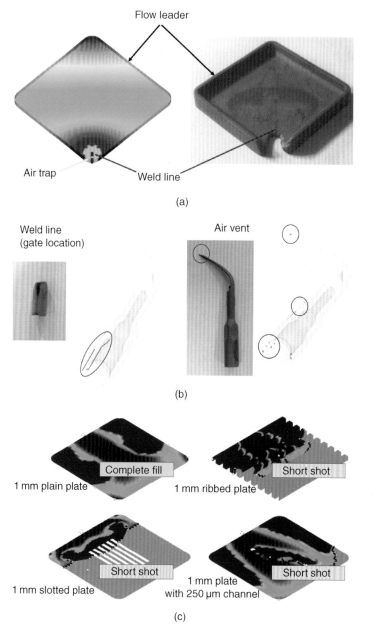

*Plate IV* Some molding defects predicted by using a CAE tool for PIM process: (a) plate;[19] (b) dental scaler tip;[20] (c) geometry evaluation of micro features.[21]

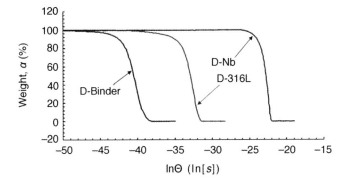

*9.12* An illustration of catalytic effect of metal powders on decomposition behavior.[22]

*Effect of metal powders*

Figure 9.12 shows the MDC of solvent debound samples of pure binder (D-Binder), Nb feedstocks (D-Nb), and 316L stainless steel feedstock (D-316L). It can be observed that for a particular binder wt%, a lower temperature (here higher work of decomposition, $\Theta$) is required for the feedstocks (D-Nb and D-316L) than just the binder without any powder. This seems reasonable as the presence of powder will enhance the debinding process and a higher work of decomposition (lower temperature) will be needed, which results from the catalytic effect or faster heat transfer due to higher thermal conductivity of the metal powder (54 W/m K for Nb, 12 W/m K for 316L, versus 0.1 W/m K for typical polymer). Another difference can also be considered between MDCs of D-Nb and D-316L. This could also explain the effect of powder shape on the debinding process. D-Nb is a niobium feedstock which has irregularly shaped Nb particles whereas D-316L is a 316L feedstock having spherical shaped powder particles. Both feedstocks have the same solids loading (57 vol%). Spherical powder particles have a lower surface area as compared to the irregular shaped powders.[21, 23] It is also evident from the MDC (Fig. 9.12) that lower work of decomposition is required for debinding D-Nb (irregular) than D-316L (spherical shape). Finally, it is concluded that the catalytic, thermal conductivity, and particle shape effects are combined in decomposition behavior of the binder system with metal powder. Detailed analyses will be required for more accurate explanation.

*Weight–temperature–time plot*

Figure 9.13 shows a weight–temperature–time plot for decomposition of solvent debound 316L feedstock in hydrogen atmosphere. Based on this information, the required time hold can be predicted for a given temperature

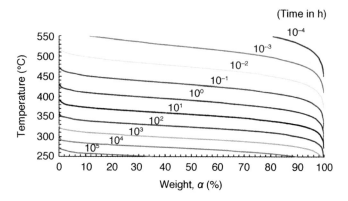

*9.13* Weight–temperature–time plot for decomposition of solvent debound 316L feedstock.[22]

and targeted binder weight loss. This will help in reducing the experimentation required to determine the optimum debinding cycle.

## 9.4 Modeling and simulation of the sintering process

Effective computer simulations of metal powder sintering are at the top of the powder metallurgy industry's wish list. Since the pressed green body is not homogeneous, backward solutions are desired to select the powder, mixing, injection molding, debinding, and sintering attributes required to deliver the target properties with different tool designs, machines, and processing conditions. In building toward this goal, various simulation types have been evaluated: Monte Carlo, finite difference, discrete element, finite element, fluid mechanics, continuum mechanics, neural network, and adaptive learning. Unfortunately, the input data and some of the basic relations are not well developed; accurate data are missing for most materials under the relevant conditions. The simulations help to define the processing window and set initial operating parameters.

### 9.4.1 Theoretical background and governing equations

The methodologies used to model the sintering include continuum, micromechanical, multi-particle, and molecular dynamics approaches. These differ in length scales. Among the methodologies, continuum models have the benefit of shortest computing time, with an ability to predict relevant attributes such as the component density, grain size, and shape.

## Constitutive relation during sintering

Continuum modeling is the most relevant approach to modeling grain growth, densification, and deformation during sintering. Key contributions have been made by Ashby and coworker,[25, 26] McMeeking and Kuhn,[27] Tikare et al.,[28, 29] Riedel and coworkers,[30–32] Bouvard and Meister,[33] Cocks,[34] Kwon and coworkers,[35,36] and Bordia and Scherer[37–39] based on sintering mechanisms such as surface diffusion, grain boundary diffusion, volume diffusion, viscous flow (for amorphous materials), plastic flow (for crystalline materials), evaporation–condensation, and rearrangement. For industrial application, the phenomenological models are used for sintering simulations with the following key physical parameters.

- Sintering stress[40] is a driving force of sintering due to interfacial energy of pores and grain boundaries. Sintering stress depends on the material's surface energy, density, and geometric parameters such as grain size when all pores are closed in the final stage.
- Effective bulk viscosity is a resistance to densification during sintering and is a function of the material, porosity, grain size, and temperature. The model of the effective bulk viscosity has various forms according to the assumed dominant sintering mechanism.
- Effective shear viscosity is a resistance to deformation during sintering and is also a function of the material, porosity, grain size, and temperature. Several rheological models for the effective bulk viscosity are available.

The above parameters are a function of grain size. Therefore, a grain growth model is needed for accurate prediction of densification and deformation during sintering.

Typical initial and boundary conditions for the sintering simulations include the following:

- initial condition: mean particle size and grain size of the green compact for grain growth and initial green density distribution for densification obtained from compaction simulations;
- boundary conditions: surface energy condition imposed on the free surface and friction condition of the component depending on its size, shape, and contact with the support substrate.

The initial green density distribution within the pressed body raises the necessity to start the sintering simulation with the output from an accurate compaction simulation, since die compaction induces green density gradients that depend on the material, pressure, rate of pressurization, tool motions, and lubrication. The initial and boundary conditions help determine the shape distortion during sintering from gravity, non-uniform heating, and from the green body density gradients.

*Numerical simulation*

Even though many numerical methods have been developed, the FEM is most popular for continuum models of the press and sinter process. The FEM approach is a numerical computational method for solving a system of differential equations through approximation functions applied to each element, called domain-wise approximation. This method is very powerful for the typical complex geometries encountered in powder metallurgy. This is one of the earliest techniques applied to materials modeling, and is used throughout industry today. Many powerful commercial software packages are available for calculating two-dimensional (2D) and 3D thermo-mechanical processes such as those found in the sintering process. To increase the accuracy and convergence speed for the sintering simulations, developers of the simulation tools have selected explicit and implicit algorithms for time advancement, as well as numerical contact algorithms for problems such as surface separation, and remeshing algorithms as required for large deformations such as seen in some sintered materials, where up to 25% dimensional contraction is possible.

### 9.4.2 Experimental determination of material properties and simulation verification

In the development of a constitutive model for the sintering simulation, a wide variety of tests are required, including data on grain growth, densification (or swelling), and distortion. These are approached as follows.

- Grain growth: quenching tests are conducted from various points in the heating cycle and the mounted cross-sections are analyzed to obtain grain size data to implement grain growth models. A vertical quench furnace is used to sinter the compacts to various points in the sintering cycle and then to quench those compacts in water. This gives density, chemical dissolution (for example diffusion of one constituent into another), and grain size as instantaneous functions of temperature and time. The quenched samples are sectioned, mounted, and polished prior to optical or scanning electron microscopy (SEM). Today, automated quantitative image analysis provides rapid determination of density, grain size, and phase content versus location in the compact. Usually during sintering the mean grain size $G$ varies from the starting mean grain size $G_0$ (determined on the green compact). A new master sintering curve concept is applied to fit the experimental grain size data to an

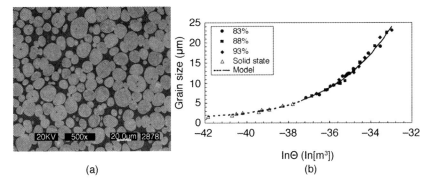

*9.14* (a) SEM of a liquid phase sintered tungsten heavy alloy after quenching; (b) the grain size model results taken from an integral work of sintering concept.[41]

integral work of sintering,[41] since actual cycles are a complex combination of heats and holds. The resulting material parameters trace to an apparent activation energy as the only adjustable parameter. Figure 9.14 shows an SEM micrograph after quenching test and grain growth modeling for a W–8.4 wt% Ni–3.6 wt% Fe mixed powder compact during liquid phase sintering (LPS).

- Densification: to obtain material parameters for densification, constant heating rate dilatometry is used for *in situ* measurement of shrinkage, shrinkage rate, and temperature. By fitting the experimental data to models that include the sintering stress $\sigma_s$ and bulk viscosity $K$ as functions of density and grain size, again relying on the master sintering curve concept,[42] the few unknown material parameters are extracted. Figure 9.15 shows the dilatometry data and model curve-fitting results used to obtain the parameters during sintering of a 316L stainless steel.[43]

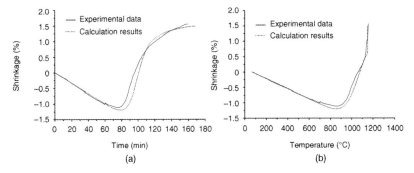

*9.15* Dilatometry data showing *in situ* shrinkage data during constant heating rate experiments and the curve-fitting results used to obtain the

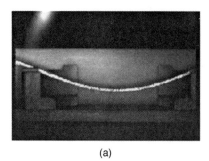 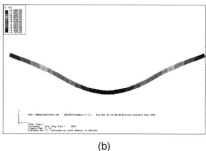

(a)                             (b)

*9.16* (a) A video image taken during the *in situ* bending test for a 316L stainless steel sample doped with 0.2 wt% boron; (b) the FEM model results used to verify the shear viscosity property as a function of time, temperature, grain size, and density during heating.[44]

- Distortion: powder metallurgy compacts reach very low strength levels during sintering. Accordingly, weak forces such as gravity, substrate friction, and non-uniform heating will induce distortion and even cracking. To obtain the material parameters related to distortion, three-point bending or sinter forging experiments are used for *in situ* measurement of distortion.[44] By fitting the experimental data with FEM simulations for shear viscosity $\mu$ with grain growth, the parameters such as apparent activation energy and reference shear viscosity are extracted. Figure 9.16 shows *in situ* bending test and FEM results for obtaining material parameters in shear viscosity for a 316L stainless steel powder doped with 0.2% boron to induce improved sintering.

Such data extraction techniques have been allied to several materials, ranging from tungsten alloys, molybdenum, zirconia, cemented carbides, niobium, steel, stainless steel, and alumina. Table 9.2 is an example set of material properties for W–8.4 wt% Ni–3.6 wt% Fe as used as input data for the sintering simulation.[45] The above experimental techniques can be used for verification of sintering simulation results.

### 9.4.3 Applications

*Gravitational distorting in sintering*

The rheological data for the sintering system allow the system to respond to the internal sintering stress that drives densification and any external stress, such as gravity, that drives distortion. When a compact is sintered to high density it is also necessary to induce a low strength (the material is thermally softened to a point where the internal sintering stress can induce densification). Figure 9.17 shows sintering simulation results for a tungsten

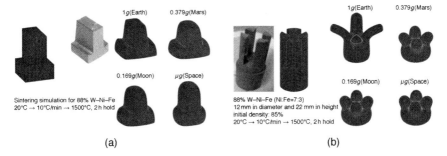

*9.17* Final distorted shape by sintering under various gravitational environments for complicated test geometries.[45] (a) T-shape; (b) joint part.

*Table 9.2* Complete set of material properties for the sintering of W–8.4 wt% Ni–3.6 wt% Fe[45]

| Sintering stress | $\sigma_s = \dfrac{6\gamma}{G}\dfrac{\rho^2(2\rho - \rho_0)}{\theta_0}$ for $\rho < 0.85$ |
|---|---|
| | $\sigma_s = \dfrac{2\gamma}{G}\left(\dfrac{6\rho}{\theta}\right)^{1/3}$ for $\rho W > 0.95$ |
| | $\sigma_s = \dfrac{(\rho_2 - \rho)}{(\rho_2 - \rho_1)}\sigma_{si} + \dfrac{(\rho - \rho_1)}{(\rho_2 - \rho_1)}\sigma_{sf}$ for $0.85 \leq \rho_W \leq 0.95$ |

| State | | Solid | Liquid |
|---|---|---|---|
| Grain growth | $\dfrac{dG}{dt} = \dfrac{k_0 \exp(-Q_G/RT)}{G^l}$ | | |
| | $k_0(m^{n+1}/s)$ | $2.8 \times 10^{-13}$ | $1.1 \times 10^{-15}$ |
| | $Q_G\,\text{kJ/mol}$ | 241 | 105 |
| | $l$ | 2.0 | 2.0 |
| Bulk viscosity | $K_i = \dfrac{\rho(\rho - \rho_0)^2}{8\theta_0^2}\dfrac{TG^3}{\alpha_i \exp(-Q_D/RT)}$ for $\rho \leq 0.92$ | $Q_D(\text{kJ/mol})$ 250 | 250 |
| | $K_f = \dfrac{\rho}{8\theta_0^{1/2}}\dfrac{TG^3}{\alpha_f \exp(-Q_D/RT)}$ for $\rho > 0.92$ | | |
| | with | | |
| | $\alpha_f = \dfrac{\theta_0^2}{\sqrt{0.08}(0.92 - \rho_0)^2}\alpha_i$ | $\alpha_i(\text{m}^6\,\text{K/s})$ $1.3 \times 10^{-17}$ | $5.0 \times 10^{-17}$ |
| Shear viscosity | $\mu_i = \dfrac{\rho^2(\rho - \rho_0)}{8\theta_0}\dfrac{TG^3}{\beta_i \exp(-Q_D/RT)}$ for $\rho \leq 0.92$ | | |
| | $K_f = \dfrac{\rho}{8}\dfrac{TG^3}{\beta_f \exp(-Q_D/RT)}$ for $\rho > 0.92$ | | |
| | with | $\beta_i(\text{m}^6\,\text{K/s})$ $1.3 \times 10^{-17}$ | $1.3 \times 10^{-12}$ |
| | $\beta_f = \dfrac{\theta_0}{0.92(0.92 - \rho_0)}\beta_i$ | | |

heavy alloy, relying on test data taken on Earth and under microgravity conditions, to then predict the expected shapes for various gravitational conditions – Earth, Moon, Mars, and in space. The results show that gravity affects shape distortion during sintering.[35] Accordingly, the computer simulations can be used to reverse engineer the green component geometry to anticipate the distortion to achieve the desired sintered part design.

*Sintering optimization*

Usually a small grain size is desired to improve properties for a given sinter density. In this illustration, the design variable is the sintering cycle. To obtain maximum density and minimum grain size, the following objective function $F$ is proposed[46]

$$F = \alpha \left[\frac{\Delta \rho}{\rho}\right] + (1 - \alpha) \left[\frac{\Delta G}{G}\right] \quad [9.34]$$

where $\alpha$ is an adjustable parameter. Figure 9.18 shows an example for maximum density and minimum grain size for a 17-4 PH stainless steel powder. For example, the minimum grain size will be 21.9 μm if the specified sintered density is 95% or theoretical. Figure 9.18(b) shows the corresponding sintering cycle by matching the value of adjustable parameter $\alpha$.

## 9.5 Conclusion

Computer simulations of the PIM have advanced considerably, and in combination with standard finite element techniques show a tremendous ability to guide process set-up. Illustrated here are the PIM concepts required to perform process optimization. Although the models are only approximations to reality, still they are of value in forcing a careful inspection of what is understood about the PIM process. In this regard, the greatest value of modeling is in the forced organization of process knowledge. There remain several barriers to widespread implementation. The largest is that traditional powder metallurgy is largely dependent on adaptive process control since many of the important factors responsible for dimensional or quality variations are not measured. The variations in particle size, composition, tool wear, furnace location, and other factors such as reactions between particles during heating, all impact the important dimensional control aspects of PIM, although nominal properties, such as strength, hardness, or fatigue life are dominated by the average component density. In this regard, especially with respect to the initial process set-up, the computer simulations are of great value. Still, important attributes such as dimensional tolerances and internal cracks or other defects are outside

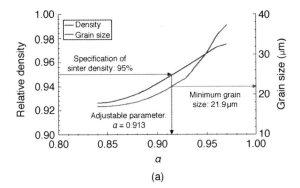

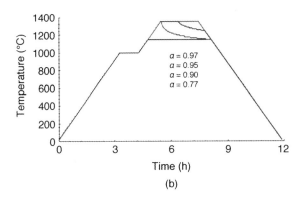

*9.18* Minimum grain size for a given final sinter density and the corresponding sintering cycle for achieving this goal in a 17-4 PH stainless steel.[46] (a) sintered density and minimum grain size; (b) sintered cycle.

the cost–benefit capabilities of existing simulations. Furthermore, the very large number of materials, processes, tool materials, sintering furnaces, and process cycles makes it difficult to generalize; significant data collection is required to reach the tipping point where the simulations are off-the-shelf. Thus, much more research and training is required to move the simulations into a mode where they are widely applied in practice. Even so, commercial software is available and shows great value in the initial process definition to set up a new component.

## 9.6 References

1. Kang, T. G. and Kwon, T. H. (2004), 'Colored particle tracking method for mixing analysis of chaotic micromixers', *Journal of Micromechanics and Microengineering*, 14, 891–899.

2. Schenk, O. and Gärtner, K. (2004), 'Solving unsymmetric sparse systems of linear equations with PARDISO', *Future Generation Computer Systems*, 20, 475–487.
3. Shannon, C. E. (1948), 'The mathematical theory of communication', *Bell Systems Technical Journal*, 27, 379–423.
4. German, R. M. (2003), *PIM Design and Applications User's Guide*, Innovative Material Solutions, State College, PA, USA.
5. Kim, S. and Turng, L. (2004), 'Developments of three-dimensional computer-aided engineering simulation for injection molding', *Modelling and Simulation in Materials Science and Engineering*, 12, 151–173.
6. Chiang, H. H., Hieber, C. A., and Wang, K. K. (1991), 'A unified simulation of the filling and post filling stages in injection molding. Part I: Formulation', *Polymer Engineering and Science*, 31(2), 116–124.
7. Hieber, C. A. and Shen, S. F. (1980), 'A finite-element/finite-difference simulation of the injection molding filling process', *Journal of Non-Newtonian Fluid Mechanics*, 7(1), 1–32.
8. Park, S. J. and Kwon, T. H. (1996), 'Sensitivity analysis formulation for three-dimensional conduction heat transfer with complex geometries using a boundary element method', *International Journal for Numerical Methods in Engineering*, 39, 2837–2862.
9. Park, S. J. and Kwon, T. H. (1998), 'Optimal cooling system design for the injection molding process', *Polymer Engineering and Science*, 38, 1450–1462.
10. Park, S. J. and Kwon, T. H. (1998), 'Thermal and design sensitivity analyses for cooling system of injection mold, Part 1: Thermal analysis', *ASME Journal of Manufacturing Science and Engineering*, 120, 287–295.
11. Ahn, S., Chung, S. T., Atre, S. V., Park, S. J., and German, R. M. (2008), 'Integrated filling, packing, and cooling CAE analysis of powder injection molding parts', *Powder Metallurgy*, 51, 318–326.
12. Hwang, C. J. and Kwon, T. H. (2002), 'A full 3D finite element analysis of the powder injection molding filling process including slip phenomena', *Polymer Engineering and Science*, 42(1), 33–50.
13. Kwon, T. H. and Ahn, S. Y. (1995), 'Slip characterization of powder-binder mixtures and its significance in the filling process analysis of powder injection molding', *Powder Technology*, 85, 45–55.
14. Rezayat, M. and Burton, T. (1990), 'A boundary-integral formulation for complex three-dimensional geometries', *International Journal of Numerical Methods in Engineering*, 29, 263–273.
15. Khakbiz, M., Simchi, A., and Bagheri, R. (2005), 'Investigation of rheological behaviour of 316L stainless steel-2 wt-% TiC powder injection moulding feedstock', *Powder Metallurgy*, 48(2), 144–150.
16. Berginc, B., Kampus, Z., and Sustarsic, B. (2007), 'Influence of feedstock characteristics and process parameters on properties of MIM parts made of 316L', *Powder Metallurgy*, 50(2), 172–183.
17. Suri, P. German, R. M., de Souza, J. P. and Park, S.J. (2004), 'Numerical analysis of filling stage during powder injection moulding: Effects of feedstock rheology and mixing conditions', *Powder Metallurgy*, 47(2), 137–143.
18. PIMSolver User's Guide, Cetatech, Ver. 2.0, 2005.
19. Hwang, C. J., Ko, Y. B., Park, H. P., Chung, S .T., and Rhee, B. O. (2007),

'Computer aided engineering design of powder injection molding process for a dental scaler tip mold design', *Materials Science Forum*, 534–536, 341–344.
20. Shi, Z., Guo Z. X., and Song, J. H. (2002), 'Modeling of binder removal from a (fibre+powder) composite pre-form', 50, 1937–1950.
21. Kissinger, H. E. (1957), 'Reaction kinetics in differential thermal analysis', *Analytical Chemistry*, 29, 1702–1706.
22. Aggarwal, G., Park, S. J., Smid, I., and German, R. M. (2007), 'Master decomposition curve for binders used in powder injection molding', *Metallurgical and Materials Transactions A*, 38A(3), 606–614.
23. Aggarwal, G. and Smid, I. (2005), 'Powder injection molding of niobium', *Materials Science Forum*, 475–479, 711–716.
24. Aggarwal, G., Park, S. J., and Smid, I. (2006), 'Development of niobium powder injection molding: Part I. Feedstock and injection molding', *International Journal of Refractory Metals and Hard Materials*, 24, 253–262.
25. Ashby, M. F. (1974), 'A first report on sintering diagrams,' *Acta Metallurgica*, 22, 275–289.
26. Swinkels, F. B. and Ashby, M. F. (1981), 'A second report on sintering diagrams', *Acta Metallurgica*, 29, 259–281.
27. McMeeking, R. M. and Kuhn, L. (1992), 'A diffusional creep law for powder compacts', *Acta Metallurgica et Materialia*, 40, 961–969.
28. Tikare, V., Braginsky, M. V., Olevsky, E. A., and Dehoff, R. T. (2000), 'A combined statistical-microstructural model for simulation of sintering', in *Sintering Science and Technology*, eds R. M. German, G. L. Messing, and R. G. Cornwall, Pennsylvania State University, State College, PA, pp. 405–409.
29. Tikare, V., Olevsky, E. A., and Braginsky, M. V. (2001), 'Combined macro-meso scale modeling of sintering. Part II, Mesoscale simulations', in *Recent Developments in Computer Modeling of Powder Metallurgy Processes*, eds A. Zavaliangos and A. Laptev, ISO Press, Ohmsha, Sweden, pp. 94–104.
30. Riedel, H., Meyer, D., Svoboda, J., and Zipse, H. (1994), 'Numerical simulation of die pressing and sintering – development of constitutive equations', *International Journal of Refractory Metals and Hard Materials*, 12, 55–60.
31. Kraft, T. and Riedel, H. (2002), 'Numerical simulation of die compaction and sintering', *Powder Metallurgy*, 45, 227–231.
32. McHugh, P. E. and Riedel, H. (1997), 'A liquid phase Sintering model: Application to Si3N4 and WC-Co', *Acta Metallurgica et Materialia*, 45, 2995–3003.
33. Bouvard, D. and Meister, T. (2000), 'Modelling bulk viscosity of powder aggregate during sintering', *Modelling and Simulation in Materials Science Engineering*, 8, 377–388.
34. Cocks, C. F. (1994), 'The structure of constitutive laws for the sintering of fine grained materials', *Acta Metallurgica et Materialia*, 45, 2191–2210.
35. Kwon, Y. S. and Kim, K. T. (1996), 'High temperature densification forming of alumina powder – constitutive model and experiments', *Journal of Engineering Materials and Technology*, 118, 448–455.
36. Kwon, Y. S., Wu, Y., Suri, P., and German, R. M. (2004), 'Simulation of the sintering densification and shrinkage behavior of powder-injection-molded 17-4 PH stainless steel', *Metallurgical and Materials Transactions A*, 35, 257–263.

37. Bordia, R. K. and Scherer, G. W. (1998), 'On constrained sintering – I. Constitutive model for a sintering body', *Acta Metallurgica*, 36, 2393–2397.
38. Bordia, R. K. and Scherer, G. W. (1998), 'On constrained sintering – II. Comparison of constitutive models', *Acta Metallurgica*, 36, 2399–2409.
39. Bordia, R. K. and Scherer, G. W. (1998), 'On constrained sintering–III. Rigid inclusions', *Acta Metallurgica*, 36, 2411–2416.
40. German, R. M. (1996), *Sintering Theory Practice*, John Wiley, New York, USA.
41. Park, S. J., Martin, J. M., Guo, J. F., Johnson, J. L., and German, R. M. (2006), 'Grain growth behavior of tungsten heavy alloys based on master sintering curve concept', *Metallurgical and Materials Transactions A*, 37A, 3337–3343.
42. Park, S. J., Martin, J. M., Guo, J. F., Johnson, J. L., and German, R. M. (2006), 'Densification behavior of tungsten heavy alloy based on master sintering curve concept', *Metallurgical and Materials Transactions A*, 37A, 2837–2848.
43. Chung, S. H., Kwon, Y. S., Binet, C., Zhang, R., Engel, R. S., Salamon, N. J., and German, R. M. (2002), 'Application of optimization technique in the powder compaction and sintering processes', *Advances in Powder Metallurgy and Particulate Materials, Part, 9*, 9-131–9-146.
44. Blaine, D., Bollina, R., Park, S. J., and German, R. M. (2005), 'Critical use of video-imaging to rationalize computer sintering simulation models', *Computers in Industry*, 56, 867–875.
45. Park, S. J., Chung, S. H., Johnson, J. L., and German, R. M. (2006), 'Finite element simulation of liquid phase sintering with tungsten heavy alloys', *Materials Transactions*, 47, 2745–2752.
46. Park, S. J., German, R. M., Suri, P., Blaine, D., and Chung, S. H. (2004), 'Master sintering curve construction software and its application', *Advances in Powder Metallurgy and Particulate Materials, Part 1*, 1-13–1-24.

# 10
# Common defects in metal injection molding (MIM)

K.S. HWANG, National Taiwan University, Taiwan, R.O.C.

**Abstract:** Inconsistent product quality, including poor dimension control, distortion, and internal and external defects, has tended to be underestimated by MIM practitioners. These defects may originate in the early processing steps, but they often do not manifest until after debinding or sintering. Thus, the solutions are difficult to provide. This chapter presents an overview of MIM defects. Explanations for these defects are described and, where available, remedies and suggestions are provided.

**Key words:** defects, powder/binder separation, dimensional control, swelling, distortion.

## 10.1 Introduction

Powder injection molding comprises several processing steps, and defects may occur in each step if care is not taken. The defects encountered could be caused by mechanical factors, such as poor mold design and mold manufacturing, or by processing related factors, such as incomplete kneading, inadequate molding pressure, injection speed, holding pressure, and non-optimized debinding and sintering parameters (Zhang *et al.*, 1989; Hwang, 1996). Some of these defects may originate in the early processing steps but are difficult to identify because they do not manifest until after debinding or sintering. In the following section, the defects that frequently occur in each processing step are examined and their causes are explained. Hopefully, with an understanding of the scientific background of the defects, exhaustive trial and error experimentation can be avoided.

## 10.2 Feedstock

### 10.2.1 Feedstock uniformity

The raw powders used for MIM are usually quite fine, mostly below 20 μm, and thus agglomeration could be serious. When hard agglomerates, which cannot be broken up during high-shear-rate kneading, are included in the feedstock, the final sintered product could then contain inhomogeneous microstructures. If the agglomerates are alloying element additions, highly alloyed areas will develop and cause lean-alloyed regions, which will affect the mechanical properties (Hwang *et al.*, 2007). Moreover, the highly alloyed region may also have low density after sintering, as shown in Fig. 10.1, since the agglomerated alloying elements, such as Mo, may not be densified at the sintering temperature. Owing to unbalanced interdiffusion,

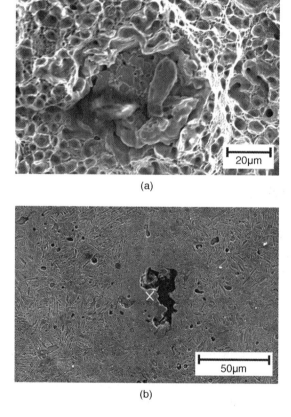

*10.1* (a) Fracture surface showing Mo agglomerates; (b) large voids in Mo-rich region.

other alloying elements, such as nickel, when too large or agglomerated, may also develop Kirkendall pores that are too large to be annihilated.

The uniformity of the feedstock constituents is always a concern in kneading. If the kneading time and shear rate are insufficient, the metal powders will not be uniformly distributed. Some pockets of organic binders may also exist and cause blistering during the subsequent thermal debinding process. In addition, with excessively long kneading time, decomposition or evaporation of low-temperature binder components will occur. The optimized kneading time can be determined by monitoring the power input to the kneader. When the power decreases and becomes steady, the feedstock is ready and should contain uniformly distributed powders and binders. This uniform distribution and lot-to-lot consistency can be confirmed using a pycnometer density meter or a capillary rheometer (Kulkarni, 1997).

## 10.2.2 Recycled feedstock

To lower the manufacturing cost of MIM products, the gate, runner, sprue, and green parts with defects are usually recycled. Two methods are usually adopted by the MIM industries (Kulkarni and Kolts, 2002). The first method is to add 30–50% recycled feedstocks to fresh materials, while the other is to use 100% recycled materials. Unfortunately, these feedstocks deteriorate as the number of recycling iterations increases. The main cause of the deterioration of feedstock is oxidation of the binder components, particularly in the case of paraffin wax, caused by the transformation of the C–C chain to the C=O chain. The backbone binder, such as polyethylene, also deteriorates during recycling (Cheng et al., 2009). These deteriorations cause decrease of the intrinsic binder strength and weakening of the bonding at the powder–binder interface and thus lower green strength. The loss of the lubrication effect of the binder among powders also results in higher viscosity. This means that the lot-to-lot dimensions and green densities will be difficult to control (Kulkarni and Kolts, 2002). Consequently, the molding parameters, including injection pressure and barrel temperature, should be re-adjusted with recycled feedstocks. In addition, as a result of the weaker bonding between powders and binders, more cracks and distortions of production parts may result during solvent debinding with further iterations of recycling, as shown in Fig. 10.2 for feedstocks recycled six and eight times.

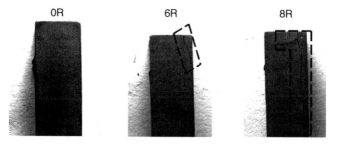

*10.2* The surface condition of the specimen after solvent debinding in 50°C heptane for 4 h. More defects can be found in the specimens prepared with highly recycled feedstocks (6R and 8R).

## 10.3 Molding

### 10.3.1 Flash

When high numbers of cavities are built in the mold, long runners are required. This requires a high molding pressure to transport materials to each cavity. When a high molding pressure is applied, flash usually occurs, as shown in Fig. 10.3, unless the molds are perfectly clamped without any clearance. Since the materials and shapes of the mold components are different, different amounts of elastic deformation and, in some cases, even plastic deformation, will occur and create clearance between tooling components. As a result, materials will be forced into the clearances. This problem becomes even more serious when large molds with large numbers of cavities and undersized molding machines are used, because the stiffness of the mold and the precision of the mold dimensions become more difficult to control.

*10.3* Flashes occur when the feedstock is forced into the clearances between mold components under high molding pressures.

## 10.3.2 Residual stress

Another defect, which is caused during molding but does not appear until the later processing steps of solvent debinding or thermal debinding, is residual stress. This stress is frozen in owing to the rapid cooling by the relatively cold mold surfaces. When this stress is relieved during heating in the debinding stage, either in solvent debinding or in the early stage of thermal debinding, distortion could occur. If a high injection pressure is applied, distortion may even occur for parts sitting at room temperature for a long time.

## 10.3.3 Binder/powder separation

Another common defect of cosmetic parts is a poor appearance around the gate. As the feedstock passes through the narrow gate into the die cavity, binder separation can occur due to powder migration from a high-shear-rate region to a low-shear-rate region. This binder/powder separation becomes more serious at the end of the filling stage and during the pressure holding stage, wherein the feedstock flows slowly and the pressure builds up rapidly. To fill in the clearance in the die cavity resulting from thermal shrinkage, more feedstock is pushed into these openings by the holding pressure. Since the movement of the metal powders in the feedstock is slow, due to the high interparticle friction, the binder, which has a low viscosity, is forced to flow through interparticle interstices. Thus, the binder content at the gate areas is increased, causing molding defects, particularly at the gate, where the shear rate is the highest. If the bonding between the powder and the binder is weak, separation will occur easily and form binder-rich areas at the surface of the part near the gate, as shown in Fig. 10.4. This gate mark remains after sintering, since the solid content in this region is low and thus the appearance is different. Post-sintering surface treatment, such as sand blasting or coating, is often applied to eliminate or alleviate this problem. Development of new binders that can reduce the gate mark is still a focus of research. Occasionally, the binder-rich area at the surface of the part may delaminate from the bulk or form a hidden delaminated void near the surface. When a post-sintering secondary operation, such as drilling, is applied, these defects will then be noticed, as shown in Fig. 10.5. In some cases, gate redesign helps to alleviate this powder/binder separation problem. If the original gate has a narrow opening and is located at a thin section, it can be enlarged and relocated to a thick section. As a result, the level of binder/powder separation will be reduced and the flow marks and weld lines will be minimized.

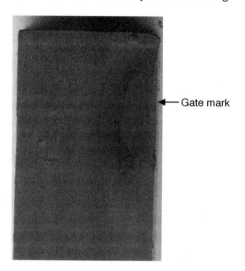

*10.4* With a high shear rate at the gate, powder/binder separation occurs and binder-rich gate marks form.

*10.5* Powder/binder separation occurs and delamination is observed after drilling.

### 10.3.4 Other defects

Other defects in MIM parts are similar to those found in conventional plastic injection molding due to improper molding parameters. When the molding pressure or holding pressure is low, low green density areas and incomplete filling can occur, as shown in Fig. 10.6. When combined with poor venting design and early solidification at long runners or sprue, internal voids may also develop in thick sections. After sintering, these regions with low density or internal voids could form concave sink marks.

Weld lines and flow marks are caused by low feedstock temperatures, low

Common defects in metal injection molding 241

*10.6* Incomplete filling occurs when the molding pressure is too low.

mold temperatures, or low barrel temperatures. When cold materials are forced by high pressure to flow over mold surfaces, flow marks are left behind, as shown in Fig. 10.7. When the flowing feedstock is separated into two streams by a post and the two separated cold fronts meet again, weld lines form, as shown in Fig. 10.8. These defects could also be related to improper mold designs, such as long runners, improper gate locations, improper venting ports, etc. The typical solution to these molding defects is to prevent cooling of the feedstock before the end of molding by increasing the barrel, nozzle, and mold temperatures and by redesigning the part to avoid partitioning of the flowing feedstock stream in the die.

Most defects encountered during molding of MIM, including the ones described above, are similar to those found in conventional plastic injection molding and are listed in Table 10.1. Also shown are the causes and remedies of these defects.

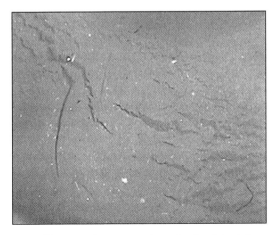

*10.7* Flow marks occur when the feedstock temperature is too low.

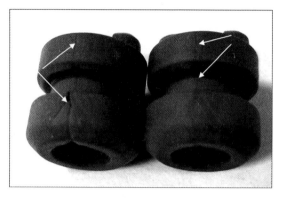

*10.8* Weld lines form when two cold fronts meet each other.

*Table 10.1* Defects frequently found in molding

| Defect type | Possible causes | Remedies |
| --- | --- | --- |
| Flash | Too high a pressure inside the die, poor flatness of mold surface along the parting line, venting channel too large | Use large tonnage machine, proper tool making, use a lower injection speed and molding pressure, optimize the switch point |
| Sticking in cavity | Too high a molding pressure, not enough thermal shrinkage, early ejection, improper mold design or making | Use a lower injection speed, molding/holding pressure, and mold temperature, increase cooling time, eliminate undercut and increase draft angle, adjust ejection area and location, redesign the binder |
| Sink mark | Thermal shrinkage, low density | Increase molding/holding pressure and injection speed, decrease mold temperature, increase gate area, add venting channels, decrease speed when passing thick sections |
| Voids | Trapped gas, absorbed moisture | Increase holding pressure, decrease injection speed, increase mold temperature, increase gate area, move gate to thick sections |
| Burn marks | Overly heated binders | Decrease injection speed and feedstock temperature, increase gate area, change gate location |

Table 10.1 (cont.)

| Defect type | Possible causes | Remedies |
|---|---|---|
| Weld lines | Cold feedstock in the die | Increase injection speed, mold temperature, and feedstock temperature, enlarge gate opening, add venting channels or overflow wells near weld line locations, move gate location, redesign parts to avoid stream partition |
| Flow mark | Cold feedstock in the die | Increase injection speed, mold temperature, and feedstock temperature, enlarge gate opening, change gate location |

## 10.4 Debinding

A successful debinding process usually consists of two or three stages, during each of which one of the binder components is removed. This ensures that the shape of the molded compact remains intact throughout the debinding process. Minor binder components, such as plasticizers, surfactants, coupling agents, and lubricants, are usually removed first. Polymeric backbone binders, which account for 30–60% of the binder, are the last type removed, and this late debinding stage is accompanied by light sintering of particles.

Several debinding processes have been identified, including the two-stage debinding process of solvent/water debinding followed by thermal debinding, straight thermal debinding, and catalytic debinding. The 100% thermal debinding is a slow process, since decomposed gas components that develop in the core section cannot escape to the atmosphere effectively through any pore channels. Defects are frequently encountered unless extremely slow heating rates and long debinding times, such as several days, are applied. Catalytic debinding is used for polyacetal-based materials, which also contain about 10% polyolefin binder components such as polyethylene and polypropylene. The polyacetal is thermally decomposed at around 135°C into formaldehyde in a diluted nitric acid gas atmosphere. This decomposition is basically a direct solid–gas reaction. No liquid is involved and thus the compact remains rigid throughout the debinding process. Moreover, since the reaction occurs only at the binder–vapor interface, there is no internal pressure build-up caused by decomposed gases. As a result, distortion or slumping can be prevented and blistering, voids, and cracks can thus be minimized. Although catalytic debinding can produce good quality debound parts in a very short debinding period, the more widely used debinding process today is the two-stage debinding process of solvent debinding followed by thermal debinding, for economic reasons. Thus, this process will be used to illustrate the examples of debinding defects in more detail.

### 10.4.1 Solvent debinding

In the solvent debinding process, cracking and slumping are frequently observed if the process is not well executed. Since the solvent is usually heated in order to increase the debinding rate, the minor components, such as paraffin wax and lubricant, become soft or even melt. As a result, warpage and slumping frequently occur, particularly when the part has a complicated shape with overhanging sections. To alleviate this problem, using a stronger backbone binder and a lower debinding temperature is usually helpful. A modification of the part shape to give better support for thin or extended sections is also frequently employed.

When parts are subject to solvent debinding, cracking may also occur, even though the part is soft. This problem is mainly caused by the swelling of polymers when solvents penetrate into the binder. The amount of expansion depends on the temperature, and the types of binder component and solvent used. Figure. 10.9 shows the *in situ* length change of a 100 mm long plate immersed in heptane at 40, 50, and 60°C using a laser dilatometer (Lin and Hwang, 1998). The part expands and the amount increases as the

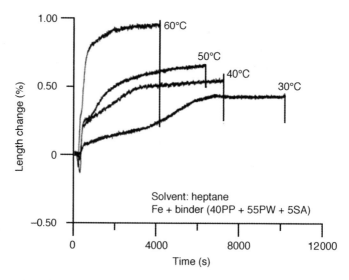

*10.9* Amount of swelling of molded specimens increases with increasing temperature.

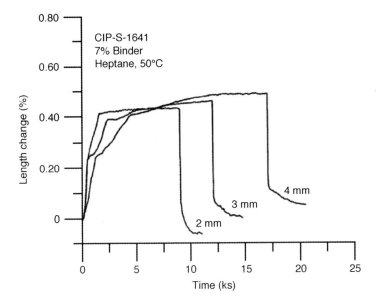

*10.10* Effect of thickness on the swelling during solvent debinding.

temperature increases, in particular when the temperature exceeds the melting point of a binder component. The main cause of the expansion is the reaction between the solvent and the backbone binder. This has been confirmed by using compacts with and without soluble binder components, which show similar amounts of swelling for both specimens (Hu and Hwang, 2000). The addition of isopropyl alcohol reduces the amount of expansion by retarding the penetration of the heptanes (Fan et al., 2008). The amount of expansion also increases as the total binder content increases. These observations suggest that the solvent debinding temperature and the amount of binder should be reduced if distortion occurs.

Parts with different cross-section thicknesses also behave differently (Hu and Hwang, 2000). Thin sections expand quickly in the early stage, while thick sections expand slowly, as shown in Fig. 10.10, owing to the constraints from the inner section of the part. As a result, distortion may occur early in the debinding stage because of the different amounts of expansion in different sections of the part.

## 10.4.2 Thermal debinding

Debinding defects are often observed after thermal debinding. However, the causes of these defects are not necessarily related to the thermal debinding process per se. They could originate from problems in the kneading, molding, or solvent debinding processes, which are amplified and manifested during thermal debinding.

Thermal debinding defects are frequently seen when the applied heating rate is too fast. In most cases, the defects are caused by the fast decomposition of the binder components. Adsorbed water has also been shown to produce defects when it is converted to steam. When the decomposed gas molecules cannot escape fast enough to the ambient through interconnected pore channels, they cause blisters or even bloating holes if the green body is plastic. When the green body is more rigid or the interparticle friction is low, cracking will occur.

It is widely recognized that the heating rate should be slow to prevent the formation of defects. This heating rate can be determined using thermogravimetric analysis (TGA), which shows the critical temperature ranges where binder components decompose. The heating rate through these ranges should be slow or have a holding period to prevent drastic binder burn-off. However, it should be noted that the debinding behavior of parts in a production furnace with a heavy loading is quite different from that used in TGA. The temperature range where slow heating or holding is required is usually higher than that measured in TGA. It is also recommended that the gas flow rate be increased and a short flow path be used to help carry away the decomposed gases. When a vacuum furnace is used, this means that a high partial pressure should be applied during thermal debinding. The entrance and exit of the gas should be designed in such a way that the flow path over parts is short in order to maintain a laminar flow, which removes decomposed gas more effectively than does a turbulent flow.

As thermal debinding proceeds to the end, only a small amount of backbone binder is left. This remaining backbone binder is mostly located at interparticle necks. The capillary force induced tends to rearrange particles and thereby to induce internal stress. A feedstock using high tap density powder and high solids loading could help to alleviate these stresses. If the content of the backbone binder is high or the percentage of the soluble binder removed is low, distortion could occur because too much liquid binder is present during thermal debinding.

For solvent debound parts, the heating rate can be quite fast, owing to the presence of interconnected open pores. With more and larger interconnected pore channels, the gases generated during the thermal debinding can escape to the ambient without causing blistering or cracking. Fan *et al.* (2009) used a model to predict the minimum amount of binder removal required during solvent debinding for producing interconnected pores in the middle section of products with different thicknesses. For example, to create open pores in the middle section of a part with 4.2 wt% soluble binder, the minimum

Common defects in metal injection molding 247

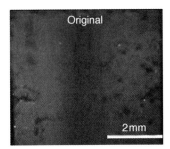 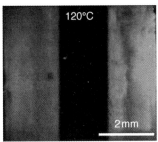

*10.11* Fluorescence dye penetration test of 6 mm thick specimens with 68% of the soluble binder removed, showing that the binder redistributes and fills open pore channels at the center of the part after being heated to 120°C.

debinding fraction needed is about 59% of the total amount of soluble binder, despite the thickness of the part, yielding a local debinding fraction of 37% and a porosity of 8.5% at the center. However, with too much residual binder and too few open pore channels, the large amounts of gases generated during thermal debinding may not be able to escape to the surface. Moreover, a subsequent study by the same group found that the remaining binders in the solvent debound parts redistribute within the compact during thermal debinding (Fan and Hwang, 2010). As the solvent debound parts are heated above the melting points of the binder components, capillarity drives binders into fine pore channels and interparticle contact areas in order to reduce the total surface energy. As a result, the open pores in the middle section are blocked again, as shown in Fig. 10.11. With too much binder and few interconnected pore channels, these regions behave similarly to those occurring when as-molded parts are subjected to thermal debinding directly, without solvent debinding. Thus, to improve the efficiency of thermal debinding, the amount of binder removal during solvent debinding must be increased so that the amount of the remaining binder is too little to flow into and fill in the pores in the middle section.

Experimental results have shown that this minimum amount of binder removal for the prevention of binder redistribution increases as the thickness of the part increases. For parts with 4.20 wt% soluble binder and a thickness of 6 mm, the suggested minimum amount of binder removal for solvent debinding is 79%, or 3.32 wt% of the total weight of the part, when a heating rate of 5°C/min is used (Fan and Hwang, 2010). This value is higher than the 59% required for creating open pores in the center of the part during solvent debinding. Since the minimum amount of the binder that needs to be removed depends on the thickness of the part and the heating rate of the thermal debinding, a general number of 90% is recommended for a safe start.

The exfoliation phenomenon, which is caused by the delamination of a skin layer from the main body, has also been reported (Woodthorpe *et al.*, 1989; Zhang *et al.*, 1989). This has been attributed to the binder-rich surface caused by incorrect molding procedures in combination with a fast heating rate during thermal debinding. One postulation is that, as the feedstock cools in the die cavity, the material shrinks owing to the volume shrinkage of the metal powder and binder components, in particular the paraffin wax. This shrinkage leaves a small clearance between the part and the die wall, which allows further penetration of the feedstock. Since the clearance is small, the easiest material to fill this gap in the pressure holding stage of the molding process is the low melting material, such as paraffin wax. The other possibility causing the binder-rich surface is the increased binder/powder ratio at the surface due to the emerging flow of the binder from the interior during thermal debinding. With a high binder content and less interparticle friction, a layer, or skin, could detach from the main body. With a fast heating rate, this phenomenon may worsen if the outward flow of the binder is accelerated by the high pressure of the decomposed gas in the core of the part.

To facilitate explanation, Table 10.2 presents a summary of the various types of defects found after solvent and thermal debinding, the possible sources of the problems, and the recommendations for their removal (Hwang, 1996).

Several techniques have been developed to alleviate the above defects of cracking, blistering, and distortion that occur during thermal debinding. They include the use of slow heating rates in the temperature range where binder components decompose or evaporate, high gas flow rates, higher binder removal percentages during solvent debinding, and supports or fixtures for overhanging and intricate sections. The use of more irregular powders also helps to increase the green strength and allows the use of a faster debinding rate. However, these powders will have an adverse effect on feedstock flowability and sinterability.

### 10.4.3 Binder residues

The binder residues left from debinding have been paid little attention in the literature (Zhang *et al.*, 1989). These residues could be critical for some structural or magnetic parts. Oxide residues from high-density polyethylene and metal stearates have been reported in several studies. These oxides may come from catalysts, which are used as Ziegler–Natta initiators for polymerization of polyethylene and isostatic polypropylene, while some other oxides may come from metal atoms in stearates. The metal atoms can

*Table 10.2* Defects frequently found in debinding

| Defect type | Possible causes | Remedies |
|---|---|---|
| Cracks (solvent debinding) | Swelling of binder components, poor bonding between binder and powder, low strength of the backbone binder, too high a molding pressure, large differences in section thicknesses | Change the type and composition of solvent or binder, use a lower injection speed and molding pressure, redesign parts with smaller differences in section thicknesses, use lower debinding temperatures |
| Bending/distortion (solvent debinding) | Residual stress from molding, lack of support for overhanging sections, entrapped air | Bake between 50 and 90°C, use fixtures, adjust molding parameters |
| Corrosion/stain (solvent debinding) | High acidity of solvent, humid environment | Replenish solvents or use new ones, leave parts in a dry atmosphere |
| Cracks/blistering (thermal debinding) | Overly fast heating, absorbed water in feedstock, insufficient binder removal for solvent debinding, poor binder distribution, low solid content | Use slow heating rates, extend solvent debinding time, use longer kneading time and adjust binder components, keep feedstock dry, use higher gas sweeping rate and shorter flow path |
| Bending/distortion (thermal debinding) | Overly fast heating, insufficient binder removal for solvent debinding, lack of support for overhanging sections, insufficient interparticle friction, too much binder | Use slow heating rates, extend solvent debinding time, use fixtures or sands for the support, use higher gas sweeping rate, use more irregular powders, increase solids loading |
| Exfoliation (thermal debinding) | Wax separation to the surface, overly high heating rate | Extend solvent debinding time, use slower heating rate, bake below 100°C |

Source: (Hwang, 1996).

usually be dissolved in the matrix during sintering. However, reactive metals such as titanium and aluminum will react with oxygen or water vapor in the metal powder or sintering atmosphere, forming stable metal oxides. These oxides or dissolved elements could influence the mechanical properties of structural parts. These residues may not always be detrimental to MIM parts. It has been demonstrated that residual Ti from high-density polypropylene helps densification and increases the strength (Lu *et al.*, 1996). Magnesium from magnesium stearate has also been shown to improve the sintered density and bending strength of injection molded alumina by forming fine and uniformly distributed magnesia in the alumina matrix (Hwang and Hsieh, 2005).

Another residue is carbon soot left from thermal debinding. With an increase in the carbon content, the corrosion resistance of stainless steels such as 316L will deteriorate due to the formation of chromium carbide and Cr-lean areas. If an excessive amount of carbon is present, the melting point decreases and may cause local distortion or melting of the part. For Kovar (Fe–29Ni–17Co), the coefficient of thermal expansion will also increase and cause adverse effects on glass–metal sealing (Tokui *et al.*, 1994). To ensure complete debinding without leaving carbon residues, a high gas sweeping rate during thermal debinding is recommended. With the use of continuous furnaces, such as pusher or walking beam furnaces, a high dew point in the debinding zone is also recommended to facilitate the removal of the carbon residues. It should be noted that carbon has a high diffusion rate into iron and that such diffusion starts at about 875°C, depending on the carbon content and the phase transformation temperature. Thus, thermal debinding and carbon removal should be completed before parts reach this temperature.

## 10.5 Sintering

### 10.5.1 Appearance and discoloration

One of the advantages of the MIM process is its capability to produce parts with shiny surface finishes that have small surface roughness. To achieve this goal, high density must be obtained, and the metal surface must be free from reaction with sintering atmospheres to avoid the formation of oxides, nitrides, or other reaction products. Consequently, the dew point or oxygen content in the atmosphere must be sufficiently low, and the amount of hydrogen or the degree of vacuum must be sufficiently high, to reduce the metal oxides on metal powder surfaces. Otherwise, fine oxide particles, such as silicon dioxides, can easily form. When nitrogen is contained in the atmosphere, such as dissociated ammonia, it will react with the chromium in stainless steels, forming chromium nitrides, which deteriorate the corrosion resistance of stainless steel. When a high vacuum is used at high temperatures, chromium will evaporate, with a resultant reduction in corrosion resistance. Thus, a partial atmosphere of argon or pure argon is often used for high-temperature sintering. These sintering concerns and solutions are similar to those for press-and-sinter parts.

## 10.5.2 Dimensional control and distortion

High density, above 95% of the theoretical density, is usually required for MIM parts. To achieve this density, fine powder and high-temperature sintering are used most of the time. In some cases, liquid phase sintering, including supersolidus liquid phase sintering, is required. Unfortunately, with the typically low solids loading in MIM parts, less than 70 vol%, preservation of the geometry is a challenge. This problem is an even greater challenge in prealloyed powders for which the solidus and liquidus lines are flat and the temperature difference between these two lines is small. These characteristics make the amount of liquid very sensitive to the temperature variation in the sintering furnace, as indicated by the phase diagram and the lever rule. A good example is the SKD11 or D2 tool steels. The temperature must be controlled to within $\pm 5°C$. If the temperature is too high, too much liquid is formed; then gravitational forces cause distortion or even slumping, while too small an amount of the liquid will not densify the compact (German, 1997).

Several solutions have been reported. A recent patent discloses prealloyed tool steel powders doped with 2–5% niobium (Soda and Aihara, 2007). The niobium forms NbC and inhibits grain growth during liquid phase sintering. With fine grains, the thickness of the liquid layer at the grain boundaries decreases, which makes particle rearrangement more difficult (Liu et al., 1999). The addition of carbides has also proven effective to open up the sintering window with the same underlying mechanism (Chuang and Hwang, 2011).

Similar to the distortion problems found in solvent debinding, overhanging sections, steps, and narrow openings, as shown in the schematic diagram of Fig. 10.12, may also distort during sintering due to gravitational force or uneven amounts of shrinkages in different regions. This problem becomes even more serious when liquid phase sintering or heavy metals such as tungsten alloys are used. The use of high solids loading and more irregular powders helps to a certain degree. But sintering fixtures are often required to prevent warping, and dummy bridges, which are removed by mechanical methods after sintering, are often designed into the part to prevent opening or narrowing of the long slots during sintering.

## 10.6 Conclusion

Although considerable time and effort has been spent to resolve the problems encountered during the fabrication of MIM parts, and the technology itself has advanced significantly since the birth of the MIM process, defects still frequently occur during the daily practices of kneading, injection molding, debinding, and sintering. Some examples are gate marks,

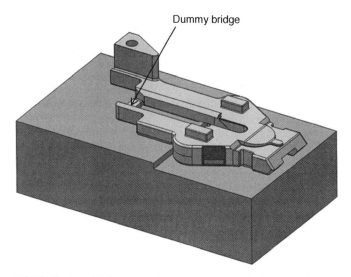

*10.12* Parts with long overhangs, steps, and slots require sintering fixtures and dummy bridges to prevent sagging and opening/closure of the slots.

sink marks, distortions, cracking, blistering, and poor dimensional control. Some of these problems have been resolved, and the underlying basics are understood, as briefly described above. However, many challenging problems remain, and in-depth understanding of the root causes and the fundamentals of the processes has yet to be attained. There is still much to learn about the residual stress in molded articles, powder/binder separation, the flow behaviors of binders during molding, binder redistribution behavior during thermal debinding, and shape retention during supersolidus liquid-phase sintering.

## 10.7 References

Cheng, L.H., Hwang, K.S., and Fan, Y.L. (2009), 'Molding properties and causes of deterioration of recycled powder injection molding feedstock', *Metallurgical and Materials Transactions A*, 40A, 3210–3216.

Chuang, K.H. and Hwang, K.S. (2011), 'Preservation of geometrical integrity of supersolidus-liquid-phase-sintered SKD11 tool steels prepared with powder injection molding', *Metallurgical and Materials Transactions A*, 42A, in press, doi: 10.1007/s11661-010-0593-8.

Fan, Y.L., Hwang, K.S., and Su, S.C. (2008), 'Improvement of the dimensional stability of powder injection molded compacts by adding swelling inhibitor into the debinding solvent', *Metallurgical and Materials Transactions A*, 39A, 395–401.

Fan, Y.L., Hwang, K.S., Wu, S.H., and Liau, Y.C. (2009), 'Minimum amount of binder removal required during solvent debinding of powder injection molded compacts', *Metallurgical and Materials Transactions A*, 40A, 768–779.

Fan, Y.L. and Hwang, K.S. (2010), 'Defect formation and its relevance to binder content and binder redistribution during thermal debinding of PIM compacts', *Proceedings of the 2010 PM World Congress*, EPMA, Shrewsbury, UK, vol. 4, pp. 595–601.

German, R.M. (1997), 'Supersolidus liquid-phase sintering of prealloyed powders', *Metallurgical and Materials Transactions A*, 28, 1553–1567.

Hu, S.C. and Hwang, K.S. (2000), 'Length changes and deformation of PIM compacts during solvent debinding', *Metallurgical and Materials Transactions A*, 31A, 1473–1478.

Hwang, K.S. (1996), 'Fundamentals of debinding processes in powder injection molding', *Reviews in Particulate Materials*, 4, 71–103.

Hwang, K.S. and Hsieh, C.C. (2005), 'Injection-molded alumina prepared with Mg-containing binders', *Journal of the American Ceramics Society*, 88, 2349–2353.

Hwang, K.S., Wu, M.W., Yen, F.C., and Sun, C.C. (2007), 'Improvement in microstructure homogeneity of sintered compacts through powder treatment and alloy designs', *Materials Science Forum*, 534–536, 537–540.

Kulkarni, K.M. (1997), 'Dimensional precision of MIM parts under production conditions', *International Journal of Powder Metallurgy*, 33(4), 29–41.

Kulkarni, K.M. and Kolts, J.M. (2002), 'Recyclability of a commerical MIM feedstock', *International Journal of Powder Metallurgy*, 38, 43–48.

Lin, H.K. and Hwang, K.S. (1998), 'In-situ dimensional changes of powder injection molded compacts during solvent debinding', *Acta Materialia*, 46, 4303–4309.

Liu, J., Lal, A., and German, R.M. (1999), 'Densification and shape retention in supersolidus liquid phase sintering', *Acta Materialia*, 47, 4615–4626.

Lu, Y.C., Chen, H.C., and Hwang, K.S. (1996), 'Enhanced sintering of iron compacts by the addition of $TiO_2$', in *Sintering Technology*, eds R.M. German, G. L. Messing, and R.G. Cornwall, Marcel Dekker, New York, USA, pp. 407–414.

Soda, Y. and Aihara, M. (Mitsubishi Steel Mfg Co. Ltd) (2007), 'Alloyed steel powder with improved degree of sintering for MIM and sintered body', US Patent 7,211,125.

Tokui, K., Sakuragi, S., Sasaki, T., Yamada, Y., Ishihara, M., Nakayama, M., Kuji, I., and Nomura, M. (1994), 'Properties of sintered Kovar using metal injection molding', *Journal of Japan Society of Powder and Powder Metallurgy*, 41, 671–675.

Woodthorpe, J., Evans, J.R.G., and Edirisinghe, M. J. (1989), 'Properties of ceramic injection moulding formulations, part 3, polymer removal', *Journal of Materials Science*, 24, 1038–1048.

Zhang J.G., Edirisinghe, M.J., and Evans, J.R.G. (1989), 'A catalogue of ceramic injection moulding defects and their causes', *Industrial Ceramics*, 9, 72–82.

# 11
# Qualification of metal injection molding (MIM)

D. F. HEANEY, Advanced Powder Products, Inc., USA

**Abstract:** The MIM process produces net-shape or near-net-shape components at a reduced manufacturing cost; however, significant process and product qualification is required to ensure that the final product is acceptable. The level of qualification depends upon the end use of the product. It could be as simple as making a prototype component or as complicated as characterizing and controlling the entire process. As a rule, medical and aerospace products require the greatest amount of qualification and less critical products such as tools and consumer products require the least. In this chapter, a guideline is presented to qualify a MIM process that meets a final specification at a minimum cost. A rationale for choosing only the most critical process control methods is presented.

**Key words:** quality, process characterization, SPC parameters.

## 11.1 Introduction

Metal injection molding (MIM) is a process for forming net-shape or near-net-shape components at a reduced manufacturing cost as compared to machining and at a higher precision level than other forming technologies, such as investment casting. However, the process is fairly complicated, requiring knowledge from various disciplines to ensure that a quality product is manufactured. Knowledge of powder handling, powder sintering, injection molding, powder/polymer rheology, polymer degradation, metallurgy, and so on must be understood and used to ensure a stable process and a quality product. Furthermore, each process step is interactive, i.e. a molding defect might only be detected under certain sintering conditions, thus, characterization of each process is essential.

In view of the complexity of the MIM process, an engineer can become lost in all the possible variables that could be implemented to attain a controlled process, and most importantly, a quality product. Often, a high level of process control is required, but at other times it is not. In this chapter, a program is presented for a design engineer to easily qualify a component vendor or a process engineer to qualify and monitor a MIM process. In this way the design engineer can understand the MIM process well enough to make intelligent decisions with regard to the use of MIM in their applications and the process engineer can ensure a controlled process for consistent production.

## 11.2 The metal injection molding process

To use and to validate a MIM process, an engineer must have a basic understanding of the process. The process must be divided into its sub-process categories. The number of process steps can be as many as nine. The number of steps depends upon the particular technology and the amount of processing that a manufacturer performs. For example, whether the manufacturer purchases feedstock or manufactures proprietary feedstock in-house. The potential process steps are as follows:

1. raw material selecting and monitoring;
2. material blending;
3. feedstock compounding;
4. injection molding;
5. solvent or catalytic debinding;
6. thermal debinding;
7. sintering;
8. secondary operations (HIP, machining, heat treating, grinding, etc.);
9. inspection and packaging.

Each process step produces a product that feeds the next process step. Thus, process control can be performed on the individual processes and also on the product of each process step to ensure that the entire process is in control. Table 11.1 lists a MIM process sequence and the inputs and output of each sub-process.

## 11.3 Product qualification method

The first question that must be asked is whether MIM can be used for an application. This is a two-sided question. First, is it technically feasible, and second, is it economically feasible? With this in mind, one must determine the economics of using the MIM process as opposed to the current fabrication technique or conventional method for an application. If the economic exercise is successful, the next step is to determine the required application properties and critical dimensions of the component. Prototype mold fabrication is then performed and candidate materials are selected.

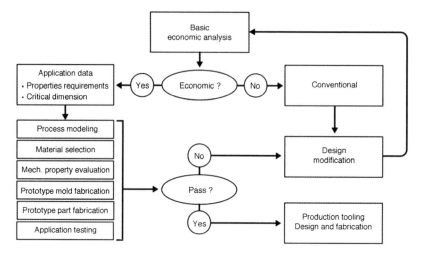

*11.1* Logic diagram to go from concept to production for MIM.

*Table 11.1* Process input and output products for comprehension and process control

| Process step | Process input | Process output |
|---|---|---|
| Blending | Powder and binder | Powder/binder mixture |
| Compounding | Powder/binder mixture | Feedstock |
| Molding | Feedstock | Green part |
| Debinding | Green part | Brown part |
| Sintering | Brown part | Finished part* |

* May require secondary operation

This is followed by near-net-shape part fabrication and application testing. Figure 11.1 shows a logic diagram that can be followed to fabricate and qualify a MIM component.

The first step in qualifying MIM for an application is to do the economic analysis. If the economics look undesirable, one must return to a conventional manufacturing technique or make design modifications in material type or part size to make the project economical. Often the size of the part will dictate whether MIM can be used, not only from a technical point of view for removal of the binder, but also because of the powder cost. Typically, parts greater than 300 g cannot be fabricated economically using contemporary MIM technology. After a successful economic analysis, the next step is to look at the technical aspects of the component – specifically the application data such as property requirements and critical dimensions. After these have been defined, the proper material or group of materials is selected for evaluation. These materials are then injection molded and evaluated for properties, such as tensile strength and/or corrosion resistance. Property evaluation can be skipped if a vendor has defined these data in previous work. Satisfactory results for properties justify the fabrication of a prototype mold. 'Prototype' is emphasized since these components are produced using low-cost tooling. The prototype components might require secondary operations that would not be required while using production tooling. Application testing is the final test to validate that MIM can be utilized for an application. If application testing is unsuccessful, the development cycle starts again with design modifications and economic analysis, provided there is continued management support.

The above methodology is a philosophy. Variations to this technique may exist for different applications, different amounts of available capital, and also considering time to market for a product. For example, one may choose to go directly to production tooling fabrication and do the development on the production tool. In this way the time to market is reduced; however, the up-front cost and the risk are higher.

## 11.4 MIM prototype methodology

Following the economic analysis, the next step for any potential MIM application is the evaluation of mechanical properties of potential materials and the production of prototype components. The production of test samples (tensile, fatigue, wear discs) or actual prototypes is done to test a new material for a particular application with the minimum of cost. Property evaluation can be as simple as locating property values in the Metal Powders Industry Federation (MPIF) standards manual, from a MIM component supplier's specifications, or from other literature that pertains to a specific material. Furthermore, mechanical properties can be MIM component supplier dependent, since all suppliers use different processing methods and raw materials. For example, vacuum sintering of a stainless steel may reduce its corrosion resistance or the feedstock binder may leave a carbon residue that could compromise properties or corrosion resistance of the sintered metal.

### 11.4.1 Material selection

Selection of the proper material for a particular application often decides the success of a MIM project. Chapter 2 lists some available metal alloys and their potential applications. Commonly, a number of materials are evaluated for an application and a decision is made based on performance, cost, and input from a MIM component supplier or expert. As a general rule, if a material is available in a fine powder form and can be sinter densified, it can be metal injection molded. A good vendor or MIM design engineer should be able to select the proper material based on application input.

### 11.4.2 Prototype production

A MIM prototype is a MIM component that has been fabricated using the MIM process; however, the tooling that is used is much lower in cost than production tooling. A prototype tool will cost less than one quarter the cost of a production tool, since expensive mold components such as slides and cams are not used. Difficult features are machined in as secondary operations, since fewer than 1000 prototype components are typically produced. The tooling is also made with easy-to-machine metals such as P20 and unhardened H13. Aluminum can also be used; however, this metal is easily dinged and mauled during prototype development. It is often valuable to have the mold vendor work with materials that they are accustomed to machining.

## 11.5  Process control

Once a MIM prototype component has passed initial qualifications for use, the next step is to ensure that a process is sufficiently in control for a particular application. Minimum process control is required if the specifications are broad and a significant amount of process control is required if the specifications are narrow. This section breaks down the MIM process to make it possible to determine which process control needs are required for a specific MIM process to achieve the required specifications at the minimum of cost.

To analyze the MIM process for process control, the process must be divided into its sub-process categories. Each of these sub-processes can be controlled to ensure a more repeatable process; however, the more a process is monitored, the more costly is the overall process. Therefore, an engineer must balance between cost and control to ensure that the MIM process is profitable for a particular application. For example, if a company is producing aerospace or medical components, where the cost of a catastrophic failure is high, there must be a high level of process documentation and control. A company that manufactures products that have less financial exposure for catastrophic failure would require less. The general concept is to produce the best possible product with the least amount of process monitoring.

Table 11.2 lists each of the process steps for MIM, and also the parameters that could be controlled for that process step. Although there are many potential parameters to control in the MIM process, not all need to be controlled. Application and process type define the required control.

*Table 11.2* Potential parameters to control for the PIM process

| Process step | Process attribute | Measurable attribute | Monitor method |
|---|---|---|---|
| Raw materials | Powder | Chemistry | Specification sheets, Chemistry analyzer |
| | | Powder size | Specification sheets, PSD analysis |
| | | Powder size distribution | PSD analysis |
| | | Density | Pycnometry |
| | | Tap density | Tap density |
| | | Moisture level (ceramic) | Hydrometer |
| | Binder | Moisture level | Hydrometer |
| | | Viscosity | Capillary rheometry |
| Compounding | Feedstock | Density (powder/binder ratio) | Pycnometer, Archimedes |
| | | Viscosity stability | Capillary rheometry |
| | | Viscosity vs. shear rate | Capillary rheometry |
| Injection molding | Switch-over pressure* | Switchover pressure stability | Machine |
| | Screw return torque | Screw return torque stability | Machine |
| | Shot size | Shot size stability | Machine |
| | Part | Part mass | Scale |
| | Defects | Blisters, voids, cracks, powder/binder separation, knit lines | Visual, X-ray |
| Solvent debinding | Part | Mass loss | Scale |
| | Defects | Cracks, blisters | Visual |
| Thermal debinding | Part | Mass loss | Scale |
| | Part | Shrinkage | Linear measurement |
| | Defects | Cracks, blisters | Visual |
| Sintering | Maximum temperature | Maximum temperature stability | Thermocouple |
| | Temperature uniformity | Temperature uniformity stability | Thermocouple |
| | Defects | Voids, cracks | Visual, X-ray |
| | Part | Shrinkage, final dimensions | Linear measurement |
| | Part | Density | Archimedes, pycnometer |
| | Part | Chemistry (carbon) | Leco or other chemistry method |
| | Part | Surface finish | Profilometer |
| | Part | Corrosion resistance | Salt spray, potentiodynamic scans |
| | Part | Properties | Application testing |
| Inspection | Part | Dimensions | Linear measurement |

* Only if switchover by position, may also monitor position if switchover by pressure.

*Table 11.3* Process auditing comparison for minimum and precision process control

| Attribute | Minimum control | Precision control |
| --- | --- | --- |
| Feedstock | • As-molded component mass auditing | • As-molded part mass auditing<br>• Feedstock density auditing (pycnometer, Archimedes)<br>• Feedstock viscosity stability auditing |
| Injection molding | • As-molded component mass auditing<br>• Position switchover and monitor switchover pressure | • Part density (pycnometer, Archimedes)<br>• Part dimension<br>• Cavity pressure switchover and monitor switchover position, closed-loop control on hold pressure |
| Solvent debinding | • Weight loss studies | • Weight loss studies<br>• Weight loss auditing |
| Thermal debinding | | • Weight loss studies<br>• Weight loss auditing |
| Sintering* | • Select part dimension auditing<br>• Component density auditing | • Select part dimensions<br>• Component density auditing<br>• Chemistry analysis, particularly carbon<br>• X-ray<br>• Crack detection<br>• Microstructure<br>• Mechanical testing |

* Many modern processes combine thermal debinding with sintering, thus thermal debinding does not get monitored in precision control.

Table 11.3 lists a comparison of process control auditing for two levels of control. One process control would be considered minimum in both cost and effort, whereas the other provides precise control for precision and high-performance components.

## 11.6 Understanding of control parameters

The following section is devoted to describing the different process controls and the reason for their use. These can also be used in the process set-up and qualification stage to ensure a stable process and reviewed periodically or when a problem arises.

## 11.6.1 Powder characteristics

*Chemistry*

Chemistry monitoring is most critical for materials that are sensitive to carbon level and oxygen level; however, other elements such as chromium for stainless steels may be important to monitor. Carbon level is required for materials where it is important for properties and heat treatment; the most common are tool steels, low-alloy steels, and martensitic stainless steel. In these materials, carbon level affects dimensional stability, sintered density, and mechanical properties. Also, materials sensitive to carbon embrittlement, such as titanium, should have the carbon monitored. Oxygen monitoring is important for materials such as titanium because it affects the elongation. Also, oxygen in combination with silicon in stainless steels may affect the elongation by the formation of silica particles.

*Powder size and size distribution*

Powder size and size distribution affects mixture viscosity and injection molding. As particle size increases, mixture viscosity decreases. This affects the molding process consistency. Also, sintered density and mechanical properties are affected by the powder size. As particle size decreases, the sintering response increases. Therefore, variability in particle size affects the part dimensions, part density, and mechanical properties.

## 11.6.2 Feedstock behavior

*Density (pycnometer, Archimedes)*

Density is a direct measure of the ratio between the powder and the binder. Improper feedstock density affects sintered size, mixing viscosity, and molding.

*Feedstock viscosity*

Improper feedstock viscosity will result in variability during molding and part quality. It also can be an indication of improper raw materials, degraded raw materials, degraded feedstock, and poorly mixed feedstock.

## 11.6.3 Injection molding

*Component mass*

A variation in component mass will result in a variation in dimensions in a sintered component. The component mass variability may be the result of the feedstock preparation step or of a molding process variation. Variability in the switchover position or hold pressure can cause mass variability and final component dimensional variability. Furthermore, the molding operation is a constant-volume process, thus, a variation in mass will result in a variation in sintered dimensions.

### 11.6.4 Debinding

*Weight loss*

Understanding the amount of binder in the component and the rate at which it is removed is critical for defect-free components. Also, if the binder is not removed correctly, excess carbon may result in the component, and affect the final mechanical properties of the components.

### 11.6.5 Sintering

*Select component dimensions*

Dimensional variability shows the most effect after sintering. As the temperature and gas flow in the furnace varies, so do the dimensions. Therefore, knowledge of the dimensions and how they vary is important to understand the sintering process and to maintain a controlled process. Furthermore, variability in component mass at molding manifests itself as variation in component dimensions at sintering. The variability of all the previous process steps will be exacerbated in the sintering step and show up as component dimension variability.

*Component mass*

The mass of a component can change if there is an alloying element that vaporizes in the furnace during sintering. For example, the loss of chromium in stainless steel during vacuum sintering is well documented.

*Component density*

The density of a component can change if the temperature or gas flow is incorrect during sintering.

*Chemistry analysis*

Material evaporation or contamination from the furnace or process gas can be detected to high accuracies using chemistry analysis.

*X-ray*

Voids and cracks are easy to detect using X-ray equipment. This could be used for the set-up of critical parts for medical or aerospace applications.

*Crack detection*

Many methods exist for the detection of cracks. These can range from acoustics to visual detection.

*Microstructure*

The evaluation of microstructure can be extremely valuable. The detection of oversintered or undersintered microstructure is simple using this method. Also, many other characteristics of the component can be detected from the microstructure. These include carbide formation, phase ratios, pore size and location, and so on.

*Mechanical testing*

Often, the components can be put through function testing for strength or evaluated for hardness on the actual component. Another method to monitor the sintering process is to have test specimens sintered with the components and evaluate the test samples for strength, elongation, or some other test.

## 11.7 Conclusion

A method to qualify metal injection molding for an application has been presented. This was done for both product and process. A thorough evaluation of process controls and monitoring that can be carried out on a MIM process has been laid out. Also, a rationale for the selection of the best process control for a particular application has been presented. In general, one should monitor only the most critical parameters that are dependent upon the application and its specifications.

## 11.8 Sources of further information

ASM (1998), *Powder Metal Technologies and Applications, ASM Handbook*, Vol. 7, ASM International, Materials Park, Ohio, USA.

European Powder Metallurgy Association (2009), *Metal Injection Moulding – A Manufacturing Process for Precision Engineering Components,* 2nd edition, updating 1st edition (2004), European Powder Metallurgy Association, Shrewsbury, UK.

German, R. M. (2003), *Powder Injection Molding – Design and Applications*, Innovative Materials Solutions, Inc., State College, PA, USA.

Heaney, D. F., (2004), 'Qualification method for powder injection molded components', *P/M Science and Technology Briefs*, 6(3), 21–27.

Heaney, D., Zauner, R., Binet, C., Cowan, K., and Piemme, J. (2004), 'Variability of powder characteristics and their effect on dimensional variability of powder injection molded components', *Journal of Powder Metallurgy*, 47(2), 145–150.

MPIF (2007), *Material Standards for Metal Injection Molded Parts*, Metal Powders Industry Federation, Princeton, NJ, USA, MPIF Standard # 35, 1993-942007 edition.

Zauner, R., Binet, C., Heaney, D., and Piemme, J. (2004), 'Variability of feedstock viscosity and its effect on dimensional variability of green powder injection molded components', *Journal of Powder Metallurgy*, 47(2), 151–156.

# 12
# Control of carbon content in metal injection molding (MIM)

G. HERRANZ, Universidad de Castilla-La Mancha, Spain

**Abstract:** The control of carbon content is a key aspect of metal injection molding (MIM). This chapter describes the main systems in which carbon control is required to ensure high-quality components are produced. The main parameters used to control the carbon content during industrial processing are discussed, especially for debinding and sintering processes, together with the influence of the carbon content on final mechanical properties.

**Key words:** carbon control, debinding, sintering, MIM.

## 12.1 Introduction: the importance of carbon control

Nearly all materials are sensitive to carbon. In fact, it is the most important issue during the manufacture of some materials, as even minor fluctuations in carbon content can alter the microstructure. Carbon contamination from the binder, which is a carbon-based chain that must be properly incinerated during the debinding step, is an inherent problem in the powder injection molding process. Its influence has a strong impact on the sintering process, microstructure and mechanical properties of metals.

The influence of residual carbon may, however, vary significantly depending on the system. The sintering temperature and carbon content are the most important process variables in tool steels since they dictate the volume fraction of the liquid phase. In general, fractions of residual carbon help the densification process, thus allowing a better control of the process and an increase in the final hardness. In contrast, in cemented carbide systems it is important to achieve an accurate control of the carbon to avoid carburization or the production of brittle phases. In other systems, such as most stainless steels, magnets (Fe–Si, Fe–Ni), or titanium alloys, the carbon content should be kept to a minimum as carbon contamination can lead to unacceptable mechanical properties and negative modification of the densification rate.

Carbon is an important aspect of the debinding process and depends strongly on the kind of binder system used. The large number of different binders therefore results in a complex approach to understanding the decomposition mechanism. A binder may be removed using any of a number of methods, and sometimes by a combination of several methods. The most commonly used process, which results in degradation of the polymer, is known as binder burnout. The binder contains a high level of carbon and therefore debinding requires an appropriate selection of the heating rate, temperature, time, and atmosphere to extract the binder and control the residues, the interactions during heating, and the formation of reaction products.

The most common tool to determine the required debinding parameters is thermogravimetrical analysis (TGA). This tool allows the monitoring of sample weight with the increase of temperature. It is possible to program different heating rates or the use of different atmospheres. Analysis of the results allows the user to design the right debinding cycle controlling the residual weight of polymer, and thus the carbon content. Theoretical models allow the decomposition behavior of polymers and the residual carbon left in the powder compacts to be predicted, thereby helping the development of optimised cycles.

True control of the carbon content occurs as a result of the atmosphere selected. Thus, the use of different atmospheres leads to the formation of different species during the debinding process. As the reducing character of the atmosphere increases, an increase in the final carbon content is observed. This fact is attributed to the reaction between the hydrogen and the oxygen on the powder preventing the reactions between the carbon and the oxygen that leads to a reduction of the carbon content. In other words, the burnout products change from carbon–oxygen predominance to hydrogen–oxygen predominance as the reducing character of the atmosphere increases. The debinding temperatures also change upon changing the atmosphere. A balance between minimum debinding time, compact integrity, and suitable carbon content must therefore be achieved during industrial processing.

Finally, it is important to correlate the carbon content to the material's properties. Several authors have found a strong influence of carbon content on microstructure development. Thus, in systems such as stainless steels, lower residual carbon favors the formation of microstructures formed by a mixture of martensite and δ-ferrite instead of austenite and martensite. Increasing residual carbon leads to a decrease in the density and the yield strength of the sintered compacts, whereas the tensile strength increases (Wu et al., 2002). Accurate control of the carbon content through combinations of solvent and thermal debinding in the WC–Co system allows hard metal

Control of carbon content in metal injection molding    267

parts with a high transverse strength to be obtained. It is therefore crucial to adjust the carbon content by varying the thermal cycle, process time, and debinding atmosphere to obtain an optimal combination of strength and ductility.

## 12.2 Methods of controlling carbon, binder elimination and process parameters affecting carbon control

### 12.2.1 Carbon source

In most alloys, carbon control has been achieved by adjusting carbon potential in the sintering atmosphere. In MIM, the presence of a binder component that is a source of carbon, iron oxide that can decarburize, the complex atmosphere composition that produces different gas–solid reactions, the $CO/CO_2$ and $H_2/CH_4$ ratios, and the different effects caused by the residual carbon in the different alloys make the carbon control more complex.

The polymeric binder decomposes into hydrocarbons ranging from $C_1$ to $C_6$. These molecules may themselves be reactants and break further into small hydrocarbon chains by radical processes. The process consists of thermolytic bond cleavage, recombination reactions and volatilization of low molecular weight products (Mohsin *et al.*, 2010).

The thermal degradation mechanisms of different types of polymers leads to the production of different amounts and kinds of residues. Without knowing the amounts of each species and the existing gases, it is very difficult to predict which reactions are occurring but it is possible to give some general ideas. For example, the products of a depolymerization type polymer such as POM (polyacetal) or PBMA (polybutyl-methacrylate) are completely different from the residues produced by random type polymer or polyolefin such as PP (polypropylene) or EVA (ethylene-vinyl acetate), as described by Kankawa (1997).

When POM or PBMA decompose, they do not produce any carbon residue in either air or nitrogen. They decompose into monomers of the main chain which are easily evaporated at a low temperature (300°C), as illustrated in Fig. 12.1(a). Monomers are sequentially lost at the chain's end. However, polyolefins or random type polymers decompose to low molecular weight compounds at high temperature and require longer time. The degradation can be described by a random scission of the chain. It could produce some carbon double bonds that require even higher decomposition temperatures (600°C), Fig. 12.1(b). In spite of these results, the use of depolymerization type polymers as a single component of the binder could lead to the production of samples with cracks and holes during thermal debinding in air because of the faster decomposition rate. A combination of

**12.1** Degradation mechanism of polymers (based on Kankawa (1997)): (a) depolymerization type (polyacetal); (b) random decomposition type (polypropylene).

both kinds of polymers seems to lead to better specimens. In this case, the best results are obtained using a sequence of catalytic debinding to eliminate the polyacetal under $NH_3$ atmosphere and leave the polyolefin, followed by a thermal debinding cycle under a controlled atmosphere. These results were demonstrated by TGA–FTIR (Fourier transform infrared spectroscopy) studies, tools that we will describe later in this chapter.

If the debinding experiments take place in $H_2$ atmosphere, the degradation can be attributed to the reaction of $H_2$ with the binder (equation [12.1]) but it is very common to observe decarburization processes following the same mechanism (equation [12.2]) (Myers and German, 1999)

$$[CH_2]n + xH_2(g) \Rightarrow CH_4, C_2H_4, C_2H_6, C_3H_8, C_4H_{10}, \text{etc.} \quad [12.1]$$

$$C + xH_2(g) \Rightarrow CH_4, C_2H_4, C_2H_6, C_3H_8, C_4H_{10}, \text{etc.} \quad [12.2]$$

If the debinding take place in $N_2$, Ar, or a vacuum, it is possible to observe other kinds of decarburization processes attributed to other reactions of the residual carbon with residual oxygen or some oxides present on the metallic powders

$$C + \tfrac{1}{2}O_2(g) \text{ or } (s) \Rightarrow CO(g) \quad [12.3]$$

$$C + O_2(g) \text{ or } (s) \Rightarrow CO_2(g) \quad [12.4]$$

$$C + CO_2(g) \Rightarrow 2CO(g)s \quad [12.5]$$

In those alloys in which it is important to keep the initial carbon content (for example $Fe_2Ni$ alloys or 4600 steel) the most successful results are obtained by an optimum hydrogen/nitrogen mixture atmosphere to ensure residue-free debinding and reduce the oxides but to avoid the severe decarburization effect of $H_2$. In industrial practice, the optimal hydrogen/nitrogen proportion has to be individually determined.

### 12.2.2 Tools for measuring carbon content

It is important to be aware of the main tools available to control the carbon content. The main tool used during carbon control is the elemental analyser designed for wide-range measurement of carbon content of metals, ceramics and other inorganic materials using the combustion and infrared detection technique. This analyser allows for carbon control after the debinding in most of the alloys. In some cases it is necessary to modify the calibration range of this method by modifying the amount of the sample and using special patterns, but it is adaptable to most alloys.

Another way to monitor the carbon content is the use of TGA, as described above, by monitoring the sample weight as the temperature increases. The weight loss as a function of time shows an almost linear decrease up to a certain value, after which the weight loss becomes slower. This deceleration of the weight loss during binder removal for nearly equiaxed shapes is mainly due to geometrical effects. The explanation could be that the binder interface advances at almost constant speed in the molding, the reaction surface decreases and hence the rate of weight loss decreases.

Thermogravimetrical analysis is suitable for studying the decomposition behavior of the polymers used as binders. Indeed, by choosing different atmospheres and a constant heating rate, it is possible to distinguish the effect of the atmosphere on the decomposition rate of each binder. As an example, Fig. 12.2 shows the TGA of a binder of two components, paraffin wax (PW) and high-density polyethylene (HDPE). The TGA was programmed using a constant rate of 10°C/min under $N_2$ atmosphere. It can be observed that the PW starts to decompose near to 220°C and at 450°C the main weight losses takes place, corresponding with the degradation of the HDPE. Based on these results it is possible to design a debinding cycle for the feedstock.

The debinding schedule comprises different steps with two main elements: heating rate ramps and isothermal stages. A combination of these elements will produce defect-free parts. In the first step of the heating cycle, a high rate of 5°C/min can be chosen before any weight loss occurs. At 220°C where the paraffin starts to decompose, an isothermal step has to be selected to favor a slow elimination of this component. Between 220 and 500°C, the

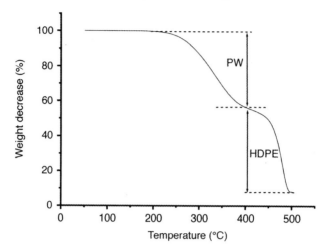

*12.2* Thermogravimetrical analysis (TGA) of a two-components binder.

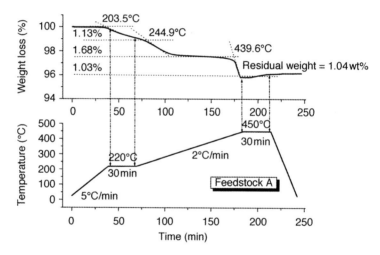

*12.3* Correlation between the debinding cycle and the thermogravimetrical analysis for a feedstock based on PW and HDPE (Herranz, 2004).

main weight loss occurs progressively and the heating rate is reduced by 2°C/min. These heating rates were optimized because high heating rates produce external blisters and cracks, and slow heating rates extend the process. The last isothermal step is considered as the 'maximum debinding temperature' and it determines if there is some residual binder left over. The optimal thermal debinding cycle has to be applied directly on the feedstock to check the viability because the presence of metal powder could slightly modify the transition temperatures. Figure 12.3 shows the correlation between the debinding cycle and the TGA applied directly over the feedstock, showing what a maximum debinding temperature left, which in this case is 1%wt of residual carbon in the sample.

Further information regarding this process can be obtained by analysing the degradation gas by FTIR and mass spectrometry (MS). This approach allows the simultaneous evaluation of bond breaking and product formation during decomposition of the polymeric binder because, during the decomposition process, the gases are collected and analysed. These systems allow the residual weight of each kind of binder to be obtained, essentially the residual carbon, as well as an in-depth understanding of the decomposition mechanism. The time–temperature–gas relationship can be also obtained from these studies (Kankawa, 1997; Mohsin *et al.*, 2010).

### 12.2.3 Debinding mechanisms

Among the debinding variables, the debinding route, thermal cycle, debinding atmosphere and gas flow rate are key issues when it comes to carbon control. Indeed, use of a combination of different debinding processes generally leads to a better control capability than a single-stage process such as thermal debinding. The large number of different binders and different debinding methods (solvent, thermal, wick, catalytic vacuum, etc.) results in a complex approach to the decomposition mechanisms. The most commonly used combination to ensure appropriate carbon control involves a solvent debinding step, which results in partial removal of the binder by the solvent, followed by a thermal cycle to burn out the residual binder under a protective atmosphere. The solvent extraction consists of diffusion of solvent into the binder, dissolution of the binder, interdifussion of the solvent and the binder, and the outward diffusion of the solution to the surface (German, 1987; Hwang, 1996).

In the case of thermal debinding, the binder can be removed using different mechanisms (Wright *et al.*, 1989):

- evaporation (observed in the case of low molecular weight binders);
- thermal degradation (which takes place in the depolymerization of the polyolefins that transform into low molecular weight substances, which diffuse to the surface and finally evaporate);
- oxidative degradation (in which the binder reacts with oxygen and is eliminated).

The binder residues left from debinding could be critical for structural, physical or magnetic properties, as will be shown in this chapter. Owing to incomplete debinding, carburization can take place, but decarburization can occur as well, because the carbon reacts with the oxygen of the particles (Jinushi *et al.*, 2002). The residual components, mainly low-weight species or carbon, may not always be detrimental to MIM components. These residues are usually very fine and they are homogeneously distributed through the matrix; their effect depends on the quantity of the residue and the kind of alloy.

Debinding is affected by the nature of the individual binder components, the interaction between the different components and the interaction between the binder and the powder particles (Angermann and Van der Biest, 1995). When taking into account all these aspects in order to understand the source of the residual carbon after the debinding process, it becomes clear that binder removal by the first mechanism, namely evaporation, does not leave carbon residues. In contrast, thermal degradation occurs all over the compacts and the amount of residue depends on the molecular weight of the binder components. It is important to note, however, that if the components of the binder, or the binder and the powder, interact physically and/or chemically, the amount of residual carbon content can change completely.

Oxidative mechanisms are used less frequently, as most metal powders undergo severe oxidation at the debinding temperatures. Some companies apply debinding in air, up to the temperature at which the TGA curve shows weight gain of the powder as long as the TGA curve shows weight loss of the binder. This process helps to reduce the long debinding time cycles because oxidative degradation lowers the decomposition temperature of the polymers and accelerates their removal. The use of an oxidative atmosphere could lead to an excess of carbon as well. This effect is observed because in air the excessive oxide formation on the powder surface may close the channels for the decomposition gases, producing incomplete thermal debinding. In these cases, debinding in air is restricted to a maximum temperature. Therefore, when the atmosphere used is non-oxidative, binder removal is mostly restricted to evaporation and thermal degradation processes (Angermann and Van der Biest, 1995; Grohowski and Strauss, 2000).

One of the various models proposed to predict the residual carbon content after debinding (Lin, 1997) suggests that the residual carbon content is determined by the isothermal holding time at the 'maximum debinding temperature'. Low molecular weight materials (such as wax) are easily removed by evaporation and therefore leave only minimal amounts of residual carbon. Polymer can degrade in a number of ways, including (Lewis, 1997; Lee et al., 1999):

- chain depolymerization;
- random scission;
- side-group elimination.

Chain depolymerization produces volatile molecular fragments that are practically monomeric. Polymers that decompose into volatile alkanes and

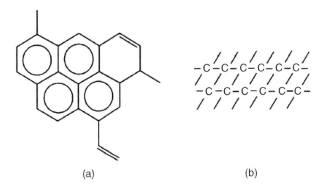

*12.4* Schematic illustration of the chemical structure of carbonaceous residue: (a) pseudographitic and (b) cross-linked (Lewis, 1997)

alkenes (such as polymethyl methacrylate (PMMA)) appear to be some of the cleanest binders.

Random scission (as found for HDPE, polyethylene glycol (PEG) or EVA) produces a wide spectrum of molecular fragments that contain little monomer, therefore they frequently result in residual carbon formation. This is often due to cross-linking reactions between initially degraded chains. Both this kind of reaction and cyclization have to be considered as they both lead to the formation of non-volatile carbonaceous residues. The residual carbon resulting from these processes generally has one of two forms: pseudo-graphitic carbon, which is composed of multiple polycyclic groups (similar to graphite), or highly branched carbon, which is a non-aromatic carbon residue (see Fig. 12.4). The residual carbon content increases linearly with an increase in the logarithm of the molecular weight. Although this type of residue is easily eliminated in an oxidizing atmosphere, the limits on the use of oxygen in the presence of metallic powders lead to the need to choose a binder that degrades into volatile components, has a complex composition that allows its progressive elimination (for example, PP/low-density polyethylene (LDPE)/PW; HDPE/PW/stearic acid (SA); PEG/PMMA; HDPE/PP/PW/EVA/SA; cellulose acetate butyrate (CAB)/PEG; EVA/PW/HDPE/SA) or includes additives that help to volatilize the polymer components in inert atmospheres to ensure low carbon residue contents in those materials where the carbon residue must be kept to a minimum.

The final mechanism, namely side-group elimination (found in polyvinyl butyral (PVB) or polyvinyl acetate (PVA), for example), also produces appreciable amounts of non-volatile carbonaceous residues. Besides water or butanal, such polymers typically produce long chains containing some unsaturation and conjugation, which can subsequently undergo cyclization and cross-linking reactions that result in significant amounts of residual carbon. A suitable binder must therefore be chosen on the basis of the requirements of the metallic system to be produced. For example, non-carbon containing materials such as copper-based alloys can be processed in hydrogen gas and eliminate any of this residual carbon as long as the metal does not react with the hydrogen to form hydrides.

### 12.2.4 Additives

Coupling agents or surface active agents are widely use to improve the interaction between the metallic or ceramic powder and the polymer components. Some additives, such as carnauba wax, act as lubricants at the die walls, decreasing the total viscosity of the feedstock. Other additives, such as stearic acid, have been described as dispersant, reducing the viscosity of the mixtures when it is added directly to the feedstock. However, owing to the chemical nature of the stearic acid (it has a polar end and a non-polar end) this additive is able to create chemical bonds between the binder and the powder surfaces (Tseng et al., 1999; Johansson et al., 2000; Herranz, 2004), which frequently leads to an increase in residual carbon in the form of other coupling agents such as aluminate, silane and titanate, frequently used for improving the rheology of the mixtures (mostly in ceramic systems) (Takahashi et al., 1992). To create the chemical bonds between the binder and the powder surfaces, it is necessary that stearic acid ($C_{17}H_{35}COOH$) is adsorbed onto the metal powder in a previous treatment in a concentrated solution. The metal powder is able to interact with the polar end of the stearic acid molecule and the binder interacts with the non-polar part.

The formation of monolayer coverage can be detected using X-ray photoelectron spectroscopy (XPS) analysis because the carbon peak intensity increases. To determine the optimum amount of SA adsorbed onto the metal powder surface, measure the SA melting enthalpy of the observed peaks in the differential scanning calorimetry (DSC) curves. It can be observed that melting enthalpy increases with the SA concentration of the solution, achieving the saturation at a particular concentration of SA and the optimum percentage of SA. This is indicative of a complete monolayer surface coverage.

There is one final factor to be considered, namely that the presence of a metallic powder can alter the mechanism or the rate of binder elimination. TGA studies must therefore be performed in the presence of this powder as the shape and temperatures of the curve could suffer important modifications. Indeed, in some cases, burnout may be retarded as the small size of the particles hinders binder evaporation (German and Bose, 1997), or the presence of a powder can enhance the thermal degradation. This catalytic effect may be due to reactions between the metal powder surface and polymer. For example, the activation energy for decomposition of EVA decreased in the presence of spherical powder 316L (Atre *et al.*, 2008). Besides, certain oxide surfaces affect the decomposition behavior of polymers (Grohowski and Strauss, 2000). In this sense, it has been reported that purer powders and larger surface areas (smaller particle size) increase the catalytic activity, thereby reducing the debinding time. Great efforts have been made to simulate these processes in order to anticipate the behavior of binder systems in the presence of metallic or ceramic powders (Lewis, 1997; Krug *et al.*, 2002; Atre *et al.*, 2008). It has also been reported that the interaction between the binder component and the powder surface leads to increased residual carbon. The activity of the powder surface can be quantified by the iso-electric point (IEP). In several investigations the carbon content was smallest for powder with iso-electric point close to 7 (neutral). The more acidic or basic a powder's surface, the more interaction between the powder and the binder and the higher the amount of residual carbon (Angermann and Van der Biest, 1995).

The cost of removing the binder is very high, and several aspects of the polymer removal process and predictions of the residual carbon left behind by different binder systems remain to be investigated in order to optimize this step of the MIM process. It is important to consider some particular factors that have to be analysed in any case, such as the size of the part that dictates the ability of the atmosphere to interact with the binder. This fact means that thermogravimetric data from one specific system cannot be directly applied to another.

## 12.2.5 Process parameters affecting carbon control

The true control of the carbon content comes from the combination of atmosphere and temperature. The thermal cycle is a critical process because carburization or decarburization can take place depending on several factors, mainly the atmosphere used during the process. The use of different atmospheres during the debinding process leads to the formation of different species changing generally from carbon–oxygen dominance to hydrogen–oxygen and hydrogen–carbon dominance as the reducing character of the atmosphere increases. The best results are obtained when equilibrium between the gas species is achieved by a combination of carburizer and decarburizer gases. However, although the chosen atmosphere may have a good equilibrium between $CO/CO_2$ or $CH_4/H_2$, the production of gases upon decomposition of the binder requires the ratio of the gas mixture to be varied continuously, thereby making it difficult to ensure a constant composition. Debinding temperatures are also observed to change as a function of the atmosphere, thus meaning that a balance between minimum debinding time, compact integrity and suitable carbon content has to be achieved during industrial processing. In general, the use of pure $H_2$ leads to a minimal carbon content and lower decomposition temperatures, although this is sometimes counterproductive as it may result in severe decarburization of the resulting parts and hydride formation (German and Bose, 1997).

Another relevant industrial parameter that must be controlled during the debinding process is the temperature. Thus, while it may be more interesting industrially to use high temperatures to reduce the debinding time as the elimination rate increases considerably, an appropriate balance must be found. Indeed, higher temperatures can easily produce internal stresses due to vaporization of the binder occurring too quickly, which could cause the component to blister or crack. Moreover, higher temperatures could lead to distortion of the components due to viscous flow under gravity as the binder softens. It is therefore important to choose heating rates according to the decomposition stages of the different components of the binder in order to prevent the onset of defects.

In short, the debinding cycle must be tailored to the binder. A model proposed to predict the residual carbon content after debinding under $N_2$–$H_2$ (Lin, 1997) suggests that the isothermal holding time at the maximum debinding temperature is one of the most important parameters. The main reason behind this is that this stage involves the above-mentioned thermal decomposition and therefore determines the residual carbon left in the component. Industrial processes have to balance a minimum debinding time and compact integrity with a correct selection of the atmosphere according to the metallic system produced. The combination of different debinding processes usually reduces distortion of the components while allowing the debinding stage to be minimized.

## 12.3 Control of carbon in particular materials

The influence of residual carbon may vary markedly depending on the system. A description of the most important systems will therefore be provided in this section. The sintering temperature and carbon content are the most important process variables in tool steels since they dictate the volume fraction of the liquid phase. In general, higher fractions of residual carbon help the densification process, thereby allowing for better control of the process and an increase in the final hardness. In contrast, it is important to achieve close control of the carbon in cemented carbide systems to avoid carburization or production of the brittle $\eta$ phase. In other systems, such as stainless steels, magnets (Fe–Si, Fe–Ni) or titanium alloys, the carbon content should be kept to a minimum as such contamination could lead to undesired mechanical, electrical and magnetic

# Control of carbon content in metal injection molding

properties. The most important aspects of each system are carefully described in this section.

## 12.3.1 High-speed steels (HSS)

High-speed steels have unique physical and mechanical properties that make them good candidates for the production of parts with an optimal combination of high strength, wear resistance, toughness and hardness. Their production by powder metallurgy (PM) techniques results in parts with a uniform distribution of carbides and therefore isotropic mechanical properties. The major disadvantage of this production method for HSS is its moderate sensitivity to sintering parameters such as temperature and atmosphere (Jauregi et al., 1992). Moreover, the optimal conditions are determined by the composition of the HSS, with the carbon content having a particularly pronounced influence on the microstructure evolution and sintering temperature (Wright et al., 2000). The initial studies of Shepard et al. (1973) showed that the carbon and oxygen contents of the starting material undergo major changes during the sintering process.

The critical dependency of the properties of these steels on carbon content led to the development of a technique to compensate for the loss of carbon during manufacture. The most reliable way of obtaining good results was found to involve blending elemental carbon (graphite) with the metal (Shepard et al., 1973), as it was demonstrated that this process can not only alter the composition but also enhance the sintering kinetics (Price et al., 1985; Maulik and Price, 1987). Finally, it proved possible to explain the sintering mechanism that governs these results. The sintering of these materials takes place by a supersolidus liquid phase sintering (SLPS) process that allows a near full density to be achieved (German, 1990). The important difference between SLPS and traditional liquid phase sintering (LPS) is that SLPS produces liquid at the grain boundaries, interparticle boundaries and inside the grains, whereas LPS produces liquid only at the interparticle boundaries. The temperature and carbon content are the most important variables as they dictate the volume fraction of liquid phase that appears during the densification process.

The high mechanical properties and uniform microstructures obtained by PM, and the possibility easily to achieve near full density by accurately controlling the composition, temperature and atmosphere, make HSSs good candidates for production by MIM. Moreover, the processing of HSSs by injection molding to obtain near-net-shape parts avoids costly machining operations. The main challenge in PIM processing of such materials therefore lies in achieving an accurate control of the carbon content. Control of these parameters by controlling the atmosphere is very difficult

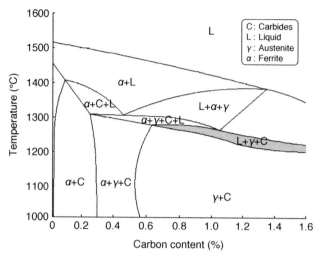

*12.5* Pseudobinary M2 phase diagram based on Hoyle (1988).

as the atmospheric composition, $CO/CO_2$ ratio and the dew point must be controlled and measured accurately throughout the sintering process (Zhang, 1997).

The first method for maintaining accurate carbon control is to program a presintering step after the initial debinding under hydrogen atmosphere. The temperature of this presintering process has to be such that the hydrogen will only attack the free carbon coming from the binder, while leaving the combined carbon unchanged, otherwise the parts suffer a heavy loss of carbon which results in a low density and poor mechanical properties. After presintering, the part could be sintered under high vacuum or hydrogen to achieve full densification (Zhang, 1997) at temperatures similar to those used in conventional PM, for example, near to 1240°C for M2 HSS. This method avoids the distortion that could arise from a non-homogeneous distribution of admixed carbon. The resulting samples contain amounts of carbon similar to those of the starting powder (0.8% of carbon in the case of M2, the most common HSS studied because of its industrially interesting applications). If presintering is performed under nitrogen at 700°C, the parts retain only a very small amount of residual carbon (1.02%). This, together with the smaller particle size used in PIM, explains why it is possible to sinter the parts at lower critical temperatures of around 1210°C (Myers and German, 1999; Liu *et al.*, 2000a). The sintering window remains very narrow, at less than 10°C, in all cases, as shown in Fig. 12.5.

The debinding process inherent in MIM could allow a further possibility suggested by several researchers, although one which has apparently been relatively unsuccessful to date, namely to vary the residual carbon coming

Control of carbon content in metal injection molding 279

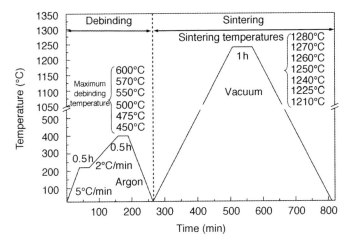

12.6 Debinding and sintering cycles applied to study the effect of incomplete debinding processes in HSS (Herranz et al., 2007).

from the debinding process. Some of the first researchers to explore this option used a modified MIM process based on a thermosetting binder and M2 as HSS powder (Levenfeld et al., 2002). These authors decided to increase the carbon content to prevent carbon loss and found that incomplete debinding led to a notable reduction of the sintering temperature, with a very wide sintering window, for samples containing 3% of residual carbon. This behavior is a consequence of the residual carbon produced upon binder degradation. This expansion of the sintering window allows a microstructural study of the carbide evolution that occurs during the sintering process to be performed, thus giving some idea of the complex sintering process that takes place in this system. The estimation of mechanical properties was made by hardness measurements and the results are in agreement with the values found in similar systems with lower carbon content. The microstructures suggest changes in other properties that have not been evaluated, but the observed changes in the sintering temperature are undoubtedly interesting.

The use of incomplete debinding processes by modifying the maximum debinding temperature, as in Fig. 12.6 (Herranz et al., 2007), permits study of the behavior of steels with different carbon contents in the range between the carbon content of the initial powder and 3 wt%. In industry, the accurate control of incomplete debinding can be really difficult from part to part. However, samples do not suffer from distortion in this range of carbon content. The residual carbon in the incompletely debound samples reduces the optimum sintering temperature, with this temperature decreasing as the residual carbon content increases. The fully debound samples achieve a maximum densification at 1270°C and the partially debound parts achieve

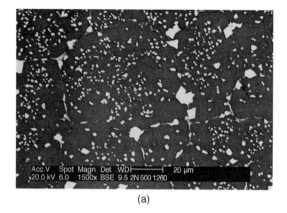

(a)

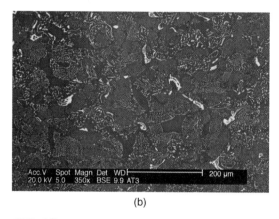

(b)

*12.7* Microstructure comparison between (a) a M2 HSS component with 2 wt% of residual carbon with an optimal distribution of carbides and (b) component with > 3 wt% of residual carbon with the formation of undesirable eutectic carbides (Herranz, 2004).

quasi-full density at 1240°C. A carbon content of more than 2wt% does not seem to affect the results, whereas carbon contents above 3wt% lead to the formation of eutectic carbides at very low temperature, which results in a worsening of the mechanical properties, heterogeneous microstructures and distortion of the components.

The microstructures of a correctly sintered HSS component and a component with an excess of residual carbon are compared in Fig. 12.7 (Herranz, 2004). The second effect is the widening of the sintering window by more than 20°C, as can be seen in Fig. 12.8 (Herranz *et al.*, 2007), which is much wider than normally found for these kinds of steels (Wright and Ogel, 1993; Liu *et al.*, 2000b). Careful control of the carbon content could help the industrial processing of these steels, although the relationship between the carbon content and the atmosphere should be studied in depth

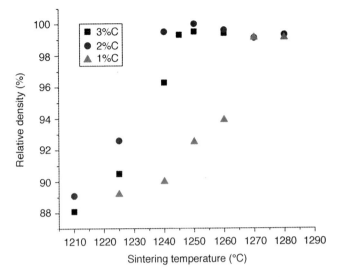

*12.8* Sintering curves for samples debound at different temperatures, and then different residual carbon contents (Herranz *et al.*, 2007).

for each particular case. The size of the samples and the reproducibility of the process are the main factors to extract any industrial advantages from these facts.

In general, the use of $H_2$-rich atmospheres leads to decarburization of the samples, therefore the use of vacuum or $H_2$–$N_2$ mixtures is often more appropriate. Indeed, the use of a $N_2$-rich atmosphere is of particular interest from a technological and economic perspective as it permits the continuous production of HSSs as well as other important benefits, such as changes in the final microstructure. The presence of $N_2$ during sintering results in the substitution of MC carbides by fine MX coarsening-resistant carbonitrides while leaving $M_6C$ carbides unaffected (Jauregi *et al.*, 1992; Miura, 1997). This effect has rather greater consequences in the case of high-vanadium HSS grades. Thus, as described by several researchers (Palma *et al.*, 1989; Aguirre *et al.*, 1999), the great affinity of vanadium for nitrogen results in a sequence of reactions [MC + N → M(C,N) + C → MN + C] that leads to the precipitation of vanadium nitrides dispersed through the matrix, which is increased if the sintering takes place at a higher nitrogen pressure (Gimenez *et al.*, 2008). This implies a further factor that must be considered when designing the sintering process for HSS components in order to be able to control the volume fraction in the liquid phase (German, 1985) as, if these reactions take place, the amount of carbon available increases and therefore the sintering conditions change.

High-speed steel has been used as a cutting material and frequently competes with cemented carbides in some applications (Hoyle, 1988). The addition of several types of reinforcement to HSS during PM generally results in a decrease in the sliding wear rate (Oliveira and Bolton, 1999; Gordo et al., 2000), thus allowing for a broader field of applications. These findings led to the production of HSS-based metal matrix composites (MMCs) by MIM. This production route offers many unique advantages for the mass production of small and complex parts. The main purpose of steel-based composites is to improve wear resistance; therefore the addition of carbides and nitrides to feedstock compositions to meet this objective has notable consequences on the carbon control of the system, as described briefly below.

The addition of carbides both improves the mechanical properties and helps to obtain a homogeneous distribution of the liquid phase thanks to the addition of a carbon source (Zhang et al., 2001); in industry, this is more reproducible than the use of residual carbon. Recent results concerning the production of reinforced M2 HSS demonstrate that carbides have a

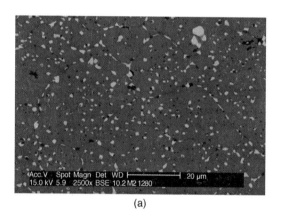

(a)

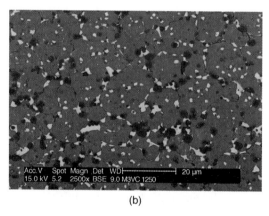

(b)

*12.9* Microstructure of (a) M2 HSS sintered under $N_2$–$H_2$ atmosphere at 1280°C and (b) M2 HSS reinforced with 3 wt% of VC sintered at 1250°C (based on Herranz et al. (2010)).

# Control of carbon content in metal injection molding 283

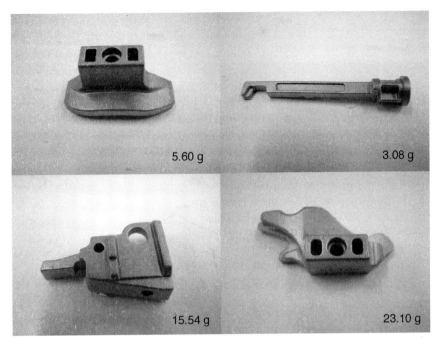

*12.10* M2 HSS components for the textile industry with hardness > 64 HRC (courtesy of MIMECRISA (Ecrimesa Group)).

powerful grain growth inhibiting effect and that these reinforcements decrease the optimum sintering temperature and broaden the sintering window. Furthermore, it has been demonstrated that the inhibiting capacity differs depending on the type of carbide and its reactivity with the steel matrix. Thus, in the case of the addition of VC, Fig. 12.9., which has a great affinity for nitrogen, the reduction of the sintering temperature, the broadening of the sintering window and the precipitation of vanadium nitrides are especially significant and are therefore of great interest for many tool applications (Herranz *et al.*, 2010). Besides the increase in wear resistance of the components, the most important aspect of the microstructural evolution seems to be the relationship between the carbon content and the nitrogen absorption. Some actual M2 HSS components produced after a strict control of carbon content during debinding and sintering are shown in Fig. 12.10.

## 12.3.2 Stainless steels

Stainless steels are the most common metal produced by MIM, followed closely by iron–nickel steels. Stainless steel parts in both the 300 (austenitic) and the 400 (ferritic or martensitic) series are produced from powders. The 300 series austenitic alloys are typically used in applications that require good corrosion resistance (303L, 304L, 316L), whereas ferritic grades are used in applications that require magnetic properties or good thermal conductivity and/or durability and in applications that involve thermal cycling (409, 410, 430). In the martensitic form, these steels are used in wear-resistant applications, although the martensitic grade has the lowest corrosion resistance of all PM stainless steel grades (420, 440). Two special grades, namely duplex stainless steel with a ferritic–austenitic microstructure, and phase-hardened (PH) stainless steel, which can be strengthened by solution treatment, are used very frequently. Duplex stainless steels have higher mechanical strength and improved resistance to stress corrosion cracking than austenitic grades. Their toughness is higher than that of the ferritic steels but slightly lower than that of the austenitic steels. The PH stainless steel (14-4PH, 17-4PH) presents high strength and good corrosion resistance. This combination of properties makes these steels very common in MIM and has widespread applications.

The corrosion resistance of sintered stainless steels, which is a key property for the majority of uses, depends on a number of factors, several of which are related to sintering. The optimal mechanical properties of MIM components require densities in the range between 97% and 98% of the theoretical value. These high densities are achieved by using extremely fine spherical powders and by sintering at high temperature. For stainless steels, these temperatures range between 1120 and 1350°C for times of between 30 to 120 min under a protective atmosphere or vacuum. During sintering, it is vital to avoid contamination and minimize the presence of precipitates of chromium carbide, chromium nitride and silicon oxide in the microstructure, as well as to control the formation of surface oxides and/or nitrides on cooling. To avoid these detrimental processes the carbon content should be very low as its presence lowers corrosion resistance (Klar and Samar, 1998). The main reason behind this phenomenon is the possibility of intergranular corrosion at the grain boundaries of stainless steels. The solubility of carbon and nitrogen into the matrix at room temperature is much lower than at high temperature. Carbon and nitrogen therefore precipitate as carbides and nitrides at the grain boundaries during cooling. Additionally, the diffusion rates of carbon and precipitated carbides are higher than the diffusion rate of chromium in the matrix. The great affinity of carbon for chromium therefore leads to the formation of chromium carbides, thereby producing a decrease in chromium concentration in those areas adjacent to the carbides.

These areas are susceptible to corrosion. This fact suggests that it is important to remove the lubricant in conventional PM to prevent diffusion of carbon into the part. The carbon content is much more critical in MIM processing and is therefore controlled during the debinding process. Binders can add up to 5% carbon to the part after molding, whereas the maximum allowed carbon content in the final sintered part can be 0.03% or less for sintered stainless steel (Klar and Samar, 1998). There is a correlation between the degree of carbon control and the oxygen content during binder elimination as oxygen assists in carbon extraction via formation of CO and $CO_2$. However, the grain surfaces can be oxidized by the binder and the resulting oxides reduced again at sinter temperatures by carbon diffusing to the surface oxide. This process therefore has to be strictly controlled.

The main stainless steels used in commercial MIM applications are 316L and 17-4PH. 316L is an austenitic steel known for its corrosion resistance and 17-4PH is a precipitation hardening grade with reasonable corrosion resistance and a much higher strength than the austenitic stainless steels. Another promising stainless steel, namely nickel-free austenitic stainless steel, arose from the need to avoid the release of nickel into the organism during biomaterial applications. The effect of residual carbon on each grade will be described below.

Despite the fact that the carbon content in the austenitic steel 316L must be kept very low in order to ensure maximum corrosion resistance, a very small quantity of carbon could be beneficial for reducing the oxide on the powder surfaces, frequently silicon oxide. At this point, the debinding atmosphere is very important to control the residual carbon. The most used atmospheres are hydrogen, nitrogen plus hydrogen, argon, nitrogen or a combination of inert atmosphere with small percentages of air or oxygen to contribute to the burnout of the binder. The interaction between the atmosphere and the powder could produce unpredictable results. Although the presence of hydrogen should lead to the reduction of the residual carbon in normal conditions, if the powder presents oxides (very common in brown parts) the hydrogen reacts with the oxygen which is not available to react with the residual carbon. The carbon content in parts debound under inert atmosphere could be lower, which is especially relevant for stainless steels.

In general, the best results are obtained upon presintering at 800°C under a hydrogen atmosphere, followed by sintering at a maximum temperature of between 1300 and 1390°C under hydrogen or vacuum (Suri *et al.*, 2005; Shu *et al.*, 2006). Industrial experience suggests that 316L with a carbon content above 0.06% contains large pores and is less corrosion-resistant (Fig. 12.11.)

The residual carbon content has a strong influence on the microstructure of injection-molded 17-4PH. Indeed, it has been demonstrated that carbon from the debound parts remains even after sintering. A relationship between carbon content and volume fraction of austenite has also been reported

*12.11* Microstructure appearance of 316L component after industrial processing when the carbon content is above 0.06 wt% (courtesy of MIMECRISA (Ecrimesa Group)).

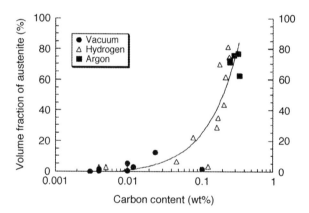

*12.12* Relationship between the volume fraction of austenite and the carbon content for sintered compacts in various atmospheres (Baba et al., 1995).

(Baba et al., 1995). It can be seen in Fig. 12.12 that the volume fraction of austenite increases with carbon content, and increases abruptly over 0.1 wt% of carbon. There is also a relationship between the amount of austenite and the final mechanical properties of the sintered and aged compacts, which present lower hardness at higher carbon contents, with an abrupt decrease above 0.1 wt% of carbon. Changes in the microstructures are detected when the residual carbon content is varied. As the carbon content decreases, the microstructure changes from predominantly austenite and martensite to martensite and δ-ferrite. These results lead to analysis of the contribution of δ-ferrite in the observed variations of properties. These variations arise as a result of the δ-ferrite phase formed during sintering (Wu et al., 2002). Thus,

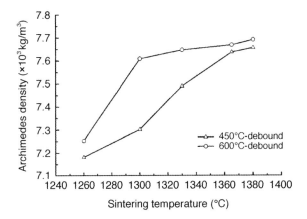

*12.13* Archimedes densities of PIM 17-4PH sintered at different temperatures debound at 450°C and 600°C to modify the residual carbon content (Wu *et al.*, 2002).

a higher amount of δ-ferrite is detected as the residual carbon content decreases as a result of the transformation γ → γ + δ that occurs at lower temperatures. The presence of this δ phase also has an important effect on the densification process. Thus, the δ/γ interphase boundaries contribute to the mass transport and this phase also increases the overall atomic diffusivity, both of which contribute to an increase in the densification of the compacts as the residual carbon decreases, as shown in Fig. 12.13, where the maximum densities are achieved at the highest debinding temperature. A similar beneficial effect of δ-ferrite, as a consequence of residual carbon, has been reported in the sintering of duplex stainless steels (Puscas *et al.*, 2001). Furthermore, 17-4PH components with carbon contents above 0.08% have much lower mechanical properties and the weldability is decreased. An example of an optimized component for the medical industry made of 17-4PH with a carbon content under 0.08 wt% is shown in Fig. 12.14.

The production of high-carbon martensitic stainless steel, such as 440C, by MIM has rarely been reported, mainly as a result of the rapid sintering densification that occurs during SLPS. Indeed, this process is often limited by the narrow processing window available to attain densification without distortion. Recent studies (Li *et al.*, 2010) have shown that it is possible to obtain distortion-free components by MIM by accurately controlling the oxygen and carbon content. These steels have a higher amount of carbon that allows for conventional heat treatment by quench and temper. In this case, carbon has been found to decrease the liquid temperature and to enhance the sintering kinetics, as more carbon means more liquid in the sintering phase. This liquid film could subsequently provide a viscous resistance to grain movement, thereby contributing to the rigidity of the structure during SLPS. A problem arises, however, when the distribution of

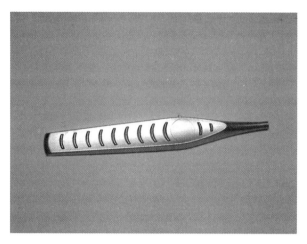

*12.14* Medical device made of 17-4PH stainless steel by MIM (14 g) (courtesy of MIMECRISA (Ecrimesa Group)).

the liquid is not uniform, as the range of carbon contents throughout the sample leads to non-uniform shrinkage. The oxidation–reduction reactions between the carbon and oxygen present in the components during sintering decrease the carbon content and make the carbon distribution non-uniform. The reduction of oxygen content that occurs as a result of the deoxidation process after debinding in $H_2$ allows components with better shape retention to be obtained. The content of carbon in this family of steels has an important effect on the shape retention.

Although the production of nickel-free austenitic stainless steels by MIM may appear unusual, the development of such materials is vital in order to obtain implants with the mechanical properties and price of stainless steel but with the ability to be used in long-term implantation. The elements that stabilize that austenitic microstructure and could substitute nickel include cobalt, carbon, nitrogen, manganese and copper. As this chapter is dedicated to the effects of carbon on steels, it is of interest to highlight the widely differing opinions available in the literature, which indicate that this matter is only likely to be resolved with more experience in the use of this special stainless steel. One such view suggests that the carbon content should be kept to a minimum (as is the case for other stainless steels), whereas the contrasting point of view notes that the activity of carbon decreases significantly with nickel content, therefore the austenitizing effect is more remarkable (Sumita *et al.*, 2004). The effect of residual carbon in such novel steels therefore remains to be clarified.

## 12.3.3 Cemented carbides

Cemented carbides are one of the most important groups of sintered tool materials due to their high hardness and outstanding wear resistance. This combination of properties is a result of their composite nature, whereby brittle refractory transition metal carbides (WC, TiC, TaC, $Cr_3C_2$ or $Mo_2C$) are combined with a tough binder metal (generally Co, but sometimes Ni or Fe). Carbon control is the most important issue during manufacture of WC–Co cemented carbides by PIM as it determines both their mechanical properties and dimensional stability.

The carbon present in these alloys is critical as even small fluctuations in carbon content can alter the microstructure of cemented carbides, producing changes in the properties as well as distortion. Thus, low carbon content gives rise to the formation of a brittle η phase, whereas excess dissociated carbon appears as graphite; both these new phases decrease the strength and hardness. The carbon content of these alloys can be measured using conventional carbon analysers using the combustion technique and infrared detection, reducing the quantity of the sample to increase the detection limit of the equipment.

Studies in this field have tended to analyse the effect of the atmosphere used during the manufacturing process to control the carbon content (Spriggs, 1970). Debinding and sintering processes are especially long in the production of cemented carbides. Different atmospheres, such as a protective argon atmosphere during debinding, a presintering treatment in vacuum and sintering under argon, $N_2$, $H_2$ or a mixed atmosphere, could be combined in conventional processing methods (Upadhyaya et al., 2001). However, an $N_2$ atmosphere would not remove the binder completely and an $H_2$ atmosphere could result in decarburization if the debinding temperature exceeds 450°C, which in turn would result in inconsistent properties and dimensional control, as carbon control is much more difficult under such an atmosphere. At temperatures above 450°C the carbon in the brown parts could react with existing oxygen to form $CO_2$ and further react with $H_2$ to form $CH_4$, producing decarburization of the samples (Fan et al., 2007)

$$WC + 2H_2 \Leftrightarrow CH_4 + W \qquad [12.6]$$

Some studies have found that thermal debinding under a 75% $N_2$/25% $H_2$ atmosphere balances the decarburizing effect of $H_2$ and the carburizing effect of $N_2$, thereby resulting in an appropriate carbon control (Baojun et al., 2002).

Another observed effect when the debinding is made under an $H_2$ atmosphere is the reactivity between the carbide and the wet $H_2$ atmosphere.

In these conditions the loss of carbon increases quite markedly (Spriggs, 1970). At temperatures between 400 and 450°C the WC, for example, would react with $H_2O$ and form CO or $CO_2$, producing the decarburization of the specimens

$$WC + H_2O \Leftrightarrow CO + H_2 + W \qquad [12.7]$$

The decarburization can be more serious if small amounts of W react with $H_2O$ producing $WO_2$. Also, loss of carbon could occur if there is any oxide in the mixtures because at temperatures around 600°C it is possible to observe reduction of these species

$$MeO + C \Leftrightarrow Me + CO \qquad [12.8]$$

Although several steps can be used during thermal debinding to shorten the debinding process, it has been shown that the best carbon content control is achieved using a combination of solvent-based and thermal debinding (better than only thermal debinding) under 75% $N_2$/25% $H_2$ or vacuum debinding while maintaining the process at a lower temperature for a longer time. This method makes it easier to manufacture injected hard metal parts with high transverse strength that require a stoichiometric quantity of carbon content in a narrow margin, usually close to 6%, without carburization and decarburization (Qu *et al.*, 2005; Fan *et al.*, 2007).

One of the main current research topics (Fernandes and Senos, 2011) in the field of cemented carbides concerns the development of new composites with partial or total substitution of the traditional cobalt binder by other cheaper and less toxic materials. A deeper knowledge of the phase diagrams of these new composites is therefore essential to obtain the desired final phase composition and to select the appropriate sintering cycle conditions, with carbon content being one of the main issues to be clarified. These advances will allow, in the near future, the development of new materials to be produced by MIM technique.

## 12.3.4 Magnets

Although there are several different types of magnets, PM in general, and MIM in particular, are effective for applications in which complex magnetic parts would otherwise require considerable machining. Furthermore, as densities approaching the theoretical density are possible when the components are processed by MIM, magnetic inductions close to the saturation magnetic induction are also possible. This means that magnetic inductions equivalent to those of wrought alloys are possible for alloys with a similar composition. In addition, it is possible to develop alloys by adding new alloying elements that cannot be considered when using wrought fabrication technology.

Magnetically soft materials that are produced in large quantities include high-purity iron, low-carbon steels, silicon steels, iron–nickel alloys, iron–cobalt alloys, and soft magnetic ferrites. The magnetism of these materials is exhibited only while an external magnetic field is applied. Thus, in general, when a piece of iron is placed near a permanent magnet or in the magnetic field generated by an electrical current, the magnetization induced in the iron by the applied field is described by a magnetization curve obtained by plotting either the intensity of magnetization or the magnetic induction, $B$, as a function of the applied field, $H$. The behavior of any magnetic material can be defined by its hysteresis loop and $B/H$ ratio, which is termed the permeability. This value indicates the relative increase in flux or magnetic induction caused by the presence of a magnetic material (Moyer, 1998).

Three elements, namely the ferromagnetic materials, iron, nickel and cobalt, and their respective alloys, are truly magnetic. Indeed, the high magnetic saturations of iron and cobalt, together with their availability and price, mean that the commercial soft magnetic alloys used in PM and MIM processing are commonly produced from high-purity iron or various ferrous alloy types such as Fe–2Ni, Fe–3Si, Fe–0.45P, Fe–0.6P and 50Ni–50Fe.

Gas-atomized powders are normally selected to fabricate these parts because they are purer and finer, although iron and nickel powders produced by the carbonyl process are also widely used. Permeability, coercive field and hysteresis loss are strongly affected by impurities within the alloy, with the impurities that are most harmful to these alloys including carbon, nitrogen, oxygen and sulfur. For this reason, careful elimination of the binder is required to ensure minimal carbon content.

Most companies that produce magnetic parts use dedicated furnaces for sintering magnetic parts and select protective atmospheres such as $N_2$–$H_2$, vacuum or argon, that do not include carbon to avoid contamination, as even a small excess of carbon content (around 0.03 wt% C) seriously degrades the magnetic properties (Salak, 1995).

It is important to note that the formation of structural defects is common when significant densification occurs prior to complete binder burnout. The binder used in this system should decompose and the channels should be opened below 500°C, as densification starts above this temperature and the carbon content after sintering remains close to the residual carbon after debinding (Fig. 12.15). The process of closing the pore channels reduces the reaction area during densification, thereby reducing the decarburization rate.

Taking into account that the initial particles do not present significant

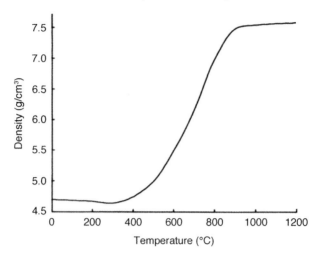

*12.15* Densification of Fe–2Ni as a function of sintered temperature in hydrogen (Ho and Lin, 1995).

oxidation, atmospheres containing high hydrogen concentrations (around 40%), which could help to remove the excess carbon, can be used to ensure a carbon content similar to the initial value. This atmosphere allows maximum values of hardness and strength to be achieved with a microstructure close to that of pure pearlite (Ho and Lin, 1995; Zhang and German, 2002).

The decarburization process is influenced by other parameters such as the gas flow rate and the holding time, which are critical not only because the carbon content changes but also because the carbon distribution through the sample may not be uniform when the gas flow rate is too low, for example. The use of vacuum conditions is also common, in which case commercial carbonyl iron grade powder coated with $SiO_2$, such as the BASF-OS grade carbonyl iron powder, is frequently used (Lin, 2011). This $SiO_2$ surface coating layer reacts with the residual carbon content to form CO or $CO_2$ during sintering, thereby essentially depleting it. Furthermore, although $SiO_2$ depletes the carbon content of iron during sintering, it also improves the mechanical properties of MIM components by inhibiting grain growth, which leads to finer grains within the components.

Other types of magnets include permanent magnets, which are normally used in a single magnetic state and have stable magnetic fields. Permanent magnets are characterized by a high coercive force ($Hc$) and residual induction ($Br$) and are generally manufactured using a variety of metals, intermetallics and ceramics, including certain steels, Alnico, CuNiFe, Fe–Co alloys containing V or Mo, Pt–Co, hard ferrites and cobalt-rare earth alloys. Permanent magnets are usually obtained by casting or sintering.

Thus, cast magnets have a higher residual induction but have the disadvantage of being strongly dependent on the cooling rate, which may result in defects that limit the production of small and complex pieces. In contrast, sintering allows the production of more complex designs, and the use of MIM guarantees close tolerances and the possibility to obtain large series of small components. Some of the most common permanent magnets, including Nd–Fe–B, Alnico and other rare-earth compositions such as Sm(Co, Fe, Cu, Zr)z, are very attractive for high-temperature applications because of their large energy products and performance at such temperatures.

The first Nd–Fe–B magnets were successfully sintered in 1984 (Sagawa et al., 1984), although their main disadvantage is their maximum operating temperature of only 170°C. Furthermore, it is critical to avoid oxidation and carbonization during production of these components due to the binder. Indeed, to obtain magnets, the sintered parts must contain less than 0.1 wt% of carbon, and to optimize the magnetic properties less than 0.08 wt% is required for such magnets (Yamashita, 1998).

Detailed studies have been undertaken to estimate the influence of the residual carbon content on the magnetic properties as it has been demonstrated that an excess of carbon produces critical changes in the microstructure of the magnets. Indeed, magnets have a uniform cellular microstructure when the carbon content is low. However, a comparison of the cellular microstructures of magnets with different carbon contents has shown that the cell size increases with increasing carbon content and that *Br* and *Hc* are close to zero when the carbon content reaches a certain value (for example, 0.43 wt% in the Sm(Co, Fe, Cu, Zr)z magnets (Tian et al., 2007)). Furthermore, carbon could react with some components of the magnets to produce secondary phases that impede the sintering process. This is the case of the reaction between carbon and Zr, which produces a ZrC component that reduces the liquid phase of the magnet during sintering, thereby drastically impeding the densification. All these effects produce a dramatic worsening of the magnetic properties of the components, as can be seen in Fig. 12.16.

The production of other permanent magnets, such as Alnico, by MIM is yet to be optimized; therefore such systems have received increasing interest in the last few years. Such magnets are hard and brittle, which makes them difficult to mechanize, and they have lower coercivities than SmCo or NdFeB magnets. However, the very high Curie temperature of Alnico magnets means that they can be used in systems with higher operating temperatures (about 540°C, compared with 180°C for NdFeB), and they also exhibit high remanence and very good corrosion and oxidation resistance. As is the case with the other magnets described herein, the MIM process requires the use of pure starting components and an appropriate

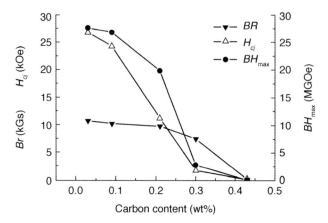

*12.16* Magnetic properties of Sm(Co, Fe, Cu, Zr)z magnets with different carbon contents (Tian *et al.*, 2007).

debinding step to extract the majority of the residual carbon. The best results are obtained under a hydrogen atmosphere, although further optimization of the debinding, sintering and thermomagnetic treatment to achieve the same magnetic properties as pressed samples is still required (Zlatkov *et al.*, 2009).

Control of the oxygen content during the MIM process affects the same factors as the carbon content, namely the binder, debinding and sintering atmospheres, therefore both contents are usually monitored simultaneously. It is common to apply antioxidation treatments to the powders, such as an agent that covers the powder surface, thereby preventing direct contact with the atmosphere, to avoid the oxidation of permanent magnets (Shengen *et al.*, 2006).

### 12.3.5 Titanium alloys

Owing to their attractive properties such as low density, high strength and good resistance to corrosion and oxidation, titanium alloys have always been considered to be important materials. However, their applications have been limited to date because of the high production costs of the raw material and the manufacturing process. Fortunately, recent research has led to a reduction in the cost of titanium products. MIM, for example, offers a low-cost alternative to otherwise expensive machining processes. Since the preliminary research in this field in the 1990s, the numerous parameters that can be varied to ensure the successful production of titanium parts have been defined (German, 2009). Consequently, the production of Ti by MIM is practised by several companies around the world. Despite this current production capacity, however, the MIM of titanium alloys remains a field of significant research interest owing to its complexity and numerous variables.

It has been demonstrated that the key attributes of the powder are its particle size, shape, and distribution, and its oxygen and carbon contents (German, 2009). The most common MIM alloys are pure titanium (CP Ti) and Ti6Al4V, although a number of groups are working on other systems, such as those obtained by replacing vanadium (toxic for the human body) with niobium, T6Al7Nb or Ti4Fe7Cr, Ti4Fe, Ti5Fe5Zr, Ti5Co, Ti6Al4Zr2Sn2Mo and intermetallics such as TiAl or the shape-memory alloy NiTi.

Titanium powders are available in different morphologies, sizes and compositions. However, the production cost of the initial powders remains an important issue. Many trials are in progress on novel production routes (Froes and German, 2000). For example, the use of titanium hydride powder, which has been explored since the 1990s, is an attractive alternative for breaking the cost barrier. Nevertheless, only recently is the number of works describing processing conditions and properties of sintered parts increasing (Li *et al.*, 2006; Carreño-Morelli *et al.*, 2010).

As far as carbon control is concerned, there is a general trend amongst titanium alloys to minimize the carbon content as, even at low concentrations, interstitial carbon severely degrades the mechanical properties of such alloys (Ouchi *et al.*, 1998). The characteristics of the starting powder with regards to carbon content are important as the elimination of residual carbon during MIM processing is complicated.

In addition to traditional thermoplastic and thermosetting binders, several types of binder have been designed in the search for easier ways to eliminate carbon residues and facilitate their decomposition. The most popular such systems are based on paraffin wax, polyethylene, polypropylene and, on occasions, stearic acid. Indeed, binder systems based on aromatic compounds have provided some interesting results (Weil *et al.*, 2006). However, although the initial solvent debinding proved to be successful, the second, thermal debinding step was more difficult. The best option therefore appears to be a vacuum process during heating while at the same time leaking argon into the vacuum chamber to favor the extraction of the binder residues. Although an accurate control of the carbon content is indispensable, experience has shown that oxygen control is more difficult in this system. Control of carbon, oxygen and nitrogen as well, is critical to success with titanium because these elements have substantial effects on properties (Conrad, 1981). Besides, the effects of these impurities are related to each other. For example, the yield strength of sintered titanium depends on density ($\rho$) and oxygen content ($X_o$) (German, 2009).

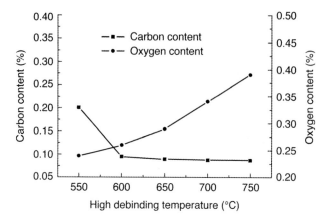

*12.17* Correlation of maximum debinding temperature during 1 h and carbon and oxygen content of the specimens during thermal vacuum debinding of Ti6Al4V. Starting carbon level of 0.056 wt% and oxygen level of 0.192 wt% (Guo et al., 2006).

$$\sigma_y = (420 + 970\, X_o)\, \rho \qquad [12.9]$$

where $\sigma_y$ is yield strength in MPa; but it is necessary to take into account that the oxygen equivalent depends on nitrogen and carbon

$$X_o = (\text{wt\% O} + 2\,(\text{wt\%N}) + 0.5\,(\text{wt\%C}) \qquad [12.10]$$

This inter-dependence and the high sensitivity to these impurities contents could explain why the reported properties of Ti PIM alloys are scattered and highly variable. The binder elimination has an important role on the final levels of these impurities.

Thus, the detailed study of both parameters carried out by Guo *et al.* (2006) on the Ti6Al4V system demonstrated that although an argon atmosphere results in lower residual carbon content, a vacuum atmosphere favors the reduction of the oxygen content. This finding gives some idea of the importance of studying the system as a whole. Indeed, measurements of carbon and oxygen content after different debinding times and temperatures under vacuum have shown that, whereas the carbon content was reduced to 0.095 wt% after 1 h at 600°C, the oxygen content increased continuously during debinding (Fig. 12.17), thereby suggesting that the debinding process should not be extended for too long. Recent studies have described the use of PEG- and PMMA-based binders that can be eliminated completely by a combination of solvent debinding in water and thermal debinding under an argon atmosphere (Sidambe *et al.*, 2010). It is preferable to remove binders at the lowest possible temperature so that titanium does not react with the oxygen and carbon. Most of the binders decompose by about 450°C and it is observed that the impurity contents increase with higher processing temperatures. Once titanium carbides or oxides are formed, they cannot be reduced during sintering. Furthermore, the use of higher sintering temperatures leads to more contamination (mostly increased oxygen content) and reduces the ductility of the final sample. Lower temperature (in the range of 1250°C) and shorter sintering time (around 3 h) is common since containerless hot isostatic pressing is used to reach full density (a treatment between 850 and 1100°C).

Finally, although the sintering process can be performed under argon to maintain a low carbon and oxygen content, in some cases this has been found to result in gas becoming trapped in the pores, thus resulting in low densifications. The best procedure therefore involves sintering under vacuum at a maximum temperature of between 1100 and 1450°C, depending on the system. These general conditions maintain the oxygen content at slightly over 0.2 wt% (near to the starting oxygen content of a gas atomized powder), the nitrogen content at below 0.03 wt% and the carbon content at 0.04 wt% or less (depending on the debinding process), and are therefore useful for even the most demanding applications. For some industrial and consumer products a low ductility is often suitable and it is possible to allow oxygen levels up to 0.5 wt% (German, 2009).

## 12.4 Material properties affected by carbon content

Finally, it is important to correlate the carbon content with material properties. Carbon is the element with the largest influence on the mechanical or magnetic properties. Thus, whereas other elements are not modified intentionally during the MIM process, the carbon content is strongly affected by the raw materials used in the feedstock and the various steps involved in MIM, ranging from debinding and sintering to heat treatments. In Table 12.1 some practical examples are shown from research studies where special attention has been paid to the residual carbon content.

Table 12.1 Practical examples of material properties affected by carbon content

| Material | Examples | Debinding conditions | Sintering conditions | Carbon content | Mechanical properties or magnetic properties | References |
|---|---|---|---|---|---|---|
| High-speed steels | SKH10 | Solvent in heptane, 85°C + thermal 600°C in $H_2$–$N_2$ | 1200–1300°C $H_2$–$N_2$ + heat treatment of quenching and triple tempering | 1.7% | Hardness: 70 HRC TRS: 3200 MPa | Miura (1997) |
| | M2 | Catalytic depolymerisation at 120°C + thermal up to 700°C in $N_2$–$H_2$ | 1275–1287°C, 30 min in $N_2$–$H_2$ | 0.66–0.77% | Hardness: 57–64 HRC TRS: 1909–2315 MPa | Myers and German (1999) |
| | M2 | Thermal in vacuum up to 250°C for 10–40 h + presintering in $H_2$ 800°C, 1 h | 1243°C, 1 h, vacuum | 0.79% | Hardness: 43 HRC UTS: 800 MPa | Zhang (1997) |
| | M2 | Thermal in argon up to 500°C | 1250°C, 1 h, high vacuum | 0.88–0.91% | Hardness: 620 HV | Herranz et al. (2005) |
| | M2 | Thermal in argon up to 450°C | 1100–1350°C, high vacuum | 0.81–3.2% (brown parts) | Hardness: 650 HV | Levenfeld et al. (2002) |
| Stainless steels | 17-4PH | Thermal in $H_2$ up to 450°C, 2 h | 1380°C, 1 h in $H_2$ | 0.203% | TS: 980 MPa | Wu et al. (2002) |
| | | Thermal in $H_2$ up 600°C, 2 h | 1340°C, 1 h in $H_2$ | 0.130% | TS: 940 MPa | |
| | 17-4PH | Solvent in heptane 5 hours 75°C + thermal in $H_2$ up to 1050°C | 1250–1350°C, 1 h, vacuum or $H_2$ | <0.1% | Hardness: 25 HRC TS: 1100 MPa (After aging treatment) Hardness: 43.5 HRC TS: 1335 MPa | Baba et al. (1995) |
| | 440C | Solvent in methylene dichloride 37°C, 6 h + | 1200–1260°C, 30 min in | 1.05% | Hardness: 57.7 HRC TS: 876 MPa | Li et al., 2010 |

| | | | | | |
|---|---|---|---|---|---|
| Cemented carbides | WC-8Co | thermal in Ar up to 600°C + presintering up to 950°C vacuum + aging treatment | 1400°C, 80 min vacuum | 5.63–5.65% | TRS: 2300–2500 MPa | Fan et al., 2007 |
| | WC-TiC-Co | Solvent in heptane, 40°C + thermal 450°C in H$_2$ | Vacuum | 6.20–6.35% | Hardness: 90–93 HRA<br>TRS: 2000–2100 MPa | Qu et al., 2005 |
| | WC-8Co | Solvent in heptane, 30°C, 2 h + thermal in vacuum | 1400°C, 1 h | 5.6–5.7% | Hardness: 90 HRA<br>TRS: 2500 MPa | Baojun et al., 2002 |
| Low-alloy steel | AISI 4600 | Solvent + thermal in N$_2$–H$_2$ up to 600°C, 1.5 h | 1200–1300°C in H$_2$–N$_2$ (different proportions) + heat treatment of reheating + quenching + tempering | 1.2–0.01% | TS: 1400 MPa<br>Hardness: 33 HRC<br>Elongation: 9% | Miura, 1997 |
| | AISI 4100 | Thermal up to 400°C in H$_2$–N$_2$ + solvent in heptane 85°C + thermal 600°C in H$_2$–N$_2$ (different proportions) | | 0.4–0.5% | TS: 1250 MPa<br>Hardness: 39 HRC<br>Elongation: 4% | Miura, 1997 |
| Titanium alloys | Ti6Al4V | Solvent in heptane, 80°C, 5 h | 1250–1280°C in vacuum | 0.073% | EM: 122 GPa<br>YS: 865 MPa<br>UTS: 955 MPa<br>Elongation: 12% | Zhang et al., 2008 |
| Magnets | Alnico 8 | Solvent in acetone, 45°C, 18 h + thermal in H$_2$ | Sintering in H$_2$ 1300–1325°C + thermomagnetic treatment | 0.183% | Br=0.772T<br>Hcb= 85.19 kA/m<br>(BxH)max=22.95 kJ/m$^3$ | Zlatkov et al., 2009 |
| | Nd-Fe-B | Thermal 300°C, 30 min H$_2$ + 30 min vacuum | 1080–1120°C, 4 h in vacuum + annealed in vacuum 500°C, 2 h | 610–790 ppm | Br=1.268T<br>iHc= 0.68–1.10 MA/m<br>(BxH)max=306–287 kJ/m$^3$ | Yamashita, 1998 |

## 12.5 References

Aguirre, I., Gimenez, S., Talacchia, S., Gomez-Acebo, T. and Iturriza, I. (1999), 'Effect of nitrogen on supersolidus sintering of modified M35M high speed steel', *Powder Metallurgy,* 42(4), 353–357.

Angermann, H. H. and Van der Biest, O. (1995), 'Binder removal in powder injection molding', *Reviews in Particulate Materials,* 3, 35–70.

Atre, S. V., Enneti, R. K., Park, S. J. and German R. M. (2008), 'Master decomposition curve analysis of ethylene vinyl acetate pyrolysis: influence of metal powders', *Powder Metallurgy,* 51(4), 368–375.

Baba, T., Miura, H., Honda, T. and Tokuyama, Y. (1995), 'High performance properties of injection molded 17-4 PH stainless steel', in *Advances in Powder Metallurgy and Particulate Materials,* Metal Powder Industries Federation, Princeton, NJ, USA, pp. 6271–6278.

Baojun, Z., Xuanhui, Q. and Ying, T. (2002), 'Powder injection molding of WC–8% Co tungsten cemented carbide', *International Journal of Refractory Metals and Hard Materials,* 20, 389–394.

Carreño-Morelli, E., Krstev, W., Romeira, B., Rodriguez-Arbaizar, M., Girard, H., Bidaux, J.-E. and Zachmann, S. (2010), 'Titanium parts by powder injection moulding of TiH$_2$-based feedstocks', *PIM International,* 4(3), 60–63.

Conrad, H. (1981), 'Effect of interstitial solutes on the strength and ductility of titanium', *Progress in Materials Science,* 26, 123–403.

Fan, J. L., Li, Z. X., Huang, B. Y., Cheng, H. C. and Liu, T. (2007), 'Debinding process and carbon content control of hard metal components by powder injection moulding', *PIM International,* 1(2), 57–62.

Fernandes, C. M. and Senos, A. M. R. (2011), 'Cemented carbide phase diagrams: A review', *International Journal of Refractory Metals and Hard Materials,* doi: 10.1016/j.ijrmhm.2011.02.004.

Froes, F. H. and German, R. M. (2000), 'Cost reduction prime Ti PIM for growth', *Metal Powder Report,* 55(6), 12–21.

German, R. M. (1985), *Liquid Phase Sintering,* Plenum Press, New York, USA.

German, R. M. (1987), 'Theory of thermal debinding', *International Journal of Powder Metallurgy,* 23(4), 237–245.

German, R. M. (1990), 'Supersolidus liquid phase sintering. Part 1. Process review', *International Journal of Powder Metallurgy,* 26(1), 23–34.

German, R. M. and Bose, A. (1997), *Injection Molding of Metals and Ceramics,* MPIF, Princeton, NJ, USA.

German, R. M. (2009), 'Titanium powder injection moulding: a review of the current status of materials, processing, properties and applications', *PIM International,* 3(4), 21–37.

Gimenez, S., Zubizarreta, C., Trabadelo, V. and Iturriza, I. (2008), 'Sintering behaviour and microstructure development of T42 powder', *Materials Science and Engineering A,* 480, 130–137.

Gordo, E., Velasco, F., Anton, N. and Torralba, J. M. (2000), 'Wear mechanisms in high speed steel reinforced with (NbC)p and. (TaC)p MMCs', *Wear,* 239, 251–259.

Grohowski, J. A. and Strauss, J. T. (2000), 'Effect of atmosphere type on thermal debinding behavior', in *Proceedings of PM2TEC 2000, International Conference*

on *Powder Metallurgy and Particulate Materials*, New York, USA, pp. 4.137–4.144.
Guo, S., Qu, X., He, X., Zhou, T. and Duan, B. (2006), 'Powder injection molding of Ti–6Al–4V alloy', *Journal of Materials Processing Technology*, 173, 310–314.
Herranz, G. (2004), *Development of New Binder Formulations Based on HDPE to Process M2 HSS by MIM*, PhD thesis, Universidad Carlos III de Madrid, Spain.
Herranz, G., Nagel, R., Zauner, R., Levenfeld, B., Várez, A. and Torralba, J. M. (2004), 'Powder surface treatment with stearic acid influence on powder injection moulding of M2 HSS using a HDPE based binder', in *Proceedings of PM 2004 Powder Metallurgy World Congress*, vol. 4, pp. 397–402.
Herranz, G., Levenfeld, B., Varez, A. and Torralba, J. M. (2005), 'Development of new feedstock formulation based on high density polyethlene for MIM of M2 high speed steels', *Powder Metallurgy*, 48(2), 134–138.
Herranz, G., Levenfeld, B. and Várez, A. (2007), 'Effect of residual carbon on the microstructure evolution during the sintering of M2 HSS parts shaping by metal injection moulding process', *Materials Science Forum*, 534–536. 353–356.
Herranz, G., Rodríguez, G. P., Alonso, R. and Matula, G. (2010), 'Sintering process of M2 HSS feedstock reinforced with carbides', *PIM International*, 4(2), 60–65.
Ho, Y. L. and Lin, S. T. (1995), 'Debinding variables affecting the residual carbon content of injection-molded Fe-2 Pct Ni steels', *Metallurgical and Materials Transactions A*, 26A, 133–142.
Hoyle, G. (1988), *High Speed Steel*, Butterworth and Co., Cambridge, UK.
Hwang, K. (1996), 'Fundamentals of debinding processes in powder injection molding', *Reviews in Particulate Materials*, 4, 71–104.
Jauregi, S., Fernández, F., Palma, R. H., Martínez, V. and Urcola, J. J. (1992), 'Influence of atmosphere on sintering of T15 and M2 steel powders', *Metallurgy Transactions A*, 23A, 389–400.
Jinushi, H., Kyogoku, H., Komatsu, S. and Nakayama, H. (2002), 'Stoichiometry between carbon and oxygen during sintering process in Cr–Mo steel by injection molding'. *Advances in Powder Metallurgy and Particulate Materials*, 10, 183–198.
Johansson, E., Nyborg, L. and Niederhauser, S. (2000), 'Rheology of carbonyl iron MIM powder plastisols-effect of stearic acid additive', in *Proceedings of the PM World Congress 2000*, Part 1, pp. 262–266.
Kankawa, Y. (1997), 'Effects of polymer decomposition behavior on thermal debinding process in metal injection molding', *Materials and Manufacturing Processes*, 12(4), 681–690.
Klar, E. and Samar, P. K. (1998), 'Powder metallurgy stainless steels', in *ASM Handbook, Powder Metal Technologies and Applications, vol. 7*, ASM International Ohio, USA.
Krug, S., Evans, J. R. G. and Maat, J. H. H. (2002), 'Reaction and transport kinetics for depolymerization within a porous body', *AIChE Journal*, 48(7), 1533–1541.
Lee, S. H., Choi, J. W., Jeung, W. Y. and Moon, T. J. (1999), 'Effects of binder and thermal debinding parameters on residual carbon in injection moulding of Nd (Fe, Co)B powder', *Powder Metallurgy*, 42(1), 41–44.
Levenfeld, B., Várez, A. and Torralba, J. M. (2002), 'Effect of residual carbon on the sintering process of M2 high speed steel parts obtained by a modified metal

injection molding process', *Metallurgical and Materials Transactions A*, 33A, 1843–1851.
Lewis, J. (1997), 'Binder removal from ceramics', *Annual Review of Materials Science* 27, 147–173.
Li, Y., Chou, X. M. and Yu, L. (2006), 'Dehydrogenation debinding process of MIM titanium alloys by $TiH_2$ powder', *Powder Metallurgy*, 49(3), 236–239.
Li, D., Hou, H., Liang, L. and Lee, K. (2010), 'Powder injection molding 440C stainless steel', *International Journal of Manufacturing Technologies*, 49, 105–110.
Lin, S. T. (1997), 'Interface controlled decarburisation model for injection moulded parts during debinding', *Powder Metallurgy*, 40(1), 66–68.
Lin, K. (2011), 'Wear behavior and mechanical performance of metal injection molded Fe–2Ni sintered components', *Materials and Design*, 32, 1273–1282.
Liu, Z. Y., Loh, N. H., Khor, K. A. and Tor, S. B. (2000a), 'Microstructure evolution during sintering of injection molded M2 high speed steel', *Materials Science and Engineering A*, 293, 46–55.
Liu, Z. Y., Loh, N. H., Khor, K. A. and Tor, S. B. (2000b), 'Sintering of injection molded M2 high speed steel', *Materials Letters*, 45, 32–38.
Maulik, P. and Price, W. J. C. (1987), 'Effect of carbon additions on sintering characteristics and microstructure of BT42 high speed steel', *Powder Metallurgy*, 30(4), 240–248.
Miura, H. (1997), 'High performance ferrous MIM components through carbon and microstructural control', *Materials and Manufacturing Processes*, 12(4), 641–660.
Mohsin, I. U., Lager, D., Gierl, C., Hohenauer, W. and Danninger, H. (2010), 'Thermo-kinetics study of MIM thermal de-binding using TGA coupled with FTIR and mass spectroscopy', *Thermochimica Acta*, 503–504, 40–45.
Moyer, K. H. (1998), 'Magnetic materials and properties for powder metallurgy part applications', *ASM Handbook, Powder Metal Technologies and Applications*, vol. 7, ASM International, Ohio, USA.
Myers, N. and German, R. M. (1999), 'Supersolidus liquid phase sintering of injection molded M2 tool steel', *International Journal of Powder Metallurgy*, 35(6), 45–51.
Oliveira, M. M. and Bolton, J. D. (1999), 'High-speed steels: increasing wear resistance by adding ceramic particles', *Journal of Materials Processing Technology*, 92–93, 15–20.
Ouchi, C., Iizumi, H., and Mitao, S. (1998), 'Effects of ultra-high purification and addition of interstitial elements on properties of pure titanium and titanium alloy', *Materials Science Engineering A*, 243, 186–195.
Palma, R. H., Martínez, V. and Urcola, J. J. (1989), 'Sintering behaviour of T42 water atomized high speed steel powder under vacuum and industrial atmospheres with free carbon addition', *Powder Metallurgy*, 32(4), 291–299.
Price, W. J. C., Rebbeck, M. M. and Wronski, A. S. (1985), 'Effect of carbon additions on sintering to full density of BT1 grade high speed steel', *Powder Metallurgy*, 28, 1.
Puscas, M., Molinari, A., Kazior, J., Pieczonka, T. and Nykiel, M. (2001), 'Sintering transformations in mixtures of austenitic and ferritic stainless steel powders', *Powder Metallurgy*, 44(1), 48–52.

Qu, X., Gao J., Qin M. and Lei, C. (2005), 'Application of a wax-based binder in PIM of WC–TiC–Co cemented carbides', *International Journal of Refractory Metal and Hard Materials*, 23, 273–277.

Sagawa, M., Fujimura, S., Togawa, N., Yamamoto, H. and Matsuura, Y. (1984), 'New material for permanent magnets on a base of Nd and Fe', *Journal of Applied Physics*, 55, 2083.

Salak, A. (1995), *Ferrous Powder Metallurgy*, Cambridge International, UK.

Shengen, Z., Jianjun, T. and Xuanhui, Q. (2006), 'Antioxidation study of Sm (Co, Cu, Fe, Zr )Z sintered permanent magnets by metal injection molding', *Journal of Rare Earths*, 24, 569–573.

Shepard, R. G., Harrison, J. D. L. and Rusell, L. E. (1973), 'The fabrication of high-speed tool steel by ultrafine powder metallurgy', *Powder Metallurgy*, 16(32), 200–219.

Shu, G. J., Hwang, K. S. and Pan, Y. T. (2006), 'Improvements in sintered density and dimensional stability of powder injection–molded 316L compacts by adjusting the alloying compositions', *Acta Materialia*, 54, 1335–1342.

Sidambe, A. T., Figueroa, I. A., Hamilton, H. and Todd, I. (2010), 'Metal injection moulding of Ti–6Al–4V components using a water soluble binder', *PIM International*, 4(4), 56–62.

Spriggs, G. (1970), 'The importance of atmosphere control in hardmetal production', *Powder Metallurgy*, 13, 369–393.

Sumita, M., Hanawa, T. and Teoh, S. H. (2004), 'Development of nitrogen-containing nickel-free austenitic stainless steels for metallic biomaterials–review', *Materials Science and Engineering C*, 24, 753–760.

Suri, P., Koseski, R. P. and German, R. M. (2005), 'Microstructural evolution of injection molded gas- and water-atomized 316L stainless steel powder during sintering', *Materials Science and Engineering A*, 402, 341–348.

Takahashi, M., Hayashi, J. and Suzuki, S. (1992), 'Improvement of the rheological properties of the zirconia/polypropylene system for ceramic injection moulding using coupling agents', *Journal of Materials Science*, 27, 5297–5302.

Tian, J., Zhang, S. and Qu, X. (2007), 'Behavior of residual carbon in Sm(Co, Fe, Cu, Zr)z permanent magnets', *Journal of Alloys and Compounds*, 440, 89–93.

Tseng, W. J., Liu, D. and Hsu, C. (1999), 'Influence of stearic acid on suspension structure and green microstructure of injection-molded zirconia ceramics', *Ceramics International*, 25, 191–195.

Upadhyaya, A., Sarathy, D. and Wagner, G. (2001), 'Advances in sintering of hard metals', *Materials and Design*, 22, 499–506.

Weil, K. S., Nyberg, E. and Simmons, K. (2006), 'A new binder for powder injection molding titanium and other reactive metals', *Journal of Materials Processing Technology*, 176, 205–209.

Wright, J. K., Evans, J. R. G. and Edirisinghe, M. J. (1989), 'Degradation of polyolefin blends used for ceramic injection moulding', *Journal of American Ceramic Society*, 72(10), 822.

Wright, C. S. and Ogel, B. (1993), 'Supersolidus sintering of high speed steels. I: Sintering of molybdenum based alloys', *Powder Metallurgy*, 36(3), 213–219.

Wright, C. S., Wronski, A. S. and Iturriza, I. (2000), 'Development of robust processing routes for powder metallurgy high speed steels', *Materials Science and Technology*, 16, 945–957.

Wu, Y., German, R. M., Blaine, D., Marx, B. and Schlaefer, C. (2002), 'Effects of residual carbon content on sintering shrinkage, microstructure and mechanical properties of injection molded 17-4PH stainless steel', *Journal of Materials Science*, 37, 3573–3583.

Yamashita, O. (1998), 'Magnetic properties of Nd–Fe–B magnets prepare by metal injection moulding', *International Journal of Powder Metallurgy*, 34(7).

Zhang, H. (1997), 'Carbon control in PIM tool steel', *Materials and Manufacturing Processes*, 12(4), 673–679.

Zhang, H., Heaney, D. and German, R. M. (2001), 'Effect of carbide addition on sintering of M2 tool steel'. *Advances in Powder Metallurgy and Particulate Materials*, 4, 66.

Zhang, H. and German, R. M. (2002), 'Sintering MIM Fe–Ni alloys', *International Journal of Powder Metallurgy*, 38, 51–61.

Zhang, R., Kruszewski, J. and Lo, J. (2008), 'A study of the effects of sintering parameters on the microstructure and properties of PIM Ti6Al4V alloy', *PIM International*, 2(2), 74–78.

Zlatkov, B. S., Bavdek, U., Nikolic, M. V., Aleksic, O. S., Danninger, H., Gierl, C. and Erman, A. (2009), 'Magnetic properties of Alnico 8 sintered magnets produced by powder injection moulding', *PIM International*, 3(3), 58–63.

# Part III

Special metal injection molding processes

# 13
# Micro metal injection molding (MicroMIM)

V. PIOTTER, Karlsruhe Institute of Technology (KIT), Germany

**Abstract:** Although miniaturization represents a global trend of our time, development of adequate manufacturing processes for metal and ceramic materials is still widespread. A promising solution is micro powder injection molding (MicroPIM), which has already reached a remarkable technological level and first applications have successfully entered the market. Smallest details in the micrometer range have been achieved with different kinds of metal (steel, iron alloys, copper, tungsten etc.) and ceramic powders. Powder particle size has been identified as one of the most influential factors concerning surface quality and accuracy. By way of further progress in micro-manufacturing, two-component and in-mold-labeling MicroPIM are under development. Both variants allow for the fabrication of multi-material, and thus multi-functional products with the additional benefit of reduced assembly costs.

**Key words:** micro powder injection molding, MicroPIM, MicroMIM, MicroCIM, two-component powder injection molding, 2C-MicroPIM in-mold-labeling powder injection molding, IML-MicroPIM microstructured tools.

## 13.1 Introduction

This chapter discusses the state of research and development of micro-component powder injection molding of metals and, to a lesser degree, ceramics. Ceramics are referred to where innovative developments have taken place which have not yet been tried with metal materials, but will most probably be attempted in the near future.

The particular micro-specific features of powder injection molding are discussed in the following sections. Section 13.2 considers the contribution of powder injection molding to microsystems technology. Since it is necessary to review micro-specific tool making, Section 13.3 describes the various methods needed to produce microstructured mold inserts. Section

13.4 covers the special features of micro powder injection molding (MicroPIM), the powders used, molding and thermal process steps, and the current capabilities of the process. Particular variants of the process that are under development are described in Section 13.5. Section 13.6 deals with the simulation of MicroPIM, focusing on the challenges and possibilities for process improvement using modified material models. Section 13.7 summarizes immediate research and development requirements and discusses the most promising developments in the field of MicroPIM.

## 13.2 Potential of powder injection molding for micro-technology

Microsystems technology is considered one of today's most promising future technologies. Its innovations have been used in various markets, including information technology, life sciences, automotive engineering and power engineering, in the white and brown goods industries, in machine construction, and in physical and chemical process engineering, to name but a few examples. The most successful microsystems products are largely manufactured from silicon or plastics, rather than metals or ceramics.

In addition to the above sectors, there are some fields, for example chemistry, telecommunications and biology, and some products, for example midget gears or counter mechanisms, that require highly resistant components made of metals or ceramics (see also Table 13.2 later in Section 13.4). The potential of metal- or ceramic-based materials is well known within precision mechanics applications that are subjected to high forces, corrosion, wear, high temperatures, or that demand low thermal expansion, biocompatibility or sterilizability. It is necessary therefore to enhance manufacturing methods for products using metal-based components in micro-dimensions. Emphasis must be placed on the development of mass production methods to obtain profitable medium- and large-scale batches of complex micro-components (Piotter *et al.*, 2008). Powder injection molding is expected to become increasingly popular for the precision manufacture of complex micro-parts. Being a suitable method for use with nearly any powder material (German, 1996), it offers a wide range of processible materials (soft and hard magnetic materials, refractory metals, functional ceramics). Often, there are hardly any other ways to process and apply these materials economically, other than by MicroPIM (Rota *et al.*, 2004). A comprehensive survey exploring market volume trends of MicroPIM can be found in (Petzoldt, 2008; Yin *et al.*, 2008; German, 2010).

Metal injection molding (MIM) should be compared to other manufacturing technologies in terms of the economic viability and production capability of the process. As a method that can be replicated and scaled up,

MIM has advantages with regard to medium- or large-scale production when compared to typical machining processes such as milling, drilling, electrical discharge machining (EDM) and grinding. The same holds true for MIM when compared to laser structuring, particularly as less scrap is produced. When compared with well-established, large-scale processes like stamping, embossing or powder pressing, MIM shows a much better capability for near-net-shape fabrication of complex geometries, including significantly reducing the need for finishing processes. These advantages, however, have to be judged against the higher processing costs caused by the addition of steps such as feedstock preparation and debinding.

## 13.3 Micro-manufacturing methods for tool making

### 13.3.1 Manufacture of micro-tools – general issues

The molding of components with structural details in the micron or even submicrometer range requires appropriate tools and mold inserts. However, as the overall dimensions of the respective parts may range from hundreds of micrometers to several centimeters, special methods are required to manufacture these tool components. These specially manufactured micro-structured mold inserts are usually incorporated into injection molding tools as interchangeable parts. Unlike mold inserts, injection molding tools can be manufactured using conventional precision tool construction methods.

If micro-parts with high aspect ratios ($\geq 5$) have to be manufactured, certain processes like core evacuation or variothermal temperization have to be used. The large variety of micro-parts means that nearly an equally large variety of single- or multi-component micro-injection molding tools have to be produced. These tools may have, for example, two- or three-plate molds or may be designed with or without hot runners.

### 13.3.2 Options for manufacturing microstructured mold inserts

As described above, the tools typically used for micro-injection molding are manufactured using conventional methods. The tools are then fitted with microstructured mold inserts that can be produced in various ways, including lithographic methods, laser ablation, erosion, optimized precision mechanics processes, and various other techniques. For a more detailed discussion of production methods see (Yang *et al.*, 2009; Fu *et al.*, 2010). Micro-machining and micro-discharge machining are currently the most popular and most established methods (Brinksmeier *et al.*, 2008). Finally, two or more structuring methods can be combined to manufacture

Table 13.1 Optional methods for manufacturing microstructured mold inserts

| Structuring process | Geometrical degrees of freedom | Typical aspect ratios | Minimum dimensions (μm) | Dimensional accuracies (±μm) Lateral | Dimensional accuracies (±μm) Vertical | Roughness, $R_a$ (nm) | Typical mold materials |
|---|---|---|---|---|---|---|---|
| Silicon etching + (electroforming)[a] | 2.5D | 0.1–50 | 1–5 (typically) | 0.02[c] | 0.033[c] | 10 | Si (Ni, Ni alloys)[a] |
| UV (SU-8) lithography + electroforming | 2.5D | 1–4 (20[d]) | 30 nm[b] >2 | 2 | 1–5 | | Ni, Ni alloys |
| X-ray lithography (PMMA resist) + electroforming (LIGA) | 2.5D | 10–100 | 0.2 | 0.1–1[i] | >5 | 10–50 | Ni, Ni alloys |
| Electron beam lithography + electroforming | 2.5D | 2[e] 4[f] | 2 (grooves)[e] 0.05[f] | n.a. 10 nm[f] | n.a. 4 nm[f] | n.a. n.a. | Ni, Ni alloys |
| Laser micro caving | 3D | 1–10 | 10 | 5–20 | 3–10 | >200 | Mainly steel |
| Laser-LIGA | 3D | 1–10 | 200–400 nm | 1 | 0.5 | >50 nm | Ni, Ni alloys |
| Micro-machining (milling, drilling, etc.) | 3D | 1–10 (50[g]) | Prominent structures: <10  Sunken structures: 15 | 2 | 3–10 | 300 | Steel, brass, aluminum |
| Micro ECM | 3D | <40 | 25 | 2 | | <200 | Almost all electrically conductive materials |
| Micro ECF | 3D | ≤10 | 200 nm[h] – 10 | | | | Certain types of steel |
| Micro EDM | 3D | 10–100 | 50 | 1–3 | | ≥400 | Almost all electrically conductive materials |
| Laser sintering | 3D+[j] | 10 | 50 | 1–10 | <10 | >500 | Steel (H13, 316L, CoCr4), Inconell, TiV4 |

[a] optional process step; [b] feasibility boundary; [c] tolerances vary according to etching depth; tolerances will be 10–50 times larger if etching depth considerably exceeds 10 μm; [d] geometry-dependent; [e] intermediate mask; [f] shim; [g] depending on geometry; [h] under restricted conditions only; [i] considering line widths from 1 μm to 10 μm; [j] 3D plus hollow and cut-back features

microstructured mold inserts. Table 13.1 surveys the main manufacturing methods and major parameters.

Issues of wear resistance, which have not yet been thoroughly investigated for macro-scale injection molding, become very important when looking at micro-injection molding. An overview of interesting studies (Bergstrom *et al.*, 2001; Schneider *et al.*, 2005) describes experiments performed in special test stands, which compared mold insert materials according to their respective wear resistances. It turned out, rather unexpectedly, that rates of wear are lowest for materials of a lesser hardness, for example nickel, while being highest for dispersion-hardening steels whose precipitated grain particles are virtually washed away by the action of the feedstock as it flows past (Schneider *et al.*, 2008). Much research will focus on mold wear as a major issue for future PIM technologies.

It is beyond the scope of this chapter to describe all the methods listed in Table 13.1 in detail. We will confine ourselves instead to discussing the LIGA method, as described in the following section.

### 13.3.3 The LIGA method

For certain applications, lithographic and electroforming methods are combined to manufacture mold inserts with intricate structural details and high demands on the side wall roughnesses and the aspect ratios. This method is referred to as LIGA. This German acronym stands for LIthographie, Galvanoformung, Abformung (lithography, electroforming, molding). Over the last few decades research institutes around the world have used this process.

At the Karlsruhe Institute of Technology the LIGA mold inserts are manufactured by gluing an aligned small plastic plate (mostly polymethyl methacrylate (PMMA)) onto a high-precision copper substrate with a properly pretreated, highly polished surface. The surface of the resist is also highly polished. The resist is structured by emitting hard X-radiation through a gold absorber mask with the desired structure. The irradiated resist areas are then dissolved through wet-chemical development (X-ray lithography). After optional partial or full-surface metallization through vaporization (as electroforming starter layers), the pattern of the plastic structure is invertedly transferred into metal (mostly nickel or nickel alloy) through electroforming. Wire-cut EDM is applied to give the tool its outer shape prior to or after substrate separation and final finishing of the mold insert and its complex surface by wet-chemical etching and wet-chemical ultrasonic cleaning. Examples for mold inserts produced by ultraviolet (UV)-based lithography are shown in Fig. 13.1(a) and (b).

The LIGA mold inserts are used, for example, as injection molding tools for series manufacturing of hollow waveguide micro-spectrometers or the

312    Handbook of metal injection molding

(a)

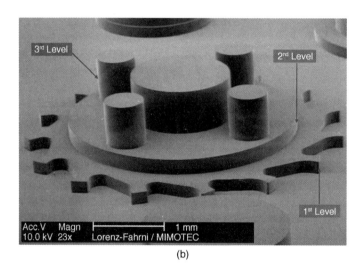

(b)

*13.1* (a) Typical LIGA mold inserts made by UV irradiation of SU-8 resist and subsequent electroplating. Both inserts carry inverse structures for replication of microfluidic functional units, the upper insert is made of nickel while the surface of the lower one consists of gold (the yellow micromold is an intermediate step of the final part; the next step consists in dissolving the gold layer (yellow) and then the final nickel cavity is obtained). (b) SEM detail view of a further UV-LIGA master in photoresist SU-8; note the smooth surfaces and the different levels of microstructures generated by a three-step irradiation process. Electroforming onto this master will generate the micromold (both figures courtesy of Mimotec, Switzerland).

manufacturing of optical or mechanical components, such as gearwheels or lenses. Such inserts are subject to very high demands relating to the material and surface properties, as well as the lateral structural and outer dimensions. To meet these demanding requirements, various improvements in handling, production sequencing and tooling have been developed in recent years and have considerably enhanced the safety of the LIGA process. LIGA technology, of course, has its limitations. In the case of narrow standing structures, detrimental overlapping effects might occur during electroplating, while filigree features, which have a large side-wall surface area, may get damaged during resist removal.

## 13.4 Powder injection molding of micro-components

Powder injection molding of metal- or ceramic-filled feedstocks is well established in industry for the production of macroscopic parts which are used in products such as industrial machines or automobiles, white goods and brown goods, micro-electronic and medical devices (German and Bose, 1997; German, 2008). High-precision, narrow-tolerance nozzles for bonding wires and ferrules serving as ceramic guide elements in optical fiber plugs are examples of micro-electronic applications that come very close to MicroPIM. It seems only natural, therefore, that PIM technology, throughout the world, is being developed for the manufacture of micro-components.

### 13.4.1 Feedstocks for micro-component powder injection molding

*Binder systems*

Micro-component powder injection molding requires the careful selection of feedstock components and compounding techniques. Classical powder injection molding of hard metal or ceramic components uses medium-size powder particles of around 0.5–10 µm. Macroscopic MIM is performed with particles sized up to 20 µm or even larger (Hartwig *et al.*, 1998). MicroPIM, however, requires particle sizes in the micro- or even submicrometer range in order to meet the specific requirements regarding surface roughness, true-to-detail design, and the mechanical properties of the sintered part. In this context it should be mentioned that surface roughness plays a much more important role than in macroscopic applications. One significant reason for this is the higher ratio between ground elevations and the structure size itself. A considerable argument for using powders that are as small as possible in MicroMIM is so that grain sizes are kept to a minimum after sintering in order to maintain a polycrystalline morphology. One option to try and ensure this outcome is to start the process sequence with the smallest possible powder particles.

MicroPIM needs feedstocks of a very low viscosity for rapid filling in the presence of high flow-length-to-wall-thickness ratios and to avoid the premature setting of the melt on account of the high thermal conductivity of the feedstocks. The binder systems, being responsible for the viscosity of the feedstocks, play an important role in compounding (Checot-Moinard et al., 2010). A further issue covers the interaction of chain length and constitution with the capability of the binder to disperse and keep the powder particles separated effectively (Hanemann et al., 2010).Viscosity is not the only criterion for determining the suitability of feedstocks for micro-molding purposes. Since micro-components are subject to high demolding forces due to relatively large surface-to-volume ratios and small load-bearing cross-sections, equal importance is attached to the strength and stability of the green parts. For example, polyoxymethylene-based binders usually have high viscosities excluding them, at first glance, from the list of binders suitable for MicroPIM. The above-average green strength, however, displaces this disadvantage if the high viscosity is compensated by, for example, a variothermal process conduct. In addition, it must be mentioned that the micro-cavities cannot be cleaned easily once they have become blocked with feedstock.

Besides requiring low feedstock viscosities and high strengths of the green parts, MicroPIM, in the same way as classical PIM technology, depends on factors such as feedstock homogeneity, shelf life and recyclability, easy and environmentally friendly debindering, and controllable shrinkage behavior (Kipphut and German, 1991; Baek et al., 2006). The most popular binder systems for commercial and scientific uses are categorized as follows:

- polymer compounds consisting of waxes and/or thermoplastics;
- thermoplastic-based binders;
- thermoplastic-reinforced polyethylene glycol-based binders;
- water-based or coagulating binders.

While thermoplastic-based feedstocks and several variants of wax-based binders have become widely accepted, there are few examples of the application of thermoplastic-reinforced and water-based types. However, as these two latter binder systems have advantages, such as the environmental friendliness of debindering in uncritical aqueous media, they are expected to be used increasingly in the near future.

The workability of feedstocks is improved by the inclusion of additives to the binders. These additives may consist of, for example, low-molecular-flow agents and dispersants with cross-linking properties that are intended to ensure an optimum distribution of the particles in the binder and to prevent re-agglomeration. The dispersants have to be selected taking into account both the surface characteristics of the powder and the chemical structure of the binder substances (polarity). In addition, care must be taken to provide for an optimum powder–binder coupling to ensure higher filling degrees which, in turn, reduce the sintering shrinkage and minimize dimensional inaccuracies. High-grade MicroPIM feedstocks are also treated with such additives. The chemistry of these is not discussed below because it is similar

*Table 13.2* Metal materials currently applied for MicroMIM, typical powder sizes

| Material | Mean particle size, $d_{50}$ (μm) | Typical aspect ratio (AR) | Min. lateral dimensions (μm) |
|---|---|---|---|
| Stainless steel 316L (1.4404) | 1.5–5 (up to 12) | 1–5 (up to 10) | 50 (down to 5*) |
| Stainless steel 17-4PH (1.4542) | 3–5 (up to 12) | 1–5 (up to 10) | 50 (down to 20) |
| Carbonyl iron | 1.5–5 | Up to 15 | Down to 10 |
| Nickel–iron alloy (NiFe) | | | ≤ 60 |
| Titanium alloys | ≥ 20 | | |
| Copper | 0.5–2 | (Up to 100) | Down to 10 |
| Copper–diamond | | 6 | 250 |
| Tungsten–copper alloy (WCu) | 1.5–3 | | ≤ 30 |
| Tungsten alloys | 0.5–6 | | |
| Hard metal (WC–xCo) | 0.5–4 | Up to 10 | 50 (down to 20) |
| Alumina ($Al_2O_3$) | 0.4–1.5 | Up to 10 | ≤ 30 |
| Zirconia ($ZrO_2$) | < 0.1–0.8 | > 10 | ≤ 10 |

* AR <1

to that of the additives used in the feedstocks of most macroscopic applications.

*Metal powders*

MicroPIM is performed using mainly metal powders from the standard PIM steels 17-4PH (1.4542, X5 CrNiCuNb 17 4) and 316L (1.4404, X2 CrNiMo 17 13 2), non-ferrous metals such as copper, and the recently developed titanium, tungsten and tungsten alloy powders (Zeep et al., 2006). Zirconium ceramic and aluminum oxide ceramic powders are mostly used for molding optical fiber ferrules and wire bond nozzles. Nitride ceramics, for example silicon nitride ceramics, are still at an early stage of development. An overview covering the range of the most important types of metal powders used for MicroPIM to date is provided by Table 13.2.

Micro-molding is mainly performed using metal powders that consist of particles with medium-sized diameters. As a general rule, the powder particle sizes should amount to no more than one-tenth or, preferably, one-twentieth of the size of the smallest detail of the cavity. The $d_{50}$ values of the steels are typically in the range of 1.5–4.5 μm, but fractions up to 10 μm or above are also used. The major MicroMIM steels 316L and 17-4PH are available as finest < 5 μm fractions. Powders of still finer structures or other types of steel can be obtained by, for example, air separation. Pure iron powders, just as the steels above, are available as fractions with particle sizes of 1.5 μm and below.

Looking at the non-ferrous metals, while fine fractions are available for copper and nickel, the minimum $d_{50}$ values of titanium amount to approximately 20 µm, which is rather large for the purpose of micro-component molding. On the other hand, it must be mentioned that due to the expressive pyrogenic nature of titanium, the powder particle size cannot be reduced infinitely.

Pre-alloyed powders are recommended to avoid the high shear rates that often occur during MicroPIM molding and cause rapid separation of the individual alloying substances. Furthermore, pre-alloyed powders show advantages with respect to dimensional details and surface quality as a result of a better homogenized mixture. Micro-molding works best with gas-atomized powders consisting of globular or spherical particles that ensure high powder filling rates and acceptable feedstock viscosities.

### 13.4.2 Micro powder injection molding (MicroPIM) – molding procedure

Micro powder injection molding techniques have been developed using special machines and tools designed for polymer-based micro-injection molding. This special basic equipment will be introduced below. It should be noted that the expression 'micro-part' covers a lot of quite different devices. These range from tiny singular parts with a weight of only a few milligrams to so-called microstructured parts which are large, mostly flat base plates with microstructural features on the surface. As a consequence, different machine types have to be applied. In the case of microstructured parts, precise but still commonly configured injection molding machines with clamping forces in the range of 15–50 t are sufficient. To achieve the best accuracies, electrical driven units are preferable.

For the filigree singular parts, however, special machines had to be developed. To obtain the smallest injection volumes, they are usually not equipped with standard screw units but with one or two plunger injection facilities. As the plunger diameters can be much smaller than those of the screws, even the lowest shot volumes can be measured out and injected precisely. With respect to the relatively high injection pressures, and to avoid flashing effects, the clamping force in general shall be chosen higher than calculated by the projected area of the micro-parts.

Several micro-injection molding machines have been developed and are now available on the market. They require only a few additions in order to adapt them for MicroMIM. For example, they should have the advantage of better wear resistance, i.e. coated steel units or even hard metal should be used. As the typical micro-molding machines have been designed for polymer materials, the plastisizing units should be modified for a good homogenization of PIM feedstocks, for example by an optimized geometry of the extruder screw.

Mold inserts for MicroPIM are characterized by their filigree detail dimensions. It is not surprising, therefore, that additional features had to be developed to adapt injection molding processes to the requirements of micro mold insert manufacturing. Two relevant examples, evacuation and variothermal tempering of the injection molding tools, will be discussed below.

Cavities in a typical micro-molding tool are 'blind holes', i.e. they are impermeable to gas at the bottom. The compressed, heated air that would develop if the heated feedstock was pressed into such cavities would cause burning of the organic share of the binder. While this so-called 'diesel effect' can be remedied when occurring in macroscopic injection molding processes, it cannot be avoided in microstructures whose cavities cannot be provided with common ventilation slots as used in laminate tools or bores. This problem can be adequately solved by evacuating the tool directly prior to injection using a tool-connected vacuum pump. As a rule, pressures of 1 bar or less are achieved in the cavities.

Variothermal temperature control is performed in case of tools with high flow-length-to-wall-thickness ratios. During this special process, temperatures close to the melting point of the feedstock have to be reached prior to injection, thus ensuring that the viscosity of the latter remains high enough for molding the filigree structural details. After the mold has been filled completely, the tool and feedstock are cooled down to temperatures which ensure that the strength and stability of the green part are high enough to allow clean demolding. Since the temperature increases and decreases prolong the lengths of cycles, and may take several minutes in the case of non-optimized processes, doubts are sometimes raised regarding the method's efficiency.

Thorough tool design and process optimization can help to address these doubts. For example, the lengths of cycles can be reduced. Additional major reductions in the lengths of cycles can be achieved by the use of novel heating methods such as inductive heating.

Another point worth mentioning is that MicroPIM procedures require high-precision tool movements owing to the extreme sensitivity of the molded micro-details. To achieve these, the mold opening and ejection velocities of micro-product injection molding machines may be adjusted to values of approximately 1 mm/s or less. Finally, the control software of the injection molding machines must be retrofitted for automation of the respective processes. For many years, some machine manufacturers have been providing complete software packages to meet that need.

In 1995, development work on MicroPIM was taken up by the former Forschungszentrum Karlsruhe, based on the micro-injection molding process and commercial feedstocks available at that time. The knowledge and expertise gained over decades in the use of metallic and ceramic

318    Handbook of metal injection molding

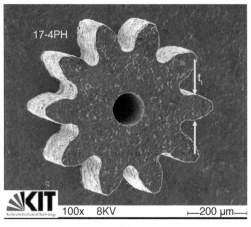

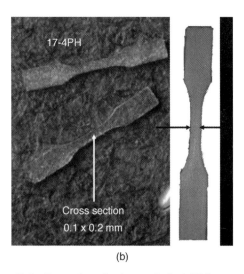

*13.2* Examples of microtechnical PIM components. (a) Gearwheel made of 17-4PH steel with outer diameters of approximately 580 μm and a tooth width of approximately 110 μm; note the rough surface due to the relatively coarse metal powders; (b) micro tensile specimen of steel with minimum cross-section of approximately 110 μm × 220 μm.

materials had been focused primarily on assessing inherent process limitations relative to the respective feedstocks and binder systems that had increasingly been developed in-house. Figure 13.2 shows examples of metallic micro-components produced at the laboratory scale so far.

Micro-components, as compared to macro-parts, are characterized by high surface-to-volume ratios that ensure lower temperature gradients between core and surface areas, thus facilitating effective debindering and

Micro metal injection molding 319

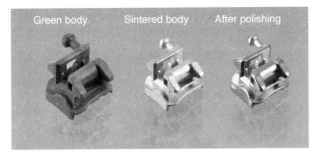

*13.3* Tooth brackets made of Ni-free steel alloy (courtesy of Bernhard Förster GmbH, Pforzheim, Germany).

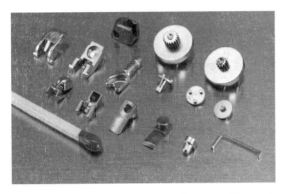

*13.4* MIM components for watch gears. The parts consist of either 17-4PH, 316L or NiFe alloys (courtesy of PARMACO AG, Switzerland).

sintering procedures. Heat treatment, however, must be given special attention. Debindering of the injection-molded parts subsequent to micro-molding can be done by melting out, pyrolysis, and dissociation in a solvent or by catalytic decomposition of polymers in an acid atmosphere. Two-stage processes may consist, for example, of pre-debindering in a solvent followed by thermal debindering as the main procedure.

The binder-free micro-parts are then sintered into finished parts. Metallic components (see Figs 13.3 and 13.4) are sintered under reduced conditions, mostly in pure hydrogen. During that process, the molded parts shrink by the volume assumed before by the binder. Since MicroPIM feedstocks contain relatively high shares of binders, sintering shrinkage may amount to as much as a linearly increasing share of 16 to 22%. The sintered parts are characterized by what is referred to as the 'best achievable precision' relative to the nominal dimensions. This significant parameter is given below as standard deviation from the nominal measure. Precision is determined, among other things, by the condition of the powders and the homogeneity of the feedstocks. A detailed analysis of the influence of the injection

*Table 13.3* State of micro-injection molding in terms of the main parameters. For comparison, equivalent values for MicroCIM and polymer micro-injection molding have been added. It is evident that the powder particle size is of extreme importance to the structure molding and surface qualities

| Material | Min. lateral dimension (µm) | Min. details (µm) | Aspect ratio, prominent structures | Aspect ratio, recessed structures | Tolerances (± %) | Surface roughness,* $R_{max}/R_a$ (µm) |
|---|---|---|---|---|---|---|
| Metals | 10 | 5 | 10 | > 10 | 0.5 | 7 / 0.8 |
| Ceramics | ≤ 10 | ≤ 3 | < 15 | 15 | 0.4 (0.1**) | <3 / 0.2 |
| Polymers | 1 | < 0.1 | 20 (200*) | 25 | 0.05 | 0.05 / 0.05 |

*Note*: AR = aspect ratio = flow-length-to-wall-thickness ratio; *depending on the type of mold insert; **after thorough process optimization.

molding parameters on the precision of the molded parts is given, for example, in (Billiet, 1985; Zauner et al., 2004; Greene and Heaney, 2007; Li et al., 2007; Beck et al., 2008; Nishiyabu et al., 2008). Depending on the powder particle sizes and filling rates, the reproducibility of component dimensions is ±0.3 to ±0.5% for metals, ±0.5% for alloyed powders, and ±0.2 to ±0.5% for ceramics.

Enhancements in materials and processes may considerably reduce tolerances and yield reproducibilities of ±0.1%. Such outstanding values, of course, can only be achieved if all other effects, such as lot-to-lot variations and environmental influences, are kept to a minimum. Table 13.3 shows the main parameters obtained from research and development carried out at the Karlsruhe Institute of Technology. These results have been equalized with values given by several PIM-related publications so that the numbers in Table 13.3 can be regarded as the current geometrical limitations of MicroPIM.

### 13.4.3 Debinding and sintering

As with macroscopic PIM, the removal of the binder is a critical step in the MicroPIM process, both from the technical and the economic points of view (German, 2005). In general, it can be performed by thermal melting or degradation, solvent extraction or catalytic decomposition of the binder components, or by a combination of these methods.

In the case of so-called microstuctured parts (i.e. devices of relatively large overall dimensions but with micro-scale features on the surface), the debinding times, which depend on the wall thickness of the part, have the potential to limit the economic viability of the whole process. For micro-parts with wall thicknesses significantly below 1 mm, however, debinding times play a minor role. Here the interaction of the material with the sintering tray and the shape conservation of filigree details are more important than shortening the debinding cycle. Furthermore, one run in a typical debinding furnace usually produces a large number of parts, meaning that operational capacity is very high even if debinding times are quite long.

Again, similarly to macroscopic PIM (German, 1996), the sintering conditions depend mainly on the type of metal processed and parameters differ little from the usual ones. Owing to the high specific surface of the fine powders for metal injection molding, holding times, as well as peak sintering temperatures, can be reduced for micro-parts. On the other hand, to avoid exorbitant grain growth, the heating and cooling procedures usually have to be accelerated. Heating and cooling rates, however, can only be increased to values which will not cause distortion effects.

Batch furnaces running under reducing atmospheres ($H_2$ or $N_2/H_2$) or vacuum are quite common. During the sintering process the parts experience a linear shrinkage of up to 23% due to the comparably lower powder loadings. Values for densities reached are in the range 95–99% of theoretical density.

### 13.4.4 Metrology and handling

Metrology and quality assurance are crucial matters in micro-technology. When measuring the outer dimensions of green bodies and sintered parts one can rely on the measurement systems developed for application in micro-electronic or micro-electro-mechanical systems/micro-opto-electro-mechanical systems (MEMS/MOEMS) fabrication (Lanza et al., 2008). For example, test stands based on coordinate measuring machine (CMM) units, or even white light interferometry and atomic force microscopy (AFM) are in use for geometrical inspection. These test systems are quite expensive. If an automated system is built up it can represent a useful quality inspection system. Reliable measurement and quality inspection is an important matter for micro-technology. As a result, a lot of research and development approaches are working in this field and no additional efforts are necessary for MicroPIM.

Much more complicated is the 'view into the body' to determine micro-cracks, cavities, and areas of powder/binder segregation (Hausnerova et al., 2010). This internal inspection has to be carried out rapidly, preferably on-line, to avoid excessive failure production. Therefore, the classical way of cutting and grinding, and the subsequent optical investigation of the cross-section, is not the most effective method of testing. Alternative methods, such as ultrasonic inspection and/or thermographical testing, show much more promise in terms of performance potential. As they are already under development, and are already in use for macroscopic PIM, they are not described in detail here.

Finally, the two-dimensional and three-dimensional inspection methods based on X-ray irradiation should be mentioned. MicroMIM parts profit by their small thicknesses, meaning that they can be irradiated without thorough energy dissipation and beam widening. A good description of the different approaches and their capabilities can be found in (Jenni *et al.*, 2009). Using monochromatic synchrotron radiation a three-dimensional profile of the powder distribution over a whole MicroMIM sample can be generated and the determination of powder/binder segregation phenomena becomes possible (Heldele *et al.*, 2006). Nevertheless, such investigations are costly and time consuming, meaning that faster and less complex variants have to be derived for industrial applications. More efficient methods have only recently begun to be developed (Albers *et al.*, 2008). It is essential that this development continues as the down-scaling of functional test procedures has the potential to optimize the production of micro-parts and microsystems.

As in macroscopic PIM, handling, automation and the interfaces of the relevant production facilities play an important role in MicroMIM. Existing or soon to be developed tools can be used for MicroMIM on the condition that they are adapted to the desired small dimensions. This is a challenge for the whole micro fabrication world and considerable research and development efforts are underway from which MicroPIM will also benefit.

In the case of singular micro-parts, the precise positioning of gripper to part is essential as tolerances are 1 μm or less. This positioning of gripper to part has to be considered thoroughly during process planning (Freundt *et al.*, 2008). This is also true for automated quality assurance. In the case of MicroPIM, the relatively low green strength, which might cause problems if mechanical grippers are used, has to be considered. Similarly, the higher weights due to the powder loading can be a disadvantage in the case of vacuum grippers. On the other hand, metal-filled components reveal some advantages for handling. For example, unlike plastic ones, they are not charged electrostatically. Thus, MicroMIM has an advantage over polymer micro-injection molding as regards the easier gripping or moving of parts.

### 13.4.5 Leading and innovative trends

Micro-component powder injection molding is an attractive method both from the technical and economic points of view. It is not surprising, therefore, that major research efforts are not limited to a closely restricted circle of scientists. In recent years there has been an increase in MicroPIM research in East Asia (Tan, 2007; Jang *et al.*, 2010; Ko *et al.*, 2010). Although MicroPIM research is carried out across Asia, the Osaka Prefectural College of Technology (now at Kinki University) and the

(a)

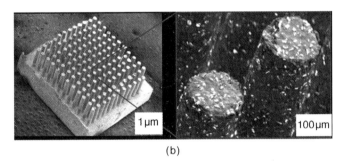

(b)

*13.5* (a), (b) Peltier elements made by MicroMIM using Cu–diamond powders. Diameter of pillars 250 μm, height 1500 μm (courtesy of Kinki University, Osaka).

Nanyang Technological University in Singapore are examined more closely here.

The researchers in Osaka used PMMA lost cores to manufacture complex PIM components of 316L stainless steel. The PMMA molds have been fabricated using lithographic (nanoimprint lithography) or other methods (Nishiyabu *et al.*, 2006; 2010). Furthermore, research and development has been dedicated to investigating micro-/nano-powder compounds for use in MicroPIM (Nishiyabu and Tanaka, 2008). For micro-electronic applications, heat sinks have been injection molded using Cu–diamond filled feedstocks. The reason for this material development was to obtain high thermal conductivity combined with a coefficient of thermal expansion (CTE) similar to that of silicon (Fig. 13.5). Similar goals are pursued by other study groups in Japan (Yasui *et al.*, 2006) and Korea (Baek *et al.*, 2006), among others.

As the second example, the Nanyang Technological University in

324  Handbook of metal injection molding

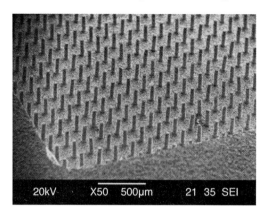

*13.6* MicroMIM samples made of 316L stainless steel. Injection molding was performed using a variothermal temperature profile. Diameter of columns 60 μm, height 190 μm (courtesy of Nanyang Technological University, Singapore).

Singapore has been engaged in development and manufacture of micro gearwheels made of 316L steel powders (Loh *et al.*, 2007) using wire-eroded or silicon-etched (Tay *et al.*, 2006) mold inserts (Fig. 13.6). Moreover, micro-specific auxiliary injection molding devices were developed and launched (Fu *et al.*, 2007) and the conditions relevant for the sintering of micro-parts were investigated (Liu *et al.*, 2006).

Among the relatively few research activities on MicroPIM in North America during recent years were analyses of ultrasonic PIM feedstock flow behavior (Cheng *et al.*, 2007). Economic studies and MicroCIM experiments, with an emphasis on the molding of wire-eroded mold inserts using commercial aluminum oxide feedstocks (Catamold AOF), have also been carried out (Atre *et al.*, 2006). The pronounced powder–binder segregation that occurred during these experiments may be assumed to have been due to the neglect of variothermal process control. Comprehensive studies on MicroPIM, for example on the production of medical components, have also been performed at Los Alamos National Laboratory (Heaney, 2006a; 2006b; 2006c).

In Europe, intensive research on MicroPIM has been carried out at Fraunhofer Institut IFAM in Bremen, Germany, including on the sintering behavior of single- and multi-component MicroMIM specimens (Simchi *et al.*, 2006). The researchers developed microstructured metal components for heat sinks. WCu and MoCu had been chosen due to their high thermal conductivity and low coefficient of thermal expansion (Schmidt *et al.*, 2009). Further research and development approaches deal with the processing of biocompatible materials for implant applications. For example, a special 316L powder mixture has been created by the addition of ultra-fine and nano iron fractions. This mixture enabled the replication of microstructured surfaces with hemispheres of 5 μm diameter (Haack *et al.*, 2010). Aiming for medical applications, Ti-based feedstocks partially doped with nano powder fractions have been developed for bone implants. Among other effects, the influence of binder composition on processibility and contamination has also been studied (Friederici *et al.*, 2010).

FOTEC Research and Development GmbH in Wiener Neustadt, Austria, has developed an advanced X-Cooler to be used for central processing unit (CPU) temperization. The design is characterized by thin Cu tubes located on a carrier plate, all manufactured by MIM using pure Cu feedstocks. The wall thickness of the Cu tubes is 0.3 mm at a height of 30 mm, thus, a record-holding calculative aspect ratio of 100 has been achieved (Zlatkov and Hubmann, 2008; Zlatkov *et al.*, 2008a). More material related, and not exclusively focused on MicroMIM, are investigations for processing of hard magnetic powders (Zlatkov, 2009). It should be noted here that powder injection molding of filigree precision parts or micro-parts is not confined to research but has already been adopted by the industry.

## 13.5 Multi-component micro powder injection molding

### 13.5.1 Approaching 2C-MicroPIM

The joining and assembly of singular micro-components into complex products is carried out through extremely intricate and quite expensive processes that risk causing damage to the often filigree structures of the individual parts. Problems of this kind can be reduced considerably by combining single-component molding with a joining step, as is done in multi-component micro-injection molding (2C-MicroPIM) (Moritz, 2008; Ruh *et al.*, 2008; Li *et al.*, 2010). It is a particular challenge of multi-component powder injection molding that, in addition to joining two different materials during the molding process, a defined bond between these materials must be maintained during debinding and sintering. To ensure optimum procedures, the debinding and sintering behaviors of the feedstocks and powders must be coordinated and adapted precisely regarding quantities, morphologies, and thermal and chemical behavior (Oerlygsson *et al.*, 2003; Petzoldt, 2010).

With the intention of adapting powder injection molding techniques to the requirements of micro-dimensions, a joint research project was carried out in Germany some years ago: '2K-PIM' was aimed at developing multi-component powder injection molding for the manufacture of micro-parts or micro-components from two types of metal or ceramics and with corresponding different local properties and functions. Both partners, the Karlsruhe Institute of Technology (formerly Forschungszentrum Karlsruhe)

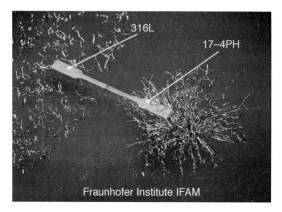

*13.7* 2C-MicroPIM demonstrator consisting of non-magnetic 316L and magnetized 17-4PH steel (courtesy of Fraunhofer Institute IFAM, Bremen, Germany).

and Fraunhofer Institut IFAM, had to cope with the task of developing suitable materials and process technologies, placing emphasis on the identification of appropriate material combinations and feedstocks for micro-application, and the development of an injection molding, debindering and sintering method suited for multi-component parts. Within 2K-MIM, the composites 17-4PH/316L (Fig. 13.7) and iron/316L, two soft magnetic/non-magnetic composites, were developed successfully and used to manufacture two demonstrator specimens. The results obtained for iron/316L showed that different materials, with strongly differing sintering characteristics, can be tailored for joint sintering.

### 13.5.2 Fixed and movable structures

2C powder injection molding allows the manufacture of both fixed and movable structures if use is made of the powders' different sintering behaviors. The necessary preconditions can be provided by adapting the powder contents and selecting either sinter-active powders or powders that are inert to sintering. In addition, care must be taken to choose temperature programs that are adapted to the requirements of the respective materials and to use sintering underlays that are suited, for example, to free sintering of movable structures.

While fixed structures require equal sintering shrinkages and temperatures to avoid residual stresses or even self-destruction during compaction, movable structures require separation of the partial volumes during sintering as well as different powder contents and sintering temperatures. This ensures that shrinkage of one component sets in prior to that of the other component. Macroscopic movable 2C-PIM components have been developed for some time (Maetzig and Walcher, 2006).

### 13.5.3 Micro in-mold labeling with the use of PIM feedstocks

Typical two-component injection molding that blends at least two different melt flows in a cavity is not the only method available for PIM manufacturing of multi-component parts. Equally promising results can be achieved combining in-mold labeling with powder injection molding processes (IML–MicroPIM). PIM feedstock is injection-molded around a green film placed in an injection molding tool and filled with metal or ceramic particles. Both partial volumes are debindered and sintered at the same time to obtain a material composite. The material and process conditions are similar to those selected for 2C-MicroPIM of fixed structures.

While macroscopic methods of that kind have been developed for some time, the novel microscopic applications offer further possibilities. Extremely fine (nano-) particles can be added to the film feedstocks without the workability being affected by the increase in viscosity. This means that functional particles or nanoparticles can be applied to surfaces of metal or ceramic components while the components' three-dimensional bodies themselves are manufactured at lesser costs by means of common powder injection molding.

Micro in-mold labeling using PIM feedstocks is presently being developed within the framework of the EU Multilayer project (FP7-NMP4-2007-214122). Early interesting results were obtained with the successful injection molding of zirconium oxide-filled ceramic films on green parts of another zirconium oxide type. The solid, mostly pore-free structures of the material composites were maintained during debindering and sintering. The microstructures, imprinted in the outer surface of the film through the injection pressure, remained intact after debindering and sintering (Vorster *et al.*, 2010). The micro in-mold labeling tests described above were performed using mainly ceramic and metal/ceramic materials. Future research and development activities may assess whether the same processes can be used on purely metallic films or feedstocks.

### 13.5.4 Sinter joining

Composites can also be produced by sinter joining; that is, by joining green parts after cooling and demolding. Basically, two or several green parts are assembled, debindered and, finally, sintered into a fixed structure. Since green parts can be assembled in various different ways (for example automatically by use of modular computer-assisted robots), this method is of considerable geometric flexibility. In addition, sinter joining may serve to manufacture components with undercuts or hollow parts with inside inserts. The technology is discussed with a focus on micro applications in (Fleischer et al., 2008) and (Munzinger et al., 2010).

One disadvantage of sinter joining is that the sintering of green parts, especially those of different powders or materials, may not be easy once they have been cooled and demolded. It is advisable, therefore, to provide subcomponents with joining elements that ensure positive locking. If necessary, the green parts that have been assembled may be weighted during sintering to obtain tight joints. Another drawback of sinter joining as opposed to multi-component injection molding is that the adjustment and assembly of the parts is more intricate.

## 13.6 Simulation of MicroMIM

As the challenges and solutions related to a meaningful simulation of the PIM process are thoroughly described elsewhere in this chapter, at this point only the micro-specific considerations shall be discussed. In general, designing a MicroPIM part follows the same basic rules as for macroscopic powder injection molding. It should be mentioned here that, since it is very difficult to modify or alter the microstructured molding tools, computer programs would be extremely useful for avoiding errors at the design stage. Further remarkable differences between the macro- and micro-worlds are the increasing heat losses due to the larger surface-to-volume ratio and the (usually detrimental) influence of relatively high shear rates. Additionally, the limitations given by micro mold insert production and powder size have to be considered. A description of design aspects related to MicroPIM can be found in (Albers et al., 2007).

Although widely in use, commercial simulation programs have weaknesses that limit their reliability for predicting typical MicroPIM effects. Owing to the single-phase material models, the phenomena of segregation cannot be simulated unless additional features are provided. Moreover, certain properties of the metal particles, for example higher inertia due to higher density, cannot be assessed sufficiently and the typical effects of PIM, such as the disproportionate formation of strands and folds, wall friction, and yield points, are often quite difficult to define correctly.

Since shear rates are often above average and changes in cross-sections of the flow channels can be quite abrupt, micro injection molding applications are more affected by the weaknesses described above than macroscopic PIM. Moreover, the higher surface/volume ratios of micro-components must be assumed to aggravate all surface-dependent effects. With this in view, the principles described below are particularly relevant for MicroMIM. It must also be mentioned that commercial software tools do not consider any of the special micro-injection molding methods such as variothermal process control or tool evacuation. Modified or even new simulation tools, adapted to the needs of micro-technology, must be developed in the future.

Tests had been performed at Karlsruhe Institute of Technology to find out whether it makes sense to use the popular commercial programs to simulate micro-injection molding processes. Components with certain test geometries were defined and a suitable injection molding tool was manufactured to obtain and compare the calculated values and experimental process data for an exemplary part. The first experiments using data not specific to the feedstocks revealed considerable discrepancies between the quantitative simulation data and the measured values. The pressure that is actually required for complete molding, for example, was found to be above the calculated pressure by up to 100%. Results turned out to be much better and to deviate by less than 10% when measuring and using specific feedstock data. It was also found that fluid mechanical processes, such as mold filling, splitting and convergence of melt flows, cannot always be calculated qualitatively in advance. With this in view, profound knowledge and experience are required in order to interpret simulated calculations. Although the process of injection molding is not described exhaustively, the specifically determined material data and experienced interpretation of the results obtained already allow for a limited utilization of simulation tools within MicroPIM development projects.

Thorough and reliable simulations of MicroPIM processes are unlikely to be achieved without multi-phase material models. Several research facilities have developed approaches that may turn out to be attractive. A quite promising approach for simulating MicroMIM is currently under development at IMTEK, University of Freiburg. The scientists in the group of Professor Korvink generated a new material model based on the smoothed particle hydrodynamics (SPH) method (Monaghan, 2005). To enhance this model for sufficient MicroPIM prediction they incorporated two new features, an inherent yield stress and shear induced powder segregation. The first has been performed by means of a bi-viscosity approach (O'Donovan and Tanner, 1984). For the simulation of shear-induced powder segregation during injection, three important effects have to be considered: the migration of powder particles towards areas of lower concentration, which is caused by different collision frequencies; the migration of powder particles towards areas of lower shear rates; and the migration of powder particles towards areas of lower viscosities caused by viscosity inhomogeneities. After incorporating these approaches into the SPH model a correct prediction of powder distribution in MicroMIM parts was achieved as proven by computer tomography (CT) measurements using synchrotron radiation (Kauzlaric et al., 2008; Greiner et al., 2011). A similar objective, the creating of a PIM-specific simulation routine, is being pursued at the

University of Besancon through the study of novel approaches, which are described in detail in Barriere *et al.* (2003; 2009) and Kong *et al.* (2010).

## 13.7 Conclusion and future trends

In summary, MicroPIM is a key technology that is suited to the manufacture of medium-sized to large quantities of heavy-duty microcomponents of differently complex geometries by use of metals, metal alloys, hard metals or ceramics. It is primarily due to the higher availability of finest ceramic powders that high aspect ratio micro-components are predominantly made from ceramics. Metal powders made up of micrometer- or submicrometer-sized particles, however, can also be used for manufacturing precision mechanical components or micro-parts that require little or no reworking. The first industrial applications are already available on the market.

Probably the most important demand for research and development is the comprehensive investigation of the material–parameter–properties interdependencies. This covers many singular effects including powder–binder segregation, jetting and wrinkling, grain growth and size distribution, and distortion during sintering.

Directly related to the exploration of how materials and parameters effect green and sintered parts performance is the improvement of analytical methods including the detection of defects such as micro-cuts and cavities. Equally as important is the demand for the monitoring of powder distribution in the green bodies to detect areas in danger of distortion during thermal treatment as early as possible.

Much as in the case of macroscopic PIM, improvements in dimensional accuracy and surface quality are necessary optimizations in the case of MicroPIM. Mostly, one tries to achieve enhancements by reducing the size of particles through use of particularly fine (for instance screened) powder fractions. Since the material costs are low enough that they do not significantly raise production costs, MicroPIM can be assumed to have advantages over the macroscopic methods. Dimensional reproducibility through minimization of the tolerances between the sintered part and nominal dimensions is known to play a significant role in industrial series production. In this regard, much research goes into optimizing feedstocks, placing emphasis on studies of suitable dispersants for coupling powder particles and organic binders. Investigations on the wear behavior and long-term stability of tools and mold inserts are scarce and have to be expanded. Among the future trends observed at present for MicroPIM is the enlargement of the range of suitable materials for the development of future fields of application. Tungsten or titanium are typical examples. These trends are driven by the lack of adequate alternative procedures and by material savings compared to the ablating methods. Now as before, steel, mainly 316L and 17-4PH, seems to be the material mostly used for micro-mechanical applications. At the same time, the use of copper or copper–diamond is expected to increase owing to the material's high electrical and thermal conductivity. The latter, of course, might result in unintended short shots. Nevertheless, MicroMIM using Cu-filled feedstocks is reported in, among others, Zlatkov *et al.* (2008b).

Other important trends are the multi-component manufacturing methods including two-component micro-injection molding, lost-core technology, micro in-mold labeling or sinter joining of metal or ceramic micro-parts. Further intensive research and development is expected due to these methods' particular attractiveness to microsystems technology. 2C-micro-injection molding has, with the use of polymers, achieved a high technical standard which has already led to the first industrial applications and can therefore rely on efficient equipment and machinery in the future.

Both macroscopic and microscopic production increasingly demand effective and realistic simulation tools enabling the modeling and calculation of multi-component material systems for powder injection molding and allowing the prediction of segregations and sintering deformations. Moreover, the relevant material models should admit unconventional flow profiles and consider typical PIM properties such as extreme strand formation. It is opportune that such software routines are already being developed to support the commercial programs and that novel research approaches are being taken to develop multi-phase material models. Finally, it should be mentioned that holistic design rules, material data bases, standardized peripheric equipment and the set-up of PIM-specific standards are yet to be developed and it is hoped that these areas will open up opportunities for future research and investigation.

## 13.8 Sources of further information and advice

Additional information can be found under the following web addresses:

www.fhg-ifam.de
www.iam.kit.edu/wpt/
www.ntu.sg
www.epma.eu
www.apmi.com
www.pulverspritzgiessen.de (in German)

Books and review articles include:

Loehe, D., Hausselt J., et al. (2005), *Advanced Micro and Nanosystems, Vol. 3, Microengineering of Metals and Ceramics*, eds H. Baltes, O. Brand, G. K. Fedder, C. Hierold, J. Korvink, and O. Tabata, Wiley VCH-Verlag, Weinheim, Germany, Ch 10–12.

Petzoldt, F. (2008), 'Micro powder injection moulding – challenges and opportunities', *Powder Injection Moulding International*, 2(1), 37–42, Inovar Communications Ltd.

German, R. M. (2010), 'Materials for microminiature powder injection molded medical and dental devices', *International Journal of Powder Metallurgy*, 46(2), 15–18.

## 13.9 References

Albers, A., Burkardt, N., Deigendesch, T., and Merz, J. (2007), 'Micro-specific design flow for tool-based microtechnologies', *Microsystem Technologies*, 13, 303–310.

Albers, A., Deigendesch, T., Enkler, H.-G., Hauser, S., Leslabay, P., and Oerding, J. (2008), 'An integrated approach for validating micro mechanical systems based on simulation and test', *Microsystem Technologies*, 14, 1781–1787.

Atre, S. V., Wu, C., Hwang, C. J., Zauner, R., Park, S. J., and German, R. M. (2006), 'Technical and economical comparison of micro powder injection molding', in *Proceedings of 2006 Powder Metallurgy World Congress*, Korean Powder Metallurgy Institute, ISBN 89-5708-121-6 94580, vol. 1, pp. 45–46.

Baek, E.-R., Supriadi, S., Choi, C.-J., Lee, B.-T., and Lee, J.-W. (2006), 'Effect of particle size in feedstock properties in micro powder injection molding', *Proceedings of 2006 Powder Metallurgy World Congress*, Korean Powder Metallurgy Institute, ISBN 89-5708-121-6 94580, vol. 1, pp. 41–42.

Barriere, T., Liu, B., and Gelin, J. C. (2003), 'Determination of the optimal process parameters in metal injection molding from experiments and numerical modeling', *Journal of Materials Processing Technology*, 143, 636–644.

Barriere, T., Quinard, C., Kong, X., Gelin, J.C., and Michel, G. (2009), 'Micro injection moulding of 316L stainless steel feedstock and numerical simulations', in *Proceedings of EuroPM 2009 Conference*, Kopenhagen, European Powder Metallurgy Association, ISBN 978 1 899072 07 1, pp. 325–330.

Beck, M., Piotter, V., Ritzhaupt-Kleissl, H.-J., and Hausselt, J. (2008), 'Statistical analysis on the quality of precision parts in ceramic injection moulding', in *Proceedings of the 10th Euspen Conference 2008*, Zurich, ISBN 978-0-9553082-5-3, vol. 2, pp. 179–183.

Bergstrom, J., Thuvander, F., Devos, P., and Boher, C. (2001), 'Wear of die materials in full scale plastic injection moulding of glass fibre reinforced polycarbonate', *Wear*, 251, 1511–1521.

Billiet, R. (1985), 'The challenge of tolerance in P/M injection molding', in *Proceedings of the Annual Powder Metallurgy Conference*, San Francisco, California, ed. H. I. Sanderow, Metal Powder Industries Federation, Princeton, NJ, USA, pp. 723–742.

Brinksmeier, E., Gäbe, R., Riemer, O., and Twardy, S. (2008), 'Potentials of precision machining processes for the manufacture of micro forming molds', *Microsystem Technologies*, 14, 1983–1987.

Checot-Moinard, D., Rigollet, C., and Lourdin, P. (2010), 'Rheological characterization of powder and micro-powder injection moulding feedstocks', in *Proceedings of Powder Metallurgy 2010 World Congress*, EPMA, ISBN: 978 1 899072 13 2, vol. 4, pp. 563–571.

Cheng, C.-C., Ono, Y., Whiteside, B. D., Brown, E. C., Jen, C. K., Coates, P. D. (2007), 'Real-time diagnosis of micro powder injection molding using integrated ultrasonic sensors', *International Polymer Processing*, XXII (2), 140–145, Hanser Publishers, Munich.

Fleischer, J., Munzinger, C., and Dieckmann, A.-M. (2008), 'Thermal micro-sinter-joining for realizing a shaft to collar connection of μPIM bevel gears', in *Proceedings of Euspen 10th International Conference*, Zürich, ISBN 13: 978-0-9553082-5-3, vol. 2, pp. 236-239.

Freundt, M., Brecher, C., and Wenzel, C. (2008), 'Hybrid universal handling systems for micro component assembly', *Microsystem Technologies*, 14, 1855–1860.

Friederici, V., Bruinink, A., Imgrund, P., and Seefried, S. (2010), 'Micro MIM process development for regular surface patterned titanium bone implant materials', in *Proceedings of Powder Metallurgy 2010 World Congress*, EPMA, ISBN: 978 1 899072 13 2, vol. 4, pp. 785–790.

Fu, G., Tor, S., Loh, N. H., Tay, B., and Hardt, D. E. (2007), 'A micro powder injection molding apparatus for high aspect ratio metal micro-structure production', *Micromechanics and Microengineering*, 17, 1803–1809.

Fu, G., Tor, S. B., Loh, N. H., and Hardt, D. E. (2010), 'Fabrication of robust tooling for mass production of polymeric microfluidic devices', *Journal of Micromechanics and Microengineering*, 20, 085019.

German, R. M. (1996), *Sintering Theory and Practice*, John Wiley and Sons, New York, USA, ISBN 0-471-05786, pp.147–155.

German, R. M. (2005), *Powder Metallurgy and Particulate Materials Processing: The Processes, Materials, Products, Properties, and Applications*, Metal Powder Industries Federation, Princeton, NJ, USA.

German, R. M. (2008), 'Divergences in global powder injection moulding', *Powder Injection Moulding International*, 2(1), 45–49.

German, R. M. (2010), 'Materials for microminiature powder injection molded medical and dental devices', *International Journal of Powder Metallurgy*, 46(2), 15–18.

German, R. M. and Bose, A. (1997), *Injection Molding of Metals and Ceramics*, Metal Powder Industries Federation, Princeton, NJ, USA, ISBN 1-878-954-61-X, p. 80ff.

Greene, C. D. and Heaney, D. F. (2007), 'The PVT effect on the final sintered dimensions of powder injection molded components', *Materials and Design*, 28 (1), 95–100.

Greiner, A., Kauzlaric, D., Korvink, J. G., Piotter, V., Hanemann, T., Weber, O., and Haußelt, J. (2011), 'SPH-simulation of micro powder injection moulding', *European Ceramic Society*, 2011, submitted.??

Haack, J., Imgrund, P., Hein, S., Friederici, V., and Salk, N. (2010), 'The processing of biomaterials for implant applications by powder injection moulding', *Powder Injection Moulding International*, 4(2), 49–52.

Hanemann, T., Heldele, R., Mueller, T., and Haußelt, J. (2010), 'Influence of stearic acid concentration on the processing of $ZrO_2$-containing feedstocks suitable for micro powder injection molding', *International Journal of Applied Ceramic Technology*, doi: 10.1111/j.1744-7402.2010.02519.x.

Hartwig, T., Veltl, G., Petzoldt, F., Kunze, H., Scholl, R., and Kieback, B. (1998), 'Powders for metal injection molding', *Journal of the European Ceramic Society*, 18, 1211–1216.

Hausnerova, B., Marcanikova, L., Filip, P., and Saha, P. (2010), 'Wall-slip velocity as a quantitative measure of powder–binder separation during powder injection moulding', in *Proceedings of Powder Metallurgy 2010 World Congress*, EPMA, ISBN: 978 1 899072 13 2, vol. 4, pp. 557–562.

Heaney, D. (2006a), 'Mass production of micro components utilizing lithographic tooling and injection molding technologies', in *Proceedings of the First International Conference on Micro-Manufacturing, ICOMM 2006*, no. 61.

Heaney, D. (2006b), 'Injection molding of micro medical components utilizing MEMS technologies for tooling', in *Proceedings of Materials Science and Technology 2006*, 15–18 October, Cincinnati, OH, USA, ASM–TMS.

Heaney, D. (2006c), 'Sintering of medical microcomponents', *Heat Treat Progress*, July/August, 35–36.

Heldele, R., Schulz, M., Kauzlaric, D., Korvink, J. G., and Haußelt, J. (2006), 'Micro powder injection molding: process characterization and modeling', *Journal of Microsystem Technologies*, 12, 941–946, doi 10.1007/s00542-006-0117-z.

Jang, J. M., Lee, W., Ko, S.-H., Seo, J.-S., and Kim, W.-Y. (2010), 'Sintering of MIMed part with various micro features', in *Proceedings of Powder Metallurgy 2010 World Congress*, EPMA, ISBN: 978 1 899072 13 2, vol. 4, pp. 471–475.

Jenni, M., Zauner, R., and Stampfl, J. (2009), 'Measurement methods for powder binder separation in PIM components', in *Proceedings of EuroPM 2009 Conference*, Kopenhagen, European Powder Metallurgy Association, ISBN 978 1 899072 07 1, pp. 141–146.

Kauzlaric, D., Lienemann, J., Pastewka, L., Greiner, A., and Korvink, J. G. (2008), 'Integrated process simulation of primary shaping: multi scale approaches', *Microsystem Technologies*, 14, 1789–1796.

Kipphut, C. M. and German, R. M. (1991), 'Powder selection for shape retention in powder injection molding', *International Journal of Powder Metallurgy*, 27(2), 117–124.

Ko, S.-H., Lee, W., Jang, J. M., and Seo, J.-S. (2010), 'Effects of debinding and sintering atmosphere in microstructure of micro MIMed part', in *Proceedings of Powder Metallurgy 2010 World Congress*, EPMA, ISBN: 978 1 899072 13 2, vol. 4, pp. 485–492.

Kong, X., Barriere, T., and Gelin, J. C. (2010), 'Micro-PIM process with 316L stainless steel feedstock and numerical simulations', in *Proceedings of Powder Metallurgy 2010 World Congress*, EPMA, ISBN: 978 1 899072 13 2, vol. 4, pp. 669–676.

Lanza, G., Schlipf, M., and Fleischer, J. (2008), 'Quality assurance for micro manufacturing processes and primary-shaped micro components', *Microsystem Technologies*, 14, 1823–1830.

Li, S. G., Fu, G., Reading, I., Tor, S. B., Chaturvedi, P., Yoon, S. F., and Youcef-

Toumi, K. (2007), 'Dimensional variation in production of high-aspect-ratio micro-pillars array by micro powder injection molding', *Applied Physics, A*, 89, 721–728.

Li, He, Y., H., Wang, G., and Deng, Z. (2010), 'Effect of delay time on material distribution of metal co-injection moulding', in *Proceedings of Powder Metallurgy 2010 World Congress*, EPMA, ISBN: 978 1 899072 13 2, vol. 4, pp. 511–517.

Liu, L., Loh, N. H., Tay, B. Y., Tor, S. B., Murakoshi, Y., and Maeda, R. (2006), 'Micro powder injection molding: Sintering kinetics of microstructured components', *Scripta Materialica* 55, 1103–1106.

Loh, N. H., Tor, S. B., Tay, B. Y., Murakoshi, Y., and Maeda, R. (2007), 'Fabrication of micro gear by micro powder injection molding', *Microsystem Technologies*, 14, 43–50.

Maetzig, M. and Walcher, H. (2006), 'Assembly moulding of MIM materials', in *Proceedings of Euro PM 2006 – Powder Metallurgy Congress and Exhibition*, 23–25 October 2006, EPMA, ISBN 978-1-899072-33-0, vol. 2, pp. 43–48.

Monaghan, J. J. (2005), 'Smoothed particle hydrodynamics', *Reports on Progress in Physics*, 68, 1703–1759.

Moritz, T. (2008), 'Two-component CIM parts for the automotive and railway sectors', *Powder Injection Moulding International*, 2(4), 38–39.

Munzinger, C., Dieckmann, A.-M., and Klimscha, K. (2010), 'Research on the design of sinter joined connections for powder injection moulded components', *Proceedings of Powder Metallurgy 2010 World Congress*, EPMA, ISBN: 978 1 899072 13 2, vol. 4, pp. 477–484.

Nishiyabu, K. and Tanaka, S. (2008), 'Small is better if testing MIM nano theories', *Metal Powder Report*, 3, 28–32.

Nishiyabu, K., Kanodo, Y. T., and Tanaka, S. (2006), 'Innovations in micro injection molding process by lost form technology', in *Proceedings of 2006 Powder Metallurgy World Congress*, Korean Powder Metallurgy Institute, ISBN 89-5708-121-6 94580, vol. 1, pp. 43–44.

Nishiyabu, K., Andrews, I., and Tanaka, S. (2008), 'Accuracy evaluation of ultra-compact gears manufactured by the microMIM process', *Powder Injection Moulding International*, 2(4), 60–63.

Nishiyabu, K., Tanabe, D., Kanoko, Y., and Tanaka, S. (2010), 'Micro metal injection moulding by NIL lost form technology and using nanopowder', in *Proceedings of Powder Metallurgy 2010 World Congress*, EPMA, ISBN: 978 1 899072 13 2, vol. 4, pp. 445–453.

O'Donovan, E. J. and Tanner, R. I. (1984), 'Numerical study of the Bingham squeeze film problem', *Non-Newtonian Fluid-Mechanics*, 15, 75–83.

Oerlygsson, G., Piotter, V., Finnah, G., Ruprecht, R., and Hausselt, J. (2003), 'Two-component ceramic parts by micro powder injection moulding', in *Proceedings of the Euro PM 2003 Conference*, Valencia, Spain, pp. 149–154.

Petzoldt, F. (2008), 'Micro powder injection moulding – challenges and opportunities', *Powder Injection Moulding International*, 2(1), 37–42.

Petzoldt, F. (2010), 'Multifunctional parts by two-component powder injection moulding (2C-PIM)', *Powder Injection Moulding International*, 4(1), 21–27.

Piotter, V., Bauer, W., Hanemann, T., Heckele, M., and Mueller, C. (2008),

'Replication technologies for HARM devices: status and perspectives', *Journal of Microsystem Technologies,* 14, 1599–1605.
Rota, A., Imgrund, P., and Petzoldt, F. (2004), 'Micro MIM – a production process for micro components with enhanced material properties', in *Proceedings of Euro PM 2004,* European Powder Metallurgy Association, ISBN 1899072 15 2, pp. 467–472.
Ruh, A., Dieckmann, A.-M., Heldele, R., Piotter, V., Ruprecht, R., Munzinger, C., Fleischer, J., and Haußelt, J. (2008), 'Production of two-material micro assemblies by two-component powder injection molding and sinter-joining', *Microsystems Technology,* 14, 1805–1811.
Schmidt, H., Rota, A. C., Imgrund, P., and Leers, M. (2009), 'Micro metal injection moulding for thermal management applications using ultrafine powders', *Powder Injection Moulding International,* 3(2), 54–58.
Schneider, J., Iwanek, H., and Zum Gahr, K.-H. (2005), 'Wear behaviour of mould inserts used in micro powder injection moulding of ceramics and metals', *Wear,* 259, 1290–1298.
Schneider, J., Kienzler, A., Deuchert, M., Schulze, V., Kotschenreuther, J., Zum Gahr, K.-H., Löhe, D., and Fleischer, J. (2008), 'Mechanical structuring, surface treatment and tribological characterization of steel mould inserts for micro powder injection moulding', *Microsystem Technologies,* 14, 1797–1803.
Simchi, A., Imgrund, P., and Rota, A. (2006), 'An investigation on the sintering behavior of 316L and 17-4PH stainless steel powders for graded composites', *Materials Science and Engineering, A,* 424, 282–289.
Tan, L.-K. (2007), 'MIM technology set to transform the design and production of heat sinks', *Powder Injection Moulding International,* 1(4), 27–30.
Tay, B. Y., Liu, L., Loh, N. H., Tor, S. B., Murakoshi, Y., and Maeda, R. (2006), 'Injection molding of 3D microstructures by µPIM', *Microsystem Technologies,* 11, 210–213.
Vorster, E., Piotter, V., Plewa, K., and Kucera, A. (2010), 'Micro inmould labelling using PIM-feedstocks', in *Proceedings of Powder Metallurgy 2010 World Congress,* EPMA, ISBN: 978 1 899072 13 2, vol. 4, pp. 505–510.
Yang, I., Park, M. S., and Chu, C. N. (2009), 'Micro ECM with ultrasonic vibrations using a semicylindrical tool', *International Journal of Precision Engineering and Manufacturing,* 10(2), 5–10.
Yasui, N., Satomi, H., Fujiwara, H., Ameyama, K., and Kankawa, Y. (2006), 'The influence of powder size on mechanical properties of small MIM parts', in *Proceedings of 2006 Powder Metallurgy World Congress,* Korean Powder Metallurgy Institute, ISBN 89-5708-121-6 94580, vol. 1, pp. 39–40.
Yin, H., Jia, C., and Qu, X. (2008), 'Micro powder injection molding – large scale production technology for micro-sized components', *Science in China, Series E: Technological Sciences,* 51(2), 121–126.
Zauner, R., Binet, C., Heaney, D., and Piemme, J. (2004), 'Variability of feedstock viscosity and its correlation with dimensional variability of green powder injection moulded components', *Powder Metallurgy,* 47(2), 151–156.
Zeep, B., Piotter, V., Torge, M., Norajitra, P., Ruprecht, R., and Haußelt, J. (2006), 'Powder injection moulding of tungsten and tungsten alloy', in *Proceedings of the EuroPM 2006,* Ghent, Belgium, pp. 85–90.
Zlatkov, B. S. (2009), 'The processing of advanced magnetic components by PIM:

current status and future opportunities', *Powder Injection Moulding International*, 3(3), 41–50.

Zlatkov, B. S. and Hubmann, R. (2008), 'Tube type X-cooler for microprocessors produced by MIM technology', *Powder Injection Moulding International*, 2(1), 51–54.

Zlatkov, B. S., Danninger, H., and Aleksic, O. S. (2008a), 'Cooling performance of tube X-cooler shaped by MIM technology', *Powder Injection Moulding International*, 2(3), 64–68.

Zlatkov, B. S., Griesmayer, E., Loibl, H., Aleksic, O. S., Danninger, H., Gierl, C., and Lukic, L. S. (2008b), 'Recent advances in PIM technology I', *Science of Sintering*, 40, 79–88.

# 14
Two-material/two-color powder metal injection molding (2C-PIM)

P. SURI, Heraeus Materials Technology LLC, USA

**Abstract**: Two-material powder injection molding or two-color powder injection molding (2C-PIM) combines the near net shape manufacturability of powder injection molding (PIM) with significant cost savings. This chapter reviews the technologies, processing parameters and applications of this process.

**Key words**: two-color powder injection molding (2C-PIM).

## 14.1 Introduction

Two-material powder injection molding or two-color powder injection molding (2C-PIM) has its basis in the more prevalent two-color injection molding of plastics. An extension of powder injection molding (PIM), 2C-PIM is viewed and investigated as a technology to manufacture near-net-shaped, functionally graded composites combined with the shape complexity that can be realized via PIM. Pischang et al.,[1] were among the early investigators to present exploratory study of 2C-PIM. This was followed by Alcock et al.,[2] and Pest et al.,[3] who applied and extended the ideas proposed to make a case for 2C-PIM. Since then, the concept and evaluation of 2C-PIM has been applied to numerous material systems with a variety of potential applications ranging from the size and shape complexity of a typical PIM part to Micro-PIM parts.

## 14.2 Injection molding technology

As with the PIM process, 2C-PIM starts with the preparation and rheological characterization of feedstock, followed by injection molding to form a green component. Molding of the components can be accomplished by over-molding and co-injection molding. In the over-molding variant, a

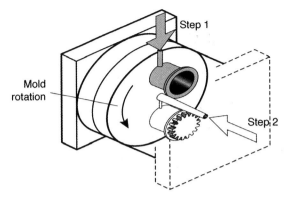

*14.1* Schematic diagram of 2C-PIM process via overmolding.[10]

molding machine that is equipped with two injection units is typically used to inject two different polymer/metal powder mixtures or compounds into the desired shape. The molded part composed of two different materials is thermally processed to remove the polymer and sintered to yield a single, integrated component. A schematic diagram of the injection molding step of this process is shown in Fig. 14.1. The concept is first to mold one part in a cavity and then rotate the tooling to form another cavity and mold around the already molded part. The component is then composed of two interlocked materials and is ejected from the mold. This is typically a manual process, where first the component is molded, cooled to room temperature, and then transferred to another mold followed by over-molding, but this is best achieved using a twin-barrel injection molding unit. Typically, over-molding is used to evaluate the process feasibility and make prototypes.

In co-injection molding, a functionally graded structure is produced using the flow behavior of the materials, through the same runner system, to produce a structured component that has a core and skin of two different materials. This is a well-established technology for plastics and has been experimentally examined for two metal powders.[2, 4, 5] Co-injection molding machines are equipped with one-, two- or three-channel systems. In the one-channel system, the feedstock is introduced sequentially into the mold by shifting a valve, as shown schematically in Fig. 14.2(a). Owing to the fluid flow characteristics, the first feedstock adheres to the cooler mold surface, forming a skin. Thickness of this skin is controlled by the injection rate, temperature and to some extent the flow compatibility of the two materials. In a two-channel system, Fig. 14.2(b), it is possible to inject two feedstocks sequentially or simultaneously. For plastics, profile injection is preferred as it provides enhanced control of the surface appearance. A three-channel system allows for simultaneous injection with a direct sprue gate, Fig. 14.2(c).

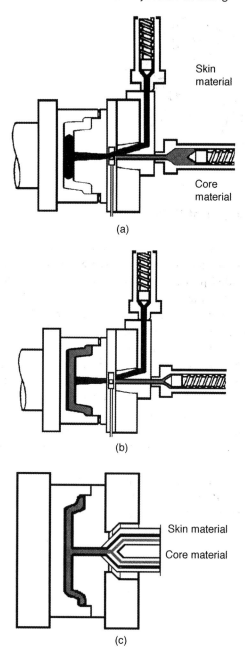

*14.2* Schematic diagram of runner system and molding in (a) one-channel, (b) two-channel and (c) three-channel co-injection molding.

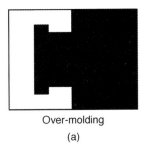 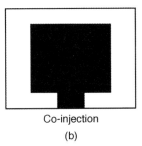

Over-molding (a)     Co-injection (b)

*14.3* Schematic diagram of the cross-sectional difference between 2C-PIM via (a) over-molding and (b) co-injection molding.[10]

Typically, the skin thickness is influenced on both sides of the part. Figure 14.3 illustrates the cross-sectional differences between over-molding and co-injection molding.

There are no known challenges to forming green parts via over-molding that are very specific to 2C-PIM. Rheological characterization and mold filling operations are a direct extension from the typical PIM cases. Different feedstock systems such as wax–polymer based, Catamold® based or water based feedstocks are used successfully.

## 14.3 Debinding and sintering

It has been largely recognized from early stages of development that the two materials should exhibit or be tailored to exhibit similar densification behavior to avoid defect formation. This aspect of 2C-PIM is by far the most challenging and will ultimately make the design and the product either successful or unviable.

During thermal processing, the injection molded part undergoes various changes and as a consequence assumes the character of material with different thermomechanical characteristics.[6–8] The polymer softens, disintegrates and burns off in the temperature range of 200–500°C, resulting in a highly porous and fragile brown part. For a typical powder characteristics of $D_{50}$ of 5–10 μm for metals and about 0.2–0.5 μm for ceramics used in PIM, the particle size is too large to exhibit any significant initial stage sintering during debinding. The brown part maintains its integrity due to interparticle friction. The strength of the brown part after debinding and before the onset of initial stage sintering is typically less than 1 MPa. At this stage, the mechanical behavior is similar to a very brittle material and differences in the thermal expansion or degradation characteristics of the binders are likely to have a large impact. Typically, the two materials utilize the same binder system and, other than small dilation due to thermal expansion, one can expect an absence of relative motion between the two components during this stage. The part is expected to survive this stage without generating any defects.[9]

Strength increases with the onset of neck growth. With a further increase in temperature and the ensuing densification, the stresses in the components and the assembly gain importance due to differences in sintering shrinkage. Porosity during the initial stage of sintering is approximately 30–45 vol%. Owing to porosity, the material continues to behave like a brittle material but with increasing plasticity with an increase in temperature. Defects formed during this stage include delamination of the interface or channeling cracks through the thickness of the component, largely attributable to the differential shrinkage and the ensuing stress state.

With the onset of the second stage of sintering, thermal softening lowers the strength and increases the plasticity of the material. It is conceivable that the differences in shrinkage behavior can be tolerated to a higher degree during the second and the final stage of sintering. In case of liquid phase sintering, experiments suggest that these tolerances are even higher with the possibility that any defects formed are effectively healed.[10]

As mentioned earlier, success of 2C-PIM depends on minimizing the differences in shrinkage during sintering. This would occur naturally in very few material systems – systems with similar compositions and particle size distribution.[11–13] For dissimilar materials, parameters that can be availed to match the sintering characteristics include a choice of material system, powder particle size, and powder volume fraction in the feedstock.

For metallurgically compatible systems, defects can be avoided if the intrinsic strength of the material exceeds the stresses induced.[8,14] Induced stresses are a strong function of component geometry and the extent of shrinkage mismatch. Qualitative measures such as the Apparent Co-Sintering Index[15] can be used as a rapid screener for 2C-PIM, but such measures restrict the choice of material without any consideration to the component design. Mapping the process window for 2C-PIM involves four distinct information streams.

1. Difference in the sintering shrinkage with temperature: difference in shrinkage during sintering is essential to understand and develop products that utilize 2C-PIM technology. Dilatometry is a very useful tool in understanding and providing a quantitative measure of material compatibility. Dilatometry curves for the materials under consideration can quickly validate the merits of 2C-PIM. Figures 14.4 (a) to (d) compare the shrinkage behavior of M2 tool steel with Fe–10Cr–0.5N and 316L SS + 0.5B to illustrate the impact of difference in shrinkage behavior and the importance of dilatometry.

2. *In situ* strength of material: *in situ* strength of the PIM component changes during different stages of processing.[6–8] This plays an

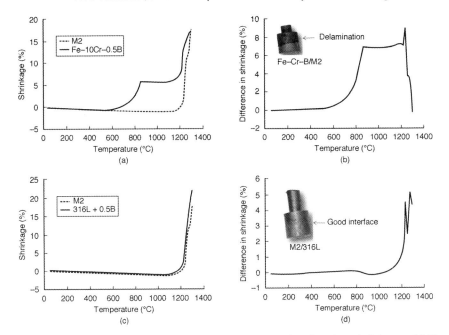

*14.4* Parts (a) and (c) are dilatometer results showing shrinkage of M2 tool steel with Fe–Cr–B and 316LS + B; (b) and (d) show the difference in shrinkage (mismatch) and its effect on the success or failure of 2C-PIM.[10]

important role in the ability of the material to withstand stresses caused by difference in the sintering shrinkage.[9] Strength of specific material can be experimentally determined by high-temperature mechanical testing. However, an approximate value will suffice during the design and proof-of-concept stage and can be obtained by simple models.[7,8] These models are found to be in general agreement with the experimental results.[16] Variation in the *in situ* strength of a powder compact is a function of its material properties, particle size, relative density of the compact, and the neck size.[7,8]

3. Thermomechanical behavior: choice of the correct thermomechanical behavior is important to model the induced stresses. These models can range from simple elastic to elasto-plastic and visco-plastic models. To predict the success of 2C-PIM geometry, an elastic model can provide a conservative estimate. These models should consider the reduction in the strength of the material due to porosity and the temperature effects.[8]

4. Stress state: difference in sintering shrinkage does not necessarily indicate material incompatibility towards 2C-PIM. The stress state of the final geometry is far more important. Despite differences in sintering shrinkage, component geometry can be designed to withstand

*14.5* Pictures of components with varying densification characteristics to induce and illustrate the defects during 2C-PIM: (a) shows cracking due to hoop stress initiated by excess shrinkage of the inner core (left) and outer core (right) and (b) shows delamination and poor bonding at the interface between metallurgically incompatible systems.[9]

or lower the induced stress.[9] Simple geometries can use closed-form solutions such as radial stress and hoop stress for circular cross-sections in either plane stress or plane strain conditions and can prove very useful in understanding the effect of different geometries.[14] Figure 14.5 shows the impact of difference in sintering shrinkage and metallurgical compatibility on the success of 2C-PIM.[9] An increase in shape complexity warrants the use of finite element method models to understand the stresses induced during sintering.

## 14.4 2C-PIM products

Figure 14.6 shows a few prototypes made via 2C-PIM. Successful prototypes made using 2C-PIM technology include a holder for angle-of-rotation sensors used in automotives,[12] tool steel inserts with stainless steel core,[10] hermetic microelectronic packages with embedded heat sinks,[17] micro-injection molded ceramic resistive heater elements,[18] and micro-shaft and gear assemblies.[19] The technology is also demonstrated on a variety of material systems and applications. For instance, copper based heat sinks with porous channels acting as heat pipes,[11] graded WC–Co composites,[13] 17-4PH/316L,[20, 21] Inconel 718/Inconel 625,[22] M2/316L,[23] and porosity-graded Co–Cr–Mo alloy,[24] Al/AlN[25] and Al/Fe–Nd–B,[26] $Al_2O_3$/TiN,[18]

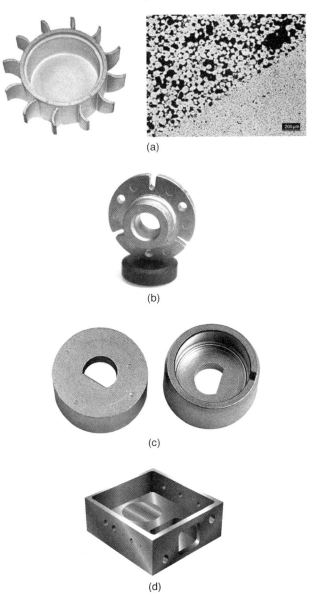

*14.6* Pictures and schematic diagrams of a few prototypes explored via 2C-PIM: (a) copper based heat sink with porous structure to act as heat pipe;[17] (b) and (c) automotive sensor holders; and (d) housing for hermetically sealed electronic package with embedded heat sink (photograph courtesy Professor Randall M. German).

$Al_2O_3/ZrO_2$[19, 27] and 3Y–TZP/stainless steel.[28] Case studies for some of the above combinations are detailed by Petzoldt.[29] Layered composites similar to low-temperature co-fired ceramics (LTCCs) or high-temperature co-fired ceramics (HTCCs) are excluded as they are not injection molded. While the challenges in sintering defect-free LTCC and HTCC components exist and are similar to 2C-PIM, there is also a possibility of using external uniaxial pressure during sintering to alter the shrinkage and densification characteristics.

## 14.5 Future trends

PIM provides the geometric shape attributes associated with plastic injection molding and the performance attributes associated with full-density powder metallurgy and ceramic sintering. The technology excels in the mass production of complicated shapes from materials that are difficult to cast or machine. PIM continues to see impressive gains in the production but also continues to be restricted to monolithic materials.

Since the initial study on 2C-PIM in 1992, there have been less than 50 unique technical articles on the subject and less than ten patents applied that utilize this method of forming. To the knowledge of the author, there have been a few prototypes manufactured to make functional material via 2C-PIM but there are no products with production volumes that can be realized by PIM.

2C-PIM aims to provide multi-functionality to the components by combining the near-net-shape manufacturability of PIM with the cost savings. Cost saving is largely realized by reducing the secondary joining operations. Success of two-color plastic injection molding is derived by functionality and economics. On the other hand, 2C-PIM is yet to mimic this success. PIM competes with other forming technologies such as die casting, investment casting and machining. To derive the competitive advantage, a successful PIM product has specific characteristics, operating within a range of component weight, shape complexity and production volume. 2C-PIM narrows down this window further owing to the constraints imposed to make a defect-free component along with other competitive joining technologies such as laser brazing, electron beam welding and fusion brazing.

Evolution of this technology to provide multi-functionality is yet to be realized. It is likely that the initial focus on reducing cost is not the appropriate strategy for 2C-PIM. Competing joining technologies, where available, are relatively mature, minimizing the incentive to investigate and develop products that yield defect-free components. While large-scale adoption is yet to be realized, it is likely to come from a combination of material systems that can only be processed via powder routes and have sufficient geometric complexity so that they cannot be processed via the press-and-sinter technique or joined by way of brazing or diffusional bonding. Similar to the ability of PIM technology to make products whose properties meet or exceed the wrought properties of equivalent material, the interfacial strength of a successful 2C-PIM product is likely to be superior to alternative joining technologies. 2C-PIM will likely find its success in niche applications, adding significant value to the component, and not as a highly productive, cost competitive technological alternative. Adoption of this technology will also require a greater integration of the design and stress analysis of the component to generate a defect-free component and provide the benefit of multi-functionality.

## 14.6 References

1.  Pischang, K., Birth, U., and Gutjahr, M. (1994), in *Proceedings of the 1994 International Conference and Exhibition on Powder Metallurgy and Particulate Materials,* Toronto, ON, Canada, 8–11 May, vol. 4, pp. 273–284.
2.  Alcock, J. R., Darlington, M. W., and Stephenson, D. J. (1996), *Powder Metallurgy,* 39(4), 252–254.
3.  Pest, A., Petzoldt, F., Hartwig, T., Veltl, G., and Eifert, H. (1997), *Proceedings of the 1st European Symposium on PIM-1997,* Munich, Germany, 15–16 October, pp. 132–139.
4.  Pest, A., Petzoldt, F., Hartwig, T., and German, R. M. (1996), *Advances in P/M and Particulate Materials,* vol. 5, pp. 19-171–19-178.
5.  Alcock, J. (1999), *Metal Powder Report,* June, pp. 30–34.
6.  German, R. M. (2001), *Materials Transactions,* 42, 1400–1410.
7.  Xu, X., Lu, P., and German, R. M. (2002), *Journal of Materials Science,* 37, 117–126.
8.  Suri, P., Heaney, D. F., and German, R. M. (2003), *Journal of Materials Science,* 38, 4875–4881.
9.  German, R. M., Heaney, D. F., and Johnson, J. L. (2005), *Proceedings of PM2TEC 2005: International Conference on Powder Metallurgy and Particulate Materials,* Montreal, Quebec, Canada, 19–23 June, pp. 41–52.
10. Heaney, D. F., Suri, P., and German, R. M. (2003), *Journal of Materials Science,* 38, 4869–4874.
11. Johnson, J. L., Tan, L. K., Bollina, R., Suri P., and German, R. M. (2005), *Powder Metallurgy,* 48, 123–128.
12. German, R. M., Heaney, D. F., Tan, L. K., and Baungartner, R. (2002), 'Fuel injectors, sensors and actuators manufactured by bi-metal powder injection molding', in *Proceedings of 2002 SAE World Congress,* 4–7 March, paper no. 2002-01-0343.
13. Li, T., Li, Q., Fuh, J. Y. H., Ching Yu, P., and Lu, L. (2009), *International Journal of Refractory Metals and Hard Materials,* 27, 95–100.
14. Johnson, J. L., Tan, L. K., Suri P., and German, R. M. (2003), *Journal of Metals,* 55(10), 30–34.
15. Simchi, A., Petzoldt, F., and Hartwig, T. *Proceedings of the Metallurgy World*

*Congress and Exhibition (PM2005)*, Prague, European Powder Metallurgy Association (EPMA), Shrewsbury, UK, pp. 357–363.
16. Gelin, J. C., Barriere, Th., and Song, J. (2010), *Journal of Engineering Materials and Technology*, 132, 011017-1–011017-9.
17. Tan, L. K. and Johnson, J. L. (2004), see http://www.electronics-cooling.com/2004/11/metal-injection-molding-of-heat-sinks.
18. Oerlygsson, G. *et al.* (2003), 'Two-component ceramic parts by micro powder injection moulding', in *Proceedings of the Euro PM 2003*, Valencia, Spain, 20–22 October, pp. 149–154.
19. Piotter, V. *et al.* (2005), in *Proceedings of the 1st International Conference on Multi-Material Micro Manufacture*, Karlsruhe, Amsterdam, 29 June–1 July, S.207-10.
20. Imgrund, P., Rota, A., Petzoldt, F., and Simchi, A. (2007), *International Journal of Advanced Manufacturing Technology*, 33, 176–186.
21. Simchi, A., Rota, A., and Imgrund, P. (2006), *Materials Science Engineering A*, 424, 282–289.
22. Simchi, A. (2006), *Metal Materials Transactions A*, 37, 2549–2557.
23. Firozdour, V., Simchi, A., and Kokabi, A. H. (2007), *Journal of Material Science*, 43, 55–63.
24. Dourandish, M., Simchi, A., and Godlinski, D., (2008), *Materials Science and Engineering A*, Vol. 472, pp. 338–346.
25. Liu, Z. Y., Kent, D., and Schaffer, G. B. (2009), *Materials Science and Engineering A*, 513–514, 352–56.
26. Liu, Z. Y., Kent, D., and Schaffer, G. B. (2009), *Metal and Materials Transactions*, 40A, 2785–2788.
27. Feng, J., Qiu, M., Fan, Y., and Xu, N. (2007), *Journal of Membrane Science*, 305, 20–26.
28. Baumann, A. A., Moritz, T., and Lenk, R. (2007), *Proceedings of European Powder Metallurgy Congress and Exhibition (EURO PM2007)*, Toulouse, France, October, vol. 2, pp. 189–193.
29. Petzoldt, F. (2010), *PIM International*, 4(1), 21–27.

# 15
Powder space holder metal injection molding (PSH-MIM) of micro-porous metals

K. NISHIYABU, Kinki University, Japan

**Abstract:** This chapter describes the metal injection molding (MIM) process for manufacturing metal components with micro-sized porous structures. The process has been achieved by applying the powder space holder (PSH) method to the MIM process. The novel PSH-MIM process is capable of net-shape manufacturing precise micro-porous metal parts by controlling pore size and porosity. This chapter also includes detailed explanations of the pore formation mechanism and optimum material selection, as well as describing examples of micro-porous MIM parts and their properties.

**Key words:** porous metal, metal injection molding, space holder method, porosity, pore size, dimensional accuracy.

## 15.1 Introduction

Porous materials are a class of materials with low density, large specific surface and a range of novel properties in the physical, mechanical, thermal, electrical and acoustical fields. The relationship between the structure and the properties of cellular solids made of natural materials and the properties of engineered materials including metals, ceramics and polymer has been summarized by Gibson and Ashby (1988). Porous materials can be categorized into closed porous and open porous structures. Closed foams can be used potentially in applications that need light-weight structural elements with better sound and impact energy absorption. Open foams can also be used for high-performance applications such as heat exchangers and heat sinks for thermal management, and also for medical implants, filters and electrodes for biological and chemical reactions. In recent years, there has been a strong focus on metal materials with higher porosity (more than around 70%) such as metal foams, cellular metals and metal sponges (Ashby *et al.*, 2000; Banhart, 2001; Wadley, 2002). Most of these materials have been developed for use as structural components.

In the field of powder metallurgy (PM), on the other hand, porous metal materials with comparatively low porosities (0–30%) have been attracting most attention in recent years. They have been used mostly for tribological applications such as oil-impregnated sintered bearings. Porous materials have not traditionally been an area of innovation in the field of PM. In the case of sintered materials used for mechanical parts, the emphasis has been more on high densification to enhance durability, strength and reliability in service. The metal injection molding (MIM) process has been developed using fine metal powders, thereby enabling the manufacture of sintered parts that are more dense and precise than those manufactured in PM (German, 1984; German and Bose, 1997).

However, open porous metals are finding many applications in high-tech products, e.g. bio-filters for medical microscopic measurements, heat exchangers for micro-devices, and micro-mist-generators for filtering air dust and bacteria, medical implants with low modulus, vapor recovery equipment for fuel cells and so on. These applications can benefit from the advantages provided by micro-porous structures. Many pores can be created easily by control of sintering in the PM and MIM process. Some beneficial characteristics of these materials are listed by Heaney *et al.* (2005). The pores are polygonal (rather than spherical) spaces between particles with high permeability.

The key challenge is how the porous structure is formed by the MIM process to create a controlled pore size and porosity. It is essential for the MIM process to remove most of the polymeric binders from the molded body while keeping the shape formed by the metal powder. Debinding technology is used to create a precise number of spaces during pore formation. MIM is also a net-shape manufacturing process without any mechanical cutting and polishing, where the pores formed on the surface are not subject to damage by machining forces. This makes the MIM process the preferred manufacturing process for net-shaped porous metal parts.

This chapter will also discuss other forming techniques for the manufacture of micro-porous metal parts, which have been recently developed by combining the powder space holder (PSH) method with the MIM process. The effects of material combination and sintering conditions on the pore formation and the physical properties of sintered porous metals are described in detail. Further advantages of the PSH-MIM method, such as excellent liquid infiltration properties, high dimensional accuracy in production and enhancement of mechanical properties by using functionally graded porous structures, are reviewed.

## 15.2 Production methods for porous metals

### 15.2.1 Types of porous metals and manufacturing

In most porous metals, the pore size tends to become smaller as the product gets smaller. Macro-sized porous structures are created by conventional machining methods such as cutting, grinding, welding and fastening. Micro-sized porous parts, however, require net-shape manufacturing in complicated shapes with high dimensional accuracy and the porous structure forming on the surface. So far, there have been few studies that have dealt with net-shape production of micro-porous metal components (Heaney et al., 2005), although these methods are very desirable when producing components with higher functionalities at a lower cost. A noted textbook on PM (German, 1984, pp. 468–469) describes the fabrication of controlled pore structures for bearings, filters, flow restrictors, sound absorbers, heat pipes and biomedical implants as a natural application for PM. Control of the pore size is achieved by using powder grains within a narrow size range and by closely controlling densification during the processing of the materials (Heaney and German, 2001). The possibility of pore structure manipulation is in fact the key reason for selecting PM to fabricate porous metals and this has been so since the early days of PM technology.

Generally speaking, sintering can make small pores easier to achieve than is possible using foaming or deposition methods (Ashby et al., 2000, p.7). Traditional PM, however, cannot produce micrometer or sub-micrometer sized pores with sufficient porosity, because there is a limit in combining pore size and porosity with this technology. The resulting pores are also interglobular interstices whose shape is not ideal for fluid infiltration.

Figure.15.1 shows the ranges of porosity and type of porous structure against pore size for typical PM production methods of porous metals. These products can be manufactured with existing commercial production methods such as (a) sintering of metal powders, (b) sintering of hollow spheres, (c) sintering of metal textiles, (d) electron deposition, (e) melt gas injection in liquid metal, (f) entrapped gas expansion and so on (Wadley, 2002). Each method can be used with a small subset of methods to create porous metals with a limited range of porosity and pore size. In practice, it is very difficult to produce porous metal components with pore sizes up to several tens of micrometers and either open or closed porous structures with a specified porosity.

At the moment few methods can in practice produce net-shaped metal components efficiently. It is even more complicated to produce porous metals with graded properties and complex shapes. The micro-porous MIM and sub-micron porous MIM processes developed by the authors have the

352    Handbook of metal injection molding

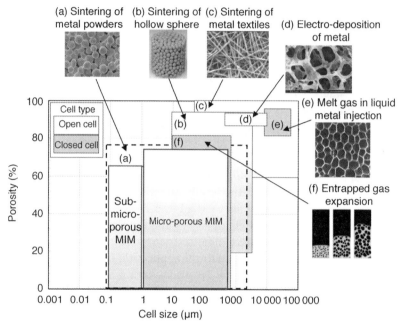

*15.1* Porosity against cell size for typical production methods of porous metals.

potential to achieve the desired production results. The process has three main advantages:

1. a wide range of independently controlled porosities and pore sizes as shown in Fig. 15.1;
2. net-shape mass-production with high dimensional accuracy;
3. various material combinations and graded structures.

## 15.2.2 Use of debinding in metal injection molding

Metal injection molding is a manufacturing method that combines traditional PM with plastic injection molding. Over the past decade it has established itself as a competitive manufacturing process for small precision components that would be costly to produce by alternative methods. It can be used to produce comparatively small parts with complex shapes from almost any type of material such as metals, ceramics, inter-metallic compounds and composites (German, 1984). Recently MIM has been studied not only for hard metals, but also for materials such as titanium, copper and aluminum (German and Bose, 1997). Unlike in the case of PM, MIM requires mixing metal powders with a large amount of polymeric binder. After this the organic constituents are removed in a debinding step

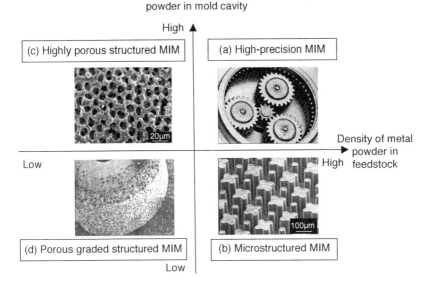

*15.2* Variously structured MIM parts determined by homogeneity of metal powder in mold cavity and density of metal powder in feedstock.

such as solvent extraction or pyrolysis. The brown body is held in the molded form only by metal powder after debinding. This debinding process and powder forming mechanism is unique to the MIM process.

A variety of MIM products produced by the authors are shown in Fig. 15.2. These suggest that various structures of MIM parts are determined by the homogeneity of the metal powder in the mold cavity and the continued solids loading of metal powder into the feedstock. To mix feedstock homogeneously the mold cavity needs to be densely filled. High-precision MIM parts and microstructured MIM parts can be manufactured where the feedstock is injection-molded into a microstructured cavity made of plastic. This process has been named micro sacrificial plastic mold insert MIM (μ-SPiMIM) by the authors.

The authors have produced metal components with a micro-sized porous structure by applying the PSH method to the MIM process (Nishiyabu *et al.*, 2004, 2005a, 2005b, 2005c). The PSH method makes use of the unique polymer-based MIM techniques developed by the authors. A homogeneous feedstock and polymeric cavities were used to make highly porous structured MIM parts. The compounds prepared by fractioning of space-holding particles were co-injection molded to make porous graded and structured MIM parts. The authors achieved some success when working with high-quality and micro-MIM processes, but sintering and mixing was poor, which caused segregation between metal powder and binder.

## 15.3 Formation of micro-porous structures by the PSH method

### 15.3.1 Powder space holder (PSH) method

The PSH method for micro-porous metals was developed based on the MIM process as illustrated in Fig. 15.3. In a conventional MIM process, the feedstock is composed of metal powder and binders; a high densification after debinding and the sintering process is very important for high-quality MIM products. However, to produce highly porous structured metals, the spaces created are required to remain stable after sintering. It is therefore useful to apply the PSH method to the MIM process. It is most important to establish the kinds of material that should be used for space holding in the process. It is preferable that the candidate materials are diffluent by water

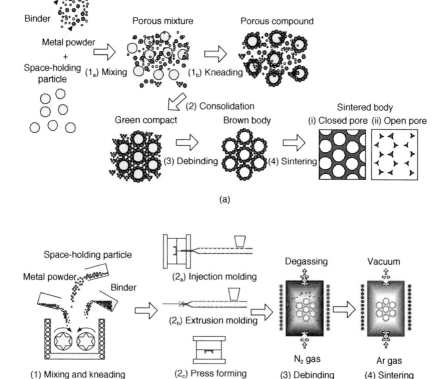

15.3 Powder space holder (PSH) MIM process for manufacturing micro-porous metal parts: (a) pore formation mechanism; (b) process flow.

and organic solvents such as sugar, salt and polymers (Lefebvre and Thomas, 2003; Bram *et al.*, 2003). There are many other requirements, such as spherical shape, availability of various particle sizes, high rigidity, thermal resistance, non-reactiveness to metal powder, reasonable cost, safety and so on. The authors have chosen mostly spherical particles made of polymethylmethacrylate (PMMA) polymer to fulfill such a requirement. PMMA polymer can be produced by monodispersed particles with a wide range of particle size from sub-micrometer to millimeters, and it has comparatively high thermal resistance and stiffness. PMMA also exhibits compatible mixing with wax and polymeric binders used in MIM feedstocks.

In the authors' process, in addition to metal powder and thermoplastic binders, extra coarse spherical materials made of PMMA were used as lost material to obtain a fine porous structure in the MIM components. As already explained, the combination of space-holding particles and metal powder together with the sintering conditions determine the porous structure. The PSH-MIM process illustrated in Fig. 15.3 is carried out in four steps as follows.

1. The first step is mixing the metal powder, binder and space-holding particles to prepare the porous compound.
2. The second step is molding the porous compound into the specified shapes by injection, extrusion molding or press forming to obtain the green bodies.
3. The third step is thermal debinding to remove the binder and space-holding particles.
4. The final step is sintering of the metal powders while keeping the dimensions of the created spherical spaces.

## 15.3.2 Mechanism of debinding in the PSH method

A key technology in the PSH method is to remove the space-holding particles which are used to make large numbers of spherical spaces. The debinding mechanism, simplified for a two-components binder system, is explained in this section. Figure 15.4(a) shows the typical debinding and sintering conditions for a green body molded using feedstock composed of 9 μm stainless steel powder and polymeric binder. Debinding is done at 600°C for 2 h, while sintering is carried out at around 1050°C for 2 h. Figure 15.4(b) shows the thermogravimetric curves for the decomposition of wax, polymer and space-holding particle (PMMA particles). Using three decomposition curves, a typical debinding mechanism for this material system can be determined. Figure 15.5 shows a schematic drawing of the debinding process at each temperature. No materials decompose below 100°C. Wax starts to decompose at 250°C, and creates numerous paths for

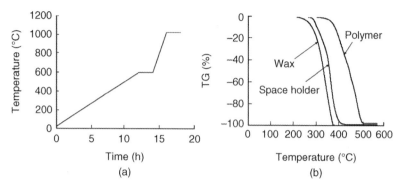

*15.4* PSH method for producing micro-porous metals: (a) debinding–sintering conditions; (b) thermogravimetric (TG) curves in debinding.

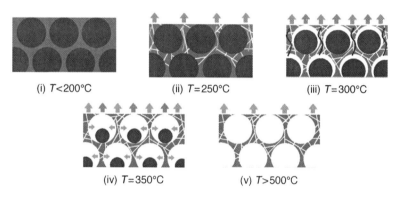

*15.5* Schematic drawing of debinding process.

degassing near the PMMA particle. Then the PMMA decomposes along with the wax at 300°C. When the temperature is further increased to 350°C, large amounts of PMMA and polymer decompose simultaneously. Finally all of the binder constituents and PMMA particles have been decomposed completely over 500°C.

### 15.3.3 Examples of micro-porous metals by PSH method

The PSH method can be applied to most kinds of metal powder, such as stainless steels, nickel, aluminum, copper, titanium and their alloys, as shown in Fig. 15.6. The pore size can be determined by diameter of space-holding particle, which can actually be prepared in a wide range of sizes, from sub-micrometers to a few hundred micrometers or larger. However, spherical pores are not completely accurately formed in all specimens, because of mismatches in size of some of the space-holding particles that are needed to match each metal powder particle.

Powder space holder MIM of micro-porous metals    357

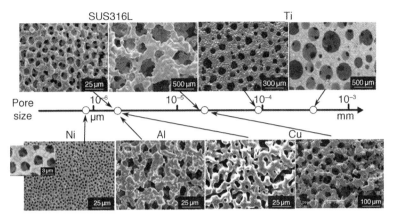

15.6 SEM images and pore sizes of micro-porous metals manufactured by PSH-MIM method.

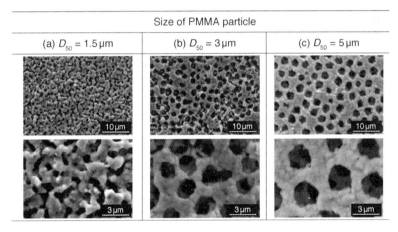

15.7 Surface structures of Ni porous specimens produced using various sizes of PMMA particles.

The surface structures of porous specimens produced using pure Ni powder ($D_{50} = 0.49\,\mu m$) and various sizes of spherical PMMA particles ($D_{50} = 1.5, 3, 5\,\mu m$), with a volumetric ratio of 60% in porous compounds, are shown in Fig. 15.7. Because the compounds are prepared using sub-micron powders with highly specific surface areas, the melt viscosity is very high, and the melt is then not easy to compact by injection molding. The molding was carried out under constant conditions where die temperature and gauge pressure was 200°C and 10 MPa, respectively. Debinding and sintering were sequentially carried out at 600°C for 2 h in $N_2$ and at 900°C for 2 h in Ar gas atmospheres in an industrial vacuum furnace to avoid oxidizing. Unsurprisingly, these specimens become reduced in pore size the smaller the PMMA particles that are used for holding spaces. As for the shape of the pores, in the cases of 3 μm and 5 μm PMMA specimens, an orderly matrix of spherical polygonal pores, with a smaller size than each diameter of PMMA particle, is visible. On the other hand, the 1.5 μm PMMA specimen lost shape in the sphere and formed numerous polygonal spaces with a size on the submicron scale. A larger difference was obtained in the 1.5 μm PMMA specimen where the pore could not hold a spherical shape because the ratio of diameter of Ni powder to PMMA particle is not large enough, so that Ni powders could not be filled into the cells among the PMMA particles. In that case, the PSH method can achieve a well-defined porous specimen, provided that the pore holds a spherical shape. In the trial production of the sintered porous metal, it was possible to achieve open pores formed homogeneously with 0.65 μm in mean pore diameter and 67% porosity. Porosity, pore size and surface area were easily controlled by optimizing the fraction of spherical materials for space-holding sintering process conditions.

Another advantage of the PSH method lies in the possibility for net-shape production of micro-porous metal components with complicated shapes and highly functionally graded structures. Figure 15.8 shows typical examples of micro-porous metal parts manufactured by extrusion molding (Fig. 15.8(a)) and injection molding (Fig. 15.8(b)) processes. The porous compound used is 9 μm 316L powder and 50 μm PMMA particle with a volumetric ratio of 60%. Longitudinal parts are better manufactured using extrusion molding, and complex-shaped parts have benefited from injection molding. The injection molded parts have a slightly denser surface than extruded ones for the reason outlined below.

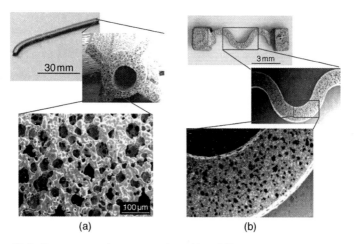

*15.8* Porous metal parts produced by different molding method using 9 μm 316L powder and 50 μm PMMA particle with volumetric percentage of 60%: (a) extrusion molding; (b) injection molding.

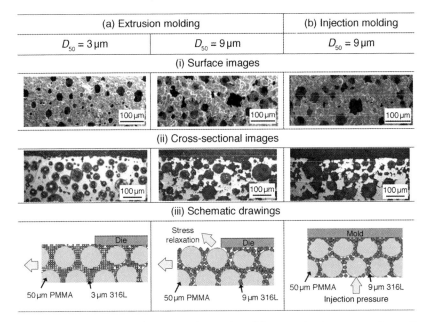

*15.9* Effects of particle size and molding methods on pore formation on the surface of porous metals.

Figure 15.9 shows the effects of particle size on pore formation on the surface of porous metals produced by injection molding and extrusion molding processes. In the case of extrusion molding, the pore size and porosity appearing on the surface were reduced in the specimen produced using finer metal powders ($D_{50} = 3\,\mu m$), which were holding the filling around the PMMA particles ($D_{50} = 50\,\mu m$). An increase of pore size and porosity, however, is visible in the specimen produced using coarser metal powders ($D_{50} = 9\,\mu m$), which were rearranged by PMMA particles that were elastic-deformed after extruding from the die. It is clear from these results that the porous structure on the surface is affected by the particle diameter of the metal powder used. In the case of injection molding, on the other hand, the pores are not changed after demolding because high molding pressure is applied in the cavity and metal powders are held in the green compact by cooling the polymeric binder before it is ejected from the mold. It is therefore clear that molding pressure and its relaxation affect the pore structure on the surface. This effect could be used to produce complex sintered porous parts with dense surfaces by one-step injection molding.

*Table 15.1* Experimental materials and fraction of constituents

| | Compositions | Mean diameter | Volume fraction | |
|---|---|---|---|---|
| | | | MIM feedstock | Porous compound |
| Metal powder | Stainless steel, 316L | 3 µm<br>9 µm | 50 vol% | 20–100 vol% |
| Binder | Wax, polyacetal | – | 50 vol% | |
| Space-holding particle | PMMA | 10 µm<br>50 µm | – | 0–80 vol% |

## 15.4 Control of porous structure with the PSH method

### 15.4.1 Materials and manufacturing conditions

The experimental materials used for porous compounds are listed in Table 15.1. The metal powders, loaded at 50 vol%, were austenitic stainless steel 316L ($D_{50}$ = 3 µm, 9 µm) produced by the water-atomization method. The binder was paraffin wax and polyacetal polymer. Spherical particles ($D_{50}$ = 10 µm, 50 µm) made from PMMA were used for holding the metal powder in position. These materials were co-mixed and pelletized with a high-pressure kneader and a plunger-type extruder. The resulting specimens are labeled specimen 3-10 or specimen 3-50. In the first case, 10 µm PMMA particles together with 3 µm sized 316L powder were used. For the second specimen, 50 µm PMMA particles were mixed with 3 µm sized 316L powder. Specimen 9-50 is produced with 50 µm PMMA particles and 9 µm 316L powder. The fraction rate of PMMA particles varied from 0 to 80 vol%, being the main experimental parameter. To keep the experiments simple, circular dishes of green compacts (40 mm diameter, 2 mm thick) were prepared from various porous compounds by hot press molding. The samples could also be produced by injection molding (Nishiyabu *et al.*, 2004, 2005a, 2005b, 2005c). Hot press molding was carried out under constant conditions where die temperature was 200°C and gauge pressure was 10 MPa. Debinding and sintering were sequentially carried out at 600°C for 2 h in $N_2$ and at 1050–1200°C for 2 h in Ar gas atmosphere to avoid oxidizing.

### 15.4.2 Surface structures

The surface structure of the sintered specimen was observed by scanning under the electron microscope. The microstructures on the surface of the sintered specimen produced by using various porous compounds are shown in Fig. 15.10. The number of pores on the surface increases as the fraction of PMMA particles increases from 0 to 80 vol%. The pore size is significantly dependent on the diameter of PMMA particles. These characteristics are

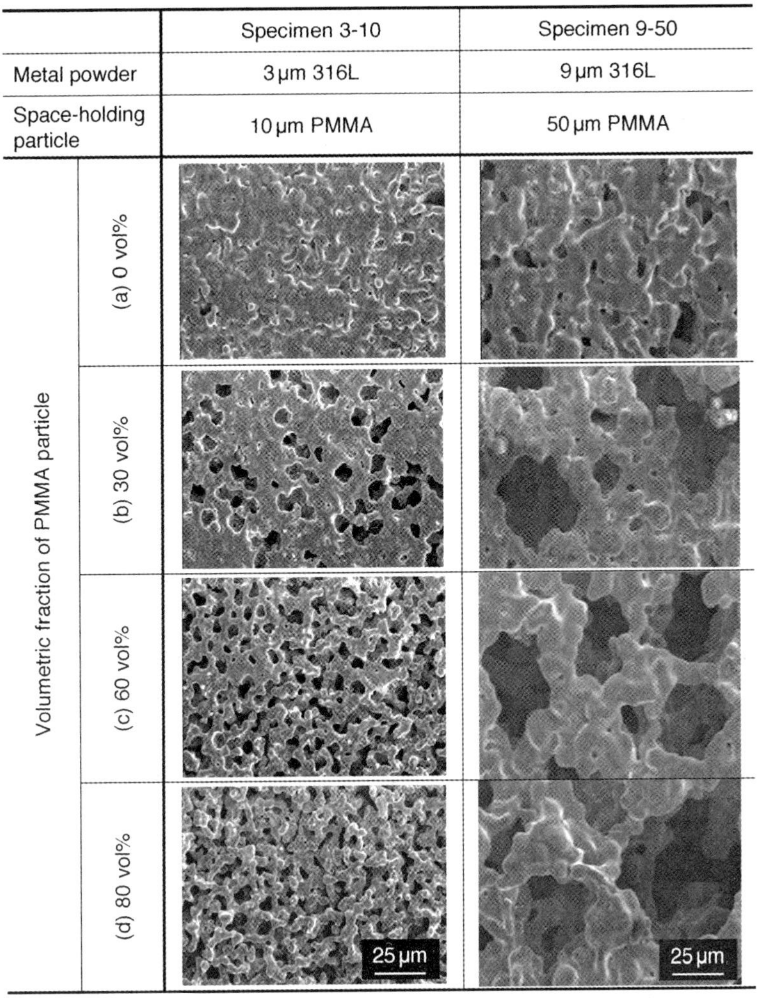

|  | Specimen 3-10 | Specimen 9-50 |
|---|---|---|
| Metal powder | 3 µm 316L | 9 µm 316L |
| Space-holding particle | 10 µm PMMA | 50 µm PMMA |

*15.10* SEM images on surface of sintered specimens with various sizes of metal powder and PMMA particle.

typical for pore formation behavior of the PSH method. To make micro-sized porous stainless steel with a much higher surface-to-volume ratio, it is better to use compounds with a combination of 3 µm 316L powder and 10 µm PMMA particles.

## 15.4.3 Sintering shrinkage and porosity

The relative density (as an inverse of the porosity) and the shrinkage of the sintered specimen were measured with micrometer calipers and an analytical balance. Figure 15.11 shows the shrinkage and the porosity of sintered

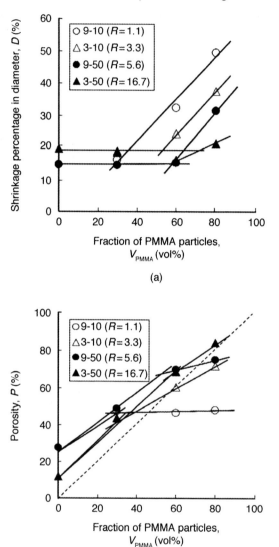

*15.11* Shrinkage and porosity as function of fraction of space-holding particles: (a) shrinkage diameter; (b) porosity.

porous specimens with various fractions of PMMA particles. Specimens 3-10 and 9-50 were sintered at 1050°C and 1200°C, respectively. For both specimens, the shrinkage is between 15 and 20 vol% and up to 50–60 vol% of PMMA particles. However, it increases rapidly when the fraction of PMMA particles becomes higher than 50-60 vol%. The transition point corresponds to the change from a closed porous structure to an open porous

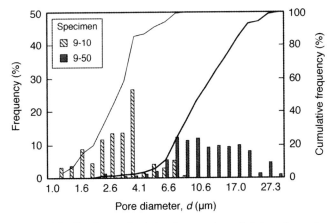

*15.12* Distributions of minimal pore size of sintered specimens with various sizes of PMMA particle.

*Table 15.2* Minimal pore size and surface area of porous specimens with various space-holding particles

|  | Specimen 9-50 | Specimen 3-10 |
|---|---|---|
| Metal powder | 9 μm 316L | 3 μm 316L |
| Space-holding particle | 50 μm PMMA | 10 μm PMMA |
| Average diameter of minimal pore | 9.58 μm | 2.39 μm |
| Specific surface | 0.04 m$^2$/g | 0.15 m$^2$/g |

one. In other words, the shrinking percentage stays constant in a closed porous structure regardless of the content of PMMA particles. In an open porous structure, the shrinking percentage will increase when the fraction of PMMA particles rises. The definite transition points are shown at the boundary between the closed porous structure and open porous structure. In the second graph, the porosity of both specimens increases as the fraction of PMMA particles increases. Here too, two regions can be distinguished, with a transition point between them. This result is similar to the first graph, where the shrinkage was plotted against the fraction of PMMA particles.

Concerning the sintered specimen described above, the pore size, pore distribution, surface area and the flow resistance against a fluid were measured with a capillary flow porometer (Porous Materials Inc., CPF-1100-AXLSP). The distributions of minimal pore size for sintered specimens with various sizes of PMMA particles are shown in Fig. 15.12. The minimal pore size and the specific surface are listed in Table 15.2. As the PMMA particle size decreases, the average minimal pore diameter decreases accordingly, but the specific surface area increases significantly. As a result, the average minimal diameter of pores was approximately equal to one quarter of the diameter of the PMMA particle.

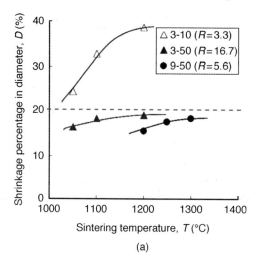

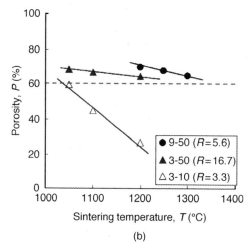

*15.13* Shrinkage and porosity as a function of sintering temperature (60 vol% PMMA): (a) shrinkage in diameter; (b) porosity.

### 15.4.4 Effects of sintering temperature

Figure 15.13 shows the shrinkage percentage in diameter and porosity for porous specimens with 60 vol% PMMA particles sintered at various temperatures. These results showed that the shrinkage increased and the porosity decreased accordingly in all specimens when the sintering temperature was increased. These phenomena agree with findings that the pore size tends to get smaller when the specimen undergoes excess sintering. The shrinkage percentage in diameter is ideally 20%, which is calculated by the volume of solid load of the MIM feedstock, i.e. 50 vol%. Therefore, when the sintering temperature is set at 1050°C for 3 μm 316L–10 μm PMMA (specimen 3-10), 1200°C for 3 μm 316L–50 μm PMMA (specimen 3-50), and 1300°C for 9 μm 316L–50 μm PMMA (specimen 9-50), the shrinkage percentage in diameter will be close to 20%. In that case a porosity could be achieved very close to the added content of PMMA particle, i.e. 60 vol%.

It can be concluded that porous structures with equally sized pores can be obtained under the sintering conditions noted above for each combination of sizes of metal powder and space-holding particles. It can be seen that sintering is more active in finer metal powder at higher sintering temperatures. Small pores form between metal particles by insufficient sintering in 50 μm PMMA specimens (specimens 3-50 and 9-50). Cellular pores resulting from space-holding particles are significantly affected by sintering temperature in the 10 μm PMMA specimens (specimen 3-10).

The effects of sintering temperature on the porous structural characteristics in the case of the 3 μm 316L–10 μm PMMA specimen (specimen 3-10) were then investigated in more detail. The distributions of the smallest pore size for specimens sintered at various temperatures are shown in Fig. 15.14. At lower sintering temperatures such as 1020°C, the mean pore size is 1.65 μm and the pore structure has a broad distribution. At higher sintering temperatures, such as 1070°C, the mean pore size was almost unchanged at 1.62 μm, but the distribution became much sharper. This is considered a result of ongoing sintering between metal powders surrounding large pores held by the space particles. When the sintering temperature is increased to 1100°C, the mean size was significantly reduced to 0.91 μm but the sharp

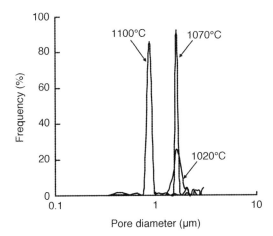

*15.14* Distributions of minimal pore size of specimens sintered at various temperatures (3 μm 316L, 10 μm PMMA (60 vol%)).

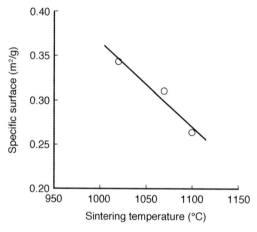

| Sintering temperature | | |
|---|---|---|
| (a) 1020°C | (b) 1070°C | (c) 1100°C |

*15.15* SEM images on surface of specimens sintered at various temperatures (3 μm 316L, 10 μm PMMA (60 vol%)).

*15.16* Specific surface as function of sintering temperature (3 μm 316L, 10 μm PMMA (60 vol%)).

distribution remained. This is due to densification and can be seen on scanning electron microscopy (SEM) images from the surface of the specimens sintered at various temperatures, as shown in Fig. 15.15. The surface area of specimens sintered at various temperatures was measured and is shown in Fig. 15.16. Specifically, the surface decreases in a linear manner with increased sintering temperature. The resistance against fluid flow after applying various temperatures was measured, and is shown in Fig. 15.17. Specifically, water flow reduced significantly as the sintering temperature increased. These results are compatible with a decreasing of pore size in specimens sintered at higher temperature, as discussed previously.

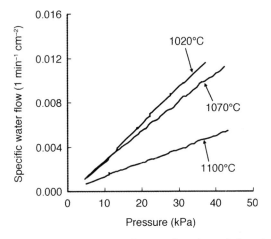

*15.17* Specific water flow as function of sintering temperature (3 µm 316L, 10 µm PMMA (60 vol%)).

### 15.4.5 Geometrical analysis

The transition point between closed porous structures and open porous structures as a function of the fraction of PMMA particles can be estimated by simple geometrical analysis. When we assume that the metal powders are uniformly located around a PMMA particle, this can be modeled as a spherical PMMA particle coated by a single layer of metal powder. When the spheres are close-packed in a face-centered cubic (fcc) structure, the volumetric fraction of PMMA particle reaches its maximum for the closed pore structure. The maximum fraction of PMMA particles, $(V_{PMMA})_{max}$, can be derived from equation [15.1] as follows

$$(V_{PMMA})_{max} = \frac{4 \times (4/3)\pi(d_p/2)^3}{[\sqrt{2}(d_p + d_m)]^3} \qquad [15.1]$$

where $d_p$ is the mean diameter of a PMMA particle and $d_m$ is the mean diameter of 316L powder. The maximum fraction of PMMA particles for a closed porous structure was estimated by geometric analysis and was then compared to the transition points as a fraction of PMMA particles obtained from experimental results. The shrinking percentage and the porosity for specimens with various size ratios of particle are shown in Figs 15.18(a) and Fig. 15.18(b), respectively. The fractions of PMMA particles at these transition points were plotted for several ratios of particle size, $R$, as shown in Fig. 15.18. For comparison the maximum fraction of PMMA particles estimated by equation [15.1] is drawn together with the curve in Fig. 15.18. The curve indicates the boundary between closed porous and open porous

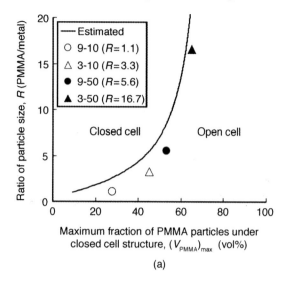

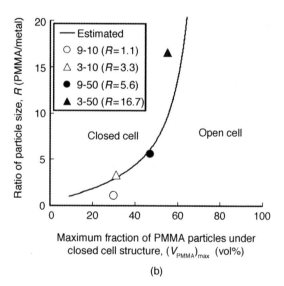

*15.18* Size ratio of PMMA particle/metal powder versus maximum fraction of PMMA particle under closed cell structure: (a) transition point in shrinkage in diameter; (b) transition point in porosity.

structures. As can be seen, the experimental results agreed well with the curve estimated simply by geometrical analysis.

## 15.5 Liquid infiltration properties of micro-porous metals produced by the PSH method

### 15.5.1 Measuring of liquid infiltration properties

This section focuses on the liquid infiltration properties of micro-porous metals produced by the PSH-MIM process. The effects of testing conditions, surface treatment and pore size on infiltration behavior were mainly investigated by the liquid infiltration test. This type of micro-porous structure is infiltrated slowly by capillary flow. The liquid can be stored in the pores and is allowed to exude slowly. The test apparatus for evaluating the liquid infiltration performance of the porous specimen was developed using an analytical balance. From the weight change during the liquid infiltration test, the factors affecting water absorption behavior of the porous specimen were shown. The results of the infiltration rate were compared to some important characteristics of the porous specimen such as mean pore diameter, liquid and gas permeability and specific surface area measured by capillary flow porosimeter. The effects of pore size on the infiltration rate were seen with a single digit micrometer in porous specimens of 15 µm constricted pore diameter.

### 15.5.2 Principle and evaluation method

A capillary flow porosimeter can be used to determine some important characteristics of porous materials such as (a) distribution of constricted part of a pore channel, $f$, (b) mean diameter, $d_m$, (c) liquid and gas permeability, $K_L$ and $K_G$ and (d) specific surface area, $S_p$ (Jena et al., 2003). The principle of this method is briefly explained as follows. The pores of a sample are filled by a wetting liquid, which is extruded by increasing the gas pressure gradually. Higher pressures are needed to empty the smaller pores. The pressure required to displace the wetting liquid from the pores can be specified. The flow curve as a function of the difference in pressure is measured for both wet and dry samples. By comparing the gas flow rates of wet and dry porous samples at the same pressures, the percentage of flow passing through the pores larger than or equal to the specified size may be calculated from the pressure–size relationship. A specific surface area is obtained by enveloping the surface area (ESA), which is based on Kozeny–Carman's equation. Otherwise the test apparatus for evaluating liquid infiltration performance of porous specimens was developed using an analytical balance as shown in Fig. 15.19. In this type of liquid infiltration test, when the porous specimen is soaked in a liquid bath at atmospheric pressure, a liquid is percolated, passing up slowly through the specimen. The weight on the analytical balance changes by buoyancy of specimen and liquid transfer.

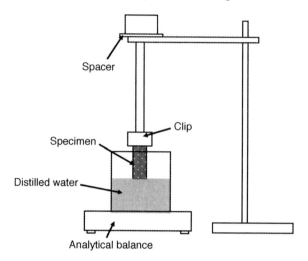

*15.19* Schematic diagram of apparatus used for evaluating the liquid infiltration performance of porous specimens.

Miyasaka *et al.* (1976) evaluated the infiltration rate and height of liquid percolated up in powders. The method is applied to the liquid infiltration test used here. The percolated height, $h$, is given by

$$h = \frac{W}{A \cdot \varepsilon \cdot \rho_L} \qquad [15.2]$$

where $W$ is weight, $A$ and $\varepsilon$ are cross-sectional area and porosity of the sample respectively, $\rho_L$ is the density of liquid. The percolated height, $h$, is given by Washburn's equation [15.3], which was developed for the capillary flow in a cylindrical pipe

$$h^2 = \frac{r^2}{4\eta}\left(\frac{2\gamma_L \cos\theta}{A \cdot \varepsilon \cdot \rho_L} + \Delta P\right)t \qquad [15.3]$$

where $r$ is radius of pipe and $t$ is time. The pressure is assumed to be constant $\Delta P = 0$, and by substituting equation [15.2] in equation [15.3], the weight change per unit of cross-sectional area of porous specimen is given by

$$\frac{W}{A} = \alpha\sqrt{t}$$

$$\alpha = \left(\frac{r \cdot \varepsilon^2 \cdot \rho_L^2 \cdot \gamma_L \cdot \cos\theta}{2\eta}\right)^{0.5} \qquad [15.4]$$

Thus the infiltration rate, $\alpha$ of liquids in porous specimens can be evaluated by its slope on the $W/A$–$\sqrt{t}$ plot.

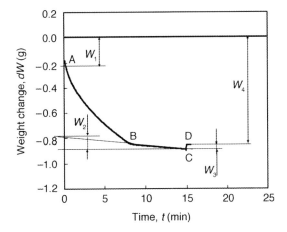

*15.20* Typical behavior of weight change during immersion of porous specimens (specimen 3-10, $l$ = 7.5 mm).

*Table 15.3* Properties of porous specimens

|  | 3-10 | 9-50 | 9-90 |
|---|---|---|---|
| Porosity, $\varepsilon$ (%) | 54.2 | 60.5 | 63.1 |
| Mean pore diameter, $d_m$ (µm) | 1.34 | 7.29 | 14.92 |
| Specific surface area, $S_p$ (m²/g) | 0.343 | 0.040 | 0.030 |
| Liquid permeability, $K_L$ | 0.105 | 3.345 | 7.518 |
| Gas permeability, $K_G$ | 0.070 | 1.564 | 3.028 |

### 15.5.3 Specimen and experimental results

Porous specimens with the physical properties shown in Table 15.3 were used. The relative density (as an inverse of the porosity) of the sintered porous specimen was measured with micrometer calipers and an analytical balance. A capillary flow porosimeter was used for measurement of pore size distribution. Galwick ($\gamma$ = 16 mN/m) is used for a wetting liquid. The typical behavior of weight change during immersion of porous specimens is shown in Fig. 15.20. Just after the specimen is immersed in water, the buoyancy and meniscus have a considerable effect on weight ($W_1$). Water infiltrates from an immersed part to a dry part between point A and point B, which causes a significant decrease in weight. Above point B, the loss of water by evaporation causes a gradual loss of weight ($W_2$). At point C, when the specimen is removed from water, the buoyancy and meniscus recover ($W_3$). As a result, the amount of water equivalent to the porosity enters the porous specimen ($W_4 - W_2$).

Figure 15.21 shows the weight change of porous specimen 3-10 measured at various immersion lengths. Increasing immersion length reduces the time

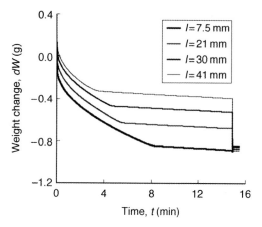

15.21 Effects of immersion length on weight change (specimen 3-10).

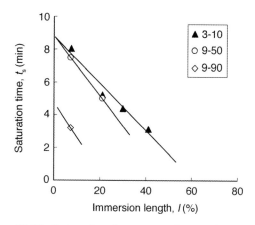

15.22 Saturation time versus immersion length.

to reach point B (namely saturation time, $t_s$), thus the saturation time decreases linearly as immersion length increases, as shown in Fig. 15.22. This is because infiltration rate in water is faster than that in air. Figure 15.23 shows the weight change of porous specimen 3-10 treated with acid cleaning or untreated. The surface treatment significantly improved water infiltration. Table 15.4 reveals that an increasing of infiltration rate by acid cleaning is due to a decreasing of contact angle. Figure 15.24 shows the effects of pore size on weight change. No significant differences in infiltration rate are seen for specimens 3-10 and 9-50, as summarized in Table 15.5, but the infiltration rate decreases considerably in specimen 9-90. The water absorption volume percentage ($V_a$) also decreases remarkably compared to the specimen with smaller pores.

From the results described – which were considering the effects of testing

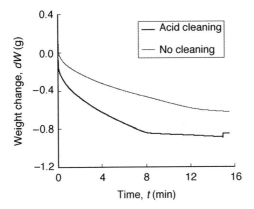

*15.23* Effects of surface treatment on weight change (specimen 3-10, $l = 7.5$ mm).

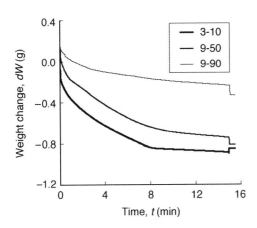

*15.24* Effects of pore size on weight change ($l = 7.5$ mm).

*Table 15.4* Effects of acid cleaning (specimen 3-10, $l = 7.5$ mm)

|  | No cleaning | Acid cleaning |
|---|---|---|
| Infiltration rate, $\alpha$ | 0.013 | 0.020 |
| Contact angle, $\theta$ | 77.4 | 58.2 |

*Table 15.5* Effects of immersion length and pore size on infiltration rate ($\alpha$) and infiltration percentage ($V_a$)

|  | 3-10 | | | | 9-50 | 9-90 |
|---|---|---|---|---|---|---|
| $l$ | 7.5 | 21 | 30 | 41 | 7.5 | 7.5 |
| $\alpha$ | 0.020 | 0.020 | 0.019 | 0.018 | 0018 | 0.005 |
| $V_a$ | 109% | 110% | 113% | 115% | 84% | 26% |

conditions, surface treatment and pore size on infiltration behavior of micro-porous stainless steels with various pore sizes produced by the PSH method and investigated by the liquid infiltration test – some conclusions can be drawn. This type of micro-porous structure is being infiltrated by a liquid slowly by capillary flow, the liquid can be stored in the pores and is leaving slowly. Some important characteristics of porous materials were measured by capillary flow porometer.

## 15.6 Dimensional accuracy of micro-porous MIM parts

### 15.6.1 Measuring dimensional accuracy

Porous materials are generally unsuitable for machining, for example cutting and polishing, because the porous structure on the surface is damaged by plastic deformation; also the machining cost is relatively high. Porous metal parts are rather only suited to net-shape manufacturing, namely powder sintering, as the only possible process. The manufacture of net-shape products with a highly dense micro-porous structure through PSH-MIM usually requires high dimensional accuracy ($\pm 0.05\%$), similar to the accuracy required in machined parts ($\pm 0.01\%$). The effects of combinations of size and fraction of space-holder particles on the dimensional error and its coefficient of variation are shown in this section. Two types of test specimen with micro-porous structures are prepared by injection molding and extrusion molding to evaluate the effect of shape complexity on the dimensional accuracy achieved.

### 15.6.2 Specimen and manufacturing method

The porous compounds used for the experimental material were prepared by mixing PMMA particles ($D_{50}$ = 10 µm, 40 µm) with MIM feedstock composed of stainless steel 316L powder ($D_{50}$ = 3 µm) and binder (paraffin wax and polyacetal polymer). As shown in Table 15.6, the fractions of PMMA particles to MIM feedstock are 0, 30 and 60 vol%, and the solids loading of MIM feedstock is 50 vol%. Using these porous compounds, two types of test specimen were fabricated by different molding methods, as shown in Fig. 15.25. A small impeller part was manufactured by injection molding (Fig. 15.25(a)) and the plate specimen was produced by extrusion molding (Fig. 15.25(b)). The dimensions of an impeller part are $D$ = 6.38 mm in outside diameter and $h$ = 2.28 mm height of the green compact. The plate specimens were cut at $L$ = 65 mm, $W$ = 65 m from longitudinal green sheet just after extrusion. The shrinkage after molding of both specimens is approximately 1–2% in each dimension (Greenc and Heaney, 2007). The debinding and sintering of these green compacts were

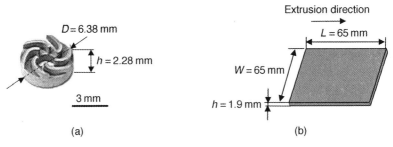

*15.25* Test specimens and locations for dimensional measurement: (a) impeller by injection molding; (b) plate specimen by extrusion molding.

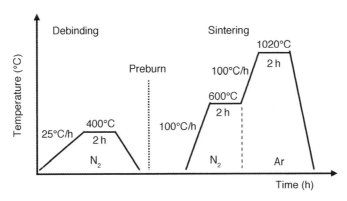

*15.26* Temperature control condition in debinding and sintering processes.

*Table 15.6* Experimental constituent materials and compositions

|  | 316L metal powder | | Binder | PMMA particle | |
| --- | --- | --- | --- | --- | --- |
|  | Particle size (μm) | Fraction (vol%) | Fraction (vol%) | Particle size (μm) | Fraction (vol%) |
| Dense | 3 | 50 | 50 | – | 0 |
| Closed cell | 3 | 35 | 35 | 10 40 | 30 |
| Open cell | 3 | 20 | 20 | 10 40 | 60 |

sequentially carried out at 400°C for 2 h in $N_2$ and at 1200°C for 2 h in an Ar gas atmosphere under the heating conditions shown in Fig. 15.26.

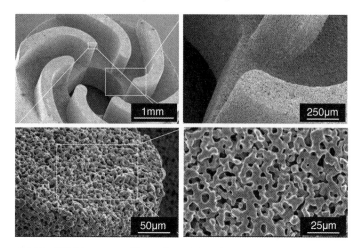

*15.27* SEM images of impeller with micro-porous structures.

### 15.6.3 Experimental results

The SEM images of the impeller produced, using the porous compound with added 60 vol% of 10 μm PMMA particles are shown in Fig. 15.27. In macroscopic observation the shape appeared accurately molded to the edges of the impeller blade. Microscopic observation also showed that the open pores were uniformly distributed on all surfaces of the sintered part and the pore size was approximately several micrometers.

Dimensional deviation and its coefficient of variance (CV) value for the outside diameter and height of the impeller specimen are shown in Fig. 15.28. The materials used for porous MIM compounds of various composition are listed in Table 15.6. In the case of closed porous specimens to which 30 vol% of 10 μm PMMA particles were added, dimensional deviations and their CV values are small and similar to dense specimens. Height directions showed higher dimensional accuracy than diametric dimensions. It is supposed that this is due to the size effect, i.e. the dimensional relation between outside diameter and height of the specimen ($D/h = 3$), but not due to the effect of flow direction and opening of the mold.

Furthermore, the dimensional deviations and their CV values of open porous specimen with added 60 vol% of 10 μm PMMA particles are a few times larger than that of closed porous ones. The main reason for this is that open porous specimens shrunk more during sintering, because a large amount of PMMA particles were added to the MIM feedstock. Consequently, the PSH-MIM process is capable of manufacturing micro-porous metal parts with controlled 20 μm pores with less than 0.4% CV value in both directions.

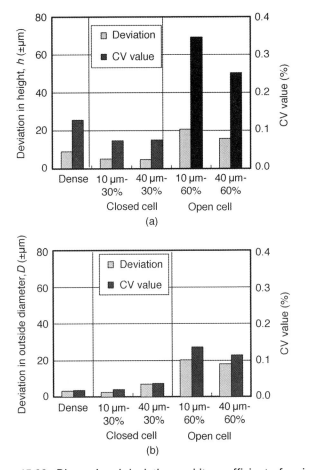

*15.28* Dimensional deviation and its coefficient of variance of impeller specimens: (a) height; (b) outside diameter.

Dimensional deviation and CV values of the plate specimens produced by extrusion molding are shown in Table 15.7. The results showed no significant differences in accuracy between width and length dimensions. This means that the extruding force had no influence on the dimensional accuracy compared to a one-directional flow of green sheet. The in-plane isotropic property is useful for manufacturing porous parts by extrusion molding. Importantly, however, a lower dimensional accuracy in height was obtained because of the size effect. Also the deviation and CV value in weight and porosity were not sufficiently small. The overall production quality, however, reached the same level as in conventional MIM, despite the green sheet being aired out from the die by extrusion and the micro-porous structure being unstable after debinding.

A comparison of deviation and CV values in dimensions and weight of

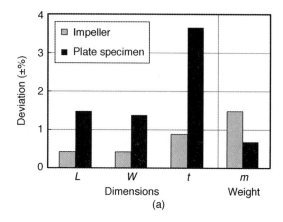

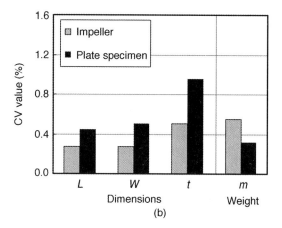

*15.29* Comparison in MIM impeller and extrusion plate: (a) dimensional deviation; (b) coefficient of variance.

*Table 15.7* Dimensional deviation and coefficient of variance of plate specimens

|  | Average | Deviation | | CV |
|---|---|---|---|---|
| Width, W | 48.15 mm | ±0.66 mm | (±1.37%) | 0.51% |
| Length, L | 47.23 mm | ±0.70 mm | (±1.47%) | 0.45% |
| Height, h | 1.37 mm | ±0.05 mm | (±3.65%) | 0.96% |
| Weight, m | 11.64 g | ±0.08 g | (±0.67%) | 0.32% |
| Porosity, P | 52.55% | ±1.51% | (±2.88%) | 1.38% |

porous specimens produced by injection molding and extrusion molding is shown in Fig. 15.29. To the porous compound, 60 vol% of 10 μm PMMA particles were added. The impeller specimen produced by injection molding obviously attained higher dimensional accuracy than the plate specimen produced by extrusion molding. This is because injection molding can manufacture green compacts with high density, as the molding pressure is applied uniformly in the cavity. However, a converse result was obtained for deviation and CV values in weight. This might be explained by the difference in shape complexity.

From the above-mentioned measuring results on dimensional accuracy, it can be concluded that the PSH method could manufacture commercially micro-porous metal components with high dimensional accuracy in both injection molding and extrusion molding. The dimensional accuracy for closed porous structured parts is less than several micrometers and therefore the same as in dense MIM parts production, while that of open porous structured parts is within several tenths of microns. The porosity can be controlled within several percent of deviation.

## 15.7 Functionally graded structures of micro-porous metals

### 15.7.1 Formation of multilayered porous structure by hot pressing

Mechanical properties of porous specimen are much lower than those of dense specimens. This is because the true cross-sectional area is incomparably lower in the porous specimen. This strength reduction is a main point of disadvantage in porous materials, and therefore an effective method should be considered for enhancing the strength of porous components.

It is well known in structural design of composites that sandwich structures show good performance in specific moduli and specific strengths. The authors tried to produce multilayered specimens with a graded porous sandwich structure. The effectiveness of the method to compensate for the deficiencies in mechanical properties was investigated for porous metals in comparison with materials with a homogeneous dense and porous structure. Co-sintering was applied to form the graded porous structures after each porous compound was consolidated by sequential hot press molding. The multilayer metals were obtained by changing the stacking sequence in the co-sintering process. The skin layer and core were formed with a highly dense structure and a micro-porous structure, respectively and conversely.

Figure 15.30 shows a co-sintering production process of multilayered metals with graded porous structures (Nishiyabu *et al.*, 2005a, 2005b, 2005c). Using dense MIM feedstock and porous compounds with various contents of space-holding particles, circular shapes of green compacts (40 mm diameter, 2 mm thick) were prepared by hot press molding for the sake of simplicity in these experiments, but could also have been produced by injection molding. The molding was carried out under constant conditions where die temperature and gauge pressure were 200°C and

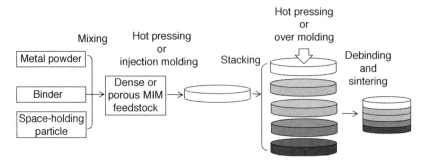

*15.30* Production process of multilayered metals with graded porous structures.

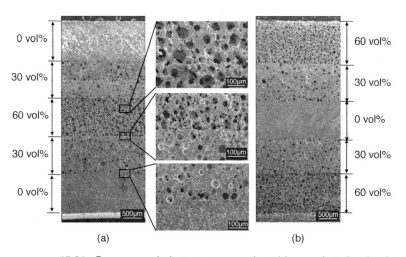

*15.31* Porous graded structures produced by co-sintering laminated green compacts: (a) dense surface graded structure; (b) porous surface graded structure.

10 MPa, respectively. The debinding and sintering were sequentially processed at 600°C for 2 h in $N_2$ and at 1050–1200°C for 2 h in Ar gas atmospheres in the vacuum furnace to avoid oxidizing.

Figure 15.31 shows SEM images of graded porous metals produced by co-sintering the laminated green compacts, which were stacked in five plates with various contents of PMMA particles in several sequences, namely (a) dense surface graded structures and (b) porous surface graded structures. These SEM images revealed the macroscopically graded porous structures and showed that no defects appeared in the interfacial region between each layer.

Figure 15.32 shows the bending stress–strain curves and bending modulus of plain porous specimens and graded porous ones. A three-point bending

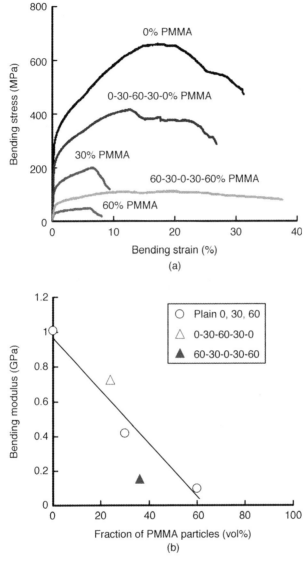

15.32 Bending properties of specimens with various fractions of PMMA particles: (a) bending stress–strain curves; (b) bending modulus.

test was carried out for short beam specimens (5 × 10 × 40 mm) and test conditions in reference to ISO178 standards. Bending strain was calculated by elastic theory using span length and height of specimen in addition to defection at loading point. In the case of the plain porous specimens, the ultimate stresses on specimens with 30 and 60 vol% PMMA particles were

| Strain \ ID | Plain porous specimens | | |
|---|---|---|---|
| | (i) 0 | (ii) 30 | (iii) 60 |
| Near fracture point | 23.1% $\varepsilon$ | 11.6% $\varepsilon$ | 11.6% $\varepsilon$ |

| Strain \ ID | Graded porous specimens | |
|---|---|---|
| | (a) 0-30-60-30-0 | (b) 60-30-0-30-60 |
| Near fracture point | 23.1% $\varepsilon$ | 23.1% $\varepsilon$ |

*15.33* Fracture aspects of specimens with various fractions of PMMA particles under bending load.

very low, and the bending strength and modulus decreased drastically with increasing fractions of PMMA particles.

The behavior can also be observed from fracture behavior shown in Fig. 15.33. Brittle fracture occurred due to rapid fracture propagation in the upper tensile side. On the other hand, both graded porous specimens show a strong deformation behavior, as with the dense 0 vol% PMMA specimen. Definite differences in bending modulus appeared between type (a) 0-30-60-30-0 and type (b) 60-30-0-30-60 specimens. The bending modulus is higher for type (a) specimen and is lower for type (b) specimen than expected. This is because the cracks occurred in the lower skin layer with the open porous structure, which was subjected to tensile load, in the type (b) specimen. Thus, definite differences were shown in mechanical properties between the plain porous structure and the sandwich structure. Therefore, it is concluded that this proposed manufacturing method is effective for material design combining material features on the micro- and the macro-scales.

### 15.7.2 Formation of multi core-in-sheath porous Ti-MIM parts by sequential injection molding

Pure titanium was chosen for the experiments as the representative materials of medical implants. The aim of this study is to demonstrate the feasibility and effectiveness of MIM in producing multilayered porous metal components. Hollow, thick-walled cylindrical structures with three layers of various porosities were molded by sequence metal powder injection molding. Pore formation and some physical properties of the sintered porous specimen were investigated. The specimens were produced using porous compounds prepared in changing fractions of space-holding particles. It was confirmed that the method proposed in this study was useful in producing the metal components with micro-sized pore and multilayered structures.

The possibility of dental implants for functionally disordered hard tissues like bone and teeth have received a lot of attention recently, in addition to substituting hard tissue instrumentations like artificial bones, artificial hip joints and artificial teeth. Pure titanium (Ti) has been the most widely used metal for orthopedic implant material to date because of its excellent combination of biocompatibility, corrosion resistance to acid, salt water and blood, and mechanical properties, such as low specific gravity and high strength (Wen *et al.*, 2001; Oh *et al.*, 2002; 2003). However, Young's modulus of pure Ti (110 GPa) is much higher than that of human bone (12–23 GPa). Critical problems caused by the mismatch of elastic modulus between implant and human bone are still unresolved. One way to alleviate the problems is, therefore, to reduce Young's modulus of pure Ti by introducing pores, thereby minimizing damage to tissues adjacent to the implant and eventually prolonging device life. According to Oh *et al.* (2003), sintered porous Ti compacts having Young's modulus close to that of human bone have been developed. Porous compacts having a porosity of 19–35 vol% were fabricated using pure Ti powder with 300–500 μm particle sizes. Also, in Oh *et al.* (2002) porosity-graded Ti compacts were investigated using Ti powders with three different particle sizes: 65, 189 and 374 μm.

Looking at practical production, Ti implants have very complicated shapes, including core structures of high durability and surface structures of high osteogenesis. It is therefore necessary to manufacture those products with high dimensional accuracy and high reliability in terms of strength. However strongly desired these micro-porous metal components may be for high functioning applications in medical implants, not many studies deal with their cost-efficient net-shape production. The sequential injection molding method was used for producing multilayer porous structured parts. In the first step, inner material was molded into the die (2 mm and 4 mm in inside and outside diameter), in the second and third steps, middle and outer materials were over-molded after having inserted the over-molded parts in turn. Three layers of structural green compacts with 2 mm layer thickness were obtained using dense MIM feedstock and porous compounds with various fractions of PMMA particles. The metal powders are pure Ti produced by the gas-atomization method. The multi-component binder is formed from polyacetal polymer and wax. Spherical particles (180 μm in mean diameter) made of polymethylmethacrylate (PMMA) were used for holding the space. These materials were co-mixed and pelletized with a high-pressure kneader and plunger-type extruder. The fraction of PMMA particles varied from 0 to 65 vol% as the main experimental parameter.

To evaluate the physical properties of sintered porous Ti, the disc shape of the green specimen was compacted with hot-press forming, using a steel die 40 mm in diameter. A sintered porous specimen with 33.4 mm diameter and 2.6 mm thick was obtained after debinding and sintering. Also, single-layer and multilayered porous pipes were prepared by injection molding using a die with a cavity size of 8 mm diameter and 24 mm high. Two types of multilayered specimen, namely 65-30-0 and 0-30-65, were prepared by changing molding sequence. Debinding and sintering were sequentially processed at 600°C for 2 h in Ar and at 1200°C for 2 h in 0.1 MPa level high-vacuum atmospheres in a metal hot zone furnace.

Figure 15.34 shows the surface structures of single-layered micro-porous Ti pipes with various fractions of PMMA particles. The highly dense sintered Ti pipe possessed high modulus and strength, but the sintered porous Ti pipe produced using 30 vol% or 65 vol% PMMA particles had closed or open pores, respectively. Figure 15.35 shows the surface structure of multilayered porous Ti pipe (65-30-0) produced by the sequential

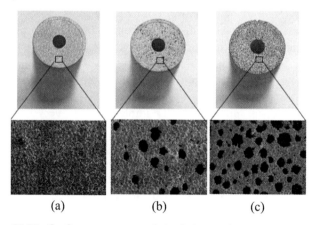

*15.34* Surface structures of single-layered porous Ti pipes with various fraction of PMMA particles: (a) 0 vol%; (b) 30 vol%; (c) 65 vol%.

Powder space holder MIM of micro-porous metals 385

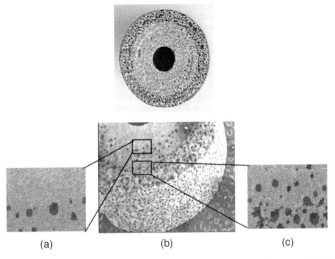

*15.35* Surface structure of multilayered porous Ti pipe: (a) 0–30 vol%; (b) overviews; (c) 30–65 vol%.

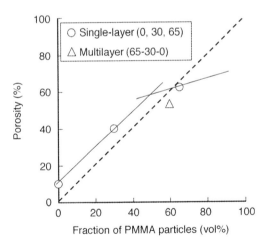

*15.36* Porosity of sintered porous Ti as a function of fraction of PMMA particles.

injection molding method. Open pores appeared in the outer cylindrical layer, and closed pores are visible in the middle layer. No defects are recognized in the interfacial region between each layer. Figure 15.36 shows the porosity of single-layered and multilayered (65-30-0) micro-porous Ti pipes as a function of the fraction of PMMA particles. As the fraction of PMMA particles increases, the porosity increases linearly up to approximately 50 vol% of PMMA particles, where it is equivalent to a transition from a closed porous structure to an open porous structure. Therefore, it

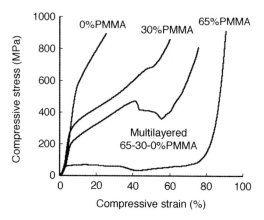

*15.37* Compressive stress against strain curves of single-layered and multilayered (65-30-0) porous Ti pipes.

was confirmed that we were able to control the porosity of sintered bodies by the fraction of space-holding particles.

Figure 15.37 shows the axial compressive stress against strain curves of single-layered and multilayered (65-30-0) porous Ti pipes. In the case of single-layered porous Ti pipes, the highly dense specimen (0 vol% PMMA) shows high yielding compressive strength, and is fractured in the shear mode, while the specimen with an open porous structure (65 vol% PMMA) deformed largely at low compressive stresses. Thus the yielding stress decreases drastically with increasing fraction of PMMA particles. On the other hand, multilayered porous Ti pipe shows a comparatively high compressive property, although an open porous structure exists on the surface.

These experimental results show that the MIM base powder space-holder method proposed in this study has great potential to realize net-shape manufacturing of micro-porous metal components with complex shape and multilayer structure. Furthermore, to advance the manufacturing method proposed, green compacts with graded content of space-holding particles were stacked by hot press molding and co-sintering was applied to them to form graded porous structures. By comparing the mechanical properties of the materials with plain, homogeneous, porous specimens and porous graded structures, it was confirmed that, first, the desired functionally graded porous structures can be produced easily, and second, that graded porous structures were effective in enhancing the mechanical properties of porous metals. When this method was applied to injection or extrusion moldings, then a near-net-shaped production of metal components with complex three-dimensional shapes could be achieved.

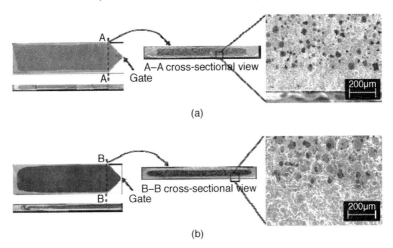

*15.38* Cross-sectional views of sandwich structure in longitudinal and lateral directions: (a) skin: 0 vol%/core: 30 vol% PMMA; (b) skin: 0 vol%/core: 60 vol% PMMA.

### 15.7.3 Formation of steamed bread-like porous structures by co-injection molding

The co-injection molding method is characterized by using a sandwich structure in porous MIM components. Co-injection molding of flat specimens applied to the hot runner system with a double gate, which could be injected sequentially, produced two kinds of materials. Figures 15.38(a) and 15.38(b) exhibit cross-sections of the sandwich structure in longitudinal and lateral directions. The core material in dark color is visible in the centre of the specimen, and the skin material, in light color, formed at about 200–300 μm in thickness. When the fraction of PMMA particles is 30 vol%, the core material forms closed pores, while when the fraction of PMMA particles is 60 vol%, it forms open pores. From these photographs, observed by SEM, it is clearly shown that the boundary region between core (porous) and skin (dense) materials of sintered specimens is free from any defects. Figure 15.39 shows the results of a bending test of the plain porous and sandwich porous specimen with various fractions of PMMA particles. The sintered specimens with sandwich porous structures showed higher bending strength than those with a plain porous structure.

It was confirmed that the PSH method was useful in producing metal components with micro-sized porous and high functionally graded porous structures. The feasibility of producing graded porous structures was confirmed when using co-sintering, showing effective compensation for the mechanical deficiencies of porous structures.

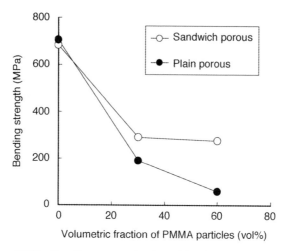

*15.39* Bending strength plotted by function of PMMA particle.

## 15.8 Conclusion

The manufacturing method for micro-porous metal components using a PSH method together with a MIM process was presented in this study. From experimental results and simple geometrical analysis, it was concluded that porous metals with micro-sized pores could be formed homogeneously when two parameters were optimized: first, the particle size of the spherical materials supporting the metal powder, and second, the sintering temperature. The porosity could be easily controlled by the fraction of PMMA particles for space holding. The pore size is dependent on the particle size of PMMA. Surface area and resistance of a fluid to flow through sintered porous metals are significantly affected by the sintering temperature. The method could be applied to injection or extrusion moldings for the net-shaped production of micro-porous metal components with complex three-dimensional shapes.

## 15.9 Acknowledgements

The author deeply appreciated the great support for the experimental work by Mr S. Matsuzaki, former research engineer, the support and understanding for the research foundation by President Dr S. Tanaka of Taisei Kogyo Co., Ltd, and by graduate students from Osaka Prefectural College of Technology.

## 15.10 References

Ashby, M. F., Evans, A., Fleck, N. A., Gibson, L. J., Hutchinson, J. W., and Wadley, H. N. G. (2000), *Metal Foams, A Design Guide*, Elsevier Science.
Banhart, J. (2001), 'Manufacture, characterization and application of cellular metals and metal foams', *Progress in Materials Science*, 46, 559–632.
Bram, M., Laptev, A., Stover, D., and Buchkremer, H. P. (2003), 'Method for producing highly porous metallic moulded bodies close to the desired final contours', US Patent 7147819.
German, R. M. (1984), *Powder Metallurgy Science*, 2nd edition, Metal Powder Industries Federation, Princeton, NJ, USA.
German, R. M. and Bose, A. (1997), *Injection Molding of Metals and Ceramics*, Metal Powder Industries Federation, Princeton, NJ, USA.
Gibson, L. J. and Ashby, M. F. (1988), *Cellular Solids – Structure and Properties*, Pergamon Press.
Greene, C. D. and Heaney, D. F. (2007), 'The PVT effect on the final sintered dimensions of powder injection molded components', *Materials and Design*, 28, 95–100.
Heaney, D. F. and German, R. M. (2001), 'Porous stainless steel parts using selective laser sintering', *Advances in Powder Metallurgy and Particulate Materials*, 8, 73.
Heaney, D. F., Gurosik, J. D., and Binet, C. (2005), Isotropic forming of porous structures via metal injection molding, *Journal of Material Science*, 40, 973–981.
Jena, A., Gupta, K., and Sarkar, P. (2003), 'Porosity characterisation of microporous small ceramic components', *American Ceramic Society Bulletin*, 82(12), 9401–9406.
Lefebvre, L. P. and Thomas, Y. (2003), 'Method of making open cell material', US Patent 6660224.
Miyasaka, K., Manabe, T., and Konishi, M. (1976), 'A modified penetration rate method for measuring the wettability of magnesium oxide powders', *Chemical and Pharmaceutical Bulletin*, 24(2), 330–336.
Nishiyabu, K., Matsuzaki, S., Ishida, M., Tanaka, S., and Nagai, H. (2004), 'Development of porous aluminum by metal injection molding', in *Proceedings of the 9th International Conference on Aluminum Alloy (ICAA-9)*, pp. 376–382.
Nishiyabu, K., Matsuzaki, S., Okubo, K., Ishida, M., and Tanaka, S. (2005a), 'Porous graded materials by stacked metal powder hot-press molding', *Materials, Materials Science Forum*, 492–493, 765–770.
Nishiyabu, K., Matsuzaki, S. and Tanaka, S. (2005b), 'Production of high functionally micro porous metal components by powder space holder method', in *Proceedings of International Symposium on Cellular Metals for Structural and Functional Applications (CELLMET 2005)*, p. 41.
Nishiyabu, K., Matsuzaki, S., and Tanaka, S. (2005c), 'Production of micro porous metal components by metal injection molding based powder space holder method', in *Proceedings of the 4th International Conference on Porous Metal and Metal Foaming Technology (MetFoam 2005)*, B55.
Oh, I. H., Nomura, N., and Hanada, S. (2002), 'Microstructures and mechanical properties of porous titanium compacts prepared by powder sintering', *Materials Transactions*, 43, 443–446.
Oh, I. H., Nomura, N., Masahashi, N., and Hanada, S. (2003), 'Mechanical

properties of porous titanium compacts prepared by powder sintering', *Scripta Materialia* 49(12), 1197–1202.

Wadley, H. N. G. (2002), 'Cellular metals manufacturing', *Advanced Engineering Materials*, 4, 10.

Wen, C. E., Mabuchi, M., Yamada, Y., Shimojima, K., Chino, Y., and Asahina, T. (2001), 'Processing of biocompatible porous Ti and Mg', *Scripta Materialia*, 45 (10), 1147–1153.

# Part IV

Metal injection molding of specific materials

# 16
Metal injection molding (MIM) of stainless steel

J. M. TORRALBA, Institute IMDEA Materials,
Universidad Carlos III de Madrid, Spain

**Abstract:** This chapter reviews the topic of stainless steels manufactured with MIM technology. The following questions are considered. What is a stainless steel? What is the place of stainless steels in the MIM world? What are the special characteristics of the different steps of MIM technology (kinds of powders used, feedstocks, debinding methods, sintering problems and new technologies)? The chapter finishes with a description of some applications.

**Key words:** stainless steels, martensitic, austenitic, phase hardened, nickel-free.

## 16.1 Introduction

When we use the term 'stainless', we are trying to describe materials with resistance to staining, rusting and pitting in air. The term 'steel' refers to an iron-based alloy, usually with carbon and other alloying elements. In fact, steel is an iron–carbon alloy. For this reason, we can say that stainless steel is the 'two lies' material because it is not a steel (usually does not have carbon) and, in some circumstances (extreme conditions depending on the material), is not stainless. What is true is that a stainless steel has a certain level of corrosion and oxidation resistance (higher than any other family of iron-based alloys), and we call them 'steels' because they have historically been held to the standards of the steel family.

What assures the level of capability of these iron-based alloys is the 12% (by weight) of chromium in the solution of the alloy. It is important take into account that this is 'in solution'. This means that if chromium is forming carbides or nitrides, the level of chromium in solution decreases and the stainless capability of the material is strongly reduced. Owing to the special manufacturing methods of powder metallurgy (PM) (in general) and MIM (in particular), when there are unusually slow cooling rates after sintering, the formation of chromium carbides in the range of temperatures between 400 and 800°C can take place (the so-called sensitisation process), leading to a lack of corrosion performance in these materials. In the same way, sintering atmospheres containing nitrogen are particularly suitable for the formation of chromium nitrides, leading to the same perverse effect.

The main consequence of the possible sensitisation of stainless steels during the cooling step after sintering makes MIM technology particularly attractive. First, the original carbon content in the powder should be very low (L grades, under 0.03% by weight), and the debinding should be complete if one desires no residual carbon in the steel, which could reduce the corrosion performance of the final part. Even the martensitic grades, which in the wrought grades have some carbon in their composition, do not have any (or very low) carbon in the MIM grades and therefore avoid this problem. In this sense, they could be considered a sort of ferritic grade with low chromium.

The main advantage of MIM in producing stainless steels is the enormous freedom to produce highly complex parts of relatively small sizes, with high dimensional tolerance and mechanical properties very close to the wrought grades (thanks to the high level of densities that can be reached by MIM). A small reduction in the corrosion performance due to the inherent characteristics of the MIM process (debinding and sintering) is rarely an issue, making the MIM product highly competitive. This is true because stainless steel is one of the first materials to be industrially produced by MIM. Today, about 50% (in Europe) to 57% (in Japan) of the total metallic parts produced by MIM are made of stainless steels,[1] so we can consider, at least in volume, that stainless steels are the most important metal produced by MIM.

Stainless steels can be divided into grades according to their microstructure. The main commercial grades belong to three families: austenitic, ferritic and martensitic. In general terms, austenitic steels are Fe–Cr–Ni alloys, with the most extended alloys being the grades with 18–20% Cr (wt) and 8–10% Ni (wt). Ferritic steels have high Cr contents (and are free of carbon), and the conventional martensitic grades used have more than 12% Cr (wt), with some amount of carbon to allow the steel to be hardened by martensitic quenching. Some powder metallurgy grades (including some MIM grades) do not have carbon, making them a sort of ferritic grade with less chromium.

There are two special stainless steel grades within the wrought and PM grades (including MIM), the so-called duplex stainless steels (meaning a ferritic–austenitic microstructure) and the precipitate hardened (PH) stainless steels. The former alloy can be strengthened by solution treatments.

Table 16.1 Orientative values for some mechanical properties of stainless steels

| Grade | Tensile strength (MPa) | Yield strength (MPa) | Elongation (%) | Hardness |
|---|---|---|---|---|
| Austenitic | | | | |
| Wrought | 520 | 210 | 40 | 90 HRB |
| PM (press and sintered) | 310 | 126 | 16 | 65 HRB |
| MIM | 415 | 170 | 38 | 80 HRB |
| Ferritic | | | | |
| Wrought | 450 | 200 | 25 | 90 HRB |
| PM (press and sintered) | 270 | 120 | 10 | 65 HRB |
| MIM | 360 | 160 | 24 | 80 HRB |
| Martensitic[a] | | | | |
| Wrought | 540 | 350 | 20 | 100 HRB |
| PM (press and sintered) | 325 | 210 | 8 | 70 HRB |
| MIM | 450 | 290 | 19 | 90 HRB |
| Duplex | | | | |
| Wrought | 600 | 550 | 15 | 30 HRC |
| PM (press and sintered) | 360 | 330 | 6 | 20 HRC |
| MIM | 480 | 440 | 14 | 27 HRC |
| Phase hardened[b] (PH) | | | | |
| Wrought | 1400 | 1200 | 10 | 45 HRC |
| PM (press and sintered) | 840 | 720 | 4 | 32 HRC |
| MIM | 1120 | 960 | 9.5 | 40 HRC |

[a] Quenched and tempered
[b] Solution hardened

We could include, thanks to the REACH regulation,[2] a new family of stainless steels, the so-called nickel-free austenitic stainless steels, especially for medical applications.

Table 16.1[3–5] displays some possible values (estimated) obtained for various mechanical properties of the different families of stainless steels. We have to take into account that a wrought stainless steel has a density near 8 g/cm$^3$ (depending on the alloying contents), with the density of a conventional PM stainless steel being about 90% of this theoretical density, and the possible density of MIM stainless steel on the order of 95% of this theoretical density (as an average value, being lower for martensitic grades and higher for PH and some austenitic grades).

As can be seen in the table, MIM mechanical features are closer to the properties of the wrought steels than conventional PM steels, especially the elongation, which is much more sensitive to the porosity than the tensile features and hardness. The corrosion performance follows a similar behaviour.

## 16.2 Stainless steels in metal injection molding (MIM)

### 16.2.1 Binders, feedstocks and debinding

Stainless steels have played an important role in the development of MIM technology, and for this reason, many of the theoretical studies regarding the technological parameters were developed with stainless steels. The reason for this was the larger range of applications for this family of materials, which was more than any other material in the period that MIM was being developed.

All of the steps of the process were investigated, including the influence of binder components[6, 7] and the rheological parameters. Of particular interest were those works that studied the effect of surface additives on the interaction with the other binder components and the rheology of the feedstock.[8, 9] Thanks to these kinds of studies, the use of surface agents like stearic acid has been widely adopted in numerous feedstock formulations.

An important set of studies was performed to understand the influence of the morphology of the powders (rounded or spherical, from water-atomised powders or gas-atomised powders)[10] and the size and granulometric distribution (which can affect the packing characteristics). The morphology of the particles is an important factor that influences the rheology, so the availability in the market of water-atomised and gas-atomised powders of different sizes and distributions allows study of this important factor.[11] Using similar sizes, the results obtained for both kinds of powders were very similar, although they were slightly better for gas-atomised powders.[12] One possible advantage of using the water-atomised powders is that, when mixed with gas-atomised powders, the shape retention during sintering and dimensional control can be improved.[13] Also, water-atomised powders are cheaper than their equivalent grades gas atomised.

Even more important to the rheology than the morphology is the size of the particles. By using different gas-atomised powders with different sizes, it was found that up to an average size of 45 µm, there is no important change in the properties, but with larger particle sizes, the packing features can be improved.[14] Large particles can produce some negative effects, like higher sintering temperature, galling in the screw check ring of molding, worse surface finish and worse shape retention, but with the benefit of lower cost. Most of the studies were performed with the most-used powders (especially the 316L grade), but there are some works related to phase-hardened steels[15] and special processes such as µMIM.[16] Powder loading is another important factor affecting rheological behaviour and technological parameters.[17]

In the MIM stainless steel feedstock, all possible binders used in any other materials can be used,[18] especially wax-based systems (paraffin wax, microcrystalline wax and natural wax) and thermoplastic-based systems

*Table 16.2* Most commonly used stainless steel grades in MIM

| Grade (AISI/SAE) | Approx. composition | Features | Code | DIN Standard | UNS Standard |
|---|---|---|---|---|---|
| AISI/SAE 420 | Fe–0.2%C–13%Cr (Mn<1%, Si<1%) | Martensitic grade/hardenable, ferromagnetic | X20 Cr 13 | DIN 1.4021 | UNS S42000 |
| AISI/SAE 430 | Fe–16%Cr (<0.08%C, <1% Mn) | Ferritic grade/ferromagnetic | X6 Cr 17 | DIN 1.4016 | UNS S43000 |
| AISI/SAE 316L | Fe–17%Cr–12%Ni–2%Mo (C<0.03%, Mn<1%, Si<1%) | Austenitic grade/non-magnetic | X2 CrNiMo 17 13 2 | DIN 1.4404 | |
| SAE J467 (17-4PH) | Fe–16%Cr–4%Ni–4%Cu–0.3%Nb (C<0.07%, Mn<1%, Si<1%) | PH grade/ferromagnetic hardenable | X5 CrNiCuNb 17 4 | DIN 1.4542 | UNS S17400 |

(polyacetals, polyethylene and polypropylene). Additionally, even thermosetting resin-based systems can be used.[19]

Today, it is possible to find any conventional stainless steel grade (of all different stainless steel families) as a powder. Different powders with different characteristics, sizes and morphologies are all commercially available. The most commonly used powders are gas-atomised with sizes under 20 μm; however, water-atomised powders are also available. Not only is it possible to find powders of most grades, but feedstocks with different characteristics are also available. Feedstocks that are catalytically debound, thermally debound and solvent (including water)-debound are also commercially available. Further, a wide range of powders and feedstocks for μMIM is now on the market. For example, Table 16.2 shows the stainless steel powders most commonly used to produce feedstocks.

There are some 'duplex' grade feedstocks available based on austenitic compositions, but with less nickel to allow for the formation for a duplex microstructure (ferritic–martensitic) when cooling from sintering temperature. The higher cost of these grades (owing to lower demand) and the fact that the duplex microstructure can be obtained by mixing ferritic and austenitic powders,[20] are the main reasons why these grades are not extended to so many applications. The presence of nitrogen, by using a nitrogen-rich atmosphere, can play an important role in the formation of a duplex microstructure,[21] with the expected reduction in the corrosion performance.

After the feedstock behaviour and its injection, binder removal is a very important topic. The debinding time, which is highly related to the industrial parameters[22, 23] and binder composition, influences all of the parameters.[24–26] Most stainless steels produced by MIM are debound by following one of the routes discussed below.

*Catalytic debinding*

The 'polyacetal binder systems' should be debound in a gaseous acidic environment (usually highly concentrated nitric or oxalic acid) at a temperature near 120°C (a temperature slightly lower than the softening of the bond system). The acid acts as a catalyst in the degradation of the polymer binder system. All of the reaction products are burned in a natural gas flame at temperatures above 600°C. This debinding method is very efficient and produces a highly interconnected porosity after short times (about 2 h or 3 h depending of the size of the component). In Europe, this is the most extended method used to produce parts. The main drawback of this method is the combination of using aggressive acids (as nitric) with relatively high temperatures, which promotes corrosion of the surrounding equipment.

*Solvent debinding*

To remove the binder using this method, the binder composition should include a constituent that can be dissolved in a liquid at a low temperature, such that a network of interconnected porosity is formed in the part while immersed in the solvent. Some possible solvents used in MIM are acetone, ethanol (and other alcohols), hexane and water. With the exception of water, all of the other options (organic options) can produce health problems if their use is not well regulated. Solvent debinding needs more time than catalytic binder removal, but the investment costs and environmental problems (especially with water) are much lower.

*Thermal debinding*

The binder system can be eliminated by heating the feedstock at temperatures where the main polymers decompose or degrade. The temperatures used can reach 800°C depending on the binder system, and the required times are longer than those of catalytic debinding. To reduce these long times, a combination of solvent and thermal debinding can be used.

## 16.2.2 Sintering aspects

After debinding, sintering is the next critical step. In this step, there are important differences between the different stainless steel grades. In the austenitic and ferritic grades, sintering should avoid any carbon or nitrogen contamination due to the possible sensitisation of the steel (especially in the austenitic grades), but in the PH and martensitic grades, additional problems have been found. By controlling the sintering temperature and the further heat treatments, different amounts of δ ferrite can be obtained in the austenitic grades;[27] at levels up to 8%, the presence of this phase improves the mechanical properties, importantly without decreasing the corrosion performance. Sintering performance of these steels can be predicted by modelling.[28, 29]

In the PH steels, sometimes the main objective is to achieve the highest tensile features, and in this sense, some carbon content can be allowed to improve these properties. The debinding process can be used for this purpose; thus, some binder left in the brown part can provide extra carbon content during sintering.[30, 31] Nevertheless, some studies have shown that an increase in carbon interferes with precipitate formation and thus lowers hardness. In this respect, carbon should be avoided in PH stainless steels; Nb is used to grab the carbon and form Nb carbides, thereby preventing the carbon from interfering in the precipitate hardening process.[32] The carbon content control can produce sintering in a δ–γ field, thereby improving the self-diffusion effect (which is much higher in ferrite than in austenite) and, as a consequence, the densification. After quenching, a duplex microstructure martensite–δ ferrite with more ductility (and slightly lower tensile strength) is produced. The same effect can be produced in austenitic grades with the addition of Mo.[33] These steels (PH) can, under good sintering conditions (up to 1380°C/90 min in an inert atmosphere), reach densities of 7.7 g/cm$^3$, ultimate tensile strength (UTS) values of 1275 MPa, and 55% elongation with good corrosion performance.[34] Non-inert atmospheres, such as $N_2$–$H_2$ mixes, can reduce the corrosion performance.[35]

The most critical stainless steel grade from the point of view of sintering is the martensitic grade.[36] Despite the fact that, with some care during sintering, it is possible to achieve a good microstructure of rounded chromium carbides in a martensitic matrix of high hardness,[37] there are inherent problems with this family of steels during this part of the process. The densification rates at temperatures below 1200°C are consistent with solid state diffusion and are, as a consequence, extremely low. However, once the liquid phase appears at higher temperatures, densification takes place quickly. The problem is that once the liquid phase appears at a relatively low temperature, the amount of this liquid phase becomes large in terms of volume percentage. A small increase in temperature leads to a significant amount of liquid phase present, which has some negative consequences, like grain growth, slumping, no uniform densification and swelling.[38] All these negative effects can be enhanced due to the presence of carbon. Therefore, the effective sintering window is very low, on the order of

10°C,[39] which causes technological problems for the industry. This sintering window can be slightly opened by using a nitrogen addition.[40] Catalytic debinding, used with polyacetal base binders, can provide this extra nitrogen. What is common to all grades is that the finer the powders are, the more enhanced is the sintering.[41]

Conventional PM stainless steels are sintered in different kinds of furnaces (either continuous furnaces, usually walking bean or pusher furnaces if high temperatures can be reached, or batch furnaces) and many types of atmospheres, even nitrogen-based atmospheres, which lead to a reduction in the corrosion performance. MIM products based on stainless steels have a higher density that those obtained by conventional press and sintering PM methods and have a slightly higher added value. As a consequence, care must be taken when sintering temperature is higher, and this process is usually conducted in batch furnaces with an extremely low dew point and in nitrogen-free atmospheres, such as hydrogen or a vacuum.

The presence of nitrogen in the atmosphere (dissociated ammonia or artificial mixes of nitrogen–hydrogen) can produce nitrogen absorption, leading to chromium nitride ($Cr_2N$) precipitation with the accompanying chromium depletion and deterioration of corrosion resistance. The formation of these chromium nitrides is promoted, especially in the range of 500–600°C, upon slow cooling from the sintering temperature. Thus, nitride formation can be decreased with very rapid cooling rates ($> 200°C$/min). However, these cooling rates are not always attainable. A hydrogen atmosphere or a vacuum represent the most suitable alternatives, and in both cases, extremely low dew points can be achieved to guarantee a good sintering behaviour of the stainless steel. The main problem when using a vacuum is the possible depletion of chromium, which can result in a reduction of the corrosion performance, but this can be reduced with a back-filling gas during sintering at some partial pressure (argon is typically used). Different works deals with this topic of using nitrogen-based atmospheres to enhance the mechanical properties[42, 43] with the risk of decreasing the corrosion performance.

The sintering temperature range used in the MIM of stainless steels varies from 1200 to 1350°C, and sintering times range from 20 to 60 min. Insufficient sintering, either too short or at too low a temperature, will result in sintered parts showing insufficient bonding, original particle boundaries, and sharp, angular pores. The sintering temperature is highly conditional on the solidus temperature of each grade. The chosen temperature should be slightly higher than the solidus temperature to avoid a massive liquid phase that can produce distortions in the shape of the sintered part. As previously explained, sintering of the martensitic grades can produce different technological problems.

### 16.2.3 The performance of MIM stainless steels

Metal injection molded stainless steels can attain higher mechanical properties than conventional PM (press and sintered) steels, mainly due to the higher density that can be achieved,[44] but their properties are slightly worse compared with their wrought counterpart grades (Table 16.1), except if post-sintering treatments are applied, like hot isostatic pressing,[39] when full density is obtained. Despite the small porosity obtained in MIM processes, the dynamic properties of these steels, in particular the fatigue properties, are strongly affected by porosity, especially over 8%.[45] The porosity of MIM stainless steels can be enhanced with space holders[46] for some special applications (such as heat exchangers, medical implants, filters, and electrodes in biological and chemical reactions). Finally, MIM stainless steels can be welded under similar conditions to wrought stainless steels.[47]

### 16.2.4 Special grades and products based on stainless steels

*Sintering improvements based on boron additions*

It is well known that in conventional PM stainless steels, boron additions can improve the sintering behaviour through liquid phase sintering, but these also cause a decrease in ductility and loss of shape.[48] There have been some related attempts in the field of MIM. Boron can be added as elemental boron, as FeB ferroalloy[49] or as nickel boride.[50] As with conventional PM, an increase in the density is produced, and as a consequence, an increase in tensile features (except the elongation and fracture toughness) is realised. The reduction in ductility is explained based on microstructural evaluations. In all the cases, a continuous brittle boride phase, depending on the addition, is formed at the grain boundaries.

*Nickel-free stainless steels*

The REACH regulation,[2] which deals with the registration, evaluation, authorisation and restriction of chemical substances (entered into force on 1 June 2007), forced the elimination of nickel, especially in powder products. This regulation promoted a great deal of research and development on new nickel-free stainless steels, including wrought steels, PM steels and MIM steels. The first attempt to reduce nickel was by introducing another gamma stabiliser element.[51, 52] Today, fully nickel-free steels are available that can compete with the PH grades and have a processing route that is quite similar because these high-nitrogen steels should be solution-annealed to obtain a fully austenitic microstructure free of irregular nitrides. Before this treatment, these steels can have a microstructure of gamma-austenite and alpha-ferrite and a network of $Cr_2N$ nitride.[53, 54]

The P.A.N.A.C.E.A. steel (protection against nickel allergy, corrosion, erosion and abrasion, Fe–17Cr–10Mn–3Mo–0.49N–0.2C) was developed for this purpose by ETH Zürich,[55] especially for MIM (polyacetal binder). This is a nickel-free austenitic stainless steel with excellent corrosion resistance and superior properties compared to 316L.

*Metal matrix composites based on stainless steels*

In conventional PM, some attempts have been made to produce metal matrix composites based on stainless steels to improve wear[56] and corrosion behaviour.[57] Metal injection molding is a suitable method for producing these kinds of composites. A composite based on 316L with improved hardness and wear resistance was developed by Gulsoy.[58, 59]

## 16.2.5 Stainless steels in MIM emerging technologies

*µMetal injection molding (µMIM)*

In the same way that stainless steels have likely been the key material in the development of MIM as a technology, they are playing the same role in µMIM. µMIM is a process to produce µ-devices at a low cost and with high performance and good dimensional control. This method allows the production of parts with no more than a few hundreds of microns in the large dimension, whereas previously only sizes of less than 100 µm in the large dimension were possible. Sometimes, some micro-parts have dimensions less than 30–40 µm;[60] thus, if the powder used has an average size of 5 µm, there can be a maximum of six particles in this direction. As can be deduced, the main constraint of this technology is the size of the particles. This is why µPIM was first developed with ceramics (that usually have a lower particle size), and one of the trends of the technology is towards using nanopowders with dimensions of less than 100 nm.[61]

This technology has many challenges that can be met by modifying some of the technological parameters that are commonly used in MIM.[62] It is essential to have a suitable binder system with low viscosity, providing for easy mold filling and higher green strength, suitable molding parameters such as higher melt and mold temperatures, and of course, particle sizes smaller than those used in conventional MIM. To improve the performance of µMIM, additional material and process development will be carried out.[63] Further, modelling tools must be involved in any development, and fine metal powders (even close to the nano level) should be used.

*Two-colour MIM*

To produce parts with the two-colour MIM technology, many technological problems must be overcome, the main one being controlling distortions during heating, sintering and cooling produced by the possible different thermal expansion coefficients of the two injected materials. In this sense, one interesting possibility that uses stainless steels of different grades is to base the mixture on steels with different magnetic characters while maintaining similar thermal characteristics. This is the development described by Imgrund et al.,[64] where two different stainless steels were injected (a combination of a non-magnetic and a ferromagnetic stainless steel, 316 L and 17-4PH). The possibilities of two-colour MIM using stainless steels is not limited to a combination of two metallic materials. It is also possible to use cemented carbides[65] or yttria-stabilised zirconia.[66]

## 16.3 Applications of MIM stainless steels

Stainless steels are the most important material (at least in terms of tonnes produced) in MIM technology, as has been described. It is not easy, however, owing to lack of information, to know what the breakdown of applications is. Analysing different sources, the breakdown can be approximated as shown in Fig. 16.1. As can be seen, the highest market share is occupied by information technologies and communications (ITC), a market which is growing very quickly. Some important parts developed in this market include some decorative applications (push buttons, rocker switches for mobile phones, original equipment manufacturer, logo, e.g. Apple, 316L polished), internal stiffening frames (mobile phones, 17-4PH), hard disk drive (computers, 17-4PH) and hinges (laptops, mobile phones, 440 and 17-4PH).

In Table 16.3, some of the features and applications of the main stainless steels grades are described. As can be seen, the breakdown of the applications includes most of the MIM applications, with a leading role in the biomedical field. Figures 16.2 to 16.8 show some of these applications.

Figure 16.2 shows a swivel hinge assembled with two sliding discs per temple, made of 17-4PH. This part greatly improves the design freedom, thanks to the construction with no edges, which allows seamless mounting.

Figure 16.5 demonstrates one clear example of how stainless steels made by MIM can fulfill the most stringent requirements in different fields of material performance. This is the case with the prosthetic knee joint shown. The MIM parts illustrated work together as a unit within a prosthetic knee joint. They protect the knee joint from accidental flexion and serve to transmit power while stretching or bending the knee. The operational demands on the components are stringent, as patient safety is critical. Given

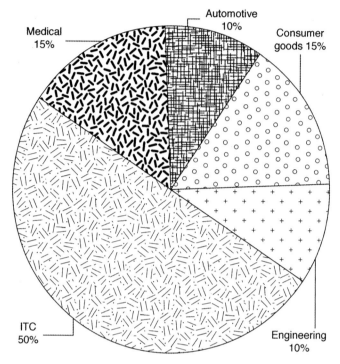

*16.1* Approximate breakdown of the stainless steel applications in MIM in the world market.

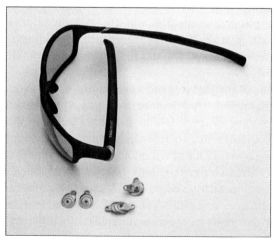

*16.2* Swivel hinge 27° assembled with two sliding discs per temple (1 g, 17-4PH) (courtesy of Ohnmacht and Baumgaertner GmbH and Co.).

Table 16.3 Main applications of stainless steels made by MIM technology

| Grade | Microstructure | Performance description | Applications |
|---|---|---|---|
| 430L | Ferritic | Magnetic stainless steel with resistance to atmospheric corrosion and general oxidation. Lower cost than nickel-containing austenitic stainless steels. | Sensors, armatures and pole pieces that require some resistance to corrosion |
| 316L | Austenitic | Non-magnetic stainless steel with excellent corrosion resistance, toughness and ductility | Medical and dental devices (mechanical joints, brackets), mobile phones (rocker key, battery lock), consumer products (watch cases, camera components, saw blade fastener, gear for electric toothbrush), spectacles (rotated spring hinge – see Fig. 16.1), defence (rotor in a safety device for missile warheads, parts in guns), marine components and non-magnetic housings (spacers for computers) |
| Duplex | Ferritic/austenitic | Weakly magnetic stainless steel providing a mix of properties between the austenitic and ferritic grades | Lock components |
| 17-4PH | Precipitation hardening | Provides an excellent combination of strength, hardness, and corrosion resistance | Ordnance components, high-strength fasteners, mobile phones (GPS antenna plug), hardware (cylinder lock casing), spectacles (eyewear swivel hinge), automotive (hydraulic connector, dashboard mounting screws, angle plug for cable connectors (see Fig.16.2), fibre optic connectors, and medical devices (mechanical joints, suturing jaws, wound forceps (see Fig.16.3), prosthetic knee joint (Fig.16.4) |
| MIM 440C | Martensitic | Provides high strength, hardness and resistance to wear with moderate resistance to corrosion | Wear plates, automotive (turbocharger vane, fuel injection nozzles), hardware (fireproof door lock encasement), medical applications (mechanical joints) and cutting instruments |
| Nickel-free | Austenitic | All the austenitic features and non-allergenic | Specially for biomedical applications |

406  Handbook of metal injection molding

*16.3* Angle plug for cable connector in the automotive industry (0.45 g, 17-4PH) (courtesy of Ohnmacht and Baumgaertner GmbH and Co.).

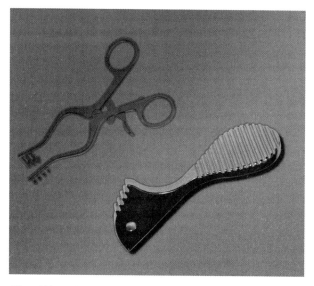

*16.4* Wound forceps (2.5 g, 1-4PH) (courtesy of Ohnmacht and Baumgaertner GmbH and Co).

these requirements, with the need for corrosion resistance, high strength and an attractive surface appearance, viable parts can only be guaranteed by MIM technology.

Figure 16.6 shows a door lock casing for domestic hardware made of 310N. This part is a relatively high weight (in the MIM field), 96 g, but provides a good demonstration of the excellent flatness and precision

Metal injection molding of stainless steel 407

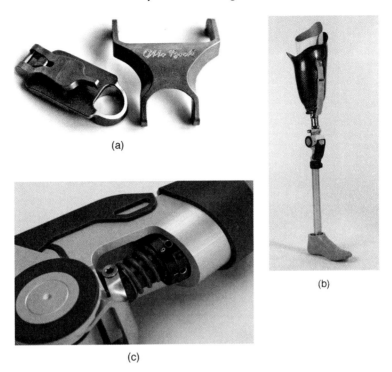

*16.5* (a) MIM parts made of 17-4 PH; (b) Otto Bock prosthetic knee joint 3R93; (c) detail (courtesy of GKN Sinter Metals).

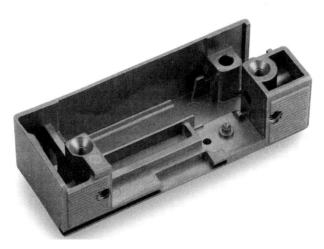

*16.6* Door lock casing for domestic hardware made of 310N (96 g) (courtesy of MimTech Alfa).

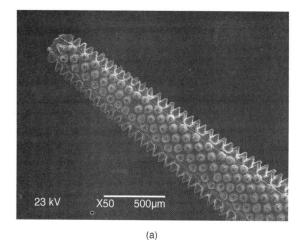

*16.7* MIM endo tips having (a) 50 μm and (b) 100 μm projections on the top section of the tip (courtesy of Ceta Tech Inc.).

achievable through this technology. This part substitutes for another made of Zamak (a die-cast zinc alloy). By manufacturing the part in stainless steel, the end-user benefits from an improved fire safety classification.

In Fig. 16.7 another interesting application in the medical sector is shown (MIM endo tips with 50 μm and 100 μm projections on the top section of the tip). This makes it possible to eliminate the need for diamond coating.

Finally, Fig. 16.8 shows the MIM option for a very difficult part (a highly precise MIM mechanical part of a quartz watch using 316L micro-grain powder). The part has a complicated shape, which requires very precise dimensions (it should be assembled with other precise parts).

Metal injection molding of stainless steel 409

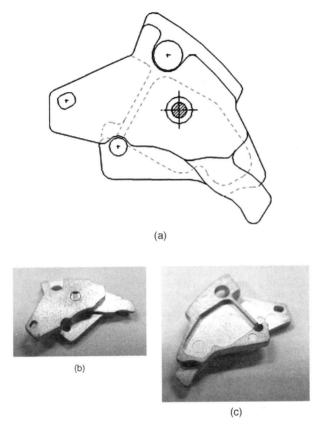

*16.8* Mechanical part of a quartz watch (micro grain powder 316L): (a) draft; (b) view from the gate side; (c) view from the ejector pin side (courtesy of EPSON ATMIX Corp.).

## 16.4 Acknowledgements

The author would like to thank the following companies for their permission to use their information in the writing of this chapter: Ohnmacht & Baumgaertner GmbH & Co, GKN Sinter Metals, Mim Tech Alfa, CetaTech Inc. and EPSON ATMIX Corp. Particular thanks are owed to Johan ter Maat, from BASF SE, for his valuable comments.

## 16.5 References

1. 'Metal injection moulding (MIM) growth slows in 2009', in *International Powder Metallurgy Directory 2010–2011*, 14th edition, Inovar Communications Ltd, Shrewbury, UK, pp. 23–25.
2. http://eur-lex.europa.eu/LexUriServ/LexUriServ.do?uri = CELEX:32006R1907:EN:NOT.
3. Davis, J. R. (ed.) (1998), *ASM Metals Handbook*, desk edition, ASM International, Materials Park, Ohio, USA.
4. MPIF (2007), MPIF Standard 35, in *Materials Standards for Metal Injection Molded Parts*, Metal Powder Industries Federation, New Jersey, USA.
5. Samal, P. K. and Klar, E. (2007), *Powder Metallurgy Stainless Steels. Processing, Microstructures and Properties*, ASM International, Materials Park, Ohio, USA.
6. Bakan, H. I., Jumadi, Y., Messer, P. F., et al., (1998), 'Study of processing parameters for MIM feedstock based on composite PEG-PMMA binder', *Powder Metallurgy*, 41(4), 289–291.
7. Huang, M. S. and Hsu, H. Ch. (2009), 'Effect of backbone polymer on properties of 316L stainless steel MIM compact', *Journal of Materials Processing Technology*, 209, 5527–5535.
8. Johansson, E., Nyborg, L., and Becker, J. (1997), 'Interactions between surface active additives and 316L MIM powder', *Proceedings of the 1st European Symposium on Powder Injection Moulding*, Munich, pp. 89–96.
9. Johansson, E., Nyborg, L., and Becker, J. (1998), 'Rheology of 316L MIM powder plastisos. Effect of surface active additives', *Proceedings of the Powder Metallurgy World Congress*, Granada, Spain, vol. 3, pp. 39–44.
10. Heaney, D., Zauner, R., Binet, C., Cowan, K., and Piemme, J. (2004), 'Variability of powder characteristics and their effect on dimensional variability of powder injection molded components', *Journal of Powder Metallurgy*, 47(2), 145–150.
11. Suri, P., Kososki, R. P., and German, R. M. (2005), 'Microstructural evolution of injection molded gas and water atomised 316L stainless steel powder during sintering', *Material Science and Engineering A*, 402, 341–348.
12. Kyogotu, H., Komatsu, S., Shinzawa, M., Mizuno, D., Matsuoka, T., and Sakaguchi, K. (2000), 'Influence of microstructural factors on mechanical properties of stainless steel by powder injection moulding', *Proceedings of the 2000 Powder Metallurgy World Congress*, vol. 1, pp. 304–307.
13. Pascoali, S., Wendhausen, P. P. A. P., and Fredel, M. C. (2001), 'On the use of gas and water atomised 316L powder blends in PIM', *Proceedings of the PM2001 Congress*, Nice, France, vol. 3, pp. 147–152.
14. Hartwig, T., Meinhardt, H., Wernicke, K., Kunert, P., Veltl, G., and Weber, M. (1997), 'MIM of 316L powders with large particle sizes', *Proceedings of the 1st European Symposium on Powder Injection Moulding*, Munich, pp. 62–68.
15. Keams, M. A., Davies, P. A., Zauner, R., and Neubauer, E. (2007), 'Influence of gas atomised powder size on mechanical properties of 15-5PM', *Proceedings of the EUROPM2007*, Toulouse, France, vol. 2, pp. 171–176.
16. Okubo, K., Tanaka, S., and Ito, H. (2010), 'The effects of metal particle size and

distributions on dimensional accuracy for micro parts in micro metal injection molding', *Microsystem Technologies*, 16, 2037–2041.
17. Li, Y. Li, L., and Khalil, K. A. (2007), 'Effect of powder loading on metal injection molding stainless steels', *Journal of Materials Processing Technology*, 183, 432–439.
18. German, R. M. (1997), *Powder Injection Moulding*, Metal Powder Industry Federation, Princeton, NJ, USA. (1997).
19. Levenfeld, B., Gruzza, A., Varez, A., and Toralba, J. M. (2000), 'Modified metal injection moulding of 316L stainless steel powder using thermosetting binder', *Powder Metallurgy*, 4, 233–237.
20. Campos, M., Muñoz, J. J., Bautista, A., Velasco, F., and Torralba, J. M. (2003), 'Ni diffusion process between austenite and ferrite in a PM duplex stainless steel obtained by powder mixing', *Materials Science Forum*, 426–432, 4343–4348.
21. Mariappan, R., Kumaran, S., and Srinivasa Rao, T. (2009), 'Effect of sintering atmosphere on structure and properties of austeno-ferritic stainless steels', *Materials Science and Engineering A*, 517, 328–333.
22. Bogan, L. E. (1997), 'Debinding of acrylic-based 316L stainless steel MIM feedstock', in *Advances in Powder Metallurgy and Particulate Materials*, Metal Powder Industries Federation, Chicago, USA, pp. 1857–1864.
23. Anwar, M. Y., Davies, H. A., Messer, P. F., and Ellis, B. (2005), 'Rapid debinding of powder injection moulded components', *Proceedings of the EUROPM'95*, Birmingham, UK, vol. 1, pp. 577–584.
24. Bialo, D., Kulesza, T., and Ludynski, Z. (2005), 'Selected problems of injection moulding of 316L type powders', *Proceedings of the EUROPM'95*, Birmingham, UK, vol. 1, pp. 626–630.
25. Karatas, C., Gokten, M., Unal, H. I., Saritas, S., and Uslan, I. (2005), 'Investigation of rheological properties of the feedstocks composed of steatite and 316L stainless steel powders and peg base resins', *Proceedings of the EUROPM'95*, Birmingham, UK, vol. 1, pp. 367–373.
26. Jorge, H., Hennetier, L., Correia, A., and Cunha, A. 'Tailoring solvent/thermal debinding 316L stainless steel feedstocks for PIM: an experimental approach', *Proceedings of the EUROPM'95*, Birmingham, UK, October 2005, vol. 1, pp. 342–348.
27. Muterlle, P. V., Perina, M., and Molinari, A. (2010), 'Mechanical properties and corrosión resistance of vacuum sintered MIM 316L stainless steel containing delta ferrite', *Powder Injection Moulding International*, 4, 66–70.
28. Kong, X., Barriere, T., Gelin, J. C., and Quinard, C. (2010), 'Sintering of powder injection molded 316L stainless steel: experimental investigation and simulation', *International Journal of Powder Metallurgy*, 46, 61–72.
29. Kwon, Y.-S., Wu, Y., Suri, P., and German, R. M. (2004), 'Simulation of the sintering densification and shrinkage behavior of powder-injection-molded 17-4 PH stainless steel', *Metallurgical and Materials Transactions A*, 35A, 257–263.
30. Wu, Y., German, R. M., Blaine, D., Marx, B., and Schlaefer, C. (2002), 'Effects of residual carbon content on sintering shrinkage of injection molded 17-4 PH stainless steel', *Journal of Material Science*, 37, 3573–3583.
31. Kyogoku, H., Komatsu, S., Nakayama, H., Jinushi, H., and Shinohara, K. (1997), 'Microstructures and mechanical properties of sintered precipitation hardening stainless steel compacts by metal injection moulding', in *Advances in*

*Powder Metallurgy and Particulate Materials,* Metal Powder Industries Federation, Chicago, USA, pp. 18135–18144.
32. Heaney, D. (2010), Personal communication, Penn State University, USA.
33. Shu, G. S., Hwang, K. S., and Pan, Y. T. (2006), 'Improvements in sintered density and dimensional stability of powder injection-molded 316L compacts by adjusting the alloying compositions', *Acta Materialia,* 54, 1335–1342.
34. Li, Y. M., Khalil, K. A., and Huang, B. Y. (2004), 'Rheological, mechanical and corrosive properties of injection molded 17-4PH stainless steel', *Transactions of Nonferrous Metal Society of China,* 14, 934–937.
35. Ye, H., Liu, X. Y., and Hong, H. (2008), 'Sintering of 17-4PH stainless steel feedstock for metal injection moulding', *Materials Letters,* 62, 3334–3336.
36. Blömacher, M., Lange, E., Schwarz, M., Weinand, D., and Wohlfromm, H. (1998), 'Powder injection molding of martensitic stainless steels', in *Advances in Powder Metallurgy and Particulate Materials,* Metal Powder Industries Federation, Princeton, NJ, USA, pp. 5131–5141.
37. Newell, M. A., Davies, H. A., Messer, P. F., and Greensmith, D. J. (2005), 'Metal injection moulding scissors using hardenable stainless steel powders', *Powder Metallurgy,* 48, 227–230.
38. Li, D., Hou, H., Liang, L., and Lee, K. (2010), 'Powder injection molding 440C stainless steel', *International Journal of Manufacturing Technologies,* 49, 105–110.
39. Wohlfromml, H. (2002), 'Powder injection molding stainless steel-produce process, performance, application', *Powder Metallurgy,* 12, 7–15.
40. Wohlfromm, H., Uggowitzer, P. J., Blömacher, M., ter Maat, J., Thom, A., Pistillo, A., and Ribbens, A. (2005), 'Metal injection moulding of a modified 440C-type martensitic stainless steel', *Proceedings of the EUROPM'95,* Birmingham, UK, vol. 1, pp. 325–332.
41. Toyoshima, H., Kusunoki, M., and Otsuka, I. (2006), 'Sintering properties of high-pressure wáter atomised SUS 316L utra fine powder', *Proceedings of the 2006 World Congress on Powder Metallurgy,* Busan, Korea, vol. 2, pp. 769–770.
42. Cheng, L. H. and Hwang, K. S. 'High-strength powder injection molded 316L stainless steel', *International Journal of Powder Metallurgy,* 46, 29–37.
43. Krug, S. and Zachmann, S. (2009), 'Influence of sintering conditions and furnace technology on chemical and mechanical properties of injection moulded 316L', *Powder Injection Moulding International,* 3, 66–70.
44. Mishra, A. K., Paul, S., Upadhyaya, A., Barriere, T., and Gelin, J. C. (2003), 'Injection molding and sintering of ausenitic stainless steel', *Indian Journal of Engineering and Materials Sciences,* 10, 306–313.
45. Yoon, T. S., Lee, Y. H., Ahn, S. H., Lee, J. H., and Lee, C. S. (2003), 'Effects of sintering conditions on the mechanical properties of metal injection molded 316L stainless steel', *ISIJ International,* 43, 119–126.
46. Özkan Gülsoy, H. and German, R. M. (2008), 'Production of micro-porous austenitic stainless steel by powder injection molding', *Scripta Materialia,* 58, 293–298.
47. Wolf, E. L. and Collins, S. R. (2002), 'GTA welding of AISI 316L metal injection molded (MIM) components', *Proceedings of the 6th International Conference on Trends in Welding Research,* Phoenix, USA, pp. 419–423.
48. Molinari, A., Kazior, J., Marcheti, F., Canteri, R., Cristofolini, F., and Tiziani,

A. (1994), 'Sintering mechanisms of boron alloyed AISI 316L stainless steel', *Powder Metallurgy*, 37, 115–122.
49. Gülsoy, H. Ö., Salman, S., and Özbek, S. (2004), 'Effect of FeB additions on sintering characteristics of injection moulded 17-4PH stainless steel powder', *Journal of Material Science*, 39, 4835–4840.
50. Bakan, H. I., Heaney, D., and German, R. M. (2001), 'Effect of nickel boride and boron additions on sintering characteristics of injection moulded 316L powder sing water soluble binder system', *Powder Metallurgy*, 44, 235–242.
51. Tandon, R., Simmons, J. W., Covino, B. S., and Russel, J. H. (1998), 'Mechanical and corrosion properties of nitrogen-aloyed stainless steels consolidated by MIM', *International Journal of Powder Metallurgy*, 34, 47–54.
52. Cui, D. W., Qu, X. H., Guo, P., and Li, K. (2010), 'Sintering optimisation and solution annealing of high nitrogen nickel free austenitic stainless steels prepared by PIM', *Powder Metallurgy*, 53, 91–95.
53. Xu, Z. W., Jia, C. C., Kuang, C. J., and Qu, X. H. (2010), 'Fabrication and sintering behaviour of high-nitrogen nickel-free stainless steels by metal injection moulding', *International Journal of Minerals Metallurgy and Materials*, 17, 423–428.
54. Cui, D. W., Qu, X. H., Guo, P., and Li, K. (2010), 'Sintering optimization and solution annealing of high nitrogen nickel free austenitic stainless seels prepared by PIM', *Powder metallurgy*, 53, 91–95.
55. Wohlfromm, H., Blömacher, M., Weinand, D., Uggowitzer, P. J., and Speidel, M. O. (1998), 'Novel stainless steels for metal injection moulding', *Proceedings of the World Congress on PM*, Granada, Spain, vol. 3, pp. 3–8.
56. Vardavoulias, M., Jeandin, M., Velasco, F., and Torralba, J. M. (1996), 'Dry sliding wear mechanism for P/M austenitic stainless steels and their composites containing Al2O3 and Y2O3 particles' *Tribology International*, 29, 499–506.
57. Abenojar, J., Velasco, F., Torralba, J. M., Bas, J. A., Calero, J. A., and Marce, R. (2002), 'Reinforcing 316L stainless steel with intermetallic and carbide particles', *Materials Science and Engineering A*, 335, 1–5.
58. Gulsoy, H. O. 'Production of injection moulded 316L stainless steels reinforced with TiC(N) particles'. *Materials Science and Technology*, 24, 1484–1491.
59. Gulsoy, H. O. 'Mechanical properties of injection moulded 316 Lstainless steels with (TiC)N additions', *Powder Metallurgy*, 60, 271–275.
60. Liu, Z. Y., Loh, N. H., Tor, S. B., Murakoshi, Y., Maeda, R., Khor, K. A., and Shimidzu, T. (2003), 'Injection molding of 316L stainless steel microstructures', *Microsystem Technologies*, 9, 507–510.
61. Supriadi, S., Back, E. R., Choi, C. J., and Lee, B. T. (2007), 'Binder system for STS 316 nanopowder feedstocks in micro-metal injection molding', *Journal of Materials Processing Technology*, 187–188, 270–273.
62. Loh, N. H., Tor, S. B., Tay, B. Y., Murakoshi, Y., and Maeda, R. (2003), 'Micro powder injection molding of metal microstructures', *Materials Science Forum*, 426–432, 4289–4294.
63. Piotter, V., Merz, L., Ruprecht, R., and Hausselt, J. (2003), 'Current status of micro powder injection molding', *Materials Science Forum*, 426–432, 4233–4238.
64. Imgrund, P., Rota, A., Petzoldt, F., and Simchi, A. (2007), 'Manufacturing of

multi-functional micro parts by two-component metal injection moulding', *International Journal of Advanced Manufacturing Technology*, 33, 176–186.
65. Simchi, A. and Petzoldt, F. (2010), 'Cosintering of powder injection molding parts made from ultrafine WC-Co and 316L stainless steel powders for fabrication of novel composite structures', *Metallurgical and Materials Transactions A*, 41A, 233–241.
66. Dourandish, M. and Simchi, A. (2009), 'Study of the sintering behavior of nanocrystalline 3Y-TZP/430L stainless-steel composite layers for co-powder injection molding', *Journal of Materials Science*, 44, 1264–1274.

# 17
Metal injection molding (MIM) of titanium and titanium alloys

T. EBEL, Helmholtz-Zentrum Geesthacht, Germany

**Abstract:** Metal injection molding of titanium and titanium alloys is still rather unusual, even though strong advantages can be achieved from this technique. High costs and insufficient mechanical properties of the sintered parts are often stated as reasons for some reservation regarding its use. This chapter shows how these problems can be overcome and how excellent properties can be achieved. The challenges of MIM processing of titanium are pointed out, followed by an overview of available powders and feedstocks. Important processing details are shown and current research is explored.

**Key words:** MIM of titanium, fatigue, grain refinement, oxygen sensitivity, reduction of powder costs.

## 17.1 Introduction

Since around 1940, when William Justin Kroll introduced the first economic and industrial process to gain metallic titanium from the ore, titanium has been used for industrial and commercial applications at a continuously progressing rate. However, compared to steel, the annually produced mass is still rather small and applications are mainly limited to five fields: the offshore industry; chemical industry; aerospace and automotive industries; medical devices and implants; and luxury goods. The listed application areas are directly linked to the specific characteristics of titanium, which are the very high strength to density ratio, the excellent corrosion resistance against salt water and most chemicals, the high biocompatibility and the warm and optically attractive surface of the material. Depending on the demands with respect to corrosion resistance, mechanical properties and also costs, either pure titanium or an alloyed material is used. High costs of raw material and processing are the main reasons for the limitation to applications where the specific properties of titanium and titanium alloys are so beneficial that higher prices can be tolerated. Today, the light-weight properties and biocompatibility in particular are features generating high levels of interest; the usage of titanium-based materials could probably be much higher if the price of the product was lower.

Against this background, MIM appears to be ideal for the processing of titanium, as typical advantages are high material utilisation and low production costs in large-quantity manufacturing. Machining of titanium is rather expensive owing to high tool costs and low processing speed. Thus, the typical geometry of machined titanium parts is rather simple and is not optimised with regard to functionality. On the contrary, even highly complex components can be manufactured by MIM without necessarily increasing costs. However, today processing titanium materials by MIM remains rare, even if an increasing interest is noticeable. In this chapter, the specific challenges of titanium MIM and methods to overcome these challenges are presented. Furthermore, the current status of MIM of titanium is reported with respect to available feedstocks and mechanical properties. Finally, current research activities are explored, with a view to prospects for the future.

## 17.2 Challenges of MIM of titanium

### 17.2.1 Interstitial elements

The most important feature of titanium that represents a challenge for PM is the high affinity to interstitial elements like oxygen, nitrogen, carbon and hydrogen. The fact that titanium is used in the vacuum technique as a getter material for purification of the atmosphere from oxygen clearly shows the problem that arises when fine powders have to be handled. According to the binary phase diagram for Ti–O, titanium is able to absorb 13 wt% oxygen on interstitial lattice sites. The great affinity to interstitial elements is combined with a strong influence on the mechanical properties, even at low contents in the range of 0.1–0.4 wt%. For illustration, Table 17.1 shows the values for the maximum content of oxygen according to ASTM standard B348-02 for wrought bars and billets and its influence on tensile properties.

Even if the standard also regulates maximum values for iron, nitrogen and carbon content, oxygen is the most important element with regard to its influence on mechanical properties, because it is picked up preferentially. The values given in the table are just minimum limits. However, it can be estimated from these numbers that in case of pure titanium an increase by just 0.22 wt% oxygen from grade 1 to grade 4 effects more than a doubling of the tensile strength. On the other hand, the ductility is reduced strongly and this is essentially the reason why the pick-up of interstitial elements

*Table 17.1* Maximum oxygen level and tensile properties according to ASTM B348-02

|  | Max. oxygen content (wt%) | Min. YS (MPa) | Min. UTS (MPa) | Min. $\varepsilon_f$ (%) |
| --- | --- | --- | --- | --- |
| Ti Grade 1 | 0.18 | 170 | 240 | 24 |
| Ti Grade 2 | 0.25 | 275 | 345 | 20 |
| Ti Grade 3 | 0.35 | 380 | 450 | 18 |
| Ti Grade 4 | 0.40 | 483 | 550 | 15 |
| Ti–6Al–4V Grade 5 | 0.20 | 828 | 895 | 10 |
| Ti–6Al–4V Grade 23 | 0.13 | 759 | 828 | 10 |

during MIM processing has to be minimised. It is rather easy to obtain a high strength in a MIM titanium component just by increasing the oxygen content, but the provision of good ductility is a challenge. In the case of titanium alloys like Ti–6Al–4V the oxygen limits are even smaller (see Table 17.1) than those of pure titanium grade 2. There is also a standard for PM products made from titanium alloy powders (ASTM B817) allowing an oxygen content of 0.3 wt% for Ti–6Al–4V. However, this standard is commonly not applied for MIM components, not even for comparison.

From the point of view of solution hardening oxygen, nitrogen and carbon have the same effect, but with different potency. Therefore it is sensible to combine these elements in the form of an oxygen equivalent $O_{eq}$. A commonly used equation is

$$O_{eq} = c_O + 2c_N + 0.75c_C \qquad [17.1]$$

where $c_O$, $c_N$ and $c_C$ represent the concentrations of oxygen, nitrogen and carbon, respectively. In the literature, the coefficient of $c_C$ differs somewhat between 0.5 (German, 2009) 0.66 (Zwicker, 1974) and 0.75 (Conrad, 1966). In addition, equation [17.1] is given in atom percentage (Conrad, 1966), but it is often applied in the literature by using weight percentage. However, in practice this difference does not really matter because the amounts of carbon and nitrogen are rather low compared to oxygen and the atomic weights of O, N and C are comparable. Even if the effect of nitrogen is twice as high as that of oxygen, again in practice the latter remains the most important interstitial element. This is mainly due to its high diffusivity, the high solubility and the fact that the oxide layer spontaneously formed at the surface impedes penetration by other elements. Experience shows that during MIM processing the pick-up of nitrogen is usually no problem. Even hydrogen commonly need not be taken into account because sintering of titanium is usually done under high vacuum at high temperatures. Under this condition possible hydrogen contents are drawn out from the material. On the other hand, carbon has to be considered, because it is present in all

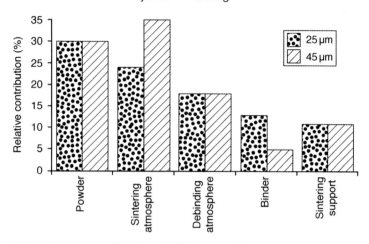

*17.1* Relative contribution of MIM processing parameters to final oxygen content of the sintered part in relation to dependence on maximum powder particle diameter. Taken from Baril *et al.* (2010).

binders and during high-temperature steps like thermal debinding it can be absorbed. Therefore it is more sensible to use equation [17.1] instead of just considering the oxygen content when discussing results from mechanical tests performed on MIM processed materials.

Baril *et al.* (2010) published a comprehensive study about the sources of interstitials, especially oxygen, during MIM processing. As shown in Fig. 17.1, several sources exist, starting with the powder and ending with the sintering support which is commonly made from ceramic materials like $Y_2O_3$ or $ZrO_2$. The ratio between these sources depends slightly on the powder particle size, but the overall pick-up is nearly the same. The study shows that storing and handling of the powder is not critical as long as the temperature is below 200°C. Above 400°C rapid oxidation starts. Even the uptake from the binder is tolerable, if the right binder components are used and debinding is performed adequately, as shown in the next sections. Most critical are the interstitial content of the powder and the sintering atmosphere, which has to be as free as possible from oxygen.

### 17.2.2 Powder and feedstock

Today, the availability of titanium-based MIM feedstock appears to be rather poor compared to stainless steel or cobalt (Co)-based materials. For many years only pure titanium feedstock of specification grade 4 was commercially available. Components made from this material show sufficient strength but the ductility does not fulfil the ASTM requirements. Recently, Ti–6Al–4V can also be bought on the commercial market, resulting in components with much higher strength and fracture elongation around

10% if processed adequately. However, it is common practice, even for commercial MIM suppliers, to manufacture the feedstock in-house by blending powder and binder components. This leads to the situation that actually no specific 'standard' feedstock exists, as in the field of MIM of stainless steel. Furthermore, titanium grade 1 or other titanium alloy feedstocks like Ti–6Al–7Nb or beta-titanium alloys are not yet available. On the other hand, manufacturing of feedstock in-house is not a serious problem and provides great freedom with regard to powders and alloys used.

*Powder*

The available powders can be divided very roughly into pure and expensive and comparably unclean and of lower cost. In this context 'pure' relates especially to the oxygen and carbon content, but also to residuals from the specific production process such as Cl, Ca, Na or Mg. Furthermore, these powders differ in geometry: the expensive ones are more spherical, the lower cost powders are typically irregular shaped and often agglomerations are found. Current techniques for the production of more expensive powders are inert gas atomisation (Hohmann and Jönsson, 1990) and plasma rotating electrode processing (Aller and Losada, 1990). Because alloys can be used as starting materials, pre-alloyed powder can also be made by these techniques. Lower cost powders are made by mechanical milling of sponge, scrap or ingots, for example as done in the hydride–dehydride (HDH) process (McCracken *et al.*, 2010); the raw material is loaded with hydrogen effecting a drastic embrittlement of the material. Thus, it can be milled easily and the hydrogen is extracted again by thermal treatment under vacuum. It is also possible to use scrap or ingot as raw material. Thus, different purities of HDH powder and also alloys are available. There is much research on reduction of the titanium oxide without Kroll process aiming at lower cost powder. Even powder production directly from the oxide is under investigation. However, at the moment HDH powders are mostly used as an economic alternative, if the properties are sufficient. Generally, HDH powders show a relatively high content of oxygen. This is because passivation steps have to be included in the processing to avoid self-ignition and burning of the powder. In addition, rather high temperatures are involved which are sufficient to effect sintering between the particles. More or less pronounced agglomeration is therefore typical. A novel process appears to be able to reduce the oxygen content after powder production (McCracken *et al.*, 2009). Liquid Ca is utilised in order to reduce the titanium to levels fulfilling the corresponding ASTM standards. However, this additional process will increase the powder costs and only the future will reveal whether this method will provide an attractive powder for the market.

Using HDH powder could even assist in keeping the oxygen level low, if the still hydrided powder is used for sintering. The hydrogen escapes during sintering and reacts with possible oxygen in the binder, powder and furnace atmosphere.

Choosing the right powder for MIM means taking into account the required mechanical properties, technical issues like powder flowability, component geometry and costs. Spherical powders produce the lowest viscosity feedstock system, thus, binder content can be kept low and sintering activity is high. Combined with high purity, using these powders usually results in the best mechanical properties. If the requirements of strength and ductility are not that high, HDH titanium powder may be sufficient. In addition, irregular-shaped powders give higher green part strength. It is also possible to mix both kinds of powders to reduce the material costs, as will be shown later. Even alloys can be produced by blended elemental technique, if the specific pre-alloyed powder is not available. Again it is possible, for example, to mix lower cost HDH titanium powder with more expensive pre-alloyed powder to gain the final composition in the sintered part.

Because of the different production techniques compared to steel powder and because of the contamination problem, which is more important when finer powder is used, for MIM of titanium commonly the size fraction < 45 µm is utilised. Accordingly, it is significantly coarser than the usual stainless steel powder. One disadvantage is the greater tendency for separation of binder and powder during injection molding. On the other hand, even with these powders, wall thicknesses of the sintered part below 0.2 mm can be realised without significant problems.

*Binder*

Several binder systems exist which are suitable to be used combined with titanium powder. Basically, the composition should provide the possibility for debinding in two steps, typically by a solvent and a thermal process; however, a binder for catalytic debinding is also available. In the latter case, for example polyacetate can be used as one component. For solvent debinding the binder often comprises paraffin wax, for water debinding polyethylene glycol. Also the use of naphthalene is possible, which evaporates slowly from the green part (Nyberg *et al.*, 2005). The second component decomposes by thermal treatment prior to sintering, usually at temperatures above 400°C. Thus, oxidation and carbonisation are possible risks at this stage. Therefore, the second component should contain as little oxygen as possible and should decompose in the right temperature range. Generally, the thermal degradation temperature should be as low as possible, because the higher this temperature is, the higher is the risk for contamination of titanium by oxygen and carbon. On the other hand, first sintering has to start before the binder has escaped completely, because of mechanical stability. Degradation temperatures between 400°C and 500°C are quite suitable. First bonding of titanium powder particles starts at about 600°C or less.

In practice, polyethylene or polypropylene, even mixed or as copolymer with ethylene vinyl acetate, are used with good success. The ratio of first and second binder component depends on the required green part strength and the powder used, and also on the required contamination level after sintering. A higher amount of the second component means basically an increase in oxygen and carbon content. Between 20 and 40 wt% second component is typically a good value. A third typical additive is stearic acid, working as a surface active agent improving the wettability of the powder and influencing viscosity and mold release. There is a wide range of percentage from 1 to 12% used. German (2009) gives a good overview on the binder systems used today.

### 17.2.3 Facilities

There is nothing really special with regard to the MIM equipment when titanium is being processed. However, the contamination problem has to be considered in the whole production chain; consequently, the precise specification of the devices is influenced. If the feedstock is made in-house, kneading under a protective atmosphere could be considered, if possible. Any device suitable for MIM of stainless steel can be used, as an injection molding machine. The mass temperature during injection ranges typically between 120°C and 180°C; the mold is heated to temperatures between 40 and 60°C. The precise parameters depend on the feedstock and on the geometry of the component.

For solvent debinding commercial products are on the market; certified for use are flammable media like heptane, hexane or acetone. The typical working temperature of the solvent ranges from 40 to 60°C. There are also trials using supercritical $CO_2$ for debinding. The motivation is to decrease debinding time and allow larger wall thickness than 10 mm of the green part (Shimizu et al., 2001; Thomas and Baril, 2010). However, the results remain ambiguous as to whether this process is really beneficial. Finally, it depends on the binder, which has to be adapted to the specific debinding method.

Thermal debinding can take place either in the sintering furnace or in a separate oven. In both cases the possibility for using sweep gas should be provided to facilitate removal of the binder gas, but also generation of a high vacuum $< 10^{-3}$ Pa should be possible. Preferentially the binder residuals are condensed in a chiller behind the furnace. If a separate furnace is used for debinding, the minimum temperature should be at least 1000°C to allow pre-sintering.

The sintering furnace should be a high-vacuum device with metallic heater (e.g. tungsten) and shielding (e.g. molybdenum). Sintering is the most critical process with regard to contamination due to the high temperature. For titanium the typical sintering temperature ranges from 1250°C to 1350°C. However, even higher temperatures up to 1500°C are necessary, when sintering, for example, intermetallic titanium aluminides.

In addition, in special cases it can be beneficial if the furnace can be run under different gaseous atmospheres, as well. For example, sintering under argon instead of a vacuum may facilitate the avoidance of oxygen pick-up, if high-purity gas is used, and sublimation of alloy elements with low melting points can be minimised (e.g. Al). Even sintering under hydrogen partial pressure may be helpful for achieving low oxygen contamination. However, this is associated with higher efforts in respect of equipment and security.

### 17.2.4 Porosity

The residual porosity of sintered titanium parts can be more critical than in the case of stainless steel. The reason is the rather high sensitivity of titanium-based materials against notches and small cracks (Evans, 1998). Especially in the case of fatigue load by stress variation, it is important that the pores are as well rounded as possible and surface defects are minimised. In a limited range these features can be influenced by proper choice of process parameters. Typically, the residual porosity amounts to a value around 3–4%. This means porosity is closed and an additional hot isostatic pressing (HIP) process can be applied without capsulation, resulting in practically 100% dense material.

### 17.2.5 Biocompatibility

Because titanium is a popular material for medical implants, application in this field is also of interest using MIM for processing. In this case, it must be ensured that possible binder residuals do not affect the biocompatibility. In the end, appropriate tests of the sintered components have to prove this. However, it is a good start to use exclusively non-toxic biocompatible binder components. In Section 17.6.1 the current state of this issue is described.

## 17.3 Basics of processing

### 17.3.1 Powder handling

The duration of open powder handling, as may occur if feedstock is made in-house, should generally be limited to a minimum. Studies show that for serious oxygen pick-up temperatures above 400°C are necessary (Baril *et al.*,

2010), which is confirmed by the current author's investigations. However, each exposition to air is a risk. In addition, titanium powder is basically burnable and explosive if it is heated to temperatures when fast oxidation occurs (above 400–500°C). As a protection gas, argon is suitable, whereas nitrogen, which is often used as an inert gas in other contexts, is picked up by titanium.

### 17.3.2 Feedstock production

During kneading binder and powder are heated to temperatures that depend on the specific binder, but usually range from 120 to 180°C. As stated above, temperatures below 200°C appear to be rather uncritical in terms of oxygen uptake, but in order to guarantee no additional oxygen pick-up kneading is done preferably under a protective atmosphere. For kneading, shear roll mixers, double planetary mixers, twin screw mixers or sigma blade kneaders are in use. Their choice depends on the binder and solids loading used. Generally, mixing under high shear is beneficial in order to provide a good homogeneity.

### 17.3.3 Injection molding

As stated previously, no special equipment is necessary. Process parameters depend on geometry and feedstock. No general advice differing from the processing of stainless steel powders can be given here. The main difference to steel feedstock is the larger powder particle size, as mentioned above, meaning that powder and binder separate more easily. On the other hand, the smaller difference in density between binder and titanium powder when compared to steel alleviates this effect. The most critical spot is the gate of the mold. The size should be generally greater than in the case of steel powder and great care should be taken concerning edges. A rounded and symmetrically constructed gate positioned in both mold parts appears to be favourable. This is also true for the runner.

It is very important to make sure that no titanium feedstock is contaminated by other materials, especially steel powder, because brittle or low-melting phases can occur in the microstructure, degrading the mechanical properties significantly. Thus, cleanliness of the machine is critically important. If the feedstock material is changed, it is necessary to clean carefully all parts of the injection molding machine in contact with the feedstock. Furthermore, attention should be paid with respect to the clearance of the back flow valve of the screw. It should be somewhat larger than the maximum size of the powder particle to avoid fretting. In the case of titanium powder, where typically particles smaller than 45 µm are used, a clearance of 0.1 mm between ring of the back flow valve and cylinder wall is usually a reasonable value.

### 17.3.4 Debinding

During thermal debinding the risk of contamination by carbon is rather high. Therefore, at that time the amount of binder should be as low as possible. This can be influenced by the ratio between first and second binder components. It should be ensured that the entire first component is removed by the first debinding stage (e.g. solvent or catalytic). This results in an open microporous structure, which facilitates the escape of the second component during thermal debinding without cracking of the part. The temperature of the thermal debinding should be as low as possible to minimise the pick-up of oxygen and carbon by titanium, as described in Section 17.2.2. Depending on the binder components used, a temperature between 400°C and 500°C is usually adequate. Furthermore, as stated above, some flow of argon gas can be used to support the transport of the binder gases out of the furnace (sweep gas). Using helium is also possible, but more expensive. Nitrogen should not be used, because it reacts with titanium. The debinding ends with heating to sintering temperature. In the case of a separate debinding furnace the parts are pre-sintered at temperatures around 900°C for about 1 h. After this treatment the parts are mechanically rigid, thus they can be handled further for final sintering.

### 17.3.5 Sintering

Sintering is the most critical production step with regard to contamination by oxygen. In addition, the resulting microstructure is defined by this process and so are the mechanical properties. Therefore great care has to be taken during sintering of titanium materials.

It is not possible to avoid oxygen pick-up completely, but it is possible to minimise it by providing a high vacuum during sintering of at least $10^{-2}$ Pa or better. It is also possible to sinter under an atmosphere of high-purity argon, but in this case the sintered density will be lower because of trapping gas in the pores. Furthermore, powder with low oxygen content should be used, for example, the extra low interstitials (ELI) variant of the Ti–6Al–4V alloy. The author's experience is that, if the whole process works well, the difference in oxygen between powder and sintered part can be limited to about 0.05 wt%. However, in practice values up to 0.1 wt% are possible, depending on initial oxygen content of the powder, number of samples in the furnace, and so on.

Choosing the right sintering parameters is always a compromise between low residual porosity and small grains (e.g. Obasi *et al.*, 2010). Generally, high sintering temperature and long hold time lead to higher density, resulting in better strength and ductility. On the other hand, grain coarsening will also be more important, affecting these properties detrimentally. Pure titanium and typical alloys like Ti–6Al–4V are sintered above the beta-transus temperature in the single phase beta region, which promotes grain growth. One should be aware that there is no chance to reduce the grain size of titanium-based materials without mechanical treatment. As Zhang *et al.* (2008) showed it can be beneficial to sinter at lower temperature for a longer time than to hold a higher temperature briefly. However, typical sintering temperature is around 1300°C combined with a holding time of 2 h. There are statistical tools helping to reduce the number of experiments for optimising the sintering parameters. Sidambe *et al.* (2009) show an example for using the Taguchi method in order to find the best combination of sintering temperature, time, heating rate and atmosphere for MIM of pure titanium and Ti–6Al–4V, respectively, for a given binder system. Another study of molding, binder and sintering parameters was performed by Shibo *et al.* (2006).

The choice of the sintering support is critical, because titanium reacts with nearly all materials. Most MIM titanium providers use $Y_2O_3$ or $ZrO_2$ or similar ceramics as supports or as the separating layer.

### 17.3.6 Further processing

The sintered parts can be surface treated by common polishing methods. Some care should be taken when applying etching techniques because of the surface porosity. Colouring methods like anodic oxidation can also be used as they are usually performed.

Hot isostatic pressing (HIP) can be applied using an argon atmosphere. Typical parameters are 850–915°C for 2 h at a pressure of 100–200 MPa. The density has to exceed 95% to ensure a closed porosity.

## 17.4 Mechanical properties

Because the mechanical properties are directly defined by the microstructure, it is important to understand the difference from wrought material. In Fig. 17.2 typical micrographs of wrought and of MIM-processed Ti–6Al–4V alloy are shown. Three obvious differences are visible.

1. The beta- and alpha-phase morphologies are different. The wrought material shows a globular structure consisting of equi-axed alpha grains with beta-phase at the grain boundary regions. By contrast, the MIM-processed material exhibits a lamellar structure with alpha–beta

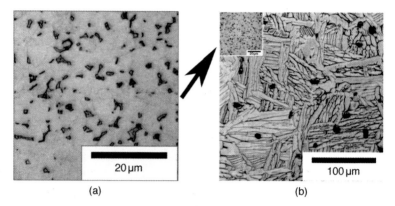

*17.2* Comparison of typical microstructures of (a) wrought Ti–6Al–4V and (b) the same material processed by MIM.

colonies. This structure is formed during cooling when passing the beta transus temperature.
2. The grain size is significantly different. As mentioned above this is due to sintering within the single-phase beta-region. In the case of the wrought material thermomechanical treatment including recrystallisation provides the small grains after the initial billet casting.
3. The MIM-processed alloy features small isolated pores with preferentially spherical shape. Depending on the sintering parameters the porosity can be influenced, but values below 2–3% have barely been realised yet.

All these differences lead to deviations of the mechanical properties from those of the wrought material. The lamellar structure is beneficial with regard to crack growth processes, but strength and ductility are generally improved by provision of a fine and globular microstructure (Lütjering, 1998). In the following section, typical tensile and fatigue properties of MIM-processed pure titanium and Ti–6Al–V4 are compared with standard values.

### 17.4.1 Tensile properties

Since 2011, a first ASTM standard for MIM-processed Ti–6Al–4V for medical applications has been in existence (ASTM F2885-11). However, usually the properties of the sintered parts are compared to the limits given by ASTM B348-02, although this standard is intended for wrought material. There are two grades applied for Ti–6Al–4V alloy: Grade 5 and Grade 23, differing primarily in the limit for oxygen. Whereas Grade 5 allows 0.2 wt% oxygen, Grade 23 (also denoted as ELI (extra low interstitials)) limits this value to 0.13 wt%. In Fig. 17.3 the tensile properties of MIM-processed Ti–

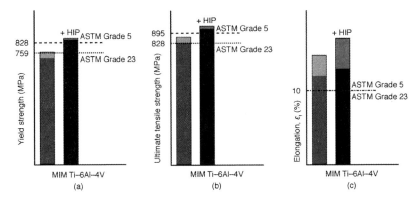

*17.3* Typical tensile properties of MIM-processed Ti–6Al–4V alloy in comparison to ASTM standards B348-02, Grade 5 and Grade 23. Additionally, the properties after an additional HIP process are shown.

6Al–4V are compared to this standard. The values are taken from three different published studies (Ferri *et al.*, 2010b; Niinomi *et al.*, 2007; Obasi *et al.*, 2010), revealing the status today, when the oxygen level is kept well below 0.3 wt%. In addition, a comparison is shown between as-sintered samples and samples exposed to an additional HIP process after sintering is done. The lighter regions of the bars represent the scatter of the values measured in the studies.

For powders, either Grade 23 pre-alloyed powder or blending of elementary and pre-alloyed powder was used. Furthermore, different binders and processing temperatures were applied in the studies. However, the results are very similar. By proper processing the tensile properties can fulfil the strength requirements for Grade 23. Closing of the porosity increases the strength so strongly that even Grade 5 requirements are met. In all cases the ductility of the specimens is significantly higher than the standards demand. This is a sign that the amount of interstitials is kept under certain limits, which are not yet precisely known. In two of the cited studies oxygen contents between 0.20 and 0.23 wt% were gained. It is also obvious that between the studies the scatter of elongation is significantly larger than that of strength. This again proves the difficulty and the need to control the processes very accurately with regard to oxygen and carbon pick-up. On the other hand, an internal investigation showed that the tensile properties do not deteriorate, if the oxygen equivalent according to equation [17.1] remains below around 0.4 wt%. As a matter of fact, in that study an oxygen equivalent of 0.37 wt% led to yield strength of 784 MPa, ultimate tensile strength of 901 MPa and elongation to fracture of 14.8%. In this case the oxygen content was 0.26 wt%. Figures. 17.4 and 17.5 show the results for tensile strength and plastic elongation in relation to the interstitial equivalent.

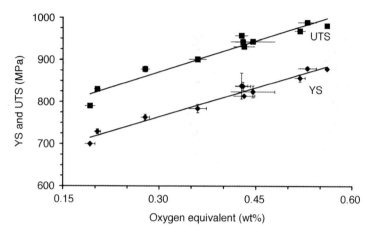

*17.4* Dependence of yield strength (YS) and ultimate tensile strength (UTS) on oxygen equivalent, measured on MIM-processed Ti–6Al–4V samples.

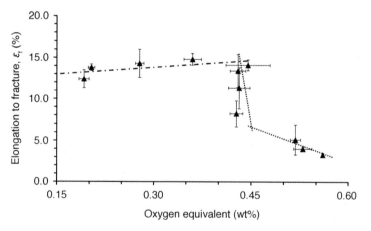

*17.5* Dependence of plastic elongation of MIM-processed Ti–6Al–4V on oxygen equivalent.

Baril (2010) evaluated the data from all studies on MIM of Ti–6Al–4V today with regard to oxygen equivalent and tensile properties. He concluded that $O_{eq}$ should not exceed 0.34 wt% and the porosity should be kept under 3% to match the mechanical requirements for surgical implants. This is in accordance with the present author's study if a certain security margin is considered.

However, a comprehensive study is still missing to confirm the impression that more oxygen than the maximum value of 0.20 wt% given by the ASTM standard is beneficial for the tensile properties of MIM-processed Ti–6Al–4V. In particular, a possible negative effect on, for example, stress corrosion cracking or fatigue properties has to be evaluated before an unconditional recommendation can be given.

Such an apparent independence of elongation within a certain range of interstitials as visible in Fig. 17.5 was not detected in the case of pure titanium as investigated by Kursaka *et al.* (1995). In their study an increase of oxygen content from 0.2 to 0.4 wt% meant a linear loss of ductility from 20% to 8% while UTS increased from 560 to 720 MPa. On the other hand, this study shows that even pure titanium is processable by MIM and good mechanical properties can be achieved if proper raw material is used.

Results from MIM processing of other alloys like, for example, Ti–6Al–7Nb exist too. One overview of mechanical properties is given by German (2009). However, it is rather difficult to compare these studies, because different powders, binders and processing parameters were used. Some samples were even additionally applied to a hot isostatic pressing process. Figures 17.6 and 17.7 display the large scatter of ultimate tensile strength and elongation resulting from these studies performed on unalloyed titanium, Ti–6Al–4V and Ti–6Al–7Nb. The diagrams make clear how important are proper processing and adequate powders for guaranteeing high quality and reproducible production. Well-balanced combinations of ultimate tensile strength and elongation in these studies were 640 MPa /20% for unalloyed titanium (Kursaka *et al.*, 1995) and 830 MPa /11% for

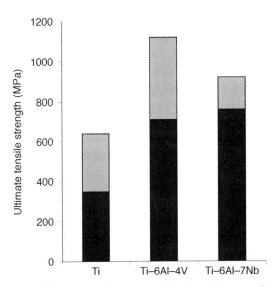

*17.6* Range of ultimate tensile strength according to experiments collected by German (2009) for unalloyed titanium, Ti–6Al–4V and Ti–6Al–7Nb. All samples were produced by MIM, but using different powders, binders and process parameters. Some samples were additionally subjected to a HIP process.

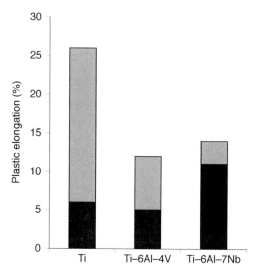

*17.7* Range of elongation according to the experiments collected by German (2009).

Ti–6Al–7Nb (Itoh *et al.*, 2009). As shown before in the case of Ti–6Al–4V even more than 14% elongation combined with 900 MPa ultimate tensile strength are possible.

The studies show that it is possible to receive highly ductile components with reasonable strength comparable to wrought material. However, if optimal tensile properties are required, closing of the pores by a HIP process is mandatory.

### 17.4.2 Fatigue properties

Studies on fatigue of MIM-processed titanium materials are very rare. In Ferri *et al.* (2009) and Niinomi *et al.* (2007) Ti–6Al–4V was investigated with regard to fatigue under bending and axial load, respectively. In both cases the endurance limit defined for $10^7$ cycles was between 350 and 400 MPa for the as-sintered samples. Muterlle *et al.* (2010) found similar fatigue resistance in a study about the effect of shot peening. Thus, the fatigue properties of MIM-processed titanium materials are better than those of cast, but inferior to those of wrought material. In the latter case there is a large scatter of values which can be found in the literature depending strongly on the actual microstructure. A range between 450 to 800 MPa can be assumed for wrought Ti–6Al–4V. The reason for the lower endurance limit of MIM Ti–6Al–4V alloy was studied in depth in (Ferri *et al.*, 2009; Ferri *et al.*, 2010a; Ferri *et al.*, 2010b). It is shown that, first, as typical for fatigue behaviour, the surface is essential. For example, by simple shot peening the endurance limit can be improved by around 50 MPa due to surface smoothing and introduction of compressive stress in the surface near region. Starting sites for crack growth are the pores and defects on the surface. Therefore, the fatigue behaviour can also be influenced by injection molding parameters and binder (Ferri *et al.*, 2010a).

Interestingly, closing of the residual porosity has a much greater effect on the tensile properties than on the fatigue properties. The best configuration investigated and exposed to an additional HIP process led to an endurance limit of 500 MPa, meaning the lower end of wrought material. On the other hand, the size of the microstructure turned out to be the most crucial parameter after surface properties. As pointed out above, the grain size of the MIM microstructure is typically much larger than that of wrought material.

In order to generate a finer microstructure of MIM Ti–6Al–4V, in Ferri *et al.* (2010b) 0.5 wt% of elementary boron powder was added to the alloy powder. During sintering needle-shaped TiB was formed, which is known as a grain refiner during casting. In fact, the addition of boron effects a drastic change of the microstructure as shown in Fig. 17.8. In order to identify the actual grains more clearly in addition to SEM images backscattered electrons (BSE) mode), EBSD measurements were made. Each gray tone represents one specific grain orientation. It is obvious that the microstructure of the boron-added alloy is completely different. Only few lamellar regions are visible and the grain size changed from around 150 μm to 20 μm.

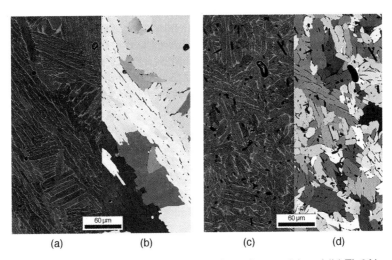

*17.8* Microstructures of MIM-processed specimens: (a) and (b) Ti–6Al–4V; (c) and (d) Ti–6Al–4V–0.5B. (a) and (c) images are made by scanning electron microscopy (SEM) in BSE mode. (b) and (d) images are evaluated by EBSD (electron backscatter diffraction), the same gray scale means the same grain orientation. The alpha phase is shown.

**432** Handbook of metal injection molding

In addition, the boron addition improves the sintering process resulting in a residual porosity of only 2.3%. The finer microstructure effects an increase in fatigue strength to 640 MPa, which is equivalent to that of high-performance wrought material. Yield stress and UTS amount to 787 MPa and 902 MPa, respectively, in connection with a plastic elongation of 11.8%. This means all mechanical requirements for Grade 23 are fulfilled by this material. This example shows that it is worthwhile to develop specific alloys optimised for MIM processing as has long been practiced in the case of materials intended to be processed by casting or forging.

## 17.5 Cost reduction

Although MIM of titanium is developed quite well today and excellent properties can be achieved, widespread application is still missing. Probably a significant cost reduction could change this situation. Many different activities are under way aiming at a reduction of the powder or component price. They can be divided into:

- more cost-efficient powder production techniques or replacement of the Kroll process;
- replacement of expensive powder by lower-cost powder.

### 17.5.1 Novel production techniques

A high portion of the overall costs for titanium is caused by the reduction of the ore, which is commonly performed by applying the Kroll process. This is a multi-step process in which first the oxide is converted into chloride, and then the chloride is reduced to metal. Several attempts have been made to replace this cost-intensive technique by other methods or by developing new powder production techniques (Froes *et al.*, 2007). Examples are described below.

*Plasma-quench process*

The raw material $TiCl_4$ is used, which is dissociated thermally by means of a plasma arc. By rapid quenching powder particles are formed. However, $TiCl_4$ is rather expensive and the process is hard to control because several reactions are involved.

*MHR, metal hydride reduction*

This technique (Froes, 1998) applies calcium hydride for the direct reduction of titanium oxide according to the chemical formula $TiO_2 + 2CaH_2 \rightarrow Ti$

+ 2CaO + 2H$_2$. The advantages are that only one step is needed, the powder is chloride-free and it is possible also to produce TiH$_2$, which can be used as a raw material, as mentioned in Section 17.5.2.

*Armstrong process*

Basically, this is a modification of the Hunter process, working with sodium: TiCl$_4$ + Na (molten) → Ti + NaCl. By a continuous process, production of both sponge and powder is possible. The drawback is again the rather high costs of the raw materials.

*FCC-Cambridge process*

The FCC (Fray–Farthing–Chen)-Cambridge process is an electrolytic process where a cathode pressed from TiO$_2$ pellets and a graphite anode are placed in a CaCl$_2$ bath. Oxygen ions diffuse from the cathode and CO$_2$ is formed at the anode: TiO$_2$ + C → Ti + CO$_2$. Sponge, which can be crushed to powder, can be made, but the cathode production is rather cost intensive. In addition, high energy consumption has to be considered. However, this technique is considered as being most suitable for providing a significant cost reduction compared to Kroll processing. By adding oxides of other chemical elements it is also possible to produce alloys.

To date, all processes still suffer from a high risk of impurity pick-up and the costs for large-scale production are not clear in all cases. But development continues and there is reasonable hope of producing lower cost powder of high quality in the future.

## 17.5.2 Powder blending

The flexibility of powder metallurgy opens the possibility to blend low-cost and expensive powders in order to reduce the costs of the raw material. For example, HDH powder of pure titanium can be mixed with gas-atomised alloyed powder or alloys can be made by blending of elemental powder, if pre-alloyed powder is not available or is too expensive for a given application. Powder blending is not only related to MIM but also to PM in general. Blending elemental powders for the PM production of alloys intended for subsequent conventional processing is the subject of many studies. Besides the reduction in costs, better and more homogeneous properties of the material compared to fabrication by ingot metallurgy can also be the reason for the application of the PM route and new alloys can be tailored rather easily (Liu *et al.*, 2006).

However, when using lower-cost powders such as HDH processed powder, it is important to bear in mind that this powder usually has a higher degree of contamination and interstitials than, for example, gas-atomised powder, as mentioned in Section 17.2.2. Thus, parts produced with such powders tend to possess rather low ductility. However, the blending with purer powder can be a good compromise between costs and mechanical properties, which is more than sufficient for many applications. Because the powders are available in different purities there is also a large range of possibilities for the ratio between price and achievable properties.

Alloys which are difficult to attain as pre-alloyed powders like Ti–6Al–7Nb or NiTi, both used in medicine, can be made by blended elemental powder metallurgy, as shown in different studies. Bolzoni et al. (2010) and Itoh et al. (2009) produced Ti–6Al–7Nb by the blended elemental approach in order to compare the mechanical results in relation to the powders used, meaning elemental or made from master alloy. It was shown that good properties including reasonable ductility can be achieved (ultimate tensile strength of 830 MPa and elongation of 11%), but in general the processing is more tricky than using pre-alloyed powder, which still leads to the best properties. In the case of NiTi, elemental powders are used because pre-alloyed powder is scarcely available and very expensive. Several studies are performed; some even under application of MIM, mostly aiming at the production of porous bone implants (e.g. Hu et al., 2008; Ismail et al., 2010). Furthermore, nearly dense samples were produced with ultimate tensile strength of 1000 MPa at a strain of 17% (Imgrund et al., 2008). In this case very fine pre-alloyed powder with $d_{50}$ of 11 µm was used.

Cost reduction is also one of the ideas behind using titanium hydride powder for PM processing. Sun et al. (2010) observed green part densities up to 90% using fine $TiH_2$ powder. They suggest that in addition to the lower costs its brittleness appears to be beneficial. At a pressure of 700 MPa they found crushing of the powder leading to better compaction. However, it is not clear whether using ductile powder could lead to a similar result due to plastic deformation. The dehydration process is combined with the sintering process providing clean particle surfaces, leading to excellent sintering conditions. In addition, hydrogen is regarded as a getter for the oxygen, thus, additional pick-up of oxygen can be eventually avoided. $TiH_2$ powders are also used in MIM studies and high sintering densities of 98% can be achieved (Carreño-Morelli et al., 2009). However, the purity of the available powders is still a concern and low ductility is usually the consequence.

There is much development in the field of finding lower-cost powders or more efficient technologies for production of powder and parts. MIM benefits from the general trend to use PM for the production of raw material, even for conventional processing, and most probably the next few years will bring significant improvement in this field.

## 17.6 Special applications

### 17.6.1 Medical application

MIM of titanium has been introduced to the field of medical devices and used for several years, mostly for the manufacture of handles for surgical instruments or other parts of endoscopic or other instruments and devices. Here the great freedom in geometry provided by injection molding is exploited in connection with the low weight and basic biocompatibility of titanium. For these applications, which are not high load bearing, the use of Ti Grade 4 feedstock is absolutely sufficient and using MIM means usually a significant cost advantage compared to machining or casting. The progress in processing of titanium alloys, as mentioned previously, during recent years has led also Ti–6Al–4V or Ti–6Al–7Nb components becoming available to withstand even high loads of 700 MPa or more. These alloys are suitable for MIM manufacturing of medical implants. Today, permanent implants made by MIM of Ti–6Al–4V are approved and available on the market, for example, the port system in Fig. 17.9 which is used for cancer therapy as a drug delivery device. An overview is given by Ebel (2008).

Using MIM for the production of implants immediately raises the question of biocompatibility. In fact, MIM is a very clean process compared to conventional manufacturing technologies. Only the possible binder residuals could represent potential candidates as toxic substances. Even if the single binder constituents may be harmless, such as paraffin or polyethylene, decomposition at high temperatures could generally lead to

*17.9* Commercial permanent implant (port system) produced by MIM of Ti–6Al–4V (courtesy of Tricumed Medizintechnik, Germany). Anodic oxidation enables the parts to be produced in different colours.

new toxic substances. However, these substances would be trapped in pores after sintering and would cover the surface. In the first instance, they would have no contact with the body and in the latter case, they could be removed if necessary. Actually, to the author's knowledge no study shows any toxicity of MIM fabricated implants. Furthermore, the direct comparison with machined alloys shows usually a better cell growth on the MIM surface, which is probably due to the rougher surface. However, scientific publications of comprehensive studies are still lacking. The great interest of the medical industry in MIM of titanium is manifested in the fact that the first standard on MIM of Ti–6Al–4V, ASTM F2885-11, focuses on medical applications. The possibility of low-cost manufacturing of anatomically shaped implants is one of the motivations for this development.

Metal injection molding for the fabrication of medical implants is also interesting because of the possibility of generating porous components. These are beneficial with regard to bone in-growth. Pore sizes between 50 and 500 μm are usually regarded as adequate, depending on a possible demand for vascularisation of the implant. Instead of applying a second process to coat the implant with a porous layer, as it is done with hip implants, the implant can be manufactured as porous in a single MIM process. Coarse powders and appropriate sintering parameters are used leading to high porosity up to 35%, for example, as shown by Oh *et al.* (2003) and Ebel *et al.* (2010). Alternatively, space-holders are added to the powder prior to feedstock production (e.g. Chen *et al.*, 2009). These space-holders can be dissolved during or prior to debinding, for example, NaCl can be dissolved by water. The latter technique is more complex, but allows a large range of possible and well-defined porosities to be realised. It should be noted that in both cases the injection molding process is more difficult compared to using standard feedstock. The large particles (powder or space-holders) influence the viscosity and even the diameter of gates, runners and so on must be taken into account. In biomedical applications the desired pore size can be 0.5 mm as mentioned above. This means that the powder or space-holder dimensions also have to be in this range.

Besides improved osseointegration, porous components also show a reduction in the elastic modulus. Typically the value of the elastic modulus is proportional to the relative density, as shown in Fig. 17.10. Thus, it is possible to adjust the stiffness to that of bone to avoid the so-called stress shielding effect. If the stiffness of the implant is higher than that of the bone, most of the load is focused on the implant. This leads to a degeneration of the bone, resulting in loosening of the implant. The elastic modulus of cortical bone is in the range 5–20 GPa. According to Fig. 17.10, MIM-processed Ti–6Al–4V possessing a porosity of around 35% shows a similar value. On the other hand, the strength of the material is reduced significantly, as Fig. 17.11 reveals.

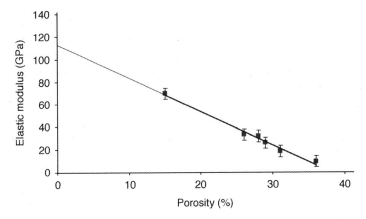

*17.10* Dependence of the elastic modulus of MIM-processed Ti–6Al–4V on porosity (Ebel *et al.*, 2010).

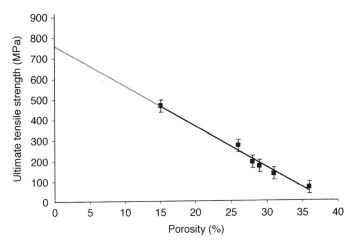

*17.11* Dependence of the ultimate tensile strength of MIM-processed Ti–6Al–4V on porosity (Ebel *et al.*, 2010).

By two-component (2C)-MIM it is even possible to realise implants with gradient porosity using two different feedstocks, one for a dense core and a second with space-holders in order to provide a porous surface. Such a technique is shown by Bram *et al.* (2010) where prototypes of a spine implant were made by 2C-MIM.

There are also studies on the combination of porous Ti–6Al–4V implants with hydroxyapatite (HA) made by MIM (e.g. Thian *et al.*, 2002). While the titanium skeleton provides mechanical stability, HA is responsible for the bioactivity. An open porosity around 50% was created by this technique and the pores were filled by HA. Such implants should be excellent in contact with bone.

As mentioned in Section 17.5.2 processing of NiTi by MIM is possible too (Bram *et al.*, 2002; Krone *et al.*, 2004; Schüller *et al.*, 2004). It is a shape memory alloy and is mostly used because of its pseudo-elasticity providing a very large region of elastic deformation, which can be exploited, for example, in the field of surgical instruments. It is also used for stents in blood vessels, where the shape memory effect accomplishes the dilation of the stent by thermal activation, when it is situated in the right position during operation. NiTi is difficult to process conventionally and MIM can offer new possibilities for shaping desired geometries. The alloy is also thought to be an ideal candidate as bone implant material because of its inherent low elastic modulus. On the other hand, the high Ni content is the subject of discussion, because Ni is an element with a high allergenic potential.

## 17.6.2 MIM of titanium aluminides

Titanium aluminides are known as novel intermetallic light-weight alloys for high-temperature applications. Owing to their low density, high specific strength and stiffness, and a temperature capability up to 800°C, titanium aluminides are ideal materials for rotating and oscillating parts in combustion engines and gas turbines. In contrast to conventional titanium alloys they consist exclusively of the intermetallic phases $\alpha_2$ and $\gamma$ and around half of the atoms are aluminium atoms. The composition of these alloys is commonly given in atomic percentage. Because they consist of intermetallic phases, titanium aluminides are very temperature stable and resistant against oxidation, but also rather brittle. At room temperature plastic elongation is around 0.2% and at 700°C this is only increased to about 2%. These properties mean that forging, machining and even casting of this alloy class are extremely demanding. Thus, at least for the production of relatively small and complex-shaped parts, MIM is regarded as a suitable and competitive production technique. On the other hand, because of the high affinity of titanium for oxygen and carbon, as discussed above, the MIM processing of fine TiAl powders is also rather sophisticated. Two main challenges have to be overcome (Ebel *et al.*, 2007).

1. Sintering temperatures have to be close to the solidus line because of low diffusivity, which is typical for materials that are stable at high temperatures. This means grain growth and loss (e.g. of Al) by evaporation can be severe problems. In addition, titanium aluminides are very sensitive to the oxygen level. Commonly, a value around 0.12 wt% is regarded as the maximum.
2. The microstructure of titanium aluminides is very sensitive to the exact path through the phase diagram during sintering and cooling. By means of different cooling profiles the mechanical properties can be influenced

strongly. Accurate process control is necessary. The phase diagram of the Ti–Al system is rather complicated and several phase changes during cooling have to be considered. In comparison Ti–6Al–4V is uncritical with regard to process variations because sintering takes place in the wide pure beta phase region.

There is little in the literature regarding MIM of TiAl (e.g. Gerling *et al.*, 2006; Katoh and Matsumoto, 1995; Kim *et al.*, 2007; Limberg *et al.*, 2009; Shimizu *et al.*, 2000; Terauchi *et al.*, 2001; Zhang *et al.*, 2009), but the latest studies show that good properties are achievable if the powders are processed adequately. The current state is that properties comparable to those of cast material can be obtained. Figure 17.12 shows typical results from tensile tests on samples from one of the present author's studies on Ti–45Al–5Nb–0.2B–0.2C (at%), measured at room temperature (RT) and 700°C. The samples were sintered at 1500°C under high vacuum ($< 10^{-2}$ Pa) for 2 h in a ceramic-free furnace and polished before tensile testing. Oxygen content was 0.12 wt% and the porosity only 0.5%. Table 17.2 compares the properties of the samples made by MIM with cast material. The MIM samples were surface polished. The cast samples were HIPed and polished.

In Zhang *et al.* (2009), MIM of Ti–45Al–8.5Nb–(W,B,Y) was investigated and ultimate tensile strength of 382 MPa and plastic elongation of 0.46% were determined. These values are also below those of cast material. Possible reasons the authors assume could be the porosity of 3.8% and the oxygen content of 0.18 wt%.

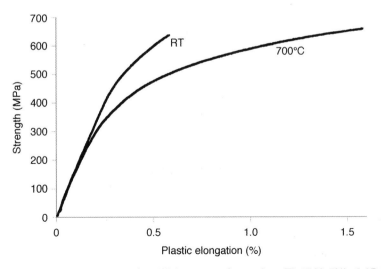

*17.12* Typical curves of tensile tests performed on Ti–45Al–5Nb–0.2B–0.2C (at%) processed by MIM, determined at room temperature (RT) and at 700°C.

*Table 17.2* Tensile properties of MIM processed Ti–45Al–5Nb–0.2B–0.2C (at%) alloy in comparison to cast material, determined at room temperature (RT) and 700°C. Both materials were surface polished; the cast samples were HIPed additionally

| Temperature | Sample | UTS (MPa) | $\varepsilon_f$ (%) |
| --- | --- | --- | --- |
| Room temperature | MIM | 630 | 0.2 |
|  | Cast | 745 | 0.1 |
| 700°C | MIM | 650 | 1.0 |
|  | Cast | 720 | 1.4 |

MIM of titanium aluminides is still young and the present results are quite encouraging. Thus, MIM appears to be a serious alternative to casting or forging.

## 17.7 Conclusion and future trends

Metal injection molding of titanium and titanium alloys is still a novel process and under continuous development. In particular, the high affinity to oxygen and carbon uptake complicates the whole production chain from powder production to sintering. On the other hand, if adequate powders and adapted binders are used and the sintering process is conducted with consideration of the special requirements of titanium, components with excellent properties can be produced, even fulfilling the demands of common standards. All materials and facilities necessary for such processing are readily on the market. Nevertheless, to date, there are no standard powders, feedstock or facilities for MIM of titanium. Even if the potential MIM provider can consult the large quantity of published information, some time and money for development should be budgeted before production can begin.

The growing interest in MIM of titanium alloys during the last years and the fact that a new ASTM standard is established now show that research is on the right track to provide MIM suppliers with missing data. Focus will be laid now and in future on:

- determination and optimisation of fatigue properties;
- development of titanium alloys adapted to the special requirements of MIM;
- development and application of new technologies for production of lower-cost powders;
- provision of standard binder systems and feedstock;
- creation of MIM standards for application in automotive and aerospace fields.

There is a great opportunity to participate in the emerging market of MIM of titanium and titanium alloys, if MIM suppliers and users do not hesitate for too long, but become active partners of researchers now. Cooperation between industry and research is basically the best prerequisite for fast and purposeful development.

## 17.8 Sources of further information

A good overview of different titanium alloys, their properties and special challenges is given by Leyens and Peters (2003).

In order to start with MIM of titanium it is helpful to read review articles, such as those written by German (2009), Baril (2010) and Ebel (2008). After that, for deeper understanding, the references given in those articles should be followed.

Information about facility suppliers, case studies and basic properties can be received via www.empa.com, www.mpif.org, www.mimaweb.org, www.jpma.gr.jp and mim-experten.de. However, the sections about titanium are usually small. On the other hand, contacting current research groups or MIM of titanium suppliers will be usually be very helpful, because there is great interest in distributing this technology further.

A visit to a PM congresses is also recommended; these are mostly organised by the European Powder Metallurgy Association (EPMA) or the Metal Powders Industries Federation (MPIF). Today, often a special session about MIM of titanium is offered and the current activities and developments in this field can be followed. Many of the references cited in this chapter are based on talks at such congresses.

## 17.9 References

Aller, A J and Losada, A (1990), 'Rotating atomization processes of reactive and refractory alloys', *Metal Powder Report,* 45(1), 51–55.

Baril, E, Lefebvre, L P, and Thomas Y (2010), 'Interstitals sources and control in titanium P/M processes', in *PM2010 Proceedings Volume 4,* PM2010 World Congress and Exhibition, Florence, 10–14 October 2010, European Powder Metallurgy Association, Shrewsbury, UK, pp. 219–226.

Baril, E (2010), 'Titanium and titanium alloy powder injection moulding: Matching application requirements', *PIM International,* 4(4), 22–32.

Bolzoni, L, Esteban, P G, Ruiz-Navas, E M, and Gordo, E (2010), 'Biomedical Ti–6Al–7Nb titanium alloy produced by PM techniques', in *PM2010 Proceedings Volume 4,* PM2010 World Congress and Exhibition, Florence, 10–14 October 2010, European Powder Metallurgy Association, Shrewsbury, UK, pp. 805–812.

Bram, M, Ahmad-Khanlou, A, Heckmann, A, Fuchs, B, Buchkremer, H P, and

Stöver, D (2002), 'Powder metallurgical fabrication processes for NiTi shape memory alloy parts', *Materials Science and Engineering A*, 337, 254–263.

Bram, M, Köhl, M, Barbosa, A P C, Schiefer, H, Buchkremer, H P, and Stöver, D (2010), 'Powder metallurgical production and biomedical application of porous Ti implants', in *PM2010 Proceedings Volume 4*, PM2010 World Congress and Exhibition, Florence, 10–14 October 2010, European Powder Metallurgy Association, Shrewsbury, UK, pp. 689–696.

Carreño-Morelli, E, Krstev, W, Romeira, B, Rodriguez-Arbaizar, M, Bidaux, J-E, and Zachmann, S (2009), 'Powder injection moulding of titanium from $TiH_2$ powders', in *Euro PM2009 Proceedings*, Euro PM2009 Congress and Exhibition, Copenhagen, 12–14 October 2009, European Powder Metallurgy Association, Shrewsbury, UK.

Chen, L, Li, T, Li, Y, He, H, and Hu, Y (2009), 'Porous titanium implants fabricated by metal injection molding', *Transactions of the Nonferrous Metals Society of China*, 19, 1174–1179.

Conrad, H (1966), 'The rate controlling mechanism during yielding and flow of titanium at temperatures below $0.4\ T_M$', *Acta Metallica*, 14, 1631–1633.

Ebel, T, Limberg, W, Gerling, R, Stutz, L, and Bormann, R (2007), 'Optimisation of the sintering atmosphere for metal injection moulding of gamma TiAl alloy powder', in *EURO PM 2007, Proceedings of the International Powder Metallurgy Congress and Exhibition, Volume 2*, Toulouse, 15–17 October 2007, European Powder Metallurgy Association, Shrewsbury, UK, pp. 203–208.

Ebel, T (2008), 'Titanium and titanium alloys for medical applications: opportunities and challenges', *PIM International*, 2(2), 21–30.

Ebel, T, Akaichi, H, Ferri, O M, and Dahms, M (2010), 'MIM fabrication of porous Ti–6Al–4V components for biomedical applications', in *PM2010 Proceedings Volume 4*, PM2010 World Congress and Exhibition, Florence, 10–14 October 2010, European Powder Metallurgy Association, Shrewsbury, UK, pp. 797–804.

Evans W J (1998), 'Optimising mechanical properties in alpha + beta titanium alloys', *Materials Science and Engineering A*, 243(1–2), 89–96.

Ferri, O M, Ebel, T, and Bormann, R (2009), 'High cycle fatigue behaviour of Ti–6Al–4V fabricated by metal injection moulding technology', *Materials Science and Engineering A*, 504, 107–113.

Ferri, O M, Ebel, T, and Bormann, R (2010a), 'Influence of surface quality and porosity on fatigue behaviour of Ti–6Al–4V components processed by MIM', *Materials Science and Engineering A*, 527, 1800–1805.

Ferri, O M, Ebel, T, and Bormann R (2010b), 'Substantial improvement of fatigue behaviour of Ti-6Al-4V alloy processed by MIM using boron microalloying', in *PM2010 Proceedings Volume 4*, PM2010 World Congress and Exhibition, Florence, 10–14 October 2010, European Powder Metallurgy Association, Shrewsbury, UK, pp. 323–330.

Froes, F H (1998), 'The production of low-cost titanium powders', *JOM*, September, 41–43.

Froes, F H, Gungor, M N, and Imam, M A (2007), 'Cost-affordable titanium: The component fabrication perspective', *JOM*, June, 28–31.

Gerling, R, Aust, E, Limberg, W, Pfuff, M, and Schimansky, F P (2006), 'Metal injection moulding of gamma titanium aluminide alloy powder', *Materials Science and Engineering A*, 423, 262–268.

German, R (2009), 'Titanium powder injection moulding: A review of the current status of materials, processing, properties and applications', *PIM International*, 3(4), 21–37.

Gopienko, V G and Neikov, O D (2009), 'Production of titanium and titanium alloy powder', in Neikov, O D, Naboychenko, S S, Murashova, I V, Gopienko, V G, Frishberg, I V and Lotsko, D V (eds), *Handbook of Non-Ferrous Metal Powders, Technologies and Application*, Elsevier.

Hohmann, M and Jönsson, S (1990), 'Modern systems for production of high quality metal alloy powder', *Vacuum*, 41(7–9), 2173–2176.

Hu, G, Zhang, L, Fan, Y, and Li, Y (2008), 'Fabrication of high porous NiTi shape memory alloy by metal injection molding', *Journal of Materials Processing Technology*, 206, 395–399.

Imgrund, P, Petzoldt, F, and Friederici, V (2008), 'μMIM: Making the most of NiTi', *Metal Powder Report*, May, 21–24.

Ismail, M H, Sidambe, A T, Figueroa, I A, Davies, H A, and Todd, I (2010), 'Effect of powder loading on rheology and dimensional variability of porous, pseudo-elastic NiTi alloy produced by metal injection moulding (MIM) using a partly water soluble binder system', in *PM2010 Proceedings Volume 4*, PM2010 World Congress and Exhibition, Florence, 10–14 October 2010, European Powder Metallurgy Association, Shrewsbury, UK, pp. 347–354.

Itoh, Y, Miura, H, Uematsu, T, Sato, K, Niinomi, M, and Ozawa, T (2009), 'The commercial potential of MIM titanium alloy', *Metal Powder Report*, May, 17–20.

Katoh, K and Matsumoto, A (1995), 'Powder metallurgy of Ti-Al intermetallic compounds by injection moulding', in *Proceedings of the 6th Symposium on High-Performance Materials for Severe Environments*, Tokyo, 20–21 November 1995, (partly in Japanese) pp. 49–55.

Kim, Y C, Lee, S, Ahn, S, and Kim, N J (2007), 'Application of metal injection molding process to fabrication of bulk parts of TiAl intermetallics', *Journal of Materials Science*, 42, 2048–2053.

Krone, L, Schüller, E, Bram, M, Hamed, O, Buchkremer, H-P, and Stöver, D (2004), 'Mechanical behaviour of NiTi parts prepared by powder metallurgical methods', *Materials Science and Engineering A*, 378, 185–190.

Kursaka, K, Kohno, T, Kondo, T, and Horata, A (1995), 'Tensile behaviour of sintered titanium by MIM process', *Journal of the Japanese Society of Powders and Powder Metallurgy*, 42, 383–387.

Leyens, C and Peters, M (eds) (2003), *Titanium and Titanium Alloys*, Wiley–VCH Weinheim, Germany.

Limberg, W, Ebel, T, Schimansky, F-P, Hoppe, R, Oehring M, and Pyczak F (2009), 'Metal injection moulding (MIM) of titanium-aluminides', in *Euro PM2009 Proceedings*, Euro PM2009 Congress and Exhibition, Copenhagen, 12–14 October 2009, European Powder Metallurgy Association, Shrewsbury, UK, pp. 47–52.

Liu, Y, Chen, L F, Tang, H P, Liu, C T, Liu, B, and Huang, B Y (2006), 'Design of powder metallurgy titanium alloys and composites', *Materials Science and Engineering A*, 418, 25–35.

Lütjering, G (1998), 'Influence of processing on microstructure and mechanical properties of (α+β) titanium alloys', *Materials Science and Engineering A*, 243 (1–2), 32–45.

McCracken, C G, Robinson, J W, and Motchenbacher, C A (2009), 'Manufacture of HDH low oxygen titanium–6aluminium–4vanadium (Ti–6-4) powder incorporating a novel powder de-oxidation step', in *Euro PM2009 Proceedings*, Euro PM2009 Congress and Exhibition, Copenhagen, 12–14 October 2009, European Powder Metallurgy Association, Shrewsbury, UK.

McCracken, C G, Barbis, D P, and Deeter, R C (2010), 'Key titanium powder characteristics manufactured using the hydride-dehydride (HDH) process', in *PM2010 Proceedings Volume 1*, PM2010 World Congress and Exhibition, Florence, 10–14 October 2010, European Powder Metallurgy Association, Shrewsbury, UK, pp. 71–77.

Muterlle, P V, Molinari, A, Perina, M, and Marconi, P (2010), 'Influence of shot peening on tensile and high cycle fatigue properties of Ti6Al4V alloy produced by MIM', in *PM2010 Proceedings Volume 4*, PM2010 World Congress and Exhibition, Florence, 10–14 October 2010, European Powder Metallurgy Association, Shrewsbury, UK, pp. 791–796.

Niinomi, M, Akahori, T, Nakai, M, Ohnaka, K, Itoh, Y, Sato, K, and Ozawa, T (2007), 'Mechanical properties of α+β type titanium alloys fabricated by metal injection molding with targeting biomedical applications', in Gungor, M N, Imam, M A, and Froes, F H (eds), *Innovations in Titanium Technology*, 2007 TMS Annual Meeting and Exhibition, Orlando, 25 February–1 March 2007, Wiley, pp. 209–217.

Nyberg, E, Miller, M, Simmons, K, and Weil, K S (2005), 'Microstructure and mechanical properties of titanium components fabricated by a new powder injection molding technique', *Materials Science and Engineering C*, 25, 336–342.

Obasi, G C, Ferri, O M, Ebel, T, and Bormann, R (2010), 'Influence of processing parameters on mechanical properties of Ti–6Al–4V alloy fabricated by MIM', *Materials Science and Engineering A*, 527(16–17), 3929–3935.

Oh, I-H, Nomura, N, Masahashi, N, and Hanada, S (2003), 'Mechanical properties of porous titanium compacts prepared by powder sintering', *Scripta Materialia*, 49, 1197–1202.

Schüller, E, Bram, M, Buchkremer, H P, and Stöver, D (2004), 'Phase transformation temperatures for NiTi alloys prepared by powder metallurgical processes', *Materials Science and Engineering A*, 378, 165–169.

Shibo, G, Xuanhui, Q, Xinbo, H, Ting, Z, and Bohua, D (2006), 'Powder injection molding of Ti–6Al–4V alloy', *Journal of Materials Processing Technology*, 173, 310–314.

Shimizu, T, Kitajima, A, and Sano, T (2000), 'Supercritical debinding and its application to PIM of TiAl intermetallic compounds', in Kosuge, K and Nagai, H (eds), *Proceedings of the 2000 Powder Metallurgy World Congress*, Japan Society of Powder and Powder Metallurgy, Kyoto, Japan, pp. 292–295.

Shimizu, T, Kitazima, A, Nose, M, Fuchizawa, S, and Sano, T (2001), 'Production of large size parts by MIM process', *Journal of Materials Processing Technology*, 119, 199–202.

Sidambe, A T, Figueroa, I A, Hamilton, H, and Todd, I (2009), 'Sintering study of CP–Ti and Ti6V4Al metal injection moulding parts using Taguchi method', in

*Euro PM2009 Proceedings*, Euro PM2009 Congress and Exhibition, Copenhagen, 12–14 October 2009, European Powder Metallurgy Association, Shrewsbury, UK.

Sun, P, Wang, H, Lefler, M, Fang, Z Z, Lei, T, Fang, S, Tian, W, and Li, H (2010), 'Sintering of TiH2 – a new approach for powder metallurgy titanium', in *PM2010 Proceedings Volume 4*, PM2010 World Congress and Exhibition, Florence, 10–14 October 2010, European Powder Metallurgy Association, Shrewsbury, UK, pp. 227–234.

Terauchi, S, Teraoka, T, Shinkuma, T, Sugimoto, T, and Ahida, Y (2001), in Kneringer, G, Rödhammer, P, and Wildner, H (eds), *Proceedings of the 15th International Plansee Seminar 2000*, Plansee Holding AG, Reutte, Austria, pp. 610–624.

Thian, E S, Loh, N H, Khor, K A, and Tor, S B (2002), 'Microstructures and mechanical properties of powder injection molded Ti–6Al–4V/HA powder', *Biomaterials*, 23, 2927–2938.

Thomas, Y and Baril, E (2010), 'Benefits of supercritical $CO_2$ debinding for titanium powder injection moulding?', in *PM2010 Proceedings Volume 4*, PM2010 World Congress and Exhibition, Florence, 10–14 October 2010, European Powder Metallurgy Association, Shrewsbury, UK, pp. 315–322.

Zhang, H, He, X, Qu, X, and Zhao, L (2009), 'Microstructure and mechanical properties of high Nb containing TiAl alloy parts fabricated by metal injection molding', *Materials Science and Engineering A*, 526, 31–37.

Zhang, R, Kruszewski, J, and Lo, J (2008), 'A study of the effects of sintering parameters on the microstructure and properties of PIM Ti6Al4V alloy', *PIM International*, 2(2), 74–78.

Zwicker, U (1974), 'Titan und Titanlegierungen' (in German), in *Reine und angewandte Metallkunde in Einzeldarstellung*, vol. 21, Springer-Verlag, Berlin/Heidelberg/New York.

# 18
Metal injection molding (MIM) of thermal management materials in microelectronics

J. L. JOHNSON, ATI Firth Sterling, USA

**Abstract:** This chapter begins with an overview of heat dissipation in electronics and the need for thermal management materials. Methods of measuring thermal properties are reviewed. The characteristics and preparation of suitable powders for MIM Cu, W–Cu, and Mo–Cu are discussed. Examples of binders and mixing techniques for preparing injection molding feedstock from these powders are described. Specific issues with molding Cu, W–Cu, and Mo–Cu feedstocks into heat sink components are discussed. Processing conditions for debinding and sintering or infiltrating such components are described with a special focus on densification and control of the oxygen content, microstructure, and dimensions. Factors affecting the thermal properties, such as impurities and porosity, are discussed. Examples of MIM heat sink components are provided.

**Key words:** thermal management, metal injection molding, copper, tungsten–copper, molybdenum–copper.

## 18.1 Introduction

Metal injection molding (MIM) can provide unique thermal management solutions to dissipate heat from microelectronic devices. Heat sinks with increasing shape complexity are being designed for more efficient heat transfer. MIM allows the necessary design freedom and provides a cost-effective means of fabricating the high quantities of heat sinks needed for microelectronics. This chapter reviews the processing conditions for MIM processing of specific materials used for thermal management and provides examples of MIM heat sink components.

## 18.2 Heat dissipation in microelectronics

The increasing power requirements and decreasing size of high-performance microprocessors present heat dissipation challenges in microelectronic package designs. High thermal conductivity materials are required, and in some cases they must also have a low thermal expansion coefficient. New heat sink designs try to handle increasing thermal loads by using aerodynamic fins, heat pipes, and microchannels. Components with these structures can be readily produced in large quantities by MIM from well-known thermal management materials, including Cu, W–Cu, and Mo–Cu.

### 18.2.1 Heat sink design

The most widely used method for extracting heat from a microprocessor is to connect it to a heat sink, which might be air-cooled, either actively or passively. The active designs require fans or pumps to circulate fluids for heat extraction, often leading to heat generation, power drain, and new failure modes. Passive designs have fins for increased surface area to increase the amount of heat that they can release to the ambient air. Both cases require high thermal conductivity materials to transfer heat from the device to the fluid stream.

Many early heat sinks were extruded from Al alloys. Extrusion is a highly automated, low-cost, high-volume process, but it inherently limits the fin geometry. Further, the alloying elements added to improve extrudability lower the thermal conductivity. Machining of extruded Al alloys or diecasting can produce fins with a higher height to gap ratio, but at a higher cost. Likewise folded-fin heat sinks can give a two-fold improvement over the performance of extruded heat sinks, but also with an increased cost (Viswanath *et al.*, 2000).

Metal injection molding enables greater flexibility in design over extrusion and die casting. For example, round pin fins are easy to produce by MIM, but machining from an extruded Al alloy is limited to square fins. The performance difference between a heat sink with round pins fabricated by MIM and one machined with square pins is shown in Fig. 18.1. Other fin geometries are also possible with MIM, such as tapered pin fins and aerofoil-shaped fins, as shown in Fig. 18.2. Heat sinks are often larger than typical MIM parts. Heat sinks up to 50 mm high by 50 mm wide by 50 mm long are most amenable to MIM processing.

Further increases in heat dissipation capability are possible by integrating a heat pipe into the heat sink. These phase-change cooling devices work by vaporizing a fluid in a hot zone, which absorbs heat. The vaporized fluid travels to a cold zone where it condenses to release heat and the liquid is wicked back to the hot zone. The effective thermal conductivities of these

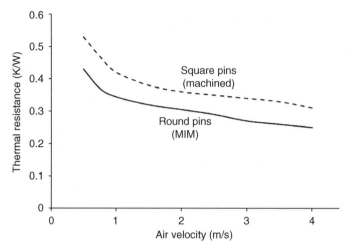

*18.1* Comparison of the thermal performance of a machined heat sink with square pins and a MIM heat sink with round pins.

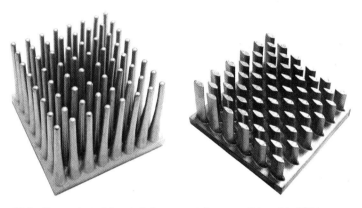

*18.2* Examples of heat sink geometries possible with MIM.

heat transfer devices can be much higher than that of Al or Cu (Swanson, 2000). Powder metallurgy techniques, including MIM, can fabricate a porous wicking structure into a heat sink to produce an integrated heat pipe (Shropshire *et al.*, 2003). Heat pipes can be miniaturized by producing them from wickless, non-circular microchannels (Babin *et al.*, 1990), which can also be fabricated by MIM.

Heat sinks are designed to remove heat from electronic packages, but packages must be designed to extract the heat directly from the semiconductor device or a heat spreader can be used to transfer heat from the back of the device to the heat sink. In both cases, high thermal conductivity is desirable, but a thermal expansion coefficient of 4–7 ppm is

# MIM of thermal management materials in microelectronics 449

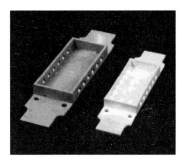

*18.3* Examples of electronic package geometries possible with MIM.

also required for compatibility with silicon to minimize thermal fatigue from off-and-on device operation.

In addition to thermal management, electronic packages also provide structural support for the semiconductor device, protection from the environment, and interconnections for power and signals (Tummala, 1991). To meet these requirements, electronic packages are usually thin-walled enclosures with holes for interconnections that can be hermetically sealed with glass. The Fe–Ni–Co alloy F-15, also known by its trademark name, Kovar® (Carpenter Technology Co., Reading, PA), is commonly used for electronic packages due to its low thermal expansion coefficient and good glass-to-metal sealing capability, but it suffers from a low thermal conductivity (Williams, 1991).

Sintered packages, such as those shown in Fig. 18.3, must have a closed pore structure to provide hermetic sealing, thus sintered densities must be over about 92% of theoretical. Dimensions typically range from 8 to 125 mm with wall thicknesses of 1–3 mm. Tolerances are very exact with typical values of $+/-25\,\mu m$ on a linear dimension (German *et al.*, 1994). Heat spreaders have similar sizes and tolerances, but lack holes. Heat spreaders or packages that are directly bonded to a semiconductor device have typical flatness requirements of $+/-0.025\,mm$ per 25.4 mm (Hens *et al.*, 1994; Ludvik *et al.*, 1991). Some surfaces can be finish ground, but others are not easily corrected for flatness. Packages are often assembled by brazing, which requires the MIM surface to be plated. The surface porosity of MIM parts can cause plating difficulties, which can be overcome by special techniques (Ludvik *et al.*, 1991).

## 18.2.2 Material selection

Copper is often used for applications that require higher thermal conductivities than possible with Al. Its lower thermal expansion coefficient also helps reduce thermal fatigue problems. Copper is more difficult to

*Table 18.1* Typical properties of heat sink materials

| Property | Cu | W–15Cu | Mo–18Cu |
|---|---|---|---|
| Density (g/cm$^3$) | 8.5 | 15.6–16.2 | 9.3–9.5 |
| Thermal conductivity (W/(m K)) | 320–340 | 180–190 | 140–160 |
| Thermal expansion coefficient (ppm/K) | 17.0 | 7.2 | 7.0 |

Source: Johnson and Tan (2004).

extrude, stamp, machine, or cast than aluminum, but it is better suited to processing with powder metallurgy techniques, including MIM.

Although the thermal expansion coefficient of Cu is lower than that of Al, it is still considerably higher than those of the silicon chip or ceramic substrate, providing challenges to mounting the heat sink. Accordingly, composite materials, such as W–Cu and Mo–Cu, which combine high thermal conductivity with a low thermal expansion coefficient, are more suitable for packaging applications. The components of these composites have widely differing melting temperatures and do not alloy, so they are usually processed using powder metallurgy techniques. Traditional methods of processing W–Cu and Mo–Cu include sheet-rolling operations and Cu infiltration of pressed and sintered W or Mo compacts (Kny, 1989; Kothari, 1982). These techniques are generally used to fabricate electrical contact materials, which have relatively high volume fractions of Cu and low degrees of shape complexity. MIM can be used to produce the porous skeleton or, in the case of W–Cu and Mo–Cu, mixed powders can be injection molded and sintered to meet heat sink design requirements.

Tungsten with 10–20 wt% Cu is a common heat sink composition. Molybdenum with similar Cu contents has a lower thermal conductivity, but its lower density makes it attractive for applications where weight is important. Typical properties of MIM Cu, W–15Cu, and Mo–18Cu are given in Table 18.1. These Cu contents for W–Cu and Mo–Cu maximize the thermal conductivity for a thermal expansion coefficient that is acceptable for most packaging applications. As discussed in later sections, theoretical thermal conductivities are not usually obtained.

## 18.2.3 Thermal property measurement

Measurement of thermal performance is not trivial. Thermal conductivity $\lambda$ is most commonly determined from the thermal diffusivity $\alpha$ as measured by the laser flash method (ASTM E1461) according to the relationship

$$\lambda = \alpha C_p \rho \qquad [18.1]$$

where $C_p$ is the specific heat and $\rho$ is the sample density. This method requires disk-shaped samples and requires great care to minimize measurement variability.

Electrical conductivity is much easier to measure and does not depend on the geometry of the sample. For Cu and other elemental metals, the electrical conductivity $\sigma$ can be measured by the four-point probe method and converted to thermal conductivity $\lambda$ using the Wiedemann–Franz relationship (Parrot and Stuckes, 1975)

$$\lambda = L\sigma T \qquad [18.2]$$

where $L$ is the Lorenz number ($2.28 \times 10^{-8}$ V$^2$/K$^2$ for copper at 25°C), and $T$ is the absolute temperature in Kelvin. Factors, such as porosity and impurities, that lower the electrical conductivity also proportionally lower the thermal conductivity. This method is not suitable for composites, such as W–Cu or Mo–Cu, whose components have different Lorenz numbers, or electrical insulators, which conduct heat via phonons.

While the thermal conductivity is an important factor, the overall thermal performance of heat sinks is better characterized by the thermal resistance. Thermal resistance in K/W is a measure of the temperature difference across the boundary between the heat sink and the ambient air as a certain amount of heat energy is passed through it. It takes into account the overall design as well as the material thermal conductivity and is affected by external factors such as air temperature and air flow. More efficient heat sinks have lower thermal resistances. The thermal resistance is often calculated in the design phase and measured for verification during testing of prototypes.

## 18.3 Copper

Unalloyed Cu powders for applications associated with heat dissipation in electronic systems have been metal injection molded for several years (Chan et al., 2005; German and Johnson, 2007; Johnson et al., 2003, 2005b; LaSalle et al., 2003; Moore et al., 1995; Shropshire et al., 2002, 2003; Terpstra et al., 1994, 1996; Uraoka et al., 1990). Many types of Cu powders are commercially available and have proven adequate for molding with conventional binder systems. The primary challenge is attaining high sintered densities and high conductivities, which requires sintering to nearly full density while reducing oxygen and other impurities to a low level. Copper is especially susceptible to hydrogen-induced swelling during sintering (Chan et al., 2005; German and Johnson, 2007; Johnson et al., 2005b; Shropshire et al., 2002; Sweet et al., 1992). Oxides in closed pores can react with hydrogen to generate trapped water vapor, which increases the gas pressure in the pores, leading to pore swelling, inhibited densification, and component blistering. Accordingly, MIM of Cu requires attention to

Table 18.2 Cu powder characteristics

| Production method | Oxide reduced | Water atomized | Gas atomized | Jet milled |
|---|---|---|---|---|
| Oxygen content (wt%) | 0.332 | 0.223 | 0.379 | 0.214 |
| Particle size distribution | | | | |
| $D_{10}$ (μm) | 5.9 | 7.8 | 4.1 | 4.7 |
| $D_{50}$ (μm) | 11 | 13 | 8.2 | 7.9 |
| $D_{90}$ (μm) | 17 | 23 | 13 | 12 |
| Pycnometer density* (g/cm$^3$) | 8.62 | 8.48 | 8.84 | 8.75 |
| Apparent density (g/cm$^3$) | 2.8 | 3.6 | 3.9 | 3.4 |
| % of pycnometer | 32% | 42% | 44% | 39% |
| Tap density (g/cm$^3$) | 3.6 | 4.4 | 4.2 | 4.3 |
| % of pycnometer | 42% | 52% | 47% | 48% |

* Theoretical density is 8.96 g/cm$^3$.

oxygen control in the initial powder and via reduction during sintering. The key requirements for processing high thermal conductivity copper components via MIM are described below.

### 18.3.1 Powders

Copper powders are produced by many processes including chemical precipitation, electrolytic deposition, oxide reduction, water atomization, gas atomization, and jet milling. Accordingly, copper powders are commercially available in a wide range of particle shapes and sizes. Electrolytic and chemical powders exhibit poor packing and poor rheology in molding, so they have been largely unsuccessful for MIM (Wada et al., 1997). Characteristics of examples of the other types are summarized in Table 18.2. Representative scanning electron micrographs are given in Fig. 18.4. These particles all have similar particle sizes but different morphologies. Typical purities reported by the manufacturers are about 99.85 wt%; however, oxygen contents can range up to 0.76 wt%. Powders are usually shipped containing desiccant and proper powder storage is essential to avoid oxidation between purchase and use.

While all of the Cu powders in Table 18.2 are suitable for MIM without further treatment, none is clearly the best choice. The jet-milled Cu powder has the lowest impurity level. The gas-atomized powder has the highest tap density, which generally translates into higher solids loadings to promote ease of MIM processing. The oxide-reduced and water-atomized powders are lower cost, which is a critical concern when competing against wrought and cast copper. Finer powders than those in Table 18.2 are more expensive, more difficult to mold, and provide little sintering advantage. Costs do not

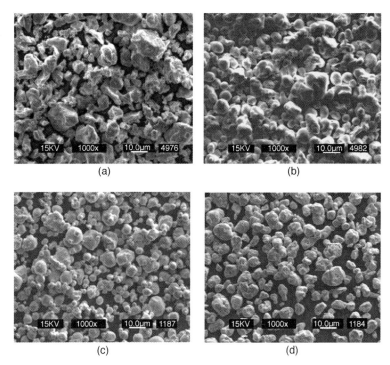

*18.4* Scanning electron micrographs of (a) 11 μm oxide-reduced; (b) 13 μm water-atomized; (c) 8 μm gas-atomized, and (d) 8 μm jet-milled copper powders.

substantially decrease with coarser powders, but give concerns with dimensional control and sintering behavior.

### 18.3.2 Feedstock preparation

Wax–polymer binders are generally compatible with Cu powders, but Cu feedstock is not available with the BASF Catamold® (Ludwigshafen, Germany) binder since the nitric acid used in debinding dissolves Cu. As an alternative, an agar-based binder system can be used (LaSalle *et al.*, 2003). The optimal solids loadings in a MIM Cu feedstock depends on the powder morphology and packing characteristics, so it can vary widely depending on the powder selection. Typical solids loadings with a polymer–wax binder are 48–52 vol% for oxide-reduced powders (Chan *et al.*, 2005; Johnson *et al.*, 2005b), 52–56 vol% for water-atomized and jet-milled powders (Johnson *et al.*, 2005b), and 65–70 vol% for gas-atomized powders ( Johnson *et al.*, 2005b; Moballegh *et al.*, 2005; Moore *et al.*, 1995). These solids loadings translate into average tooling scale-up factors of 1.24 for oxide-reduced

*Table 18.3* Examples of MIM Cu feedstock preparation

| Binder | Powder type | Particle size (μm) | Solids loading (vol%) | Mixing technique | Reference |
|---|---|---|---|---|---|
| 45% acrylate co-polymer 23% polypropylene 23% wax 9% dibutyl phthalate | Not given | 11 | 56 | Not given | Uraoka et al. (1990) |
| Paraffin-based with 2 other components | Gas atomized | 15–45 | 70 | Twin-screw | Moore et al. (1995) |
| 12% agar 12% glucose 76% deionized water + biocides | Not given | 22 | 70 | Sigma | LaSalle et al. (2003) |
| 55% paraffin wax 40% polypropylene 5% stearic acid | Water atomized | 13 | 52 | Not given | Johnson et al. (2003, 2005b) |
| 50% polypropylene 35% paraffin wax 10% polybutyl methacrylate 5% stearic acid | Oxide reduced | 7, 10, and 14 | 48 | Z-blade | Chan et al. (2005) |
| 65% paraffin 30% polyethylene 5% stearic acid | Gas atomized | 10 | 66 | Banbury | Moballegh et al. (2005) |

powders, 1.21 for water-atomized and jet-milled powders, and 1.12 for gas-atomized powders, assuming a final sintered density of 95% of theoretical.

Examples of several different Cu feedstocks are summarized in Table 18.3. Moldable feedstock has been produced with various types of mixers as well as an assortment of Cu powders. A key issue with mixing is the avoidance of contamination from previously mixed materials and wear of the mixing vessel to maintain high thermal conductivities in the final parts. Cu feedstock compounded with a wax–polymer binder is also commercially available (Zlatkov and Hubmann, 2008).

### 18.3.3 Molding

Molding Cu feedstocks is relatively straightforward, although their relatively high thermal conductivities result in faster cooling rates, making it more difficult to fill thin-walled components. As with mixing, cross-contamination must be avoided during molding to maintain high thermal conductivities. For MIM Cu, high injection speed or pressure can cause the binder to separate from the Cu powder, which can be deformed into a rigid solid due to its low yield strength. Copper powders with oxidized surfaces have better adhesion to the binder than hydrogen-reduced Cu powders (Chan *et al.*, 2005).

Gas-atomized powders have been used for molding large, complex heat sinks (Hinse *et al.*, 2007; Zlatkov and Hubmann, 2008). These parts have masses of 100–150 g and hollow structures with wall thicknesses as small as 0.3 mm. Molding such parts generally requires multiple slides, a hot-runner, and a cavity pressure transducer, but the biggest challenge is mold temperature control prior to part ejection. A combination of electrical heaters and water temperature control units has proven successful at molding tubes about 25 mm long with 0.3 mm wall thickness (Zlatkov and Hubmann, 2008). A high injection temperature and warm die required a cooling time of 5 min, but this was still much faster than heating and cooling the die, which required a cooling time of 15 min. Three-dimensional simulations that include analysis of the heating and cooling of the mold are able to accurately predict filling behavior (Hinse *et al.*, 2007).

### 18.3.4 Debinding and sintering

Copper parts can be conventionally debound using solvent and/or thermal techniques. Thermal processes require careful atmosphere control to minimize residual carbon and oxygen, which can negatively affect densification. Thermal debinding as-molded parts containing a wax-based binder in air can cause cracking due to endothermic reactions, while debinding in argon can lead to slumping. Vacuum debinding provides good dimensional control and intermediate levels of carbon and oxygen (Moore *et al.*, 1995). Solvent debinding followed by thermal debinding in hydrogen has also proven successful (Chan *et al.*, 2005; Johnson *et al.*, 2005b).

Oxygen must be reduced early in the sintering cycle to avoid entrapment of water vapor in closed pores, which is often accompanied with rapid grain growth and pore separation from grain boundaries, leading to high porosities. The addition of reactive dopants, such as Al, Cr, and Si, can help getter oxygen, but degrade conductivity (Chan *et al.*, 2005; Sweet *et al.*, 1992). Extraction of oxygen prior to final stage pore closure, which occurs near 92% density, generally requires hydrogen sintering with carefully designed thermal cycles (Hayashi and Lim, 1990; Upadhyaya and German, 1998a).

During heating, the reduction of copper oxides by dry hydrogen typically occurs in the range from 550 to 680°C (Dorfman *et al.*, 2002b). Long hold times at temperatures in this range, or higher, can eliminate swelling by reducing the copper oxides prior to final stage sintering pore closure. Maximum temperatures near the melting temperature of copper (1080°C)

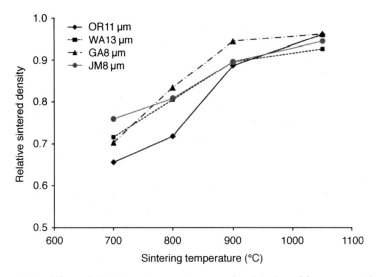

*18.5* Effect of sintering temperature on the density of four types of copper powders uniaxially pressed at 175 MPa.

are required for high sintered densities. For example, high densities have been achieved by debinding and sintering in dry hydrogen using the following thermal profile:

3°C/min to 300°C, hold for 1 h
3°C/min to 500°C, hold for 1 h
3°C/min to 600°C, hold for 1 h
5°C/min to 700°C, hold for 2 h
5°C/min to 800°C, hold for 2 h
5°C/min to 900°C, hold for 2 h
5°C/min to 1050°C, hold for 1 h

The onset of pore closure depends on the powder. The densities of the four different powders given in Table 18.2, uniaxially pressed at 175 MPa, after different stages of the above cycle are plotted in Fig. 18.5. At 700°C the density is about the same as the initial green density. Most of the sintering densification occurs during heating at temperatures between 700°C and 900°C. From 800 to 900°C, the density of the gas-atomized powder increases from 80% to greater than 90%. Thus, the oxides in this powder must be reduced below 900°C in order to prevent water vapor entrapment. The remaining powders have open porosity that allows for further reduction during heating to 1050°C. The densities after sintering at 1050°C range from 93 to 96% regardless of production method. Copper powders with mean particle sizes up to 25 μm can also achieve this level of densification at 1050°C (Johnson *et al.*, 2005b). Finer powders have higher starting oxygen

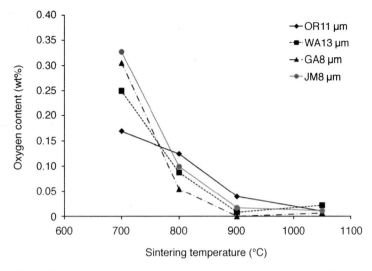

*18.6* Effect of sintering temperature on oxygen content for four types of copper powders uniaxially pressed at 175 MPa.

contents and will densify at lower temperatures, leading to a higher probability of trapping water vapor in closed pores.

Figure 18.6 confirms that the thermal cycle enables nearly complete oxide reduction of the gas-atomized powder by 900°C. The oxygen content is also 200 ppm or less for the other powders, except for the oxide-reduced powder, which contained 400 ppm oxygen. The density of the oxide-reduced powder at 900°C is 89% of theoretical, so some open porosity remains to allow for continued reduction and the escape of water vapor.

Microstructures of the water-atomized powder after sintering at 900°C for 2 h and at 1050°C for 1 h are shown in Fig. 18.7. At 900°C, the grains are small and small pores are visible at the grain boundaries. At 1050°C, both the grains and the pores have significantly coarsened with a slight increase in overall density. The large pores indicate that even with oxygen levels below 200 ppm at 900°C, sufficient oxygen remains to produce entrapped water vapor in the small pores and cause them to swell when heated to 1050°C. On the other hand, the microstructure of the oxide-reduced powder sintered at 1050°C for 1 h, shown in Fig. 18.8, consists of relatively small pores, indicating little swelling from entrapped water vapor.

## 18.3.5 Thermal properties

Thermal conductivities of MIM Cu range from 280 to 385 W/(m K), depending on porosity and impurities (Johnson *et al.*, 2005b). Porosity of 4–7% is typical, while Fe contents can range from 20 to 570 ppm. In

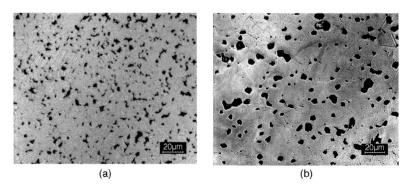

*18.7* Micrographs of a 13 μm water-atomized powder after pressing at 175 MPa and sintering in hydrogen at (a) 900°C for 2 h and (b) 1050°C for 1 h.

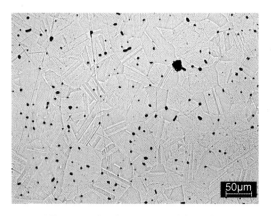

*18.8* Micrograph of a 11 μm oxide-reduced copper powder after pressing at 175 MPa and sintering in hydrogen for 1 h at 1050°C.

comparison, the thermal conductivities of commercially pure wrought Cu alloys, such as C11000, can reach 390 W/(m K) for metallic impurity levels below 50 ppm and oxygen contents as high as 0.04 wt%. Commercially pure cast Cu alloys, such as C83400, have lower thermal conductivities, usually around 340–350 W/(m K), because of the use of deoxidizers such as silicon, tin, zinc, aluminum, and phosphorus. These elements likely comprise the majority of the 0.15 wt% of impurities typically found in commercially available Cu powders.

Based on the Wiedemann–Franz relationship and Nordheim's Rule (Rose *et al.*, 1966), the predicted effect of iron impurities on the thermal conductivity of copper is plotted in Fig. 18.9 for comparison with experimental results (Johnson *et al.*, 2005b). The measured values follow the same trend as the model predictions, but for most of the samples, the measured thermal conductivities are lower than expected based on just iron

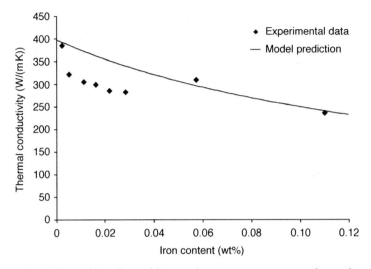

*18.9* Effect of iron impurities on the room temperature thermal conductivity of Cu in comparison to experimental results (Johnson et al., 2005b).

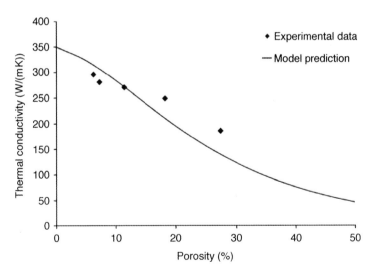

*18.10* Effect of porosity on the room temperature thermal conductivity of Cu in comparison to experimental results (Johnson et al., 2005b).

impurities. At low concentrations, Fe may be representative of the overall impurity content, and the thermal conductivity is reduced by the cumulative effects of all impurities. The highest Fe concentration likely results from contamination during processing.

In addition to impurity effects, porosity also decreases the thermal conductivity of MIM copper heat sinks as shown in Fig. 18.10.

Experimental data (Johnson *et al.*, 2005b) show slightly less dependence on porosity than predicted by the relationship proposed by Koh and Fortini (1973) assuming a value of 350 W/(m K) for the pore-free thermal conductivity (to take impurity effects into account). Impurity concentrations on the order of 0.1–0.2 wt% can be as detrimental to the thermal conductivity as 20% porosity.

### 18.3.6 Example applications

MIM Cu must compete with well-established machining and casting processes. MIM can provide a more cost-effective means of producing complex shapes than machining wrought Cu. Commercially pure Cu alloys are difficult to cast and the thermal conductivity of MIM Cu is much higher than easily castable alloys, such as C83400, owing to their high levels of alloying additions. Still, successful MIM of Cu depends on achieving the right balance of powder cost, moldability, dimensional control, sintered density, and thermal conductivity.

MIM can produce the heat sink geometries shown in Fig. 18.2 in high-conductivity Cu. As another example, a MIM Cu heat sink measuring approximately 20 mm wide by 20 mm long by 2.5 mm is shown in Fig. 18.11. This part was produced from a water-atomized powder, which was mixed at a solids loading of 52 vol% with a wax–polymer binder. After molding, the components were solvent debound to remove the wax. The remainder of the binder was burned out during heating in the sintering cycle. The sintered density was 94% with a thermal conductivity of 296 W/(m K).

MIM Cu has also been used to produce heat sinks in which the pins are replaced with thin-walled tubes for increased surface area (Zlatkov and

*18.11* An example MIM Cu heat sink. This demonstration component is approximately 20 mm wide by 20 mm long by 2.5 mm high.

*18.12* (a) A demonstration heat pipe and (b) a micrograph of the interface between the outer wall and the wick.

Hubmann, 2008). Parts have been demonstrated consisting of 96 tubes with lengths of 29.2 mm, outer diameters of 3.65 mm, and inner diameters of 3.05 mm. Sintered densities of 94% of theoretical were achieved without significant distortion of the tubes. Sintered Cu structures with large length to thickness ratios must be handled carefully to avoid deformation due to the low yield strength of fully annealed Cu (Chan *et al.*, 2005).

As another example, two-material MIM has been used to fabricate Cu heat pipe structures (German *et al.*, 2005; Johnson and Tan, 2004). In this process, a coarse Cu powder is molded to form the wick and a fine Cu powder is co-molded around it. This composite structure is then thermally processed in such a way that the two sections co-sinter together to produce a good metallurgical bond. A demonstration heat pipe is shown in Fig. 18.12 along with a cross-section showing the interface between the outer wall and the wick. This design integrates cooling fins with the outer casing, which surrounds the porous wick and a large open vapor channel. MIM enables seamless transitions between these features to eliminate thermal interface resistance.

## 18.4 Tungsten–copper

Metal injection molding of tungsten combined with 10–20 wt% copper was first patented in 1991 (Oenning and Clark, 1991), but much learning has taken place since then. As with MIM Cu, the primary challenge is attaining high sintered densities and high conductivities. Densification of W–Cu is greatly hindered by the extremely low solubility of W in Cu. Densification can be enhanced by the addition of transition metal elements, such as Ni, Co, and Fe, which have substantial solubility for W, but are detrimental to the thermal conductivity (Johnson and German, 1993a, 1993b; Johnson *et al.*, 2009, 2010). Sintering of high-purity W–Cu to near full density requires a submicron W particle size. A few W–Cu powders suitable for

MIM are now commercially available. Elemental W and Cu powders can be bought separately, but the method of producing the composite powder is critical. Alternatively, W powder can be injection molded and then infiltrated with Cu.

Successful processing of W–Cu parts via MIM requires a tailored particle size distribution for both moldability and sinterability. The key requirements for processing high thermal conductivity W–Cu components via MIM are described below.

### 18.4.1 Powders

Almost all commercially available W powders are produced from oxide reduction. Non-oxide impurities are generally less than 0.05 wt%, as required for high thermal conductivities. Sizes are typically 3–5 µm, but particle sizes below 1 µm, needed for liquid-phase sintering of W–Cu, are commercially available. Composite W–Cu powders can be produced by co-reducing tungsten oxides with copper oxides. Such powders are also commercially available and provide a homogeneous distribution of the Cu that is ready for compounding without further processing. Mixing of reduced W and Cu powders is possible but requires special care.

Oxide-reduced W powders must be deagglomerated, typically by milling, in order to produce suitable MIM feedstock. Otherwise the solids loading will be too low at a moldable viscosity. The deagglomerated W powder can be dry blended with Cu powder using a double-cone, v-cone, or Turbula® (Glenn Mills Inc., Clifton, NJ) mixer, but this adds a process step and requires a Cu particle size near that of the W powder. If the Cu particle size is much larger than the particle size of the W powder, pools of liquid form that are less effective for densification (Lee *et al.*, 1985; Moon *et al.*, 1996). High-purity Cu particles less than 10 µm are expensive. Instead, coarser Cu powders, such as the ones given in Table 18.2, can be combined with the W powder during milling.

Rod milling, ball milling, planetary, and attritor milling have all been used to prepare W–Cu powders. These methods are listed in order of increasing energy input. Coarser Cu powders can be used as the milling intensity increases. Higher energy milling processes, especially attritor milling, can embed W particles into larger Cu particles to produce a homogeneous composite powder, but the particle size and distribution depend on several milling parameters, including speed, time, and powder loading (Johnson *et al.*, 1995).

A significant obstacle to milling elemental powders is the increased likelihood of contamination from the milling media or liner. Steel media and liners are generally unsuitable, since the resulting Fe and Si contamination is highly detrimental to the thermal conductivity. Cemented carbide media

*Table 18.4* W–Cu powder characteristics

| Composition | W | Cu | Cu$_2$O | W–15Cu |
|---|---|---|---|---|
| Production method | Oxide reduced | Water atomized | Electro-chemically refined | Oxide co-reduced |
| Particle size distribution | | | | |
| $D_{10}$ (µm) | 0.4 | 1.3 | 3.6 | 1.5 |
| $D_{50}$ (µm) | 0.7 | 3.6 | 9.7 | 2.9 |
| $D_{90}$ (µm) | 1.2 | 5.5 | 17.1 | 5.6 |
| Theoretical density (g/cm$^3$) | 19.3 | 8.96 | 6.0 | 16.44 |
| Pycnometer density (g/cm$^3$) | 18.0 | 8.8 | 6.1 | 16.1 |
| Apparent density (g/cm$^3$) | 3.3 | 3.3 | 2.2 | 1.8 |
| % of pycnometer | 18% | 38% | 36% | 11% |
| Tap density (g/cm$^3$) | 4.9 | 3.7 | 3.1 | 2.6 |
| % of pycnometer | 27% | 43% | 51% | 16% |

with cemented carbide or polyethylene liners have been used more successfully, but can be cost prohibitive for production quantities.

Some or all of the Cu powder that is milled with the reduced W powder can be replaced with copper oxide, usually cuprous oxide (Cu$_2$O). Cuprous oxide powders are lower cost than reduced Cu powders, and can provide good homogeneity when milled with W powders. The copper oxide is reduced during sintering. A combination of Cu and Cu$_2$O can form a eutectic liquid at 1065°C that may enhance sintering in wet hydrogen (Jech et al., 1997). Reduction experiments have shown that the Cu$_2$O is mostly reduced in dry hydrogen below this temperature during heating (Johnson et al., 2009).

The characteristics of example components for preparing W–Cu powders are given in Table 18.4 along with those of a co-reduced W–15 wt% Cu powder. Representative scanning electron micrographs of these powders are given in Fig. 18.13. The W powder can be combined with the −10 µm Cu powder by blending or milling. It can also be milled with the Cu$_2$O powder or the Cu powders given in Table 18.2. Rod milling is not sufficiently energetic to break down Cu$_2$O particles, but can be used to obtain a homogeneous distribution with a −25 µm Cu powder as shown in the micrographs in Fig. 18.14. Tungsten particles encapsulate Cu particles in co-reduced powders which can be directly used to produce feedstock in the as-supplied condition.

*18.13* Scanning electron micrographs of (a) sub-micron W; (b) −10 µm Cu; (c) $Cu_2O$; (d) co-reduced W–15Cu powders.

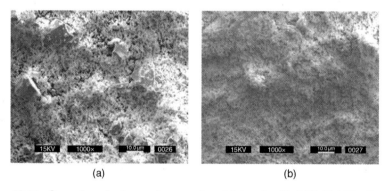

*18.14* Scanning electron micrographs of (a) rod-milled W–17.12$Cu_2O$ and (b) rod-milled W–15Cu powders.

Table 18.5 Examples of MIM W–Cu feedstock preparation

| Binder | Powder preparation | Solids loading vol% | Mixing technique | Reference |
|---|---|---|---|---|
| 39% polypropylene 49% paraffin wax 10% Carnuba wax 2% stearic acid | 1–2 µm W blended with 25 or 35 wt% Cu (8–10 µm) | 59–61 | Vacuum | Oenning and Clark (1991) |
| Wax–polymer with 40% polypropylene | Milled W blended with fine Cu, mechanically alloyed W–Cu | 52–58 | Twin-screw | Hens et al. (1994) |
| 30% polyethylene 45% paraffin wax 15% beeswax 10% stearic acid | 4 µm W blended or milled with 30 wt% Cu (various sizes) | 45–55 | Single cam mixer | Moon et al. (1996) Kim et al. (1999) |
| 35% polypropylene 60% paraffin 5% stearic acid | Submicron W milled with 2.5 wt% Cu (12 µm) | 52 | Sigma mixer | Yang and German (1997) |
| Wax–polymer | 1.5 or 3.6 µm W blended with 10, 20 or 30 wt% Cu (6.0 or 12.0 µm) | 50–63 | Double sigma | Knuewer et al. (2001) |
| Wax–polymer | 1.8 or 3.6 µm W blended with 10, 20 or 30 wt% Cu (6.0 or 13.6 µm) | 53–63 | Double sigma | Petzoldt et al. (2001) |

## 18.4.2 Feedstock preparation

Demonstrations of MIM W–Cu have been almost exclusively with wax–polymer binders (Hens et al.; 1994, Johnson et al.; 1995, Kim et al., 1999; Knuewer et al., 2001, Moon et al.; 1996, Oenning and Clark, 1991; Petzoldt et al., 2001; Yang and German, 1997). A notable exception is the use of a non-aqueous binder, such as cyclohexane, and a dispersant, such as an acrylic acid-based polyelectrolyte, which can be frozen after molding and debound by sublimation (Sundback et al., 1991). The small particle sizes required for densification have poor packing characteristics and consequently the solids loadings of these powders for injection molding are relatively low, generally ranging from 52 to 58 vol%, giving tooling scale-up factors of 1.22–1.18 for a sintered density of 95% of theoretical. Because of the dispersion difficulties, high shear rate continuous mixers are preferred for compounding (Hens et al., 1994). A summary of various binder compositions and compounding details is given in Table 18.5.

### 18.4.3 Injection molding

While any of the feedstocks shown in Table 18.5 can be used to injection mold test bars, several factors must be considered for molding complex thermal management components with thin walls and large numbers of pins or feed-throughs. Wide particle size distributions produced by lightly attritor milling a submicron W powder with a coarse Cu powder are recommended for improved molding behavior (Johnson et al., 1995). For heat sinks with pins, a higher-strength binder is generally required to prevent them from breaking upon ejection. This necessitates a high fraction of backbone polymer (polypropylene or polyethylene) or one with a high molecular weight.

For electronic packages, finding a suitable gate location for smooth flow through thin walls is often a challenge due to flatness requirements on key surfaces and the need for core pulls (Hens et al., 1994). The high conductivity of the feedstock increases the risk of premature solidification of the melt, resulting in non-fill. Mold-filling simulations can be used to optimize gate location and type, for example concentric or film gate, for specific part geometries (Petzoldt et al., 2001).

### 18.4.4 Debinding and sintering

Debinding of W–Cu is typical of other MIM materials. A combination of solvent debinding to dissolve the major binder component and thermal debinding to pyrolize the remaining backbone polymer through the open pore space is the preferred technique for the wax–polymer systems described in Table 18.5. However, thermal debinding alone (Oenning and Clark, 1991) or a combination of wicking and thermal debinding (Kim et al., 1999; Moon et al., 1996) can also be used. Hydrogen is required to reduce both tungsten oxides and copper oxides. As noted in the previous section, reduction of copper oxides occurs in the range from 550 to 680°C. Sintering in dry hydrogen reduces tungsten oxides at 800°C (Dorfman et al., 2002b). A typical debinding cycle includes a slow (about 2°C/min) ramp rate to 500°C to remove the binder and then further heating to 800–950°C for oxide reduction and to produce sufficient strength for handling.

The effects of sintering at various temperatures for 2 h in dry hydrogen on the density and oxygen content of W–15Cu are shown in Fig. 18.15. The density does not significantly increase until the temperature exceeds 1100°C, which is above the melting temperature of copper. Even at 1100°C, the oxygen content is still about 0.25 wt%, but it decreases linearly with sintering temperature. Near full density is achieved at a sintering temperature of 1300°C with an oxygen content of 150 ppm. An example micrograph showing the composite microstructure with 1 μm W grains is given in Fig. 18.16.

Liquid phases generally enhance sintering, but the extremely low

MIM of thermal management materials in microelectronics    467

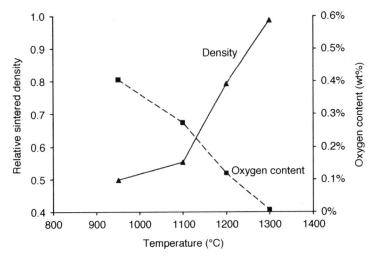

*18.15* Effect of sintering temperature on the density and oxygen content of W–15Cu.

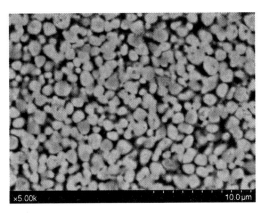

*18.16* Micrograph of W–15Cu sintered at 1300°C for 60 min.

solubility of W in Cu severely limits densification. Achievement of full density requires a submicron W particle size to promote solid-state densification at temperatures below 1400°C to avoid significant Cu evaporation (Johnson and German, 1996). The effects of W particle size and sintering temperature on the density of W–10Cu are shown in Fig. 18.17 (Johnson *et al.*, 2005a). As the Cu content increases, the dependence of the sintered density on particle size decreases, as shown in Fig. 18.18.

Solid-state densification of the W skeleton can continue even as full density is approached. Since few pores remain for the liquid copper to fill, further skeletal sintering forces it to the surface, resulting in Cu 'bleed-out' (Dorfman *et al.*, 2002a). An example of this phenomenon is shown in

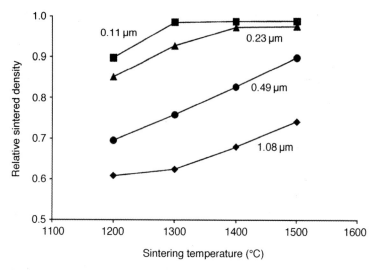

*18.17* Effect of W BET (Brunauer–Emmett–Teller) particle size and sintering temperature on the sintered density of W–10Cu after compacting at 70 MPa and sintering for 1 h in hydrogen.

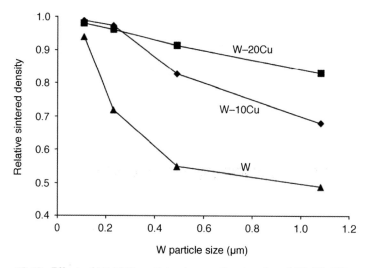

*18.18* Effect of W BET particle size on the density of W, W–10Cu, and W–20Cu after compacting at 70 MPa and sintering for 1 h in hydrogen at 1400°C.

Fig. 18.19. Although skeletal sintering increases with higher temperatures, less Cu bleed-out is seen as the temperature reaches 1400°C since liquid Cu at the surface rapidly evaporates. Over-sintering increases the W to Cu ratio and results in a Cu depletion layer at the surface. Transition metal impurities, especially Fe, can accelerate sintering of the W skeleton at lower

# MIM of thermal management materials in microelectronics

*18.19* Example of Cu bleed-out at edges of a W–Cu specimen.

temperatures (Johnson and German, 1993b), leading to greater amounts of bleed-out. Both excess Cu and depleted Cu at the surface of MIM parts are often associated with part warpage and can cause plating problems.

## 18.4.5 Infiltration

Instead of liquid-phase sintering a mixture of W and Cu powders, W–Cu composites can also be produced by placing pressed Cu powder or wrought Cu pieces in contact with a porous W preform and heating the combination above the melting temperature of Cu (Gessinger and Melton, 1977; Ho *et al.*, 2008; Kothari, 1982; Kny, 1989; Sebastian, 1981; Stevens, 1974; Wang and Hwang, 1998). In a dry hydrogen atmosphere, the liquid Cu will infiltrate the W preform. Infiltration has been used to produce electrical contacts for many decades, but these applications require a low degree of shape complexity and the preform can be produced by die pressing W powder at high compaction pressures. The preform generally gives very little dimensional change during infiltration and is infiltrated with a volume fraction of Cu approximately equal to its porosity. Since high green densities are not possible with injection molding due to the limitation of the critical solids loading, the preform must be pre-sintered to the density corresponding to the W volume fraction in the final composite. This requires a fine starting powder or high sintering temperature since sintering activators, such as Co, Ni, or Fe are highly detrimental to the thermal conductivity (Johnson and German, 1993a). A process map for the effect of W particle size and sintering temperature on density is given in Fig. 21.12 in Chapter 21.

Infiltration can be combined with pre-sintering in a single thermal cycle (Yang and German, 1997). To achieve W contents of 80 wt% or more, a submicron W particle size is required to densify the preform at temperatures below 1500 °C after the Cu melts and infiltrates it. In this case of infiltration sintering, the injection-molded W preform displays 12–14% shrinkage depending on the sintering temperature and Cu content. Densities above 99% of theoretical can be achieved.

### 18.4.6 Thermal properties

The thermal properties of MIM W–Cu depend mostly on the composition, but porosity and microstructure are also factors. Several models have been developed to estimate the thermal conductivity of composites. The rule of mixtures and the inverse rule of mixtures are the simplest and provide upper and lower boundaries on the predicted thermal conductivity. A recent model (Johnson et al., 2010) considers the effect of grain shape on the thermal conductivity of liquid-phase-sintered composites. The predictions of these models for the thermal conductivity of pure, pore-free W–Cu versus Cu content are plotted in Fig. 18.20 in comparison to experimental results (Dorfman et al., 2002a; Johnson et al., 2010). Despite best efforts to maintain high purities and high densities, most reported values are still below the predictions of the inverse rule of mixtures.

Impurities are highly detrimental to the thermal conductivity (Johnson and German, 1993a; Johnson et al., 2009, 2010). Based on the Wiedemann–Franz relationship and Nordheim's Rule (Rose et al., 1966), the predicted effect of Fe impurities on the thermal conductivity of W–10Cu and W–15Cu is plotted in Fig. 18.21. As the transition metal content increases, the effect of Cu volume fraction decreases, with little predicted difference in the thermal conductivity of W–10Cu and W–15Cu at impurity levels of about

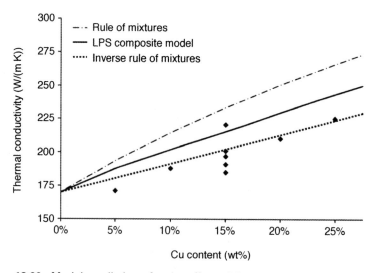

*18.20* Model predictions for the effect of Cu content on the thermal conductivity of pure, pore-free W–Cu. Experimental data from Dorfman et al. (2002) and Johnson et al. (2010) are included for comparison.

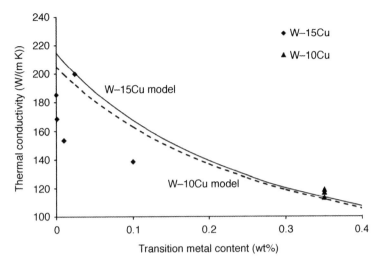

*18.21* Model predictions for the effect of transition metal impurities on the thermal conductivity of pore-free W–Cu. Experimental data (Johnson and German, 1993a; Johnson, 1994; Johnson et al., 2009) are included for comparison.

0.25 wt% or higher. These impurity levels are typically associated with intentionally added sintering aids. Decreases in thermal conductivity of 40 W/(m K) are predicted with impurity levels of 0.1 wt%, which may result from unintentional contamination.

Porosity of 1–3% is typical for MIM W–Cu parts. Model predictions (Johnson et al., 2010) for the effects of porosity on the thermal conductivity of W–15Cu are shown in Fig. 18.22. Experimental data (Dorfman et al., 2002a; Johnson and German, 1993a; Johnson, 1994; Johnson et al., 2007, 2009) are also given for comparison showing that porosity has a relatively minor role on the thermal conductivity of W–Cu, and does not explain the measurement scatter. Additional model predictions show a greater effect of an interfacial resistance with a W–W thermal boundary conductance of $10^8$ W/(m$^2$ K) for a 1 µm grain size decreasing the thermal conductivity by about 25%, in line with experimental results (Johnson et al., 2010). Thus, larger grain sizes and lower contiguities are favorable for increasing thermal conductivity, but are difficult to obtain in practice with liquid-phase sintering and are more readily produced by infiltration of a coarse W powder pre-sintered at a high temperature.

The thermal expansion coefficient is mostly dependent on the ratio of the components of the composite and the microstress between them. Porosity has virtually no effect on the thermal expansion of sintered components (Danninger et al., 2011), but the effects of microstructure are less well defined. The models of Turner (1946), Kerner (1956), and Fahmy and Ragai (1970) consider the stress-coupling of the phases and provide estimates for

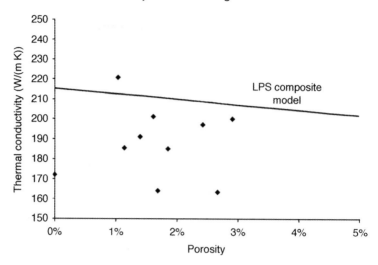

*18.22* Model predictions for the effect of porosity on the thermal conductivity of pure W–15Cu. Experimental data (Dorfman *et al.*, 2002; Johnson and German, 1993; Johnson, 1994; Johnson *et al.*, 2007, 2009) are included for comparison.

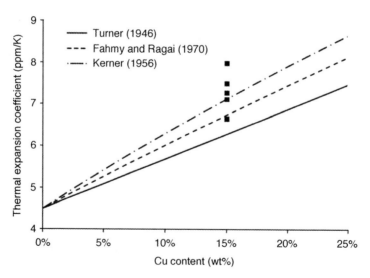

*18.23* Predicted effects of Cu content on the thermal expansion coefficient of W–Cu from three different models. Experimental data from Wang and Hwang (1998) are shown for comparison.

the thermal expansion coefficient of W–Cu for various Cu contents, as shown in Fig. 18.23. The Kerner model is closest to the average thermal expansion coefficient of W–15Cu measured by Wang and Hwang (Wang and Hwang, 1998).

MIM of thermal management materials in microelectronics 473

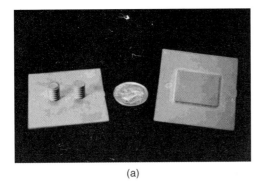

(a)

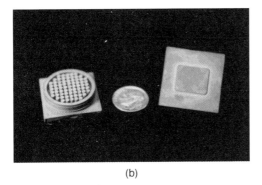

(b)

*18.24* Example W–Cu (a) heat spreaders and (b) boilers.

## 18.4.7 Example applications

Most W–Cu heat sinks and heat spreaders are produced from infiltrated sheets, but MIM enables fabrication of complex geometries, such as chip submounts and stem heat sinks for optoelectronic devices, base metals and fin-shaped heat sinks for integrated circuits, bases for multi-chipped boards, and microelectronic packages. Example heat spreaders are shown in Fig. 18.24. The heat spreader in Fig. 18.24(a) is similar to a design fabricated from infiltrated sheets, but demonstrates that threaded fasteners can be incorporated in a net shaped MIM component. The part in Fig. 18.24(b) is an example of a complex boiler designed to transfer heat from a high powered microprocessor to a liquid cooling column.

Parts that lack a flat surface require support during sintering. Since they undergo significant sintering shrinkage, fixtures must be designed to support both the green sample and the much smaller sintered part. For example, the components shown in Fig. 18.24 were supported by alumina substrates in which a rectangular section was machined to the green dimensions of the rectangular section on the bottom of the parts. The depth of this rectangular slot was identical to the thickness of the rectangular area of the sintered

474  Handbook of metal injection molding

part. In this way, the outer edges of the sintered part are supported at the end of the sintering cycle. Despite a high liquid volume fraction, the rigid W skeleton that develops early during sintering of W–Cu resists slumping. In comparison, tungsten heavy alloys, in which the liquid phase has substantial solubility for W, show significant slumping and loss of dimension precision at similar liquid volume fractions (Upadhyaya and German, 1998b), as further discussed in Chapter 21.

Pilot production of the parts shown in Fig. 18.24 was able to hold tolerances on critical dimensions to a standard deviation $+/-0.1\%$, while all dimensions could be kept within a standard deviation of $+/-0.3\%$ (Hens et al., 1994). W–Cu prototype packages similar to that shown in Fig. 18.3(a) have been produced with tolerances of $+/-0.1\%$ with a dimensional yield of 57% (Ludvik et al., 1991). A W–Cu housing for a high-frequency circuit achieved tolerances of 0.06–0.22%, 0.25–0.29%, and 0.23–0.35% for the length, width, and height, respectively (Petzoldt et al., 2001). Warpage becomes a greater concern as the sintered density increases (Knuewer et al., 2001), so densities are usually limited to about 97% of theoretical. This density is sufficient for hermeticity with little detriment to the material properties.

## 18.5 Molybdenum–copper

MIM processing of Mo–Cu is similar to that of W–Cu, but has been less researched. As with W, the low solubility of Mo in Cu hinders densification, and transition metal elements that enhance densification are detrimental to the thermal conductivity. In comparison to the W particle sizes used for W–Cu, slightly larger Mo particle sizes can be used for both liquid-phase sintering and for infiltration, but the particle sizes of Mo powders with commercial availability are limited. Because of the higher ductility of Mo and its subsequent inability to be attrited to smaller particle sizes, fewer options are available for producing composite Mo–Cu powders. Successful processing of Mo–Cu parts via MIM requires specific attention to the particle size distribution, particle morphology, and thermal cycle as described below.

### 18.5.1 Powders

Like W powders, Mo powders are generally produced from oxide reduction; however, the range of sizes available is much narrower. Particle sizes are typically 2–4 µm, but production of finer powders is difficult. Deagglomeration by milling is generally required, but high-energy processes, such as attritor milling, are less successful since Mo is more ductile than W and tends to form platelets, which pack poorly. Blending

*18.25* Scanning electron micrographs of (a) 2.5 μm and (b) 4.1 μm Mo powders.

*Table 18.6* Mo powder characteristics

| Production method | Oxide reduced | Oxide reduced |
|---|---|---|
| Particle size distribution | | |
| $D_{10}$ (μm) | 1.1 | 2.1 |
| $D_{50}$ (μm) | 2.5 | 4.1 |
| $D_{90}$ (μm) | 4.8 | 7.9 |
| Pycnometer density* (g/cm³) | 9.62 | 10.14 |
| Apparent density (g/cm³) | 2.1 | 3.0 |
| % of pycnometer | 21% | 29% |
| Tap density (g/cm³) | 3.1 | 5.0 |
| % of pycnometer | 31% | 49% |

* Theoretical density is 10.2 g/cm³.

with a Cu powder with a particle size similar to that of the Mo powder is generally required for high sintered densities (Kirk *et al.*, 1992). Co-reduction of molybdenum oxides and copper oxides is possible (Skorokhod *et al.*, 1983), but such co-reduced powders are not commercially available. In all cases, high-purity powders are needed to achieve high thermal conductivities (Johnson *et al.*, 1995).

The characteristics of two MIM grade Mo powders are given in Table 18.6. Scanning electron micrographs of the powders are presented in Fig. 18.25. The low pycnometer density of the 2.5 μm Mo powder is due to oxide impurities, which can be reduced during sintering.

## 18.5.2 Feedstock preparation

Wax–polymer binder systems have proven suitable for injection molding of both elemental Mo and Mo–15Cu compositions (Johnson and German, 1999). A plot showing the torque required to mix Mo powders at different solids loadings with a wax–polymer binder is shown in Fig. 18.26. From this

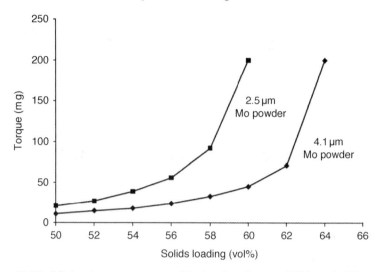

*18.26* Mixing torque versus solids loading for two MIM grade Mo powders showing critical solids loadings at 60 vol% for the 2.5 μm Mo powder and 64 vol% for the 4.1 μm Mo powder.

plot, the critical solids loading was estimated at 64 vol% for the 4.1 μm Mo powder and 60 vol% for the 2.5 μm Mo powder. The coarser Mo powder can be used for molding Mo skeletons at a solids loading of 62 vol% (tool scale-up factor of 1.15) for subsequent infiltration. The finer Mo powder is required for liquid-phase sintering of Mo dry blended with Cu, and the blend can be molded at a solids loading of 58 vol% (tool scale-up factor of 1.18). Milling techniques tend to deform the Mo particles and significantly decrease solids loadings.

### 18.5.3 Injection molding

Coarser Mo powders provide better molding behavior than blends of finer Mo powders with Cu. Heat spreaders and electronic packages have been injection molded from elemental Mo feedstock, although increased packing pressure was needed to prevent cooling cracks along the edges of the thicker section opposite the threaded studs of the component shown in Fig. 18.24(a) and at the base of the studs. The same parts could not be molded from Mo–Cu feedstock without cooling cracks (Johnson and German, 1999). Cracking was also a problem with molding Mo–Cu rectangular bars, but could be eliminated with sufficient packing pressure. The packing pressure did not transfer well to the Mo–Cu parts and this was attributed to the high conductivity of the feedstock combined with a less than optimal particle size distribution. Similar problems with molding AlN heat spreaders with the same binder system were greatly alleviated by widening the particle size distribution (Johnson et al., 1996). Thus, adjustments to the particle size distribution of Mo–Cu blends may also improve their moldability.

## 18.5.4 Debinding and sintering

Debinding of Mo–Cu is similar to that of W–Cu. Hydrogen is required during thermal debinding to reduce both Mo oxides and Cu oxides. Copper hinders the reduction of molybdenum oxides, increasing its initial reduction temperature into the range of 750–800°C (Skorokhod et al., 1983).

The solubility of Mo in Cu is much higher than the solubility of W in Cu, but it is still low enough to make liquid-phase sintering to full density difficult for the Mo contents of interest for thermal management applications. As for W–Cu, the early formation of a solid refractory metal skeleton slows the densification rate to that of solid-state diffusion (Johnson and German, 2001). In this case, the Cu contributes little to densification and simply fills pore space. The liquid Cu hinders densification since it enhances growth of the Mo grains more than for W–Cu. As the grains become larger, the driving force for densification decreases. Thus, Mo–Cu must be sintered at temperatures comparable to those used for the solid-state sintering of Mo.

For liquid-phase sintering of Mo–18Cu, sintering temperatures well above the melting temperature of Cu are required to achieve high sintered densities. These high temperatures can result in significant weight loss due to Cu evaporation. The effects of sintering temperature on the final densities and weight losses of Mo–18Cu are shown in Fig. 18.27 for the Mo powders described in Table 18.6 after liquid-phase sintering for 4 h. High sintered densities require the 2.5 µm Mo powder and a sintering temperature of 1400°C. Higher temperatures result in significant Cu evaporation and a decrease in relative sintered densities. The effectiveness of the 4 h sintering time is displayed in Fig. 18.28, which compares the sintering of Mo–18 Cu with that of elemental Mo. Even with the 2.5 µm Mo powder, a density of only 85.2% of theoretical is achieved for Mo–18Cu after 1 h at a sintering temperature of 1400°C, but the density increases to 95.7% of theoretical after 4 h at 1400°C. The slow sintering kinetics are unusual for liquid-phase sintering and are more typical of solid-state sintering of the elemental Mo powders, which continue to densify as the sintering time is increased from 1 to 4 h.

The microstructure of injection molded and liquid-phase-sintered Mo–15Cu is shown in Fig. 18.29. This sample was sintered for 3 h at 1450°C and had a density 97.1% of theoretical and a grain size of 5.7 µm. The clusters of irregular grains are representative of a slow rate of microstructural homogenization.

The high tolerances demanded by the microelectronics industry require

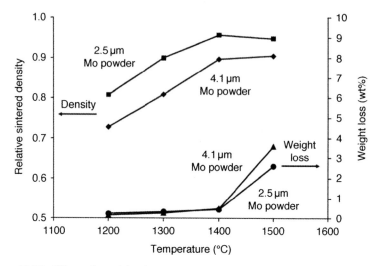

*18.27* Effect of particle size and sintering temperature on the density of Mo–18Cu. The hold time at the sintering temperature was 4 h. Weight loss is mostly due to Cu evaporation.

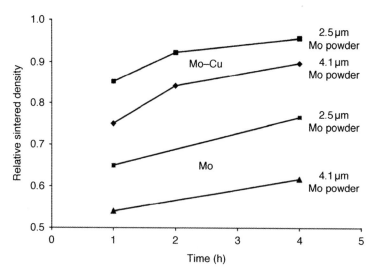

*18.28* Effect of particle size and sintering time at 1400°C on the density of elemental Mo and Mo–18Cu.

precision sintering. The heating rate and temperature gradients in the furnace can result in distortion. For example, a heating rate of 10°C/min gives warping of Mo heat spreaders. This distortion can be eliminated by decreasing the heating rate to 2°C/min. But often the dimensions of the sintered parts are related to previous processing stages, requiring tight

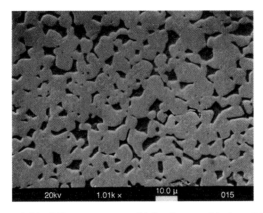

*18.29* Microstructure of injection molded and liquid phase sintered Mo–15Cu. The sample was sintered at 1450°C for 3 h in $H_2$.

quality control of every step of the injection molding process, starting from the reception of the powder.

### 18.5.5 Infiltration

Infiltration is an attractive processing route for Mo–Cu, since the sintering behavior of the Mo skeleton is improved when the Cu is not present. Also, coarser powders can be used, since sintering temperatures above 1400°C can be utilized without the concerns of Cu evaporation. Further, the Mo skeletal density can be controlled by changing the sintering cycle to allow infiltration with more or less Cu. Based on a Mo sintering model (Johnson, 2008), a process map for the effect of Mo particle size and sintering temperature on density for a starting green density of 60% of theoretical and a sintering time of 60 min is given in Fig. 18.30. For the sizes of Mo powders typically available, sintering temperatures of 1400–1800°C are needed to achieve porosity levels equivalent to the target Cu volume fractions for thermal management applications.

A 4.1 μm Mo powder has been injection molded and sintered at 1450°C for 8 h to a density of about 83% of theoretical for infiltration with 15 wt% wrought oxygen-free high-conductivity (OFHC) Cu (Johnson and German, 1999). Its microstructure is shown in Fig. 18.31. The density was greater than 97% of theoretical and the grain size was 4.3 μm. Despite the longer sintering time at 1450°C, the grain size was smaller than that of the liquid-phase sintered sample in Fig. 18.29. Thus, the presence of Cu during liquid-phase sintering contributes significantly to grain growth, although it has little effect on densification.

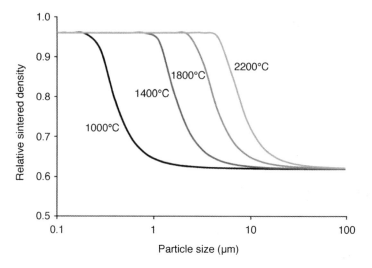

*18.30* Process map for the effect of Mo particle size and sintering temperature on density for Mo powders molded at a solids loading of 60 vol% and sintered for 1h at the sintering temperature.

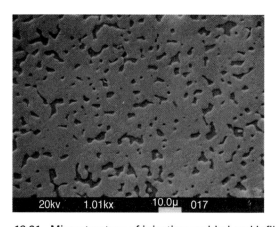

*18.31* Microstructure of injection molded and infiltrated Mo–15Cu. The

## 18.5.6 Thermal properties

The same factors of Cu content, impurities, porosity, and microstructural features that control the thermal conductivity of W–Cu also apply to Mo–Cu. The sintering temperature and cooling rate can also affect the thermal conductivity of Mo–Cu due to the higher solubility of Mo in Cu, which increases from much less than 1 wt% at 1150°C to 1.5 wt% at 1400°C (Massalski, 1986). At the higher end of this range the amount of Mo dissolved in the Cu is sufficient to lower the thermal conductivity if the cooling rate is rapid enough to prevent it from precipitating. For example, liquid-phase sintering or infiltrating at a temperature of 1400°C and furnace cooling, results in thermal conductivities of 110–130 W/(m K) despite the use of high-purity Cu and Mo powders and careful avoidance of impurities during subsequent processing steps. However, thermal conductivities of 160 W/(m K) can be achieved by infiltrating at 1150°C and furnace cooling (Johnson and German, 1999). This thermal conductivity is very close to the prediction of 169 W/(m K) for this system (German, 1993).

Cooling at a controlled rate of 1°C/min from a sintering temperature of 1400°C to 1050°C (below the melting temperature of Cu) increases the thermal conductivity from 110 W/(m K) to over 140 W/(m K), by enabling more of the Mo to precipitate. Further improvements in the thermal conductivity should be possible with a slower cooling rate or with an isothermal hold at 1150°C during the cooling cycle. Such an effect is not observed for the W–Cu system, since even at high temperatures, the solubility of W in Cu does not exceed $10^{-3}$ wt% (Ermenko *et al.*, 1976). The dissolution of 1 wt% Mo in Cu could easily decrease the thermal conductivity by 30 W/(m K), even if it is much less detrimental than Fe. For comparison, based on its effects on Cu and W–Cu, 0.1 wt% Fe is predicted to decrease the thermal conductivity of Mo–15Cu by over 30 W/(m K).

### 18.5.7 Example applications

As with W–Cu, most Mo–Cu is sold as infiltrated sheets, but MIM enables net-shape processing of more complex parts. Infiltrated Mo–15Cu heat sinks processed via powder injection molding are shown in Fig. 18.32. The density of these parts was 95% of theoretical, with practically all closed pores. No

*18.32* Picture of an infiltrated Mo–15Cu heat spreader and a small transistor package.

problems were encountered with infiltrating the threaded studs of the heat spreader. Thus, complex Mo–Cu heat sinks can be fabricated via injection molding, although an infiltration process has proven more successful than liquid-phase sintering. Less shrinkage for improved dimensional control and a lower cost Cu source are also benefits of infiltration.

## 18.6 Conclusions

High thermal conductivity materials, Cu, W–Cu, and Mo–Cu can be processed by metal injection molding. Copper can be solid-state sintered to near full density, but care is needed to avoid hydrogen swelling. Near full density can be achieved for W–Cu and Mo–Cu by either liquid-phase sintering or infiltration. The commercial availability of composite W–Cu powders with excellent sintering behavior makes liquid-phase sintering preferable for W–Cu. Because of the poor molding characteristics of fine Mo powders and their poor densification during liquid-phase sintering, infiltration techniques are recommended for Mo–Cu. With proper powder selection and good control of sintering cycles and impurities, thermal properties close to model predictions can be achieved. Metal injection molding is a net-shaping process that allows fabrication of unique geometric features that are difficult to produce with other metal-working technologies. Novel structures, such as a heat pipe with a high conductivity casing surrounding a porous wick, can be directly fabricated into complex shapes.

## 18.7 References

Babin, B R, Peterson, G P and Wu, D (1990), 'Steady-state modeling and testing of a micro heat pipe', *Journal of Heat Transfer*, 112(August), 595–601.

Chan, T Y, Chuang, M S and Lin, S T (2005), 'Injection moulding of oxide reduced copper powders', *Powder Metallurgy*, 48(2), 129–133.

Danninger, H, Gierl, C, Muehlbauer, G, Gonzalez, M S, Schmidt, J and Specht, E (2011), 'Thermophysical properties of sintered steels – effect of porosity', *International Journal of Powder Metallurgy*, 47(3), 31–42.

Dorfman, L P, Houck, D L and Scheithauer, M J (2002a), 'Consolidation of tungsten-coated copper composite powder', *Journal of Materials Research*, 17 (8), 2075–2084.

Dorfman, L P, Houck, D L, Scheithauer, M J and Frisk, T A (2002b), 'Synthesis and hydrogen reduction of tungsten–copper composite oxides', *Journal of Materials Research*, 17(4), 821–830.

Ermenko, V N, Minakova, R V and Churakov, M M (1976), 'Solubility of tungsten in copper–nickel melts', *Soviet Powder Metallurgy and Metal Ceramics*, 15, 283–286.

Fahmy, A A and Ragai, A N (1970), 'Thermal-expansion behavior of two-phase solids', *Journal of Applied Physics*, 41, 5108–5111.

German, R M (1993), 'A model for the thermal properties of liquid phase sintered composites', *Metallurgical and Materials Transactions A*, 24A, 1745–1752.

German, R M, Hens, K F and Johnson, J L (1994), 'Powder metallurgy processing of thermal management materials for microelectronic applications', *International Journal of Powder Metallurgy*, 30(2), 205–215.

German, R M and Johnson, J L (2007), 'Metal powder injection molding of copper and copper alloys with a focus on microelectronic heat dissipation', *International Journal of Powder Metallurgy*, 43(5), 55–63.

German, R M, Tan, L K and Johnson, J L (2005), 'Advanced microelectronic heat dissipation package and method for its manufacture', US Patent 6,935,022, 30 August 2005.

Gessinger, G H and Melton, K N (1977), 'Burn-off behaviour of W–Cu contact materials in an electric arc', *Powder Metallurgy International*, 9(2), 67–72.

Hayashi, K and Lim, T-W (1990), 'A consideration on incompleteness of densification of Cu, Cu–Sn, and Cu–Ni injection molding finer powders by sintering in $H_2$ gas', *PM into the 1990s, Proceedings of the World Congress on Powder Metallurgy*, Institute of Metals, London, vol. 3, pp. 129–133.

Hens, K F, Johnson, J L and German, R M (1994), 'Pilot production of advanced electronic packages via powder injection molding', in Lall, C and Neupaver, A (eds) *Advances in Powder Metallurgy*, Metal Powder Industries Federation, Princeton, NJ, USA, vol. 4, pp. 217–229.

Hinse, C, Zauner, R, Nagel, R, Davies, P and Kearns, M (2007), 'Simulation-based design for powder injection moulding', *PIM International*, 1(2), 54–56.

Ho, P W, Li, Q F and Fuh, J Y H (2008), 'Evaluation of W–Cu metal matrix composites produced by powder injection molding and liquid infiltration', *Materials Science and Engineering A*, 485, 657–663.

Jech, D E, Sepulveda, J L and Traversone, A B (1997), 'Process for making improved copper/tungsten composites', US Patent 5,686,676, November 11 1997.

Johnson, J L (1994), 'Densification, microstructural evolution, and thermal properties of liquid phase sintered composites', Ph.D thesis, Engineering Science and Mechanics, The Pennsylvania State University, University Park, PA, USA.

Johnson, J L (2008), 'Progress in processing nanoscale refractory and hardmetal powders', in Bose, A, Dowding, R J and Shields, J A (eds), *Proceedings of 2008 International Conference on Tungsten, Refractory, and Hardmaterials VII*, Metal Powder Industries Federation, Princeton, NJ, USA, vol. 5, pp. 57–71.

Johnson, J L, Brezovsky, J J and German, R M (2005a), 'Effects of tungsten particle size and copper content on densification of liquid phase sintered W–Cu', *Metallurgical and Materials Transactions A*, 36A, 2807–2814.

Johnson, J L and German, R M (1993a), 'Factors affecting the thermal conductivity of W–Cu composites', in Lawley, A and Swanson, A (eds), *Advances in Powder Metallurgy*, Metal Powder Industries Federation, Princeton, NJ, USA, vol. 4, pp. 201–213.

Johnson, J L and German, R M (1993b), 'Phase equilibria effects on the enhanced liquid phase sintering of W–Cu', *Metallurgical and Materials Transactions A*, 24A, 2369–2377.

Johnson, J L and German, R M (1996), 'Solid-state contributions to densification during liquid phase sintering', *Metallurgical and Materials Transactions B*, 27B, 901–909.

Johnson, J L and German, R M (1999), 'Powder metallurgy processing of Mo–Cu for thermal management applications', *International Journal of Powder Metallurgy*, 35(8), 39–48.

Johnson, J L and German, R M (2001), 'Role of solid-state skeletal sintering during processing of Mo–Cu composites', *Metallurgical and Materials Transactions A*, 32A, 605–613.

Johnson, J L, German, R M, Hens, K F and Guiton, T A (1996), 'Injection molding AlN for thermal management applications', *American Ceramics Society Bulletin*, 7522(8), 61–65.

Johnson, J L, Hens, K F and German, R M (1995), 'W–Cu and Mo–Cu for microelectronic packaging applications: Processing fundamentals', in Bose, A and Dowding, R J (eds), *Tungsten and Refractory Metals – 1994*, Metal Powder Industries Federation, Princeton, NJ, USA, pp. 246–252.

Johnson, J L, Lee, S, Noh, J-W, Kwon, Y-S, Park, S J, Yassar, R, German, R M, Wang, H and Dinwiddie, R B (2007), 'Microstructure of tungsten–copper and model to predict thermal conductivity', in Engquist, J and Murphy, T F (eds), *Advances in Powder Metallurgy and Particulate Materials – 2007*, Metal Powder Industries Federation, Princeton, NJ, USA, vol. 9, pp. 99–110.

Johnson, J L, Park, S J and Kwon, Y-S (2009), 'Experimental and theoretical analysis of the factors affecting the thermal conductivity of W–Cu', in Sigl, L S, Roedhammer, P and Wildner, H (eds), *Proceedings of the 17th International Plansee Seminar*, Metallwerk Plansee, Reutte, Austria, vol. 2, pp. 2/1–2/11.

Johnson, J L, Park, S J, Kwon, Y-S and German, R M (2010), 'The effects of composition and microstructure on the thermal conductivity of liquid-phase-sintered W–Cu', *Metallurgical and Materials Transactions A*, 41A, 1564–1572, 1871.

Johnson, J L, Suri, P, Scoiack, D C, Baijal, R, German, R M and Tan, L K (2003), 'Metal injection molding of high conductivity copper heat sinks', in Lawcock, R and Wright, M (eds), *Advances in Powder Metallurgy and Particulate Materials*, Metal Powder Industries Federation, MPIF, Princeton, NJ, USA, vol. 8, pp. 262–272.

Johnson, J L and Tan, L K (2004), 'Fabrication of heat transfer devices by metal injection molding', in Danninger, H and Ratz, R (eds), *Euro PM2004 Conference Proceedings*, European Powder Metallurgy Association, Shrewsbury, UK, vol. 4, pp. 363–368.

Johnson, J L, Tan, L K, Bollina, R, Suri, P and German, R M (2005b), 'Evaluation of copper powders for processing heat sinks by metal injection molding', *Powder Metallurgy*, 48(2), 123–128.

Kerner, E H (1956), 'The elastic and thermo-elastic properties of composite media', *Proceedings of the Physics Society*, 69B, 808–813.

Kim, J-C, Ryu, S-S, Lee, H and Moon, I-H (1999), 'Metal injection molding of nanostructured W–Cu composite powder', *International Journal of Powder Metallurgy*, 35(4), 47–55.

Kirk, T W, Caldwell, S G and Oakes, J J (1992), 'Mo–Cu composites for electronic packaging applications', in Campus, J and German, R M (eds), *Advances in Powder Metallurgy*, Metal Powder Industries Federation, Princeton, NJ, USA, vol. 9, pp. 115–122.

Knuewer, M, Meinhardt, H and Wichmann, K-H (2001), 'Injection moulded tungsten and molybdenum copper alloys for microelectronic housings', *Proceedings of the 15th International Plansee Seminar*, Metallwerk Plansee, Reutte, Austria, vol. 1, pp. 44–59.

Kny, E (1989), 'Properties and uses of the pseudobinary alloys of Cu with refractory metals', in Bildstein, H and Ortner, H M (eds), *Proceedings of the 12th International Plansee Seminar*, Metallwerk Plansee, Reutte, Austria, vol. 4, pp. 763–772.

Koh, J C Y and Fortini, A (1973), 'Prediction of thermal conductivity and electrical resistivity of porous metallic materials', *International Journal of Heat and Mass Transfer*, 16, 2013–2021.

Kothari, N C (1982), 'Factors affecting tungsten–copper and tungsten–silver electrical contact materials', *Powder Metallurgy International*, 14, 139–143.

LaSalle, J C, Behi, M, Glandz, G A and Burlew, J V (2003), 'Aqueous nonferrous feedstock material for injection molding', US Patent 6,635,099, 21 October 2003.

Lee, J S, Kaysser, W A and Petzow, G (1985), 'Microstructural changes in W–Cu and W–Cu–Ni compacts during heating up for liquid phase sintering', in *Modern Developments in Powder Metallurgy*, Metal Powder Industries Federation, Princeton, NJ, USA, vol. 15, pp. 489–506.

Ludvik, S, Clair, S, Krischmann, R and Clark, I S R (1991), 'Metal injection molding (MIM) for advanced electronic packaging', in Pease, L F and Sansoucy, R J (eds), *Advances in Powder Metallurgy*, Metal Powder Industries Federation, Princeton, NJ, USA, vol. 2, pp. 225–239.

Massalski, T B (1986), *Binary Alloy Phase Diagrams*, ASM, Metals Park, OH, USA.

Moballegh, L, Morshedian, J and Esfandeh, M (2005), 'Copper injection molding using a thermoplastic binder based on paraffin wax', *Materials Letters*, 59, 2832–2837.

Moon, I-H, Kim, S-H and Kim, J-C (1996), 'The particle size effect of copper powders on the sintering of W–Cu MIM parts', in Cadle, T and Narasimhan, S (eds), *Advances in Powder Metallurgy and Particulate Materials – 1996*, Metal Powder Industries Federation, Princeton, NJ, USA, vol. 19, pp. 147–156.

Moore, J A, Jarding, B P, Lograsso, B K and Anderson, I E (1995), 'Atmosphere control during debinding of powder injection molded parts', *Journal of Materials Engineering and Performance*, 4(3), 275–282.

Oenning, J B and Clark, I S R (1991), 'Copper–tungsten metal mixture and process', US Patent 4,988,386, 29 January 1991.

Parrot, J E and Stuckes, A D (1975), *Thermal Conductivity of Solids*, Pion, London.

Petzoldt, F, Knuewer, M, Wichmann, K-H and Cristofaro, N D (2001), 'Metal injection molding of tungsten and molybdenum copper alloys for microelectronic packaging', in Eisen, W B and Kassam, S (eds), *Advances in Powder Metallurgy and Particulate Materials – 2001*, Metal Powder Industries Federation, Princeton, NJ, USA, vol. 4, pp. 118–125.

Rose, R M, Shepard, L A and Wulff, J (1966), *Structure and Properties of Materials, Volume IV: Electronic Properties*, John Wiley and Sons, New York, USA.

Sebastian, K V (1981), 'Properties of sintered and infiltrated tungsten electrical contact material', *International Journal of Powder Metallurgy*, 17(4), 297–303.

Shropshire, B H, Chan, T Y and Lin, S T (2002), 'Applications of oxide-reduced copper powders in electronics cooling', *Bulletin of the Powder Metallurgy Association*, 27, 163–172.

Shropshire, B H, Klatt, K, Lin, S T and Chan, T Y (2003), 'Copper P/M in thermal management', *International Journal of Powder Metallurgy*, 39(4), 47–50.

Skorokhod, V V, Uvarova, I V and Landau, T E (1983), 'Effect of various methods of charge preparation on sinterability in the molybdenum–copper system', *Soviet Powder Metallurgy and Metal Ceramics*, 22, 185–188.

Stevens, A J (1974), 'Powder-metallurgy solutions to electrical-contact problems', *Powder Metallurgy International*, 17, 331–346.

Sundback, C A, Novich, B E, Karas, A E and Adams, R W (1991), 'Complex ceramic and metallic shapes by low pressure forming and sublimative drying', US Patent 5,047,182, 10 September 1991.

Swanson, L W (2000), 'Heat Pipes', in *The CRC Handbook of Thermal Engineering*, CRC press, New York, USA, pp. 4.419–4.430.

Sweet, J F, Dombroski, M J and Lawley, A (1992), 'Property control in sintered copper: function of additives', *International Journal of Powder Metallurgy*, 28, 41–51.

Terpstra, R L, Lograsso, B K, Anderson, I E and Moore, J A (1994), 'Heat sink and method of fabricating', US Patent 5,366,688, 22 November 1994.

Terpstra, R L, Lograsso, B K, Anderson, I E and Moore, J A (1996), 'Heat sink and method of fabricating', US Patent 5,523,049, 4 June 1996.

Tummala, R R (1991), 'Ceramic and glass-ceramic packaging in the 1990s', *Journal of the American Ceramics Society*, 74(5), 895–908.

Turner, R S (1946), 'Thermal expansion stresses in reinforced plastics', *Journal of Research NBS*, 37, 239–249.

Upadhyaya, A and German, R M (1998a), 'Densification and dilation of sintered W–Cu alloys', *International Journal Powder Metallurgy*, 34(2), 43–55.

Upadhyaya, A and German, R M (1998b), 'Shape distortion in liquid-phase-sintered tungsten heavy alloys', *Metallurgical and Materials Transactions A*, 29A, 2631–2638.

Uraoka, H, Kaneko, Y, Iwasaki, H, Kankawa, Y and Saitoh, K (1990), 'Application of injection molding process to Cu powder', *Journal of the Japan Society of Powder and Powder Metallurgy*, 37, 187–190.

Viswanath, R, Wakharkar, V, Watwe, A and Lebonheur, V (2000), 'Thermal performance challenges from silicon to systems', *Intelligent Technology Journal*, (Q3), 1–16.

Wada, N, Kankawa, Y and Kaneko, Y (1997), 'Injection molding of electrolytic copper powder', *Journal of the Japan Society of Powder and Powder Metallurgy*, 44, 604–611.

Wang, W S and Hwang, K S (1998), 'The effect of tungsten particle size on the processing and properties of infiltrated W–Cu compacts', *Metallurgical and Materials Transactions A*, 29A, 1509–1516.

Williams, C (1991), 'Design consideration for microwave packages', *Ceramic Bulletin*, 70(4), 714–721.

Yang, B and German, R M (1997), 'Powder injection molding and infiltration sintering of superfine grain W–Cu', *International Journal of Powder Metallurgy*, 33(4), 55–63.

Zlatkov, B S and Hubmann, R (2008), 'Tube type X-COOLER for microprocessors produced by MIM technology', *PIM International*, 2(1), 51–54.

# 19
# Metal injection molding (MIM) of soft magnetic materials

H. MIURA, Kyushu University, Japan

**Abstract**: Soft magnetic materials need several characteristics to realize high performance. Powder metallurgy is an effective way to produce complex-shaped parts, and also to decrease the eddy current loss at high frequency by subdividing the eddy current area by use of small grains. In particular, the metal injection molding (MIM) process allows nearly full dense and net shaping of a variety of engineering materials. The application of the MIM process to hard and brittle materials such as ferromagnetic materials demonstrates the potential of this novel process. This chapter considers the processing of three types of soft magnetic materials, namely Fe–6.5Si, Fe–9.5Si–5.5Al and Fe–50Ni alloy compacts through MIM techniques using different types of powders to obtain high performance in the soft magnetic properties.

**Key words**: metal injection molding, Fe–6.5Si, Fe–9.5Si–5.5Al, Fe–50Ni, soft magnetic properties, process conditions.

## 19.1 Introduction

In this chapter, soft magnetic properties of three different injection-molded materials will be introduced. Soft magnetic materials are used for electromagnetic applications such as motors, transformers, sensors, and so on, because of their high magnetic induction with small magnetic field. Recently, these parts have been required to be small in size, with high output and high efficiency. In order to realize these properties, soft magnetic materials need to have iron loss minimized at high frequencies. Therefore, in this chapter, three types of ferromagnetic materials which are the most attractive materials for ferromagnetic industrial applications will be discussed as follows.

The first type is Fe–6.5Si. This alloy has excellent soft magnetic properties; however, Fe–6.5Si alloy has become well known as a brittle material, so it provides poor deformation properties. Powder metallurgy is an effective way to produce complex-shaped parts, and also decreases the eddy current loss at high frequency by subdividing the eddy current area into small grains. However, it is difficult to obtain high density by conventional die-press molding of the powder. It is hoped that metal powder injection molding (MIM) could be a suitable processing technique for parts with complicated shapes, even when using hard and brittle materials. The magnetic properties of MIM compacts are expected to be better than those of conventional die-press sintered compacts, owing to the high densities achievable via MIM (Maeda *et al.*, 2005; Miura *et al.*, 1996; Shimada *et al.*, 2003). In the first section, the various magnetic properties of MIM parts of Fe–6.5Si will be discussed as a function of the type of powders, processing techniques and so on.

The second type is Fe–9.5Si–5.5Al. Fe–Ni and Fe–Si alloys are well known as soft magnetic materials (Tasovac and Baum, 1993). Specifically, Fe–9.5Si–5.5Al alloy is called a Sendust, and is suited for a magnetic head for a video tape recorder because of its high permeability and high magnetic flux density, as well as high wear resistance (Kawaguchi *et al.*, 1967). However, the applications of this alloy have been limited due to its poor workability. In the second section, the production feasibility of Fe–9.5Si–5.5Al alloys will be discussed, by MIM process, and also the soft magnetic properties of injection-molded compacts. In particular, the effects of different types of powder, grain size and impurities (such as carbon and oxygen) on the magnetic properties are investigated in detail to obtain high-performance magnetic properties by changing the process conditions; also, some factors affecting the properties will be discussed, taking the microstructures into account.

The third type is Fe–50Ni. For the MIM process of Fe–50Ni (permalloy), mixed elemental powder and prealloyed powder were used as the raw materials. In the case of the mixed elemental powder, obtaining high relative density is difficult because sintering is progressed in the area of austenite, for which the diffusion rate is low. On the other hand, high relative density can be expected with prealloyed powder because sintering is progressed in the area of ferrite. Also, the microstructure of prealloyed powder compacts appears to be more homogeneous than that of mixed elemental powder compacts. Therefore, in the final section of this chapter, the effect of mixed elemental and prealloyed powders on the magnetic properties of injection-molded permalloys will be discussed.

*Table 19.1* The characteristics of Fe–6.5Si

| Powder | $W_H$ | $G_M$ | $G_L$ |
| --- | --- | --- | --- |
| Routes | Water-atomized powder | Gas-atomized powder | Gas-atomized powder |
| Carbon (mass%) | 0.006 | 0.005 | 0.006 |
| Oxygen (mass%) | 0.31 | 0.09 | 0.004 |
| Mean particle size (μm) | 29.8 | 36.3 | 38.0 |
| Particle photo | | | |

## 19.2 Fe–6.5Si

### 19.2.1 Experimental procedure

Metal powders used in this study were three types of prealloyed Fe–6.5%Si powders. Table 19.1 shows the characteristics of those powders. Water- and gas-atomized powders with different oxygen contents were used. These powders are denoted $W_H$, $G_M$ and $G_L$ according to different atomization methods and oxygen contents. Each powder and wax-based multicomponent binder system (paraffin wax (69%), polypropylene (20%), Carnauba wax (10%), and stearic acid (1%)) were mixed at 423 K for 3.6 ks by 65% of powder loading.

In order to investigate the magnetic properties, toroidally-shaped compacts were produced by injection molding. The cavity size was 45 mm outer diameter, 30 mm internal diameter and 5.6 mm thick. Green compacts were debound at 348 K for 18 ks in an n-heptane atmosphere and this was followed by thermal debinding in a hydrogen atmosphere. Sintering was performed at 1423~1623 K for 1~3 h using electric furnaces in a hydrogen atmosphere. Figure 19.1 shows the heat pattern of thermal debinding and sintering steps, and Table 19.2 lists the sintering conditions. No secondary heat treatment was carried out.

Density measurement by Archimedes method, hardness testing (HRa), optical observation, elemental analysis for carbon and oxygen, and magnetic characterization were performed for each sintered compact.

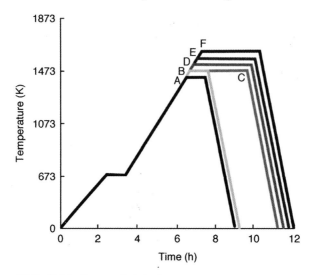

*19.1* Debinding and sintering programs.

*Table 19.2* Sintering conditions

| Pattern | Temperature (K) | Time (h) |
|---|---|---|
| A | 1423 | 1 |
| B | 1473 | 1 |
| C | 1473 | 3 |
| D | 1523 | 3 |
| E | 1573 | 3 |
| F | 1623 | 3 |

## 19.2.2 Relative density, hardness, mean grain size and chemical composition

The relationship between relative density and sintered temperature is shown in Fig. 19.2. The sintered densities of all compacts increase with increasing sintered temperature. In particular, the sintered density of $G_L$ appears higher than the others, even at lower sintering temperatures. The relationship between hardness and relative density is shown in Fig. 19.3. The hardness also increases with increasing relative density.

Figure 19.4 shows the optical micrographs of non-etched and etched cross-sections of the compacts sintered at 1623 K for 3 h. The relationship between the mean grain size and relative density is shown in Fig. 19.5. The grain size of $G_L$ appears to be about ten times larger than the other compacts. This is thought to be because the middle stage of sintering (high

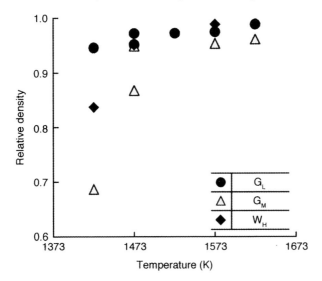

*19.2* Relationship between relative density and sintered temperature.

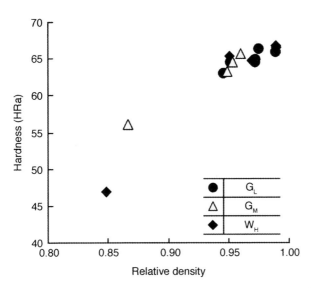

*19.3* Changes of hardness with respect to relative density.

densification) ends within a short time and the later stage of sintering (grain growth) begins because the oxygen level of the powder is low.

The relationship between the carbon content of sintered compacts and the relative density is shown in Fig. 19.6, and the relationship between the oxygen content and the relative density is shown in Fig. 19.7. In both cases, carbon and oxygen contents are not affected by relative density. Carbon

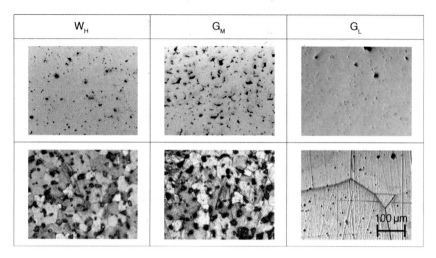

*19.4* Optical micrographs of compacts sintered at 1623 K for 3 h

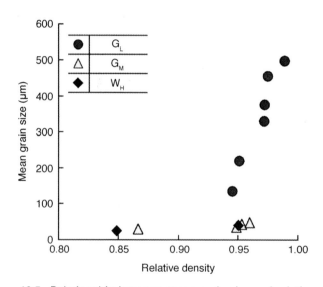

*19.5* Relationship between mean grain size and relative density.

contents of higher density compacts appear much lower than that of raw powders, as shown in Table 19.1. Oxygen contents of $W_H$ and $G_L$ are also decreased from their raw powders, but that of $G_M$ is increased and causes the lower density. Normally, the magnetic properties are mostly influenced by impurities, so that the high purity of $G_L$ is expected to show high magnetic performance.

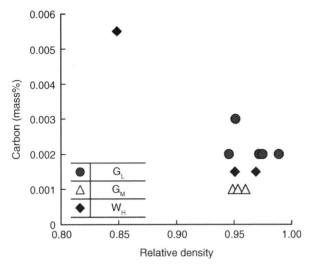

*19.6* Relationship between carbon content and relative density.

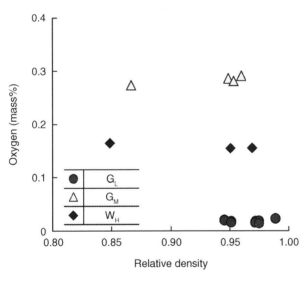

*19.7* Relationship between oxygen content and relative density.

## 19.2.3 Magnetic properties

Excellent soft magnetic material shows high magnetic induction ($B$) (Lall, 1992), high maximum permeability ($\mu_m$), low coercive force ($H_C$) and low core loss ($W_{B/F}$). Magnetic properties for wrought material of Fe–6.5%Si are compared to those of sintered compacts.

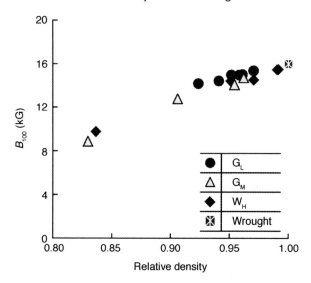

*19.8* Relationship between magnetic induction and relative density.

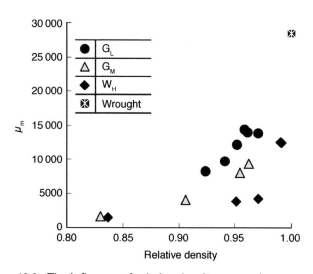

*19.9* The influence of relative density on maximum permeability.

The relationship between magnetic induction ($B_{100}$) and relative density is shown in Fig. 19.8. In all compacts, the magnetic inductions are increased in proportion to their relative densities. Moreover, the compacts with higher density show high magnetic induction, equivalent to the value of wrought material. The relationship between maximum permeability ($\mu_m$) and relative density is shown in Fig 19.9. The maximum permeability increased with increasing relative density. The relationship between coercive forces ($H_C$)

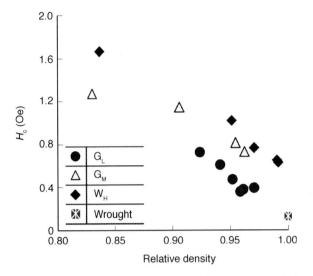

*19.10* Changes in coercive force with respect to relative density.

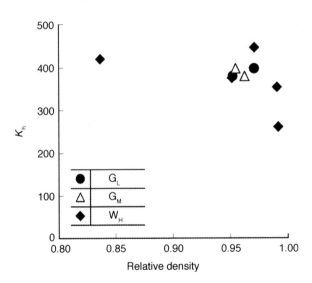

*19.11* Hysteresis loss in each specimen.

and relative density is shown in Fig. 19.10. The coercive forces are decreased with increasing relative density, and $G_L$ shows a lower value than the others. These results indicate that the sintered compacts with high density and high purity show good magnetic properties.

Figures. 19.11 and 19.12 show the hysterisis loss ($K_h$) and eddy-current loss ($K_e$) as a function of relative density. There is no apparent strong relation between either of these losses ($K_h$ and $K_e$) and density, but the eddy-

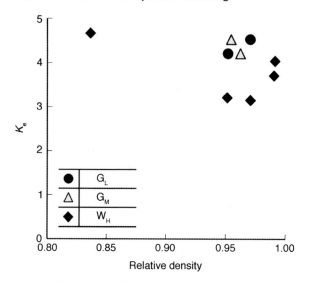

*19.12* Eddy-current loss in each specimen.

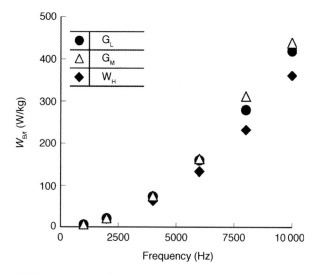

*19.13* Iron loss of selected specimens at various frequencies.

current loss of $W_H$ is lower than the others. Figure 19.13 shows the iron loss ($W_{B/F}$) at various frequencies for three types of specimens, for which the relative densities are around 0.97. Although the iron losses at lower frequency are almost the same, there are large differences at higher frequency. The iron loss of $W_H$ shows the smallest value, which is considered to be because the effect of eddy current loss on the iron loss is large. In this case, not only the purity, but also the mean grain size seems to

*Table 19.3* Chemical compositions of Fe–9.5Si–5.5A1

|  | Si | Al | Mn | P | S | C | N | O | Fe |
|---|---|---|---|---|---|---|---|---|---|
| Gas-atomized powder (mass%) | 9.5 | 6.1 | – | – | 0.001 | 0.02 | 0.002 | 0.09 | Bal. |
| Water-atomized powder (mass%) | 9.2 | 5.3 | 0.3 | 0.05 | 0.01 | 0.07 | – | 0.26 | Bal. |

have an effect on the magnetic properties. Therefore, the improvement of density and the control of grain growth are necessary to obtain exellent magnetic properties.

## 19.3  Fe–9.5Si–5.5Al

### 19.3.1 Experimental procedure

The gas-atomized powder (mean particle size: 12.7 µm) was used for the investigation. The chemical compositions of the powders are shown in Table 19.3. The powders were first admixed with wax-based multicomponent binders (paraffin wax (69%), polypropylene (20%), Carnauba wax (10%) and stearic acid (1%)) at 423 K for 3.6 ks. After injection molding, extraction debinding in condensed solvent, which was conducted at 348 K for 18 ks in n-heptane, was used to partially remove the wax components. Following this treatment, final thermal debinding was performed at 1103 K in hydrogen. Sintering was performed at 1473–1543 K for 3.6 ks in hydrogen or a vacuum. Sintered compacts were machined to the toroidal shape of 9 mm outside diameter, 6 mm inside diameter and 3 mm thick, and the compacts were annealed at 1173 K for 21.6 ks in a vacuum to eliminate the residual stresses. The maximum magnetic flux density ($B_{10}$), residual magnetic flux density ($B_r$) and coercive force ($H_c$) were measured by a B-H loop tracer. The microstructures were also examined by photomicroscopy.

### 19.3.2 Magnetic properties of gas-atomized powder compacts

Figure 19.14 shows the effect of sintering atmosphere and temperature on the relative density of injection-molded Sendust alloy compacts using gas-atomized powder. Although there is a little difference between the compacts sintered at low temperatures in hydrogen and in a vacuum, both densities reach 98% of theoretical at over 1523 K.

Figure 19.15 shows the microstructures of the compacts sintered at various temperatures in a vacuum and in hydrogen. Both compacts sintered at 1473 and 1493 K show many pores. On the other hand, the compact sintered at 1523 K shows relatively high densification and still retains a small grain size. The compact sintered at 1543 K shows clearly the grain growth

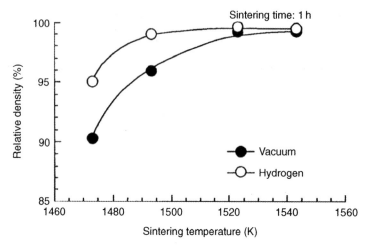

*19.14* Effect of sintering temperature and atmosphere on the density of MIM Sendust alloy compacts using gas-atomized powder.

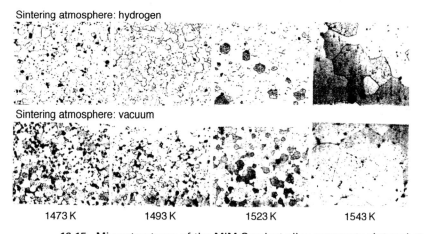

*19.15* Microstructures of the MIM Sendust alloy compacts sintered at various temperatures in a vaccum and in hydrogen.

due to the liquid phase sintering, and the compact showed some distortion. Therefore, optimum sintering temperature for the injection-molded compacts seemed to be 1523 K.

The magnetic properties such as maximum magnetic flux density ($B_{10}$), residual magnetic flux density ($B_r$), and coercive force ($H_c$) of the compacts sintered at various temperatures in a vacuum and in hydrogen are shown in Fig. 19.16. Normally, high $B_{10}$, $B_r$ and $H_c$ are suitable for soft magnetic materials. In Fig 19.16, $B_{10}$ values are little changed, but $B_r$ and $H_c$ decrease with increasing sintering temperature. Decreasing the $B_r$ and $H_c$ is a favorable tendency, but the values were relatively inferior to those of wrought materials.

Metal injection molding of soft magnetic materials 499

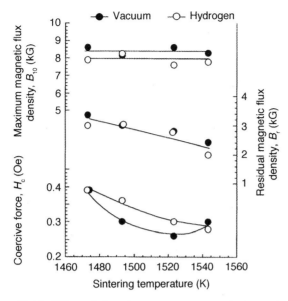

*19.16* Effect of sintering temperature and atmosphere on the magnetic properties of MIM Sendust alloy compacts using gas-atomized powder.

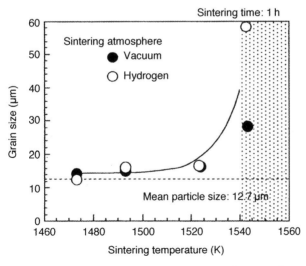

*19.17* Effects of sintering temperature and atmosphere on the grain size of MIM Sendust alloy compacts.

Generally, the magnetic properties are strongly influenced by the grain size and impurities such as carbon and oxygen (Lall, 1992). Figure 19.17 shows the grain size of compacts sintered at various temperatures. The grain sizes are little changed except in the liquid phase sintered compacts (at 1547 K). Most are similar to the original powder particle size. On the other

(a) Conventional process

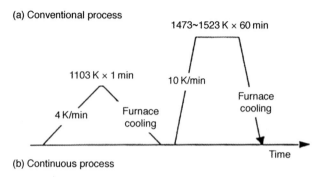

(b) Continuous process

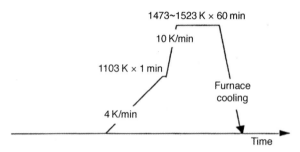

*19.18* Heating conditions for (a) conventional and (b) continuous processes.

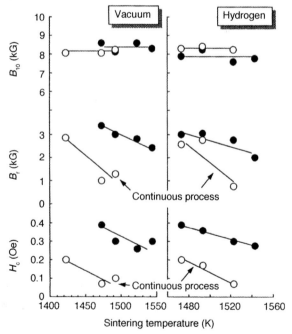

*19.19* Effect of continuous process on the magnetic properties of MIM Sendust alloy compacts using gas-atomized powder.

hand, carbon content decreased dramatically from 0.02 to 0.003 mass%, but oxygen content increased significantly from 0.09 to 0.27 mass% after sintering. From these results, a small grain size and large amount of retained oxides seem to prevent the domain wall migration, resulting in insufficient magnetic properties for soft magnetic materials.

In order to avoid the oxidation during debinding and sintering, a continuous process combining the debinding and sintering steps was performed, with the heating conditions as shown in Fig. 19.18. Figure 19.19 shows the magnetic properties of the compacts sintered by conventional and continuous processes. $B_{10}$ values are little changed even by changing the process. However, $B_r$ and $H_c$ are significantly decreased by the continuous process. In particular, the effectiveness of the continuous process seems to appear at low sintering temperature.

Figure 19.20 shows the relative densities of sintered compacts by both conventional and continuous processes. Obviously high densification occurs at low sintering temperature in the case of the continuous process. Figure

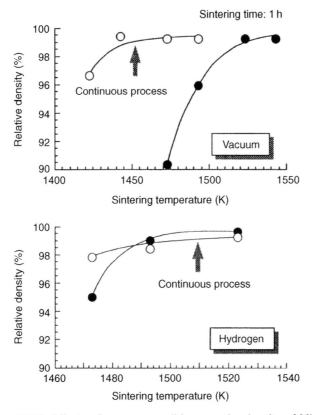

*19.20* Effects of process conditions on the density of MIM Sendust alloy compacts.

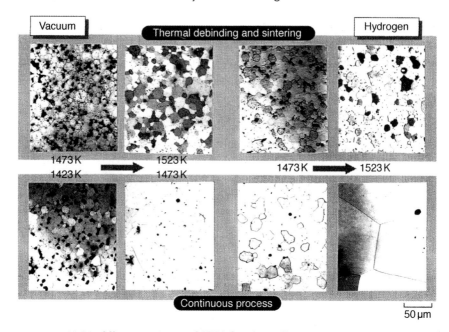

*19.21* Microstructures of MIM Sendust alloy compacts produced by different processes.

19.21 shows the microstructures of the compacts sintered by conventional and continuous processes. As would be expected, for the continuous process, there is significant grain growth at lower temperature in a vacuum and at a similar temperature in hydrogen. However, for the conventional process the grain sizes are little changed, even by increasing the temperature.

For the continuous process, low oxygen content and large grain size are obtained, as shown in Fig. 19.22. This contributes to the improvement of soft magnetic properties of injection-molded Sendust alloys.

## 19.3.3 Different magnetic properties between gas- and water-atomized powder compacts

Figure 19.23 shows the effect of sintering temperature on the density, hardness and carbon content of injection-molded compacts using water- and gas-atomized powders. Density and hardness are increased with an increase in temperature, and there is little difference between the compacts. On the other hand, magnetic properties of water- and gas-atomized powder compacts sintered at various temperatures are shown in Fig. 19.24. There is little change in $B_{10}$. However, $B_r$ and $H_c$ are very different between the powder compacts. The water-atomized powder compact shows relatively poor magnetic properties.

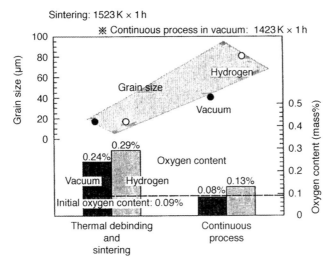

*19.22* Grain size and oxygen content of MIM Sendust alloy compacts produced by different processes.

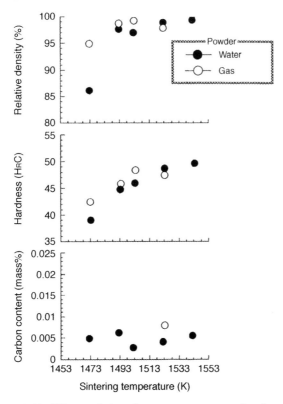

*19.23* Effects of sintering temperature on the density, hardness and carbon content of MIM Sendust alloy compacts using gas- and water-atomized powders.

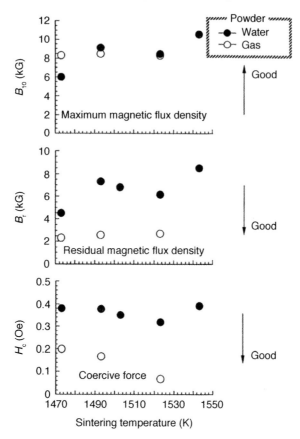

*19.24* Effects of sintering temperature on the magnetic properties of MIM Sendust alloy compacts using gas- and water-atomized powder.

Figure 19.25 shows the microstructures of both powder compacts sintered at various temperatures. The water-atomized powder compact shows fine grain size even at high sintering temperatures. Also, Fig.19.26 shows the carbon and oxygen contents of as-received powders and as-sintered compacts. After sintering, the carbon content of the water-atomized powder compact was decreased adequately under the standards. However, the oxygen content was not decreased. Specifically, the water-atomized powder compact still retained a high oxygen content. Therefore, the water-atomized powder compacts show relatively poor magnetic properties, as shown in Table 19.4. On the other hand, the gas-atomized powder compacts show excellent properties; because of the small amount of oxides and the coarse grain size, as mentioned earlier, their properties are close to that of the wrought materials.

Metal injection molding of soft magnetic materials 505

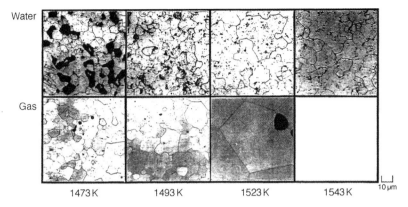

*19.25* Microstructures of MIM Sendust alloy compacts using gas- and water-atomized powders, sintered at various temperatures.

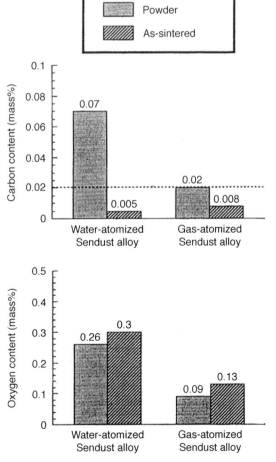

*19.26* Carbon and oxygen contents of the powders and sintered compacts.

*Table 19.4* Magnetic properties of the Sendust alloys produced by various processes

|  | MIM Gas | MIM Water | PM | Wrought |
|---|---|---|---|---|
| Relative density (%) | 98 | 99 | 92 | 100 |
| Maximum magnetic flux density, $B_{10}$ (kG) | 8.3 | 8.9 | 6.3 | 10.0 |
| Residual magnetic flux density, $B_r$ (kG) | 0.78 | 5.6 | 2.0 | – |
| Coercive force, $H_c$ (Oe) | 0.07 | 0.32 | 0.1 | 0.02 |

*Table 19.5* Chemical composition of Fe–50Ni

*Prealloyed powder*
Mean particle size: 14.2 μm

|  | Ni | Si | C | O | S | Fe |
|---|---|---|---|---|---|---|
| Mass% | 50.1 | 0.93 | 0.012 | 0.055 | 0.005 | Bal. |

*Mixed elemental powder*
Mean particle size: 4.5 μm                     Mean particle size: 5.6 μm

| Mass% | Fe | C | O | N | Mass% | Ni | C | O |
|---|---|---|---|---|---|---|---|---|
| Carbonyl Fe (L powder) | Bal. | 0.03 | 0.48 | – | Atomized Ni | Bal. | 0.003 | 0.37 |

## 19.4 Fe–50Ni

### 19.4.1 Experimental procedure

A gas-atomized powder was prepared as prealloyed powder, and the carbonyl Fe powder and atomized Ni powder were mixed in the ratio of 50 to 50 as mixed elemental powders. The composition of each powder is shown in Table 19.5. The binder consists of paraffin wax (69 mass%), polypropylene (20 mass%), carnauba wax (10 mass%) and stearic acid (1 mass%) as previously. Powders and binders were mixed in the ratio of 60 to 40 and kneaded at 418 K for 1.8 ks. Disc-type specimens (diameter: 30 mm, height: 8.3 mm) were prepared by injection-molding machine and solvent debinding was carried out to remove the wax component. Finally, thermal debinding and sintering were carried out continuously in a hydrogen atmosphere. After sintering, relative density and the amount of carbon and oxygen were measured. Ring-type specimens (outside diameter: 23 mm, inside diameter: 13 mm, thickness: 6.5 mm) were also prepared for measurement of the magnetic properties, and annealed at 1373 K for 7.2 ks in a hydrogen atmosphere to remove the work strain.

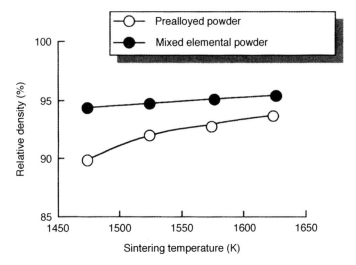

*19.27* Effect of sintering temperature on the sintered density of MIM Fe–50 mass%Ni compacts.

## 19.4.2 Magnetic properties

The sintered density of prealloyed powder compact and mixed elemental powder compact is shown in Fig. 19.27. With the increase of sintering temperature, the sintered density also increases. For all sintering temperatures, the mixed elemental powder shows higher sintered density compared with the prealloyed powder. The optical microstructure of both compacts for each sintering temperature is shown in Fig. 19.28. In the case of the prealloyed powder compact sintered at 1473 K, a lot of connected pores are observed. However, the number of connected pores decreases with increase of sintering temperature, and at 1623 K, which shows high density, circular-like pores were observed. On the other hand, the mixed elemental powder compact shows fine circular pores at all sintering temperatures. The grain size of both powders is increased to the same degree by increase of sintering temperature. For the compacts sintered at 1623 K, the mixed elemental powder compact shows large grain size. Heterogeneous microstructure of the mixed elemental powder compact could be caused by the segregation of Ni and Fe; however, it was not observed by optical microscope.

The maximum permeability, coercive force and saturation flux density are generally estimated when discussing soft magnetic properties. For soft magnetic materials, high maximum permeability and low coercive force are desirable for demagnetization. Figure 19.29 shows the results for the above characteristics. Maximum permeability of both powder compacts increases with increase of sintering temperature. However, the mixed elemental powder shows a significantly lower value of maximum permeability. Also, in

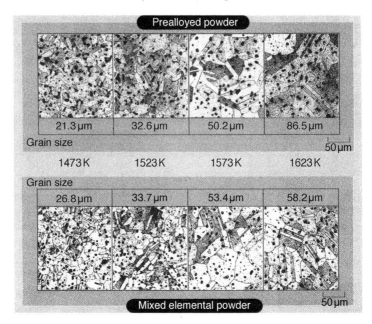

*19.28* Microstructures of MIM Fe–50 mass%Ni compacts sintered at various temperatures for 3.6 ks in hydrogen.

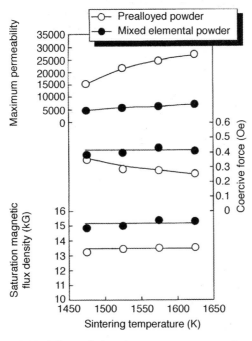

*19.29* Effect of sintering temperature on the magnetic properties of MIM Fe–50 mass%Ni compacts.

terms of coercive force, it might be interpreted that the prealloyed powder compact shows better magnetic properties. On the other hand, the saturation flux density of both powder compacts shows an approximately constant value. This proves that the mixed elemental powder compact exhibits good magnetic properties.

### 19.4.3 Improvement of magnetic properties

Among magnetic properties, the maximum permeability and coercive force are known to be susceptible to grain size and impurities (especially interstitial elements such as C, O, P and N) (Lall, 1992). Therefore, the amount of carbon and oxygen remaining after sintering was analyzed and the results are shown in Fig. 19.30. The carbon amount of both powders was suppressed at low levels; however, oxygen amount was at a significantly higher level in the mixed elemental powder compacts when compared to the prealloyed powder compacts. Therefore, to improve the magnetic properties of mixed elemental powder compacts, a further investigation was carried out on the following two points.

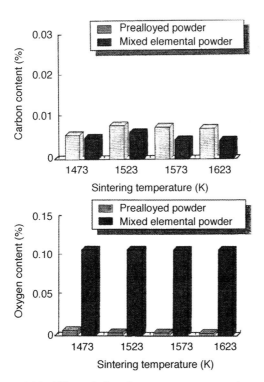

*19.30* Effect of sintering temperature on the carbon and oxygen contents of MIM Fe–50 mass%Ni compacts.

As mentioned before, the heterogeneous microstructure was not observed by optical microscope, but the diffusion degree (or concentration distribution) of each element was specifically investigated by electron probe microanalyzer (EPMA). The results are shown in Fig. 19.31. For the mixed elemental powder compact heated at 873 K, each powder particle (Fe and Ni) was observed as it is; this means that diffusion of Fe and Ni does not happen. At 1273 K, a heterogeneous microstructure was found because the diffusion of elements occurred over a partial area. However, at 1623 K, the mixed elemental powder compact shows the particles' uniform concentration distribution and homogeneous microstructure, and then the degradation of magnetic properties was not seen.

Subsequently, an investigation on the effects of oxygen decrement was carried out. First, the effect of heating rate of thermal debinding on decreasing oxygen was examined. The heating pattern shown in Fig. 19.32

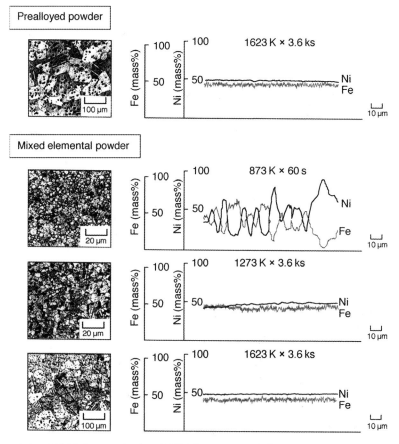

*19.31* EPMA line analysis of MIM Fe–50 mass%Ni compacts sintered under various conditions.

Metal injection molding of soft magnetic materials 511

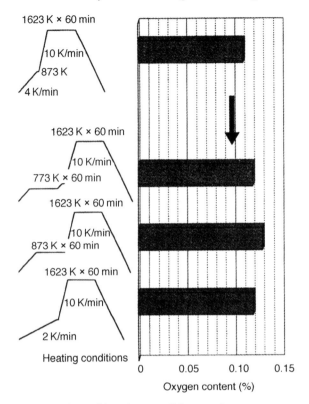

*19.32* Effect of heating condition on the oxygen content of MIM Fe–50 mass%Ni compacts.

(heating rate 4 K/min until 874 K, holding for 60 s, and then heating up at 10 K/min to sintering temperature) was applied. To promote the deoxidation in a hydrogen atmosphere, the compact was kept at 773 K and 873 K for 3.6 ks; however, a decrease of oxygen amount was not confirmed, as shown in Fig. 19.32. The compacts were also heated slowly up at 2 K/min to 873 K; however, the amount of oxygen was not changed. From the above results, it is clear that heating rate up to sintering temperature does not cause a decrease in oxygen amount.

Next, an attempt was made to promote deoxidation by using high-carbon carbonyl Fe powder, including 0.8 mass% carbon, as in Table 19.6. The results are shown in Fig. 19.33. The amount of oxygen is decreased from 0.14% to 0.01%, and it was also confirmed that the amount of carbon is not changed when compared to the low-carbon carbonyl Fe powder. The magnetic properties of all compacts are shown in Fig 19.34. Sintering was performed at 1623 K for 3.6 ks because this was the condition to obtain favorable magnetic properties of the mixed elemental powder compact. Higher maximum permeability and low coercive force were obtained by

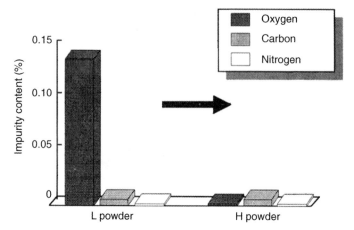

*19.33* Effect of powder type on the impurity contents of MIM Fe–50 mass%Ni compacts.

| Powder | Prealloyed powder | Mixed elemental powder | |
|---|---|---|---|
| | | H powder | L powder |
| Microstructure | | | |
| Maximum permeability | 26 600 | 12 600 | 7100 |
| Coercive force (Oe) | 0.26 | 0.22 | 0.42 |
| Saturation magnetic flux density (kG) | 13.6 | 15.5 | 15.4 |

*19.34* Microstructures and magnetic properties of MIM Fe–50 mass%Ni compacts using various powders.

*Table 19.6* Chemical composition of carbonyl Fe powder with high carbon

| Mean particle size: 4.2 μm | | | | |
|---|---|---|---|---|
| Mass% | Fe | C | O | N |
| Carbonyl Fe (H powder) | Bal. | 0.79 | 0.34 | 0.76 |

high-carbon carbonyl Fe powder; therefore, the magnetic properties could be improved by decreasing the oxygen amount of the mixed elemental powder compact. As compared to the prealloyed powder compact, a favorable coercive force was also obtained. However, the maximum permeability was not improved, thus the reason for this low maximum permeability was investigated.

| Powder | Prealloyed powder | | Mixed elemental powder | |
|---|---|---|---|---|
| Microstructure | | | | |
| | Rapid cooling | Furnace cooling | Rapid cooling | Furnace cooling |
| Maximum permeability | 27 000 | 26 600 | 19 500 | 12 600 |
| Coercive force (Oe) | 0.23 | 0.26 | 0.20 | 0.22 |
| Saturation magnetic flux density (kG) | 13.7 | 13.6 | 15.6 | 15.5 |

*19.35* Effect of rapid cooling after annealing on the microstructures and magnetic properties of MIM Fe–50 mass%Ni compacts.

The difference between the powders is the grain size, which is larger about 30 μm for mixed elemental powder, and maximum permeability is not seen to be greatly different for the small difference in grain size. It has been reported that the order lattice is formed by cooling rate in the Fe–Ni system and that the permeability decreases (Konuma, 1996), thus an X-ray diffraction test was conducted. The formation of $FeNi_3$ was confirmed in both powder compacts, because the quantification of $FeNi_3$ was difficult. Thus a higher cooling rate after annealing was attempted in order to suppress the formation of $FeNi_3$, and the effect on magnetic properties was investigated. The results are shown in Fig 19.35. The magnetic properties of the prealloyed powder compact were not much affected by cooling rate. On the other hand, the magnetic properties of the mixed elemental powder compact were significantly improved by quenching. More precisely, in the case of low cooling rate after annealing, many $FeNi_3$ phases are formed in the mixed elemental powder compact and the maximum permeability is decreased. The reason is not clear and this should be investigated specifically. Finally, supposing the heat treatment condition is optimized, the magnetic properties of the mixed elemental powder compact also could be improved as compared to those of the prealloyed powder compact.

## 19.5 Conclusion

*Fe–6.5Si:* MIM compacts from three types of prealloyed Fe–6.5Si powders with different oxygen contents were produced, and the sintered properties and magnetic properties were investigated. The relative density of low-oxygen compacts was higher than the others; however, the grain growth was remarkably higher. The compacts using low-oxygen powder also show high purity after sintering. Lower-oxygen compacts show higher maximum permeability and coercive force; however, remarkable grain growth occurred and increased the iron loss at higher frequencies. Both high density and control of grain growth are necessary to obtain the best soft magnetic properties.

*Fe–9.5Si–5.5Al:* the properties and microstructures of injection-molded Sendust alloy compacts were strongly dependent on the type of powder and the processing parameters. In particular, the grain size and retained oxides are major factors which affect the magnetic properties. A continuous process, combining the debinding and sintering steps, offered 98% of theoretical density for both the gas-atomized and water-atomized powder compacts. However, the water-atomized compacts showed fine grain size and a large amount of retained oxides, resulting in poor magnetic properties. On the other hand, the gas-atomized powder compacts showed coarse grain size and a reduction of oxides, even under the same sintering conditions, which offered excellent magnetic properties close to those of wrought materials.

*Fe–50Ni:* permalloys fabricated by MIM process using different raw powders were investigated for their magnetic properties. The prealloyed powder compact showed low relative density at about 94%, and it was possible to suppress the remaining carbon and oxygen amount at a low level. For the mixed elemental powder compact, high relative density of 96% was obtained, compared to the prealloyed powder compact, and the segregation of Fe and Ni was not observed. However, the oxygen amount was significantly increased and not such good magnetic properties were obtained. It was possible to suppress the oxygen amount after sintering in the mixed elemental powder compact by using carbonyl Fe powder including high carbon, because the carbon promotes deoxidation. However, no improvement in magnetic properties was seen. For the mixed elemental powder compact, the formation of the intermetallic compound $FeNi_3$ was a problem, and the magnetic properties were improved by quenching after annealing; it is thought that the optimization of heat treatment condition is a very important factor.

## 19.6 References

Kawaguchi, T., Tamura, K., and Yamamoto, H. (1967), 'Powder rolling of sendust alloys and their electromagnetic properties', *Journal of the Japan Society of Powder and Powder Metallurgy*, 14, 20–27.

Konuma, M. (1996), *Magnetic Materials*, Engineering Book Corporation, Tokyo Japan.

Lall, C. (1992), *Soft Magnetism: Fundamentals for Powder Metallurgy and Metal Injection Molding, Monographs in P/M Series No. 2.* MPIF, Princeton, NJ, USA.

Maeda, T., Toyota, H., Igarashi, N., Hirose, K., Mimura, K., Nishioka, T., and

Ikegaya, A. (2005), 'Development of ultra low iron loss sintering magnetism material', *SEI Technical Review*, 166.

Miura, H., Yonezu, M., Nakai, M., and Kawakami, Y. (1996), 'Influence of process condition on magnetic characteristic of soft magnetic material by MIM process', *Powder and Powder Metallurgy*, 43(7), 858–862.

Shimada, Y., Matsunuma, K., Nishioka, T., Ikegaya, A., Itou, Y., and Isogaki, T., (2003), 'Development of efficient sintering soft magnetic material', *SEI Technical Review*, 162.

Tasovac, M. and Baum, Jr, L. W. (1993), 'Magnetic properties of metal injection molded (MIM) materials'. *Advances in Powder Metallurgy and Particulate Materials*, 5, 189–204.

# 20
Metal injection molding (MIM) of high-speed tool steels

N. S. MYERS, Kennametal Inc., USA and
D. F. HEANEY, Advanced Powder Products, Inc., USA

**Abstract:** Tool steels are inherently hard and difficult to machine, thus, net shape manufacturing by MIM is a viable option for manufacturing to promote cost savings. In this chapter an introduction into the MIM of these materials, focused on high-speed steels, is presented. A review of this class of steels is presented. This is followed by a discussion of the difficulties of processing these alloys by MIM. The difficulty is primarily associated with sintering – shape retention and carbon control. The shape retention is difficult in the high-speed tool steels because these alloys sinter by liquid phase sintering and the sintering window can be as small as 15°C in the case of M2 high-speed steel. Carbon control is a universal concern for all MIM processes, particularly if hydrogen-free atmospheres are used. Finally some example processing conditions and resulting microstructures are presented.

**Key words:** high-speed steels, M2, M4, T15, sintering, microstructure.

## 20.1 Introduction

Tool steels are a family of steels that contain dispersed carbides in a hardened steel matrix. These steels are used in metal-cutting, tool and die, and many other hot and cold wear applications. The carbides typically present in these steels are vanadium-rich MC type, W- and Mo-rich $M_6C$ type, and depending on the alloy composition, Mo-rich $M_2C$, and chromium-rich $M_{23}C_6$ type. The MC carbides are the most wear resistant of all.[1–3] The $M_{23}C_6$ carbides precipitate at lower temperatures, and during heat treatment. Heat treatment includes austenitizing and quenching, similar to traditional steel; however, because of the high carbon content, retained austenite is present after the quench. Austenitizing is done near the solidus temperature, where all of the $M_{23}C_6$ and most of the $M_6C$ are taken into solution. Usually two tempering operations are performed. The first tempers the martensite formed from the first quenching, and also precipitates $M_6C$ and $M_{23}C_6$ carbides. This depletes the retained austenite of carbon, allowing it to transform to martensite during cooling. The second tempering operation will temper the newly formed martensite.[2]

Traditionally, tool steels were ingot cast, which results in segregation of alloying elements during solidification and large carbide precipitates that form stringers upon hot working. In the 1960s, hot isostatic pressing (HIP) of large, spherical-shaped, gas-atomized tool steel powders to full density was developed. The HIPed billet is hot worked into various near-net shapes. This method avoids the segregation inherent to casting and, owing to the small precipitate size in the powders, results in a finer microstructure with well-dispersed carbides. As a result of the refinement of the microstructure, HIP tool steel will possess higher toughness and similar hardness compared to an ingot cast tool steel of the same composition. The development of the HIP process also gave way to development of higher alloyed tool steels such as CPM10V (AISI A11) and CPM9V (lower carbon and vanadium content version of AISI A11) that could not be cast.[1]

Cold compacted and sintered water-atomized powders were demonstrated for several tool steel grades in the 1980s and early 1990s, allowing increased net shaping ability over HIP high-speed tool steels with microstructure and performance similar to HIP – hot worked tool steels.[4–8] Powder injection molding (PIM) of tools steels followed in the late 1990s, allowing for further increased shaping ability.[9–11] PIM tool steels can utilize the fine cuts (−325 mesh, −400 mesh) of the gas-atomized powders, which are not desirable for HIP. Common PIM high-speed tool steel grades are M2, M4, T15, and M42. MIM tool steel components find application in medium- to high-volume wear applications – such as cutting bits, crimping jaws, and other tooling components. An example MIM high-speed tool steel cutting bit application is shown in Fig. 20.1. Other tool steel grades such as A2 and S7 are in MIM production; however, their processing is different from high-speed tool steels.

## 20.2 Tool steel MIM processing

In this section a review of the MIM process is presented in the context of processing of high-speed tool steels. Where applicable a comparison with other MIM processing concerns is presented.

### 20.2.1 Feedstock

Owing to a spherical shape and size similar to stainless steel powders, feedstock formulation and injection molding of gas-atomized high-speed

*20.1* MIM high-speed steel cutting bits. Components on the left are as molded and the components on the right are heat treated, black oxided and ground sharpened.

*Table 20.1* Typical $D_{90} < 22\,\mu m$ gas-atomized high-speed tool steel powder characteristics

| Alloy | Powder size distribution (µm) | | | Pycnometer density (g/cm³) |
|---|---|---|---|---|
| | $D_{10}$ | $D_{50}$ | $D_{90}$ | |
| M2 | 5.7 | 11.2 | 18.6 | 8.02 |
| M4 | 7.8 | 14.2 | 22.8 | 8.02 |
| T15 | 7.8 | 10.7 | 18.3 | 8.18 |
| M42 | 8.2 | 15.7 | 23.6 | 8.02 |

*Table 20.2* Typical stoichiometry of a few high-speed tool steels

| Alloy | C | Mn | Si | S | P | Cr | W | V | Ni | Mo | Co | Cu | Fe |
|---|---|---|---|---|---|---|---|---|---|---|---|---|---|
| M2 | 0.80 | | 0.18 | 0.02 | | 4.10 | 6.08 | 1.93 | | 5.16 | | | Bal. |
| M4 | 1.43 | 0.41 | 0.42 | 0.13 | 0.016 | 4.42 | 5.69 | 3.9 | 0.21 | 4.5 | 0.39 | 0.10 | Bal. |
| T15 | 1.53 | 0.19 | 0.21 | 0.11 | | 4.2 | 12.0 | 4.6 | | 0.32 | 4.8 | | Bal. |
| M42 | 1.13 | 0.23 | 0.6 | 0.08 | 0.019 | 3.8 | 1.46 | 1.15 | 0.31 | 9.5 | 7.98 | 0.10 | Bal. |

tool powders are very similar to that of 316L and 17-4PH stainless steel. Typical powder properties of a few high-speed tool steel powders are given in Table 20.1. The stoichiometry of a few high-speed steels is given in Table 20.2.

Wax–polymer[9–11] and polyacetal[12] based binders have been demonstrated and are commercially available as premixed feedstock. Solids loading is typically 60–67 vol% for gas-atomized powder, and 51–63 vol% for water-atomized powder, depending on the powder size distribution and binder formulation.

Tool steels such as A2 and S7 have feedstock processing similar to the alloy steels, such as 4140 or 4605, where the base material is a carbonyl iron powder and the alloying additions are in the form of elemental additions or master alloys.

## 20.2.2 Debinding

Debinding of PIM tool steels presents some challenges in carbon control. Pure $H_2$ atmosphere results in decarburization, while $N_2$ or Ar atmospheres can result in carbon pick-up from incomplete binder burn-off, or carbon loss from reaction with oxygen in the powder or atmosphere. A mixture of 5–25% $H_2$/balance $N_2$ is often used;[9, 11–13] however, $CO/CO_2$ or $CH_4/H_2$ blends may also be used during presintering for more precise control of carbon. Temperatures of debind depend upon the temperature at which the polymer burns out in the atmosphere of interest. Thermogravimetric analysis (TGA) of the as-molded material should be performed in the atmosphere of interest to determine the proper temperature to ensure the polymer is burned out properly and not too quickly. When a polymer is burned out too quickly, it leaves carbon residue, which can drastically affect the sintering response and subsequent mechanical properties.

## 20.2.3 Sintering

Sintering of PIM tool steels is achieved by supersolidus liquid phase sintering. In prealloyed powders, liquid is formed on the grain boundaries and between particles upon heating above the solidus temperature. When a critical coverage of the grain boundaries with liquid is reached, rapid densification to near 100% density occurs. Densification may occur within as little as 10 min. This rapid densification can be clearly seen using dilatometry. A dilatometer plot of M2 tool steel is given in Fig. 20.2. Notice that little densification occurs until about 1250°C, where rapid densification occurs. This rapid densification is due to liquid formation at the grain boundaries. Too much liquid coverage of the grain boundaries will lead to distortion.[14–16] During supersolidus liquid phase sintering, grain growth is very rapid, so grain boundary area is constantly decreasing. This means that even though the liquid volume fraction may be constant once the sintering hold begins, the fraction of coverage of grain boundaries with liquid will increase during the hold time, possibly resulting in distortion.

For tool steels, microstructural problems typically precede significant distortion. $M_6C$ carbides form a eutectic liquid with the steel matrix during

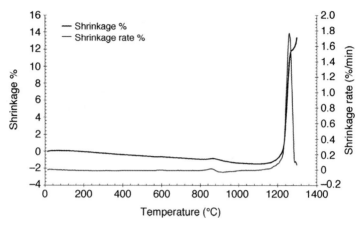

*20.2* Dilatometer plot of M2 tool steel, showing rapid densification at 1250°C.

sintering. Upon cooling, excessive liquid pools from oversintering will solidify into grain boundary carbide films or lamellar mixtures of carbide and steel, rather than discrete carbide particles. These films and lamellar structures will degrade mechanical properties by creating an easy crack propagation path along the grain boundaries. Properly sintered tool steel will precipitate discrete carbides on the prior austenite grain boundaries upon cooling. Figure 20.3 shows undersintered, properly sintered, and oversintered M2 tool steel.[12] The 'sintering window' is the allowable variation in temperature and time that will produce an acceptable density without unacceptable microstructural coarsening or distortion. Usually, for a given time, the allowable temperature variation is less than 50°C, but may be as low as 5°C. This sintering window is dictated by thermodynamics and can be illustrated by a pseudo-binary phase diagram. An example pseudo-binary phase diagram is shown in Fig. 20.4 for M2.[2] The sintering window can be approximated by the temperature range of the liquid + austenite + carbide phase region at the carbon content of M2, which is approximately 0.85%. The temperature range at this carbon content is only 13°C, from approximately 1245°C to 1258°C. Tight temperature control and short hold times are preferred for optimal processing. Highly accurate sintering control can be obtained by using well-tuned multi-zone sintering furnaces.[17]

Vacuum sintering, which aids in attaining full densification by removing gas from pores, is often used for tool steels. A furnace hot zone with graphite construction can provide the carbon potential needed to avoid decarburization; however, these furnaces are not always equipped for polymer burn-off, requiring a separate thermal debinding operation. A review of vacuum sintering configuration is found elsewhere.[17] Sintering in

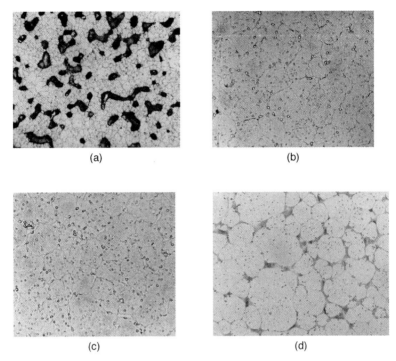

20.3 M2 tool steel: (a) undersintered; (b) properly sintered; (c) slightly oversintered, with discontinuous grain boundary carbide films; and (d) overintered, with lamellar structure.

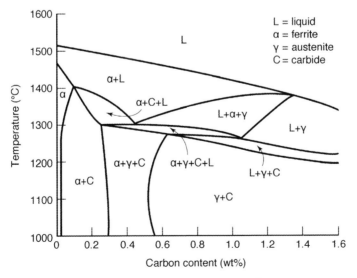

20.4 Pseudo phase diagram for M2 tool steel.[2]

*Table 20.3* Summary of published sintering temperatures in various atmospheres (given in °C)

| Sintering atmosphere | M2 | T15 | T42 |
|---|---|---|---|
| $H_2$ | $1280^4$ | $1220^8$ | $1200^8$ |
| $N_2/H_2$ | $1265–1285^4$ | $1225-1275^{4,\ 7}$ | $1215–1245^7$ |
|  | $1275–1287^{12}$ |  |  |
|  | $1270–1290^{11}$ |  |  |
| Vacuum | $1210–1220^{11}$ | $1270^8$ | $1230^8$ |
|  | $1253–1257^5$ | $1270-1285^{4,\ 7}$ | $1235–1245^7$ |

$N_2$-rich $N_2/H_2$ and $N_2/H_2/CH_4$ atmospheres have also been demonstrated, although uptake of $N_2$ into the alloy can cause formation of MX carbonitrides in place of MC carbides. This has been reported to increase the sintering window in T15, T42 and M2 tool steels during vacuum sintering, since the carbonitrides act to pin grain boundaries during sintering.[4, 7, 12] This effect is most pronounced in T15 due to the higher V content and thus a greater amount of MC is available to convert to MX carbonitride. The replacement of carbon with $N_2$ in MC carbides frees carbon to the matrix to decrease the solidus temperature, lowering the sintering temperature in steels with high V content, such as T42 and T15. Published sintering data suggest that M2 has the narrowest sintering window of the commonly used tool steels in PIM, and T15 has the widest sintering window. Table 20.3 summarizes published sintering temperatures in various atmospheres for M2, T42, and T15 tool steel. Since A2 and S7 have base carbonyl iron powders with master alloy additions, sintering is very similar to the processing of low-alloy steel.

Several researchers have demonstrated that an increase in carbon of 0.2–0.6 wt% will decrease the optimal sintering temperature by approximately 25–50°C, as well as increase the sintering window by as much as 30°C, depending on the alloy.[4–7, 13] This is achieved by a lowering of the solidus temperature with a minimal decrease in the upper phase boundary limit that leads to excessive microstructural coarsening. It should be noted, however, that a change in carbon content of this magnitude can affect the performance of the steel, and a change in grade may be a better solution.

## 20.2.4 Heat treatment

Heat treating of MIM components can be done using the same conditions as used for conventionally processed tool steels. One concern is carbon content, since the carbon content of the MIM components can vary within the specification, variability in heat treat response can occur. A best practice is to measure carbon content on each lot and adjust the heat treatment

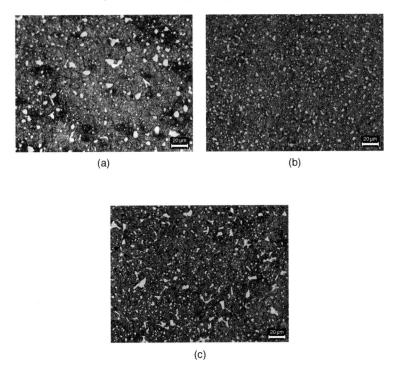

*20.5* Microstructures of (a) M4; (b) T15; and (c) M42 after heat treatment.[18]

conditions to match. Table 20.4 shows some standard heat treatment conditions for tool steels in both salt and vacuum processes. Multiple temperings are used to reduce the amount of retained austenite and increase the hardness of the tool steel. Microstructures of these high-speed tool steels after salt bath treatment are shown in Fig. 20.5.

## 20.3 Mechanical properties

The mechanical properties of MIM T15, M4, and M42 high-speed steels that experienced the heat treatment conditions shown in Table 20.4 are provided in this section.[18] These MIM samples were sintered for 30 min at 1285°C for the T15, 1260°C for the M4, and 1220°C for the M42. Table 20.5 shows the average of ten hardness measurements of the different tool steels. Notice the difference between heat treatment method and Rockwell hardness and Knoop hardness.

These same tool steels were evaluated for wear behavior using a modified ASTM G65-94 test. In this case the specimen dimensions were 6.3 mm diameter × 37 mm long and the testing conditions were 6000 revolutions at 200 r/min with a force at sample of 13 N. Wear results are shown in

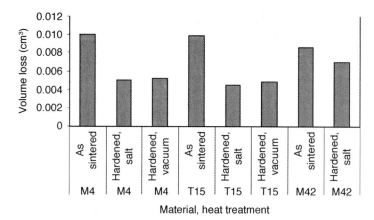

20.6 Comparative ASTM G65-94 wear results of various MIM tool steels that were heat treated using the conditions given in Table 20.4. The T15 shows the least wear in the heat treated condition.[18]

Table 20.4 Heat treatment conditions used for high-speed tool steels

| Alloy | Heat treatment | Austenitize | Quench | Temper |
|---|---|---|---|---|
| T15 | Salt bath | 1205°C, 3–5 min | 579–593°C, hold 4 min | Triple temper, 566°C, 2 h |
| T15 | Vacuum | 1177°C, 5 min | 2 bar $N_2$ to below 66°C | Double temper, 538°C, 2 h |
| M4 | Salt bath | 1205°C, 3–5 min | 579–593°C, hold 4 min | Triple temper, 566°C, 2 h |
| M4 | Vacuum | 1177°C, 5 min | 2 bar $N_2$ to below 66°C | Double temper, 538°C, 2 h |
| M42 | Salt bath | 1177°C, 3–5 min | 579–593°C, hold 4 min | Triple temper, 566°C, 2 h |

Table 20.5 Average hardness of various MIM tool steels using heat treatment conditions of Table 20.4

| Alloy | Heat treatment | As-sintered hardness, Rockwell C | HT hardness, Rockwell C | Knoop microhardness, converted to HRC |
|---|---|---|---|---|
| T15 | Salt bath | 52.0 | 60.5 | 64.0 |
| T15 | Vacuum | 50.5 | 61.5 | 64.5 |
| M4 | Salt bath | 52.5 | 62.0 | 63.0 |
| M4 | Vacuum | 52.5 | 63.5 | 65.5 |
| M42 | Salt bath | 51.0 | 63.0 | 65.0 |

Fig. 20.6. Wear resistance is slightly greater for steels heat treated in a salt bath than for steels treated in a vacuum. T15 and M4 exhibit better wear resistance than M42.

## 20.4 References

1. Dixon, R. B., Stasko, W. and Pinnow, K. E. (1998), *ASM Handbook, Volume 7, Powder Metal Technologies and Applications*, ASM International, Materials Park, OH, USA.
2. Hoyle, G. (1988), *High Speed Steels*, Butterworths, Boston, MA, USA.
3. Boccalini, M. and Goldenstein, H. (2001), 'Solidification in high speed steels', *International Materials Reviews*, 46(2), 92–115.
4. Jauregi, S., Fernandez, F., Palma, R. H., Martinez, V. and Urcola, J. J. (1992), 'Influence of atmosphere on sintering of T15 and M2 steel powders', *Metallurgical Transactions A*, 23, 389–400.
5. Wright, C. S. and Ogel, B. (1993), 'Supersolidus sintering of high speed steels: Molybdenum based alloys', *Powder Metallurgy*, 36(3), 213–219.
6. Wright, C. S. and Ogel, B. (1995), 'Supersolidus sintering of high speed steels: Tungsten based alloys,' *Powder Metallurgy*, 38(3), 221–229.
7. Urrutibeaskoa, I., Jauregi, S., Fernandez, F., Talacchia, S., Palma, R., Martinez, V. and Urcola, J. (1993), 'Improved sintering response of vanadium-rich high speed steels', *International Journal of Powder Metallurgy*, 29(4), 367–378.
8. Kar, P., Saha, B. and Upadhyaya, G. (1993), 'Properties of sintered T15 and T42 high speed steels', *International Journal of Powder Metallurgy*, 29(2), 139–148.
9. Miura, H. (1997), 'High performance ferrous MIM components through carbon and microstructural control', *Materials and Manufacturing Processes*, 12(4), 641–660.
10. Zhang, H. (1997), 'Carbon control in PIM tool steel', *Materials and Manufacturing Processes*, 12(4), 673–679.
11. Liu, Z. Y., Loh, N. H., Khor, K. A. and Tor, S. B. (2000), 'Microstructural evolution during sintering of Injection molded M2 high speed steel', *Materials Science and Engineering A*, 293(1–2), 46–55.
12. Myers, N. S. and German, R. M. (1999), 'Supersolidus liquid phase sintering of injection molded M2 tool steel', *International Journal of Powder Metallurgy*, 35(6), 45–51.
13. Dorzanski, L. A., Matula, G., Varez, A., Levenfeld, B. and Torralba, J. M. (2004), 'Structure and mechanical properties of HSS HS6-5-2 and HS12-1-5-5 type steel produced by modified powder injection moulding process', *Journal of Materials Processing Technology*, 157–158, 658–668.
14. German, R. M. (1990), 'Supersolidus liquid phase sintering, Part I: Process review', *International Journal of Powder Metallurgy*, 26(1), 23–34.
15. German, R. M. (1990), 'Supersolidus liquid phase sintering, Part II: Densification theory', *International Journal of Powder Metallurgy*, 26(1), 35–43.
16. German, R. M. (1997), 'supersolidus liquid phase sintering of prealloyed powders', *Metallurgical and Materials Transactions*, 28A, 1553–1567.
17. Heaney, D. F. (2010), 'Vacuum sintering', in *Sintering of Advanced Materials*, ed. Fang, Z., Woodhead Publishing Limited, Cambridge, UK, pp. 189–221.
18. Heaney, D. F. and Mueller, T. (2002), 'Heat treat response of metal injection molded high speed tool steels', *Proceedings of PM² TEC 2002 World Congress*, 16–21 June, Vol. 10, p 223, MPIF, Princeton, NJ, USA.

# 21
Metal injection molding (MIM) of heavy alloys, refractory metals, and hardmetals

J. L. JOHNSON, ATI Firth Sterling, USA,
D. F. HEANEY, Advanced Powder Products, Inc., USA and
N. S. MYERS, Kennametal, Inc., USA

**Abstract:** This chapter begins with an overview of the uses of heavy alloys, refractory metals, and hardmetals. The characteristics and preparation of refractory metal-based powders for metal injection molding (MIM) are discussed. Examples of binders and mixing techniques for preparing injection molding feedstock from these powders are described. Processing conditions for liquid-phase sintering MIM heavy alloys, including W–Ni–Fe and W–Ni–Cu, are described with specific emphasis on minimizing distortion while obtaining the desired properties. Solid-state and activated sintering of MIM refractory metals and alloys is reviewed with discussion of specific processing conditions for W, Re, Mo, Ta, and Nb, including the effects of alloy or oxide additions. Hardmetal compositions, primarily based on WC–Co, are reviewed and processing conditions for liquid-phase sintering are described with an emphasis on carbon control.

**Key words:** metal injection molding, heavy alloys, refractory metals, hardmetals, solid-state sintering, activated sintering, liquid-phase sintering.

## 21.1 Introduction

Refractory metals, heavy alloys, and hardmetals are typically fabricated by powder processing methods. Their main constituents are extracted from ores as powders and their high melting temperatures prevent processing by conventional melt-based forming technologies. Metal injection molding (MIM) is an attractive forming technique for these materials since it reduces the need for secondary operations. These secondary operations can prove expensive for refractory metals and heavy alloys owing to both the high cost of material scrap and the difficulty in working or machining them. Hardmetals cannot be machined and must be ground, so near-net shaping is of even greater advantage for complex shapes that cannot be formed by traditional means (Roediger *et al.*, 2000). Since processing of all of these materials usually starts with powders, MIM is not at a material cost disadvantage. In fact, MIM provides a second significant cost advantage in reduced loss and easy recycling of costly material. Thus, MIM of refractory metal and hardmetal components, even large ones that may be more technically challenging, offers a compelling cost benefit.

Metal injection molding has been demonstrated for many refractory alloys, including tungsten, tungsten heavy alloys, tungsten–copper, rhenium, molybdenum, molybdenum–copper, niobium-based alloys, titanium, titanium alloys, titanium aluminides, and several others (German, 2001). These metals, alloys, and composites possess unique combinations of strength, hardness, ductility, toughness, density, and temperature resistance, and thus they form an important class of materials that have been accepted for MIM. Applications range from ordnance components to electrical and medical electrodes.

In comparison to stainless steel MIM, where the powders are typically spherical, refractory metal, heavy alloy, and hardmetal powders have high surface area to volume ratios and irregularly shaped particles which cause the feedstock formulations to have lower quantities of powders and higher quantities of polymer for a moldable viscosity. This lower solids loading of powder requires the tooling to have greater shrinkage as compared to stainless steel and iron-based MIM. Also, these materials are sintered at a much higher temperature as compared to more conventional stainless steels and iron-based alloys. Thus, the processing is more along the lines of ceramic injection molding, where the feedstock formulations have lower solids loading and the sintering temperatures are higher. The significant difference is that refractory metals, heavy alloys, and hardmetals are typically sintered in hydrogen or vacuum, whereas ceramics are typically sintered in air.

This chapter provides an overview of the application and processing of this important class of metals, alloys, and composites. Applications, alloy additions, powder preparation, sintering, and mechanical properties are discussed.

## 21.2 Applications

As their names suggest, refractory metals are typically used in high-temperature applications; heavy alloys are used where high density is desired and hardmetals are used where high wear resistance is required. Heavy alloys and hardmetals are based on W, which is the most common refractory

*21.1* Tungsten heavy alloy disk drive counterweight.

*Table 21.1* General properties of refractory metals

| Property | Mo | Nb | Re | Ta | W |
|---|---|---|---|---|---|
| Melting point (°C) | 2617 | 2468 | 3180 | 2996 | 3410 |
| Density (g/cm$^3$) | 10.2 | 8.6 | 21.0 | 16.6 | 19.3 |
| Thermal expansion (ppm/°C) | 4.8 | 7.3 | 6.2 | 6.3 | 4.5 |
| Thermal conductivity (W/m per°C) | 142 | 52 | 71 | 54 | 166 |
| Electrical resistivity (μohm-cm) | 5.4 | 14.4 | 18.5 | 13.1 | 5.3 |
| Tensile strength (20°C, MPa) | 1030 | 550 | 1380 | 340 | 2070 |
| Modulus (20°C, GPa) | 330 | 130 | 450 | 190 | 410 |
| Crystal structure (20°C) | BCC | BCC | HCP | BCC | BCC |

Source: Lipetzky (2002).

metal. General properties of W and some other refractory metals are given in Table 21.1.

Tungsten's melting temperature of 3410°C is the highest of any metal. It has high strength, high density, high electrical and thermal conductivity, and a low thermal expansion coefficient. These properties make W useful for filaments, electrodes, electrical contacts, heat sinks, shaped charge liners, and specialty springs and fasteners. Alloys of tungsten and combinations of Ni, Fe, Co, and Cu are called tungsten heavy alloys (WHAs). They consist of W grains in a metal matrix and are used for shielding or collimating energetic x- and γ-radiation, kinetic energy penetrators, electrodes, and various types of weights. Smaller WHA components, such as cell phone vibrator weights, medical electrodes, and small inertia products are routinely injection molded. An example of a MIM WHA counterweight is given in Fig. 21.1.

Tungsten can be combined with carbon to form tungsten carbide, which has very high hardness. Tungsten carbide can be combined with metals, such as Co, Ni, and/or Cr, to produce hardmetals, also known as cemented carbides. These composites consist of WC particles 'cemented' together in a ductile metal matrix and have unique combinations of hardness and toughness, which make them widely used for tools for high-speed machining, components in earth-boring bits, road planing, and wear parts such as dies, anvils, and rolls. Some example WC–Co components that have been molded are percussive mining tips (Martyn and James, 1994), watchcase rings (Baojun *et al.*, 2002), and other complex shaped tools for wear and cutting applications (Bruhn and Terselius, 1999; Roediger et al., 2000).

Molybdenum is chemically similar to W, but its lower density is an advantage for applications where weight is important. Mo and Mo–Cu are used as electrodes, electrical contacts, shaped charge liners, and heat sinks. Mo is also used for specialty springs and fasteners.

Rhenium has the second highest melting temperature of any metal. Unlike W, Re has a ductile-to-brittle transition well below room temperature. Rhenium has the highest tensile and creep rupture strength of the refractory metals and is virtually inert to thermal shock. Its wear resistance is second only to Os, and it also has the second highest strain hardening coefficient of the refractory metals. The main disadvantages of Re are its scarcity and high oxidation rate. Owing to its high cost and the difficulty in shaping it, commercial fabrication of Re and its alloys is limited. Most applications are military, dealing with rocket systems that require high-temperature strength and high ductility.

Tantalum is the only refractory metal that is biocompatible and approved for use in long-term surgical implants as specified by ASTM F560 (ASTM, 2011). Its melting temperature and density are lower than those of W, but much higher than those of more commonly used metals. It is soft, ductile, and easy to machine. The largest application for Ta is in electrolytic capacitors owing to the dielectric properties of its oxide ($Ta_2O_5$), but it is also used for surgical implants in the forms of wire, foils, sheets, clips, staples, and meshes. It has also been considered as an implantable counterweight for some medical devices.

Niobium is chemically very similar to Ta. It has a lower melting temperature and at $8.57\,g/cm^3$, Nb has the lowest density of all the refractory metals and is thus a leading candidate for applications where the strength to weight ratio is a critical requirement, including some specialty springs and fasteners.

## 21.3 Feedstock formulation concerns

Feedstocks for refractory metals, heavy alloys, and hardmetals are similar in powder/polymer ratio to ceramic injection molding and similar in polymer formulation to other metal systems. Typically, the solids loading is in the 50–60 vol% range and the polymers can be of the wax–polymer, water-soluble, or catalytic variety. These solids loadings translate into average tooling scale-up factors of 1.18 to 1.26, assuming that the part is sintered to full density. The powder content in the feedstock is limited by the high surface area of finer particles (1–4 μm) as compared to typical gas-atomized metals (5–20 μm), which have a powder loading of up to 67 vol%.

### 21.3.1 Powder preparation

Owing to their high melting temperatures, refractory metal powders are generally produced by thermally processing chemical precursors. The nature of the chemical precursor and the temperature, time, and atmosphere of the thermal processing cycle determine the powder characteristics, which affect molding and sintering behavior.

Tungsten and Mo powders are typically produced by hydrogen-reduction of oxides. Tungsten particle sizes are typically 3–5 µm, but can range from about 0.1 to 50 µm depending on the time, temperature, hydrogen flow rate, dew point, and depth of the powder bed (Lassner and Schubert, 1999). Molybdenum particle sizes can be adjusted over a narrow range of 1 to 6 µm (Gaur and Wolfe, 2006) by the same variables that control W particle size. Composite powders can be produced by co-reducing tungsten and molybdenum oxides with other metal oxide powders.

After reduction, W and Mo powders must be deagglomerated, typically by milling, in order to produce suitable MIM feedstock. Otherwise the solids loading will be too low at a moldable viscosity. They also often need to be combined with alloying additions, such as Ni and Fe in the case of WHAs. The deagglomerated W or Mo powder can be dry mixed with other metal powders using a double-cone, v-cone, or Turbula mixer, or the reduced W or Mo powder can be co-milled with the alloying additions.

Tungsten carbide powders are produced by carburizing W powders. The characteristics, especially the particle size, of the WC powder depend primarily on the starting W particle size and the carburization temperature and time. Chemistry control is critical. Carbon contents must be held constant near the stoichiometric value of 6.13 wt%. Small amounts of oxides of vanadium and/or chromium can be added prior to carburization to control the grain size through later processing steps. VC and $Cr_3C_2$ as well as TiC, TaC, and NbC, can be added when the WC powder is milled with the metal matrix.

The conditions for milling the WC powder with the Co powder and any other additives to make hardmetal powders are key processing parameters. Two common milling techniques are ball milling and attritor milling. Both processes homogenize the mixture and can result in particle size reduction. After milling, the powders are usually spray dried. Most commercial spray-dried graded hardmetal powders contain a binder, such as wax or polyethylene glycol, which is added during milling to hold together free-flowing, spray-dried agglomerates and later to provide sufficient strength in pressed parts. Such powders can be used for MIM if the binder in the spray-dried powder is compatible with the MIM binder. Otherwise, a custom lot without a binder has to be produced or the powder has to go through a debinding cycle prior to compounding the feedstock.

Table 21.2 Characteristics of example refractory metal and hardmetal powders

| Refractory metal | W | W | Mo | Ta | Nb | WC |
|---|---|---|---|---|---|---|
| Chemistry | | | | | | |
| O (wt%) | 0.120 | – | 0.606 | 0.178 | 1.80 | 0.091 |
| C (wt%) | 0.0057 | – | 0.0017 | 0.0009 | 0.0068 | 6.14 |
| H (wt%) | – | – | – | 0.166 | 0.4000 | – |
| FSSS | 1.25 | – | – | 2.4 | – | 1.32 |
| Particle size distribution | | | | | | |
| $D_{10}$ (µm) | 0.88 | 1.5 | 2.8 | 1.8 | 3.7 | 0.9 |
| $D_{50}$ (µm) | 2.7 | 3.4 | 6.2 | 5.0 | 7.4 | 1.7 |
| $D_{90}$ (µm) | 9.8 | 6.5 | 7.0 | 7.7 | 12.4 | 3.2 |
| BET | | | | | | |
| Specific surface area (m$^2$/g) | 0.65 | 0.19 | 0.48 | 0.39 | – | 1.00 |
| Particle size (µm) | 0.49 | 1.65 | 1.23 | 0.90 | – | 0.39 |
| Pycnometer density (g/cm$^3$) | 19.0 | 19.2 | 10.1 | – | 8.4 | 15.5 |
| Apparent density (g/cm$^3$) | 3.6 | 4.1 | 1.9 | 4.7 | 2.1 | 2.7 |
| % of pycnometer | 19% | 22% | 18% | – | 25% | 17% |
| Tap density (g/cm$^3$) | 4.9 | 6.2 | 2.7 | 6.0 | 3.0 | – |
| % of pycnometer | 26% | 32% | 26% | – | 36% | – |

Hardmetal powders can also be produced from recycled scrap material via a zinc-reclaim process (Stjernberg and Johnsson, 1998) or by crushing. These 'reclaimed' powders often provide more predictable molding behavior since they have less surface area than WC powders made directly from the carburization of W.

Rhenium powders are produced by a two-stage hydrogen reduction process in which ammonium perrhenate is first reduced to an oxide before final reduction to Re metal. Different temperatures and hydrogen flow rates are used for the two stages. Typical particle sizes are 1–3 µm (Bryskin and Danek, 1991). Typical Re powders are agglomerated and have poor packing characteristics with apparent densities of 1.2–1.8 g/cm$^3$. Coarse spherical Re powders are also available for MIM; however, their sintering response is reduced owing to their lower surface area (Wang et al., 2001).

Production of Ta powders starts by reducing potassium tantalum fluoride with sodium, while production of Nb powders starts by reducing niobium oxide with aluminum (ASM, 1998a). The resulting products are subsequently melted with an electron beam, hydrided, crushed, dehydrided, and milled to produce angular particles. Alloying elements can be included in the melt. Ta and Ta alloy particle sizes range from 3 to 6 µm, while Nb and Nb alloy particle sizes range from 10 to 15 µm (ASM, 1998a). Further milling can be used to round the particles if desired.

The characteristics of some example commercial powders are summarized in Table 21.2. The most commonly used refractory metal and refractory

532     Handbook of metal injection molding

metal carbide powders have particle sizes ranging from about 1 μm to a few μm. Particle sizes below 1 μm are used in specialty applications, but increase contamination concerns owing to their high surface area. Particles coarser than a few μm can likewise be used in special situations, but their poor sintering behavior makes them less useful. Oxygen is the main impurity in refractory metal powders and can significantly lower their measured densities from their theoretical values. The agglomerated nature of the

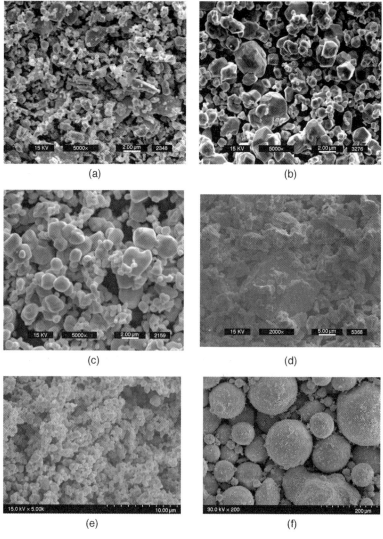

21.2   SEMs of (a) 2.7 μm W powder; (b) 3.4 μm W powder; (c) 6.2 μm Mo powder; (d) 7.4 μm Nb powder; (e) 1.7 μm WC powder; and (f) spray-dried WC–10% Co powder.

powders is indicated by their low apparent and tap densities. Their irregular particle morphologies can be seen in the scanning electron micrographs (SEMs) in Fig. 21.2, which also includes a spray-dried WC–Co powder.

## 21.3.2 Solids loading

The small, irregular particles of refractory metal and hardmetal powders result in relatively low solids loadings. As examples, Fig. 21.3 shows a plot of the torque required to mix W–Cu and Mo powders of various particle sizes at different solids loadings with a paraffin wax–polypropylene binder. From this plot the optimal solids loading for W–15Cu decreases from 58 vol% for the 0.23 μm W powder to 52.5 vol% for the 0.11 μm W powder. Similarly, the optimal solids loading decreases from 62 vol% for the 1.52 μm Mo powder to 58 vol% for the 0.47 μm Mo powder.

The optimal solids loadings for the powders in Fig. 21.3 are correlated with the green density of refractory metal powders uniaxially pressed at 430 MPa, as shown in Fig. 21.4. The predicted effects of particle size on green density for two other compaction pressures as previously modeled (Johnson, 2008) are given for comparison. Based on this plot, powders with Brunauer–Emmett–Teller (BET) particle sizes of 0.01–0.10 μm are expected to have solids loadings of 40–50 vol%.

Because of the dispersion difficulties with fine-grained powders, high shear rate continuous mixers are preferred for compounding (Hens et al., 1994; Martyn and James, 1994; Yang and German, 1998). The abrasive nature of refractory metal and especially hardmetal powders results in greater wear of mixer components, which increases processing costs. The contamination of heavy alloys and hardmetals from wear is of limited

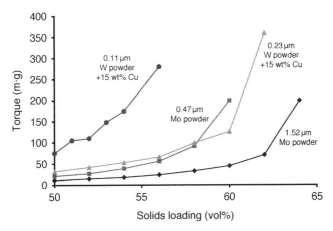

*21.3* Effect of solids loading on mixing torque for W–Cu and Mo powders with different BET particle sizes.

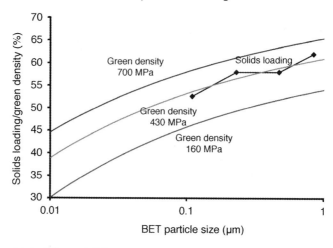

*21.4* Effect of BET particle size on the optimal solids loading in comparison to model predictions for the green density of uniaxially pressed powders.

concern since they usually contain at least small amounts of iron, but the processing and properties of refractory metals are more sensitive to contamination, as discussed in Section 21.5.2.

Examples of several heavy alloy, refractory metal, and hardmetal feedstocks are summarized in Table 21.3. Wax–polymer binders perform adequately for these materials. Solids loadings as low as 46 vol% have been reported for water-based gelling systems, wax-based low-pressure systems, and nano grain size systems (Roediger *et al.*, 2000; Youseffi and Menzies, 1997; Yunn, 2011; Zorzi *et al.*, 2003). For WHA feedstock, deagglomerating the W powder before mixing increases the maximum solids by about 3% (Suri *et al.*, 2003), but the agglomerates have no effect on the microstructure or mechanical properties of sintered parts (Suri *et al.*, 2009). For WC–Co hardmetals, the optimal surfactants are fatty acids with long carbon chain lengths, such as octadecanoic (stearic) acid (Yang and German, 1998). In general, molding and debinding are similar to other MIM feedstocks.

## 21.4 Heavy alloys

The addition of lower melting temperature transition metals such as Ni, Fe, Co, and Cu to refractory metals, especially W, can produce two-phase alloys with unique properties that can be processed at lower temperatures than commercially pure refractory metals. Limited solubility of the transition metal additions in the refractory metal and high solubility of the refractory metal in the transition metal benefit processing, but the tradeoff is decreased high-temperature performance. Still, such matrix formers enable other

*Table 21.3* Examples of heavy alloy, refractory metal, and hardmetal feedstocks

| Binder | Composition | Solids loading vol.% | Mixing technique | Reference |
|---|---|---|---|---|
| Polyethylene wax | W–4.9Ni–2.1Fe | 49 | Double planetary | Wei and German (1988) |
| Polyethylene Polystyrene Oil | W–4Ni–1Fe | 50 | Double planetary | Bose et al. (1992) |
| Paraffin wax Polypropylene | W–2.1Ni–0.9Fe | 60 | Twin-screw | Suri et al. (2003, 2009) |
| 50% microcrystalline petroleum wax 29% montan ester wax 21% synthetic hydrocarbon wax | WC–6Co | 65 | Paste mixer with trifoil blade | Martyn and James (1994) |
| 54–65% paraffin wax 30–36% polypropylene 5% octadecanoic acid | WC–11Co WC–6Co WC–15Co | 56.5 59 55 | Twin-screw | Yang and German (1998) |
| 65% paraffin wax 15% low-density polyethylene 15% vegetable oil | WC–5TiC–10Co | 57 | Roller mixer | Qu et al. (2005) |
| Paraffin wax Polypropylene Polyethylene Stearic acid | Nb | 57 | Twin-screw | Aggarwal et al. (2006) |
| Polybutene Polyethylene wax Polyethylene glycol Stearic acid | W WC WC–10Co | 55 55 56.8 | Paste mixer with impellor | Puzz et al. (2007) |
| 65% paraffin wax 10% high-density polyethylene 10% polypropylene 5% dioctyl phthalate 5% ethylene propylene diene monomer 5% stearic acid | WC–8Co | 62 | Self-made mixing device | Baojun et al. (2002) |

unique properties of refractory metals, such as high density or low thermal expansion coefficient, to be more easily utilized. In some cases, minimal alloying between the refractory metal and matrix phase is desired to produce a composite material with properties that cannot be achieved with a monolithic material, but the lack of solubility hinders densification. An example of this is the W–Cu system in which the Cu wets the W, but the W has limited solubility in the Cu matrix phase. This material system and Mo–Cu, which is chemically similar, are used for low thermal expansion heat sink applications. MIM of W–Cu and Mo–Cu is described in Chapter 18.

Tungsten alloys that are liquid-phase sintered are traditionally called tungsten heavy alloys (WHAs, as defined earlier). The most common WHAs have Ni–Fe or Ni–Cu matrices, although other transition metals such as Co, Cr, Mo, and Mn are sometimes added or substituted to improve properties or lower sintering temperatures (Bose and German, 1988a, 1990; Bose et al., 1992; Johnson and German, 2006; Srikanth and Upadhyaya, 1986). Additives that have low melting temperatures are preferred because of their low activation energies and high diffusivities at lower temperatures. The formation of intermediate compounds is generally unfavorable. High-temperature phases can lower diffusion rates, while brittle intermetallic phases that form during cooling can degrade mechanical properties.

Matrix materials increase the ductility of the alloy and also contain W in solid solution. In the case of a pure Ni addition, up to 40 wt% of W can remain in the matrix after cooling to room temperature (Massalski, 1986), however formation of the $Ni_4W$ intermetallic phase results in poor mechanical properties (Larsen and Murphy, 1965). Iron additions to W–Ni reduce both the solubility of W in the matrix and the tendency to form intermetallics. The Ni:Fe ratio should generally be in the range 1.5–4. Intermetallic formation with Ni:Fe ratios up to 15 can be reduced by resolutionizing and quenching. Ni:Fe ratios below 1.2 result in the formation of $Fe_7W_6$ intermetallics that cannot be resolutionized in a subsequent heat treatment (Caldwell, 1998).

Copper additions to W–Ni also reduce the solid solution content of W in the matrix, but are less effective at suppressing $Ni_4W$ formation (Muddle, 1984). Tungsten can also precipitate from the matrix phase during cooling, resulting in inclusions that lower strength and ductility. Rapid cooling of W–Ni–Cu alloys enables greater supersaturation of the matrix and improved mechanical properties, but the ductility is less than that of W–Ni–Fe alloys (Green et al., 1956; Muddle, 1984). The main use for W–Ni–Cu alloys over W–Ni–Fe alloys is for applications that require a non-magnetic alloy. Both types of heavy alloys can be further tailored to specific applications by adding other transition metals to adjust static and/or dynamic properties.

### 21.4.1 Liquid-phase sintering

Owing to the high diffusion rates of base (solid phase) metal atoms in a liquid, alloying additions that form a liquid phase during the sintering process often result in high densification rates, lower sintering temperatures, and greater cost effectiveness. Liquid-phase sintering (LPS) of refractory metals usually involves the addition of transition metals such as Ni (as elemental powders) that have high solubility for the refractory metal and form a thermodynamically stable second phase with a liquidus temperature below 1500°C.

Liquid-phase sintering has traditionally been broken down into initial, intermediate, and final stages (German, 1985). The initial stage, rearrangement, begins with the formation of the liquid phase. As the liquid forms an increase of the refractory metal solubility in the liquid phase enables dissolution of the solid–solid contacts that form during heating. Capillary forces due to the wetting liquid act on the solid particles and pull them into close proximity, resulting in rapid shrinkage. In the intermediate stage, solution–reprecipitation, atoms at particle contacts and other convex points dissolve in the liquid phase and diffuse to neighboring concave surfaces where they reprecipitate. As the grains change shape, they are able to pack better and release liquid to fill any remaining pores (Huppmann, 1979; Kaysser and Petzow, 1985). In addition to densification, the solution–reprecipitation process also causes grain growth via Ostwald ripening. The final stage in LPS involves continued microstructural coarsening with slower densification because of the rigidity of the solid skeleton. This stage of LPS is generally avoided in practice, particularly for the MIM process where prolonged heating can result in distortion and loss of shape retention. By this stage, densification is practically complete and further increases in grain size can degrade properties.

Most densification of WHAs occurs prior to liquid phase formation. The Ni and Fe additions, in particular, enhance the solid-state sintering of W, as discussed in Section 21.5.2. A plot showing densification as a function of temperature for W–8.4Ni–3.6Fe and W–8.4Ni–3.6Cu is given in Fig. 21.5. For W–Ni–Fe the liquid phase begins forming at 1400°C, and full densification is mostly achieved before the matrix fully melts at 1455°C (Park et al., 2006b). W–Ni–Cu is less densified when the first liquid phase forms at 1285°C, but then densifies rapidly with a second increase in densification rate occurring when the matrix fully melts at 1370°C (Sethi et al., 2009).

Once the liquid forms, the rate of grain growth increases significantly as shown in Fig. 21.6 (Park et al., 2006c). Because densification and grain growth of WHAs occur so rapidly when the liquid phase forms, they are less sensitive to the initial particle size than solid-state sintered materials. During isothermal sintering, the grain size increases in proportion to the cube root of time. The proportionality factor is called the grain-growth-rate constant and depends on the diffusivity and solubility of the W in the matrix and the volume fraction of the matrix. Higher W volume fractions result in higher

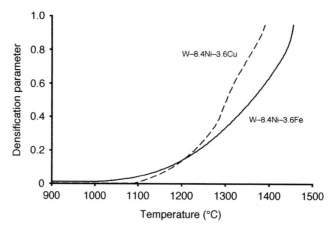

*21.5* Densification of W–Ni–Fe and W–Ni–Cu heavy alloys during heating at a rate of 10°C/min.

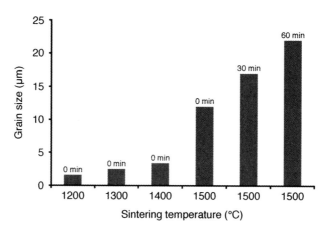

*21.6* Increase in grain size for 88W–8.4Ni–3.6Fe during heating up to 1500°C and during isothermal sintering at 1500°C. The hold times are noted. Data are from Park *et al.* (2006c). The liquid forms between 1400 and 1500°C.

grain-growth-rate constants, as shown in Fig. 21.7 (Johnson *et al.*, 2009). The matrix volume fraction has only a minor effect on densification up to about 97 wt% W, but can significantly affect distortion, as discussed later.

*Typical process cycles*

Heavy alloys are typically sintered in furnaces that are capable of 1500°C in pure hydrogen or hydrogen/nitrogen mixtures such as dissociated ammonia. Typical configurations are continuous pusher-type furnaces; however, vacuum furnaces can also be used. Dry hydrogen can be used; however,

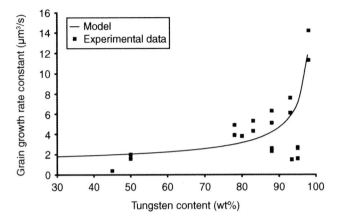

*21.7* Effect of tungsten content on the grain-growth-rate constant for LPS W–Ni–Fe heavy alloys.

wet hydrogen has become the industry standard to suppress both blister defect formation and hydrogen embrittlement. The entire process can be performed under wet hydrogen or a process that initially involves dry hydrogen prior to pore closure to obtain the maximum oxide reduction and subsequent wet hydrogen prior to pore closure and liquid formation to prevent the previously mentioned defects. Using wet hydrogen later in the sintering cycle can minimize the amount of trapped water vapor (Bose and German, 1988b; German *et al.*, 1992).

The process cycle for these alloys consists of a low temperature hold in the 900–1100°C range to obtain maximum oxide reduction and outgassing of volatiles from the powder surface prior to pore closure. Reduction of the metal oxides before pore closure is important; otherwise, water vapor can become trapped, causing residual porosity, blistering, and hydrogen embrittlement. A subsequent increase to a sintering temperature in the 1325–1500°C range with a hold of 20 min to 2 h is recommended. The time/temperature profile is a function of the alloy, desired microstructure, and size of the component. Higher sintering temperatures and longer sintering holds are recommended for higher W content alloys, the nickel–iron alloys, and property sensitive components. Higher sintering temperatures, longer sintering times, and lower W contents lead to distortion as the part slumps under its own weight (Johnson *et al.*, 1998; Park *et al.*, 2006a). Lower temperatures and shorter holds are recommended for lower W contents, nickel–copper alloys, and dimensionally sensitive components.

Micrographs of MIM microstructures are shown in Figs 21.8 and 21.9. The grains are fully rounded in Fig. 21.8 and are not fully rounded in Fig. 21.9. The slightly undeveloped microstructure of Fig. 21.9 is characteristic of a MIM component that requires superior feature definition. If a fully

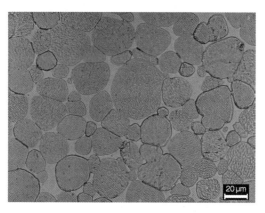

*21.8* 95W–4Ni–1Fe MIM heavy alloy sintered at 1490°C for 1 h in hydrogen gas, 35 µm grain size.

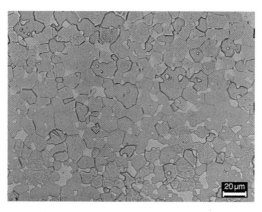

*21.9* 95W–4Ni–1Fe MIM heavy alloy sintered at 1480°C for 1 h in hydrogen gas, 10 µm grain size.

developed microstructure, as shown in Fig. 21.8, is required for particular properties, a lack of feature definition should be expected. Shape retention can also be obtained by utilization of a greater tungsten content alloy, i.e. the use of 95 wt% W as opposed to 90 wt% W.

The cooling rate also has an effect on the final properties. A slow cooling rate reduces the amount of solid solution W in the matrix, but can also result in intermetallic formation depending on the alloy composition. W–Ni–Fe alloys consisting of 90–97 wt% W are usually sintered to near full density in hydrogen at 1470–1580°C. Substituting Cu for Fe lowers the sintering temperature of the tungsten-based alloys to a range similar to the sintering temperatures of 316L or 17-4PH stainless steels. These W–Ni–Cu alloys are sintered to near full density in dry hydrogen at temperatures between 1325 and 1380°C. W–Ni–Cu alloys with lower Ni:Cu ratios form a liquid phase at lower temperatures, but have poorer densification behavior due to the lower solubility of W in the matrix.

*Distortion effects*

The high density and the liquid phase that assists densification of heavy alloys can also lead to poor dimensional stability unless sufficient contacts form between the solid grains to resist gravitational, surface tension, and frictional forces. The resistance of a LPS MIM component to distortion by these forces is determined by its viscosity during sintering. Higher amounts of liquid and larger grain sizes result in lower bulk and shear viscosities and less dimensional stability (Park *et al.*, 2006a). To avoid slumping, LPS is primarily limited to compositions with high solid volume fractions. W–Ni–Fe heavy alloys generally display distortion for W contents less than 90 wt%, corresponding to a solid volume fraction of about 0.76 (Johnson *et al.*, 1998; Kipphut *et al.*, 1988). Heavy alloys with W contents of 90 wt% or above can also slump with high sintering temperatures or long sintering times, which also results in microstructural coarsening, and property and feature fidelity degradation.

Finite element modeling can accurately predict the effects of gravity, surface tension, substrate friction, solid content, and sintering time on the shapes of sintered components (Park *et al.*, 2006a). Features to avoid with heavy alloys are cantilevers or large poorly supported sections. Larger MIM components are more susceptible to slumping due to gravity and to friction-related distortion due to sticking of the part to the substrate, while surface tension can cause rounding of features on micro-MIM parts. Figure 21.10

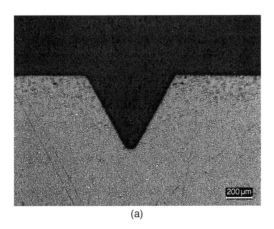

(a)

*21.10* (a) Properly sintered and (b) oversintered 90W–8Ni–2Fe MIM features. The oversintered feature shows the liquid filling the feature and rounding of the outside corners.

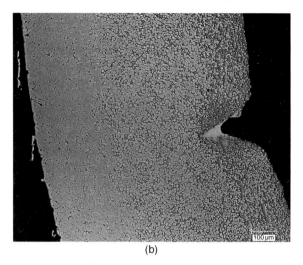
(b)

*21.10* (continued)

shows microstructures of (a) properly sintered and (b) an oversintered WHA. In the case of oversintering, the liquid phase begins to remove features owing to wetting effects at fine feature detail.

*Practical process concerns*

The primary concerns when processing heavy alloys by MIM are microstructural uniformity/ full development, distortion due to slumping, shape retention, and blistering due to an inappropriate sintering atmosphere and/or profile.

Microstructure can be affected by each process step. If the powders are not fully blended or milled, the microstructure could have regions of liquid-phase-forming elements which reduce the local properties. This microstructural behavior can also be caused by poorly deagglomerated refractory metal powder or by green state cracks that fill with liquid during sintering, giving enriched regions of refractory metal and liquid-phase-forming metal. Typically, poor milling or poor blending cause pools of liquid that are globular, whereas green state cracks produce pools of liquid that are linear in nature. Figure 21.11 shows a micrograph of this effect for a green state crack.

Distortion, which is characterized by slumping and loss of feature shape retention, occurs for all MIM systems and is primarily a function of geometry, material viscosity at sintering temperature, and density. Since heavy alloys have the greatest density of any MIM material and are liquid-phase sintered, they have the greatest propensity to distort. This distortion can be mitigated by using several techniques. The microstructure can be undersintered which minimizes the amount of liquid during sintering. This,

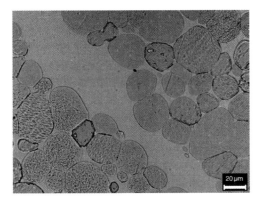

*21.11* Liquid-filled region of a tungsten heavy alloy, which had a green state crack that filled with liquid during sintering.

however, results in an undeveloped microstructure which may have inferior mechanical properties, but nominal density properties and excellent dimensional consistency. Another technique is to change the alloy to a higher quantity of solid phase, which will shift the properties, but allow for a fully developed microstructure and acceptable dimensional consistency. The final technique to reduce distortion is by the use of setters to properly support the components. Setters for heavy alloys are typically fabricated from a fiber or high alumina. This setter would be designed to support the component and minimize gravity-induced distortion or slumping. Selection of components that have minimal unsupported sections can also prevent distortion in this class of materials.

Blistering of MIM heavy alloys can be a significant issue in dry hydrogen. The primary method to eliminate these defects is to sinter in wet hydrogen to suppress the formation of water vapor voids in the microstructure, which coalesce at elevated temperatures and erupt at the surface of the component.

## 21.4.2 Properties

Tungsten heavy alloys are of engineering interest due to their uniquely high density and mechanical properties. These alloys are capable of $19\,\text{g/cm}^3$ density, 35% elongation, 43 HRC hardness, and 1380 MPa (200 ksi) ultimate tensile strength. Table 21.4 provides an overview of some typical properties.

These alloys are susceptible to hydrogen embrittlement and are exposed to hydrogen during the sintering operation, thus hydrogen outgassing can be required to enhance the ductility of the alloy. Typically, solution annealing at 900–1300°C and subsequent quenching can avoid impurity segregation and formation of intermetallic phases. Table 21.5 shows a study performed on a 95 wt% W alloy which received vacuum treatment. The

*Table 21.4* Typical heavy alloy properties

| Alloy | 90W–6Ni –4Cu | 90W–7Ni –3Fe | 95W–3.5Ni –1.5Cu | 95W–3.5Ni –1.5Fe | 97W–2.1Ni –0.9Fe |
|---|---|---|---|---|---|
| ASTM-B-777-07 | Class 1 | Class 1 | Class 3 | Class 3 | Class 4 |
| Typical density (g/cm$^3$) | 16.96 | 17.00 | 18.0 | 18.12 | 18.56 |
| Hardness (RC) | 24 | 25 | 27 | 29 | 30 |
| UTS (MPa) | 770 | 860 | 760 | 862 | 883 |
| 0.2% YS (MPa) | 517 | 610 | 590 | 620 | 586 |
| Elongation (%) | 6 | 15 | 5 | 3 | 2 |
| Elastic modulus (GPa) | 280 | 270 | 310 | 340 | 360 |
| Thermal expansion (20–400°C, ppm/°C) | 5.4 | 4.8 | 4.4 | 4.6 | 4.5 |
| Thermal conductivity (W/m per K) | 96 | 75 | 140 | 110 | 125 |
| Electrical conductivity (% IACS) | 14 | 10 | 16 | 13 | 17 |
| Relative magnetic permeability | <1.01 | 5.0–5.5 | <1.01 | 4.0–4.5 | 1.6–2.0 |

Source: Larsen and Murphy (1965).

*Table 21.5* Effect of vacuum anneal on the mechanical properties of a 95 wt% W heavy alloy

| Sintering temperature (°C) | Vacuum anneal | Density (g/cm$^3$) | YS (MPa) | UTS (MPa) | Elongation (%) |
|---|---|---|---|---|---|
| 1480 | No | 18.0 | 598 | 814 | 6.1 |
| 1480 | Yes | 18.1 | 717 | 968 | 17.3 |
| 1490 | No | 18.0 | 596 | 658 | 3.7 |
| 1490 | Yes | 18.0 | 641 | 947 | 18.7 |

material was sintered at two temperatures in pure hydrogen – 1480°C and 1490°C. After sintering, half of the samples were vacuum annealed at 1200°C for 2 h in a $10^{-6}$ torr vacuum.

## 21.5 Refractory metals

Refractory metals have many uses as described in Section 21.2, but their properties can be tailored with the addition of solution strengtheners or dispersion strengtheners. For example, W has a ductile-to-brittle transition between 200 and 500°C, but its room temperature ductility can be improved with Re additions, which increase the ability for dislocation motion by modification of lattice spacing and interatomic energies. This property improvement is known as the 'rhenium effect' (Milman and Kurdyumova, 1997) and is also observed for Mo.

Tungsten can be alloyed with Re by co-reducing ammonium perrhenate with either tungsten oxide or elemental W powder in stages at temperatures of 300–1000°C or by coating W particles (Povarova *et al.*, 1997). Rhenium can also be added in powder form to elemental W powder, but results in a less uniform Re distribution. Common alloying levels of Re range from 3 to 26 wt% for W and up to 51 wt% for Mo. Higher Re contents lead to precipitation of an intermetallic σ phase. Rhenium has a much poorer effect on group VA elements (Buckman, 1997).

Tantalum can be solution strengthened by additions of W and Hf for improved properties at temperatures over 1430°C. Niobium can also be strengthened with W and Hf additions. Most Nb and Ta alloys are relatively low-strength alloys produced by electron beam melting, high-temperature extrusion, and forging (Wojcik, 1991). MIM offers potential advantages for net-shape manufacturing of higher strength alloys, but high purities and uniform microstructures are required to attain the desired properties.

Refractory metals can be dispersion strengthened with oxides that pin grain boundaries for improved high-temperature strength. Some common examples include additions of 2 wt% $ThO_2$, $La_2O_3$, $Ce_2O_3$, or $Y_2O_3$ to W or Mo. One of the most common uses of these additions is to limit grain growth in filaments, thus, if MIM W or Mo is going to be used in a thermal application that receives cyclic heating, the use of these dopants would provide the necessary service life. A homogeneous distribution of these additives is important to achieving the desired mechanical properties. They are often added prior to hydrogen reduction of W or Mo powders. Carbides generally have lower thermodynamic stability than oxides, but can also be used to improve mechanical properties. HfC has the highest melting temperature and greatest thermodynamic stability and does not adversely affect the room temperature ductility of W (Park and Jacobson, 1997).

Refractory metals and refractory metal alloys that are solution or dispersion strengthened are usually solid-state sintered at high temperatures due to their high melting temperatures. They have been traditionally sintered by passing current through the consolidated powder to heat it directly. More current flows through the part as its density increases and its resistivity decreases. Direct sintering can achieve temperatures up to 3000°C with a few thousand amperes of current (Lassner and Schubert, 1999). Although this method is suitable for producing bars that will undergo thermo-mechanical processing for rods, wires, plates, or bars, it is not practical for sintering net-shape MIM parts. Thus, MIM refractory metals are typically sintered using radiant resistance heating, but specific conditions depend on the particular alloy, particle size, and impurities.

Sintering furnaces can also heat with microwaves instead of heating elements. The microwaves are absorbed throughout the product, which heats volumetrically. More rapid heating rates and shorter sintering cycles are possible with microwave sintering and potentially lower sintering temperatures. Tungsten has been microwave sintered to 93% of its theoretical density at a temperature of 1800°C (Prabhu et al., 2009).

### 21.5.1 Solid-state sintering

Densification of refractory metals occurs through solid-state sintering as particles in contact with each other undergo neck growth and their centers approach each other. Several sintering mechanisms can be concurrently active, but at low temperatures, grain boundary diffusion is the dominant densification mechanism of W and other refractory metals (German and Munir, 1976a). Once the porosity decreases to about 8% and the pores close, lattice diffusion begins to dominate (German and Labombard, 1982; German and Munir, 1978; Lassner and Schubert, 1999; Uskokovic et al., 1971, 1976); however, concurrent grain growth slows densification (German, 1981). In practice, densities above 96% of theoretical are difficult to achieve for solid-state-sintered refractory metals. The sintering shrinkage of all systems displays an asymptotic characteristic as the compact nears full density and can be predicted from models based on master sintering curve (MSC) concepts (Aggarwal et al., 2007; Blaine et al., 2006a, 2006b; Kwon et al., 2004; Park et al., 2006b, 2006c, 2008; Su and Johnson, 1996). These predictions can be used to construct process maps for the effects of particle size, initial density, sintering temperature, and sintering time.

Based on an MSC model (Johnson, 2008), process maps for the effect of

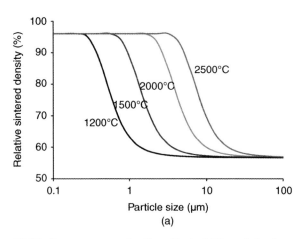

*21.12* Process maps for the effect of W particle size and sintering temperature on (a) density and (b) grain size for W powders molded at a solids loading of 55% and sintered for 10 h at the sintering temperature.

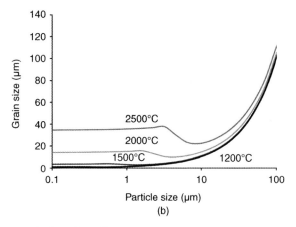

*21.12* (continued)

W particle size and sintering temperature on density and grain size for an initial density (solids loading) of 55% of theoretical and a sintering time of 600 min are given in Fig. 21.12. Even at 2500°C, W particle sizes above 10 μm show little densification or grain growth. Maximum densification can be achieved at 2000°C with particle sizes less than 2 μm with about a ten-fold increase in grain size. Although the plots show that W particles sizes of 0.5 μm or smaller can achieve near full density with little grain growth at temperatures of 1500°C or lower, the poor packing characteristics of such small particles makes achieving a solids loading of 55% very difficult.

Figure 21.13 shows the effect of increasing the solids loading on the density and grain size of a 2 μm W powder sintered at 1800°C to 2400°C for 600 min. A higher sintering temperature can partially compensate for a low

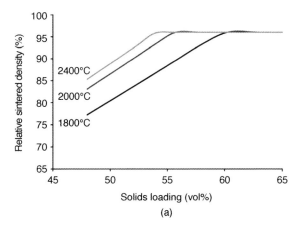

*21.13* Process maps for the effect of solids loading and sintering temperature on (a) density and (b) grain size for a 2.0 μm W powder sintered for 10 h at the sintering temperature (continued over page).

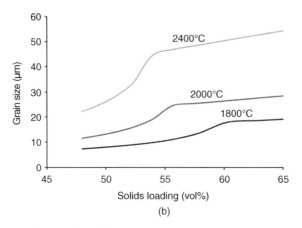

21.13 (continued)

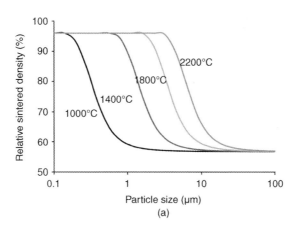

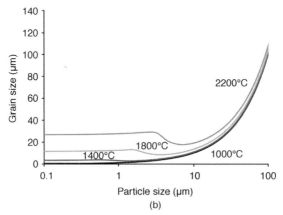

21.14 Process maps for the effect of Mo particle size and sintering temperature on (a) density and (b) grain size for Mo powders molded at a solids loading of 55% and sintered for 10 h at the sintering temperature.

solids loading, but a solids loading of at least 54 vol% is needed to achieve maximum density even at 2400°C. Grain growth becomes significant for solids loadings that enable near full density.

Using the same MSC model (Johnson, 2008), process maps for the effect of Mo particle size and sintering temperature on density and grain size for a solids loading of 55 vol% and a sintering time of 600 min are given in Fig. 21.14. The overall behavior is very similar to that of W, although the sintering temperatures are slightly lower due to the lower melting temperature of Mo and its higher diffusivity (Blaine *et al.*, 2006a).

## 21.5.2 Activated sintering

Solid-state sintering can be enhanced or 'activated' by small amounts of additives that segregate to grain boundaries and form a high-diffusivity second phase. The most effective sintering activators for W and other refractory metals are transition metals such as Co, Fe, Ni, Pd, and Pt, which have a high solubility for W and relatively low liquidus temperatures, but have limited solubility in W (German and Rabin, 1985; Petzow *et al.*, 1982; Zovas *et al.*, 1983). Additions of less than 0.5 wt% of these elements greatly reduce the sintering temperature of W and promote grain growth. Nickel and Pd are the most effective activators and can lower the sintering temperature of W by several hundred degrees Celsius (German and Ham, 1976; German and Munir, 1976b; Gessinger and Fischmeister, 1972; Li and German, 1983; Munir and German, 1977). Cobalt and Fe additions are less effective because they form intermetallic phases with W that have lower diffusivities than Ni and Pd (Li and German, 1984). Unfortunately, all of the sintering activators embrittle W.

Like W, densification of Mo is greatly enhanced by the addition of transition metals, especially Ni and Pd (Bin and Lenel, 1984; German, 1981, 1983; German and Labombard, 1982; German and Munir, 1978; Lejbrandt and Rutkowski, 1978; Zovas and German, 1984). Additions of 0.5–1 wt% Pd or Ni can lower the sintering temperature from 1800 to 1200°C (German and Labombard, 1982; German and Munir, 1978). Nickel additions to Mo form an intermetallic phase that substantially increases the diffusion rate, but is also responsible for a loss in ductility (Hwang and Huang, 2003). The ductility of Ni-doped Mo can be improved by the addition of Fe and Cu (Hwang and Huang, 2004).

The sintering of Re can be activated by Pt and Pd additions (German and Munir, 1977). Densities over 90% of theoretical have been achieved for Re with additions of 0.2 wt% Pd at temperatures as low as 1800°C (Dushina and Nevskaya, 1969). Cobalt, Fe, and Ni additions promote the densification of Nb (Samsonov and Yakovlev, 1969), although niobium oxides can act as a barrier to activated sintering up to 1600°C.

The strong effect of transition metals on the densification and properties of refractory metals necessitates special procedures to ensure that they are not inadvertently picked up from compounding mixers and from the screws and barrels of injection molding machines. These wear surfaces must be properly cleaned if an iron-based material has previously been compounded and molded. Also, the carrier polymers must be properly melted to prevent impurity pick-up from wear of these surfaces. Another concern with the use of these activators, particularly Co, Pt, and Pd, is the degradation of the carrier polymer. These elements act catalytically to break down the polymers even at room temperature. The use of antioxidants as part of the binder system is required to prevent this from occurring.

### 21.5.3 Typical process cycles

MIM refractory metal components are typically sintered at temperatures ranging from 1900 to 3050°C. The high temperatures required for sintering require the use of refractory metal setter material. For example, W sheet may be used to sinter pure W components. The tendency of refractory metals to oxidize requires reducing atmospheres or vacuum for sintering. In addition to promoting densification, the high sintering temperatures help volatilize impurities, such as alkali, earth-alkali, or transition metals, that can significantly reduce the ductility of refractory metals. Heating rates must be slow enough to prevent rapid densification and pore closure before the impurities have been evaporated. For MIM sized parts, heating rates are generally on the order of 1–4°C/min.

The process cycle for refractory metals includes removal of MIM polymers at temperatures below 600°C and a subsequent hold in the temperature range of 900–1100°C to remove oxygen and other residual impurities before heating to the final sintering temperature. One method to obtain sufficient density in a practical fashion is to debind and presinter to a near-closed pore condition in hydrogen to reduce the oxide layer from the particles and to remove other impurities. This is followed by high-temperature sintering in a vacuum. In this way, a higher temperature can be obtained in the vacuum after the hydrogen has reduced the material's particle surfaces.

Tungsten and Mo are usually sintered in 100% hydrogen with a dew point below −20°C. Tungsten components are usually sintered at temperatures of 2000–3050°C in a flowing dry hydrogen atmosphere to densities of 92–98% of theoretical with typical grain sizes of 10–30 µm (Lassner and Schubert, 1999). Molybdenum components are usually sintered in flowing dry hydrogen at temperatures ranging from 2000 to 2400°C to densities of 90–95% of theoretical (ASM, 1998b). Reduction of molybdenum oxides is important since oxygen contents above 50–200 ppm can cause brittle

## MIM of heavy alloys, refractory metals, and hardmetals

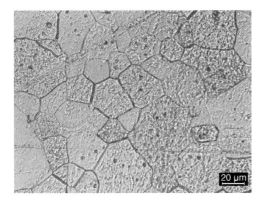

*21.15* Microstructure of pure molybdenum that was formed by MIM.

*Table 21.6* Mechanical properties of MIM Mo sintered using a hydrogen reduction up to 1400°C and vacuum sintering at 2000°C for 3 h

| Material | Density (g/cm³) | Grain size (μm) | UTS (MPa) | Elongation (%) | C (wt%) | O (wt%) |
|---|---|---|---|---|---|---|
| 1 | 92.8 | 13 | 480 | 13 | 0.01 | 0.05 |
| 2 | 95.5 | 31 | 580 | 35 | 0.01 | 0.19 |

intergranular failure. Although hydrogen is normally used, vacuum sintering at 1750°C for 10 h can result in oxygen levels as low as 170 ppm with densities of 97–98.5% of theoretical and a grain size of about 56 μm (Huang and Hwang, 2002). An oxygen partial pressure of 75 MPa or less is required for the reduction of molybdenum oxides (Huang and Hwang, 2002). A typical microstructure of MIM Mo is shown in Fig. 21.15 and typical properties are given in Table 21.6.

Net-shaped Re components can be pre-sintered at 1200–1400°C for 30 min in hydrogen to increase their handling strength followed by sintering at 2300–2700°C in hydrogen or high vacuum to densities of 97% of theoretical or higher (Leonhardt et al., 2001; Trybus et al., 2002). Sintered components can also be hot isostatically pressed to help achieve these densities. Grain sizes generally range from 25 to 50 μm. An example microstructure is given in Fig. 21.16.

Niobium and Ta form hydrides so sintering is usually performed in a vacuum of $10^{-4}$ torr or better, but can be also done in an inert atmosphere, such as Ar. They are much less tolerant of impurity pick-up than other refractory metals. Tantalum components are usually sintered in a vacuum at temperatures ranging from 2300 to 2800°C (ASM, 1998b). The high temperatures help volatilize interstitial impurities. Niobium components can be sintered to near full density at temperatures of 1900–2000°C in high vacuum (better than $10^{-7}$ torr) (Sandim et al., 2005). The effects of sintering

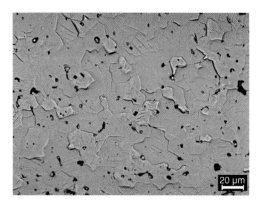

*21.16* MIM Re sintered at 2400°C, 4 h, in hydrogen and HIPed at 1660°C and 140 MPa.

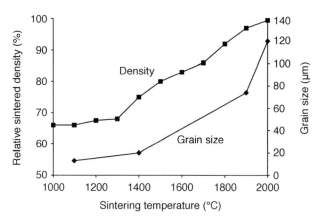

*21.17* Plot of density and grain size versus sintering temperature for Nb. Data are from Sandim *et al.* (2005).

temperature on density and grain size are shown in Fig. 21.17. As the density exceeds 90%, grain boundaries break away from pores resulting in rapid grain growth. Similar densities were achieved for MIM Nb sintered at 1800–2000°C in vacuum ($10^{-3}$ to $10^{-5}$ torr), but the resulting oxygen and carbon contents were 0.03 and 0.02 wt%, respectively, and NbC was precipitated at the grain boundaries (Aggarwal *et al.*, 2007).

### 21.5.4 Practical process concerns

The main challenge with processing MIM refractory metal components is controlling impurities and grain growth while achieving high sintered densities. Properties can be potentially improved with alloying additions, but they can increase the difficulty in sintering.

Solid-state sintering generally does not produce a fully dense component. For traditional processing of refractory metals, low sintered densities can be overcome by thermo-mechanical working, but this is not an option for net-shape MIM components. Sintering to less than 8% porosity is necessary for closed porosity to enable further consolidation by hot isostatic pressing (HIP) without external canning. Still, HIP is limited to pressures of 300 MPa at temperatures of 1700°C, which may not be sufficient for the most difficult to densify refractory metals. Other pressure-assisted consolidation techniques, such as rapid omni-directional compaction (ROC) and spark plasma sintering (SPS) have been used to densify W (Cho et al., 2006; Paramore et al., 2007), but these methods do not seem practical for net-shape MIM components.

Alloying elements can have a positive or negative effect on densification depending on specific processing conditions. For example, Re additions to W can slow densification if its particle size is larger than that of the W, but otherwise it generally promotes densification due to its slightly higher diffusivity (Solonin and Kivalo, 1982). In addition to lowering the sintering temperature of W–Re, improved homogeneity of the starting mix will also lower the temperature required to avoid inclusions from undissolved Re or σ phase precipitates. With Re coated particles, a W–5Re alloy can be completely homogenized in 30 min at 2230°C, while homogenization of the same composition produced from a blend of W and Re powders takes up to 60 h (Povarova et al., 1997). Similarly, mechanical alloying improves the homogenization of a W–25Re alloy over blending (Ivanov and Bryskin, 1997). These methods of adding Re to W can also be used to prepare other solution strengthened alloy powders such as W–Mo, Mo–Re, and Mo–W. Mo alloys such as Mo–0.5Ti, Mo–0.5Ti–0.1Zr (TZM), and Mo–1.2Hf–0.1C (MHC) can be produced by milling elemental powders or their hydrides with Mo and sintering at 1920–1980°C in hydrogen (Fan et al., 2009).

Solution strengthened Nb and Ta alloys can be processed from prealloyed hydride–dehydride powders. Preformed Nb–30Hf–9W (C-2009) can be injection molded and vacuum sintered to 94% of the theoretical density at 2250°C (Dropman et al., 1992). Full density can be achieved with subsequent containerless HIP at 1900°C for 3 h at 200 MPa with carbon and oxygen contents of 80 and 230 ppm, respectively, which are below the critical levels that cause brittle behavior. The sintering response of PIM Nb–12Hf–9W–0.1Y (C-129-Y) is much poorer, resulting in a maximum density of 84% of theoretical at a sintering temperature of 2350°C (Dropman et al., 1992). Pre-alloyed Ta–9.2W–0.5Hf powders have been HIPed at 1300–1500°C to densities of 99% of theoretical (Zhang et al., 2005b). Strengths were higher than the melted and rolled alloy, but elongations were lower. Additional pressureless sintering at 2200°C improved elongation, but decreased strength with little effect on the density.

Insoluble additions to refractory metals, such as $ThO_2$, $CeO_2$, $ZrO_2$,

$HfO_2$, $Er_2O_3$, $La_2O_3$, $Y_2O_3$, HfC, or ZrC, provide dispersion strengthening but also slow densification. These dispersoids can be added by processing appropriate precursors with the refractory metal precursor or by blending or mechanically alloying them with reduced refractory metal powders. Nitrates, such as $Th(NO_3)_4$, which can be added as aqueous solutions, are common precursors. Such precursors decompose into oxides that are thermodynamically stable even during hydrogen reduction. Hydrides, such as $ZrH_2$ can be used as precursors along with carbon additions to form *in situ* carbides. Powder forms of the dispersoids must generally have particle sizes of less than 1 μm to effectively pin grain boundaries. Typical additive amounts range from 0.4 to 4 wt%.

Sintering temperatures for dispersion strengthened W usually range from 2600 to 2800°C (Lassner and Schubert, 1999). Mo–$CeO_2$ can be sintered at 1850°C for 4 h in flowing dry hydrogen, but requires hot working (Zhang et al., 2005a). Additions of 3.5 wt% $CeO_2$ decreased the grain size to 1 μm from 17 μm for unalloyed Mo and increased the yield strength, elongation, and fracture toughness.

## 21.6 Hardmetals

The majority of hardmetals are based on combinations of tungsten carbide and cobalt. Cobalt is most commonly used in contents of 3–25 wt%, but Ni and Cr are used in applications that require enhanced corrosion resistance. The metal binder can be further modified through additional alloying. For example, Ru additions to WC–Co hardmetals significantly increase hardness without decreasing its toughness (Lisovksy, 2000; Lisovksy et al., 1991). Lower binder contents also increase hardness, but at the expense of toughness.

Hardmetal compositions can be divided into three basic types: straight grades, micrograin grades, and alloyed grades. Straight grades are primarily WC in a Co binder but may contain small amounts of grain growth inhibitors. Micrograin grades consist of WC in a Co binder with several tenths of a percent of VC and/or $Cr_3C_2$ to achieve a grain size of less than 1 μm. Alloyed grades consist of WC in a Co binder with additions of Ti, Ta, and Nb, which form separate carbide grains that have a more rounded morphology. These are commonly referred to as cubic or solid solution carbides.

Straight grades for metalworking generally contain 3–12 wt% Co. The WC grain size usually ranges from 1 to 8 μm. As the grain size decreases, hardness and strength increase, but toughness decreases. Components for rock and earth drilling tools are produced from straight grades containing 6–16 wt% Co with grain sizes ranging from 1.5 to 10 μm or larger. Straight grades for dies and punches have a medium grain size and Co contents ranging from 16 to 30 wt%.

Micrograin grades generally contain 6–15 wt% Co. The VC and/or $Cr_3C_2$ additions control grain growth during LPS, resulting in a final grain size of less than 1 µm. The fine grain size gives very high hardness and strength, which is especially useful for cutting tools for soft workpiece materials because they can be highly polished and can hold an extremely sharp cutting edge. They can also be used for machining Ni-base superalloys because of their ability to withstand temperatures up to 1200°C. Micrograin grades are also desired for rotating tools that generate shear stresses, such as drills.

Alloyed grades are primarily used for cutting steel and typically contain 5–10 wt% Co. The grain size ranges from 0.8 to 2 µm. TiC additions range from 4 to 25 wt% and reduce the tendency of WC to diffuse into steel chip surfaces. TaC and NbC additions range from 0 to 25 wt% and improve strength, cratering resistance, and thermal shock resistance. These cubic carbide additions also increase the hot hardness, which helps to avoid thermal deformation in applications where high temperatures are created at the cutting edge. Alloyed grades are also used for non-metalworking applications, primarily as wear parts. Typical grain sizes range from 1.2 to 2 µm with Co contents ranging from 7 to 10 wt%. Wear grades for applications requiring increased corrosion resistance and higher hardness are manufactured with additions of Ni and Cr.

### 21.6.1 Liquid-phase sintering

Practically all WC–Co hardmetals are processed by LPS, which was described in Section 21.4.1. Both W and C are soluble in Co, and the ternary system forms a eutectic at about 1280°C (Fernandes and Senos, 2011). Densification occurs rapidly when the eutectic liquid forms (Meredith and Milner, 1976; Snowball and Milner, 1968). The liquid phase enables achievement of densities above 99% of theoretical even with coarse WC powders and at low Co contents. Hardmetals need to be sintered to full density, since very small amounts of porosity can make them brittle. Hardmetals can be solid-state sintered to high densities, but LPS is generally required to develop the optimal microstructure.

Grain growth is controlled by the reaction at the interface of WC grains rather than by diffusion in the liquid phase, as in WHAs. Grain-growth rates for hardmetals are much lower than for WHAs, allowing for much smaller grain sizes. During isothermal sintering, the grain size increases with the square root of time. The rate of the interfacial reaction, and thus the rate of WC grain growth, can be controlled by small additions (less than 1 wt%) of other transition metal carbides, including VC, $Cr_3C_2$, NbC, TaC, TiC, and Zr/HfC, which are listed in decreasing degree of effectiveness (Morton et al., 2005).

The effect of VC additions on the grain size of WC–7Co with a 0.050 µm

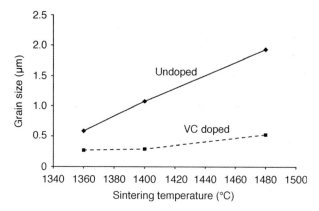

*21.18* Effect of sintering temperature on the grain size of doped and undoped WC–7Co with a starting WC particle size of 0.05 μm (Fang and Eason, 1995).

starting particle size is shown in Fig. 21.18. The undoped WC–7Co powder must be sintered below 1400°C to maintain a submicron particle size, while VC doping is able to preserve a submicron grain size even at a sintering temperature of 1480°C. Since grain growth is limited in doped hardmetals, the final grain size is generally related to the starting particle size.

Abnormal grain growth, in which a few large grains grow several times larger than the average grain size, can occur during LPS of WC–Co. In this case, a powder with a smaller average particle size can result in a larger sintered grain size than a powder with a larger average particle size. Abnormal grain growth can be avoided by using a powder with a narrow particle size distribution (Manneson et al., 2011a, 2011b). Lower sintering temperatures and the use of grain growth inhibitors also reduce the occurrence of abnormal grain growth.

## 21.6.2 Typical process cycles

Hardmetals are typically sintered in vacuum batch furnaces at 1400–1450°C, but temperatures may range from 1350 to 1600°C, depending on the Co content and the desired microstructure (Santhanam et al., 1998). The furnace hot zone is generally constructed from graphite and MIM hardmetal components are set on graphite trays coated with graphite paint. The vacuum level during sintering is usually about 0.1 Pa ($10^{-3}$ torr), but special furnaces known as sinterHIPs can backfill the furnace with argon to a pressure of up to 7 MPa while the matrix is still molten to ensure the parts are fully densified. The use of sinterHIPs has become increasingly common to virtually eliminate scrap due to residual porosity. Conventionally vacuum-sintered components can be HIPed to full density in a separate operation, but at additional cost. They can also be used in some applications without HIPing, but generally require higher sintering temperatures that can coarsen the microstructure. A smaller WC particle size and/or the addition of grain growth inhibitors can help maintain a fine grain size.

Specialized sintering cycles and adjustment of the powder chemistry can allow for enrichment or depletion of the Co at the outermost 20–30 µm of the sintered component. An enriched layer can give the component the performance of a tougher grade with the ability to resist the deformation characteristic of the lower binder content below the surface layer. A depleted layer can provide a better surface for deposition of wear-resistant coatings, such as TiC, TiN, TiCN, TiAlN, or $Al_2O_3$.

Example microstructures of a straight grade, a micrograin grade, and an alloyed grade are shown in Fig. 21.19. The straight grade contains relatively large grains in comparison to the micrograin grade, but they are still much smaller than seen in liquid-phase sintered WHAs. The fine, irregular grains make WC–Co much less susceptible to gravity-induced distortion than WHAs. The dark, blocky areas in the alloyed grade are cubic carbides.

## 21.6.3 Practical process concerns

The injection molding process requires careful control of trapped air, weld lines, and flash (German and Bose, 1997). Since carbides are much more sensitive to microstructural defects than metals, any large pores in the green body created by jetting or poor weld lines will result in unacceptable microstructures (Martyn and James, 1994). Control of flash is a challenge owing to the small particle size of the powders, allowing feedstock to enter much smaller gaps in the tooling than would a larger steel powder. If cutting inserts are to be molded, flashing at the cutting edge will limit the attainable cutting edge radius, or hone size, since green removal of the flash will result in a chamfered edge. Post-sinter honing can round this chamfer to a radius, but the hone size will exceed the thickness of the original flashing. Adding to this difficulty is the high abrasive wear imparted by the carbide feedstock to areas of the tooling that flash, which only increases the gap and limits tool life.

The avoidance of flash has led to development of low-pressure injection molding of carbides, allowing for longer tool life while using lower cost tool materials and molding equipment (Bose, 2011; German and Bose, 1997; Roediger et al., 2000). The binder and solids loading are customized to achieve a low-viscosity slurry that can be injected at lower pressures. Low-pressure injection molding may lead to difficulty in packing thick sections, resulting in sinks or voids (Zorzi et al., 2003). This can be partially compensated with a larger gate and increased mold temperature to extend

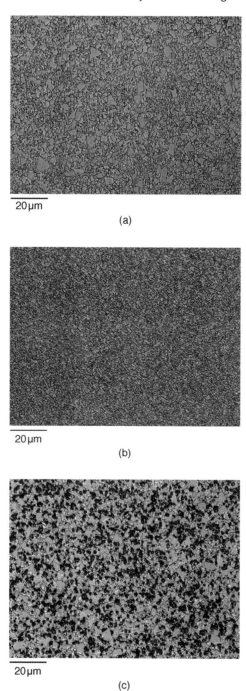

*21.19* Sintered microstructures of (a) a straight grade; (b) a micrograin grade; and (c) an alloyed grade showing cubic carbides.

the packing time. This also requires the gate to be located at or near the thickest cross-section. Injection molding of micro-sized parts is also possible, provided the rheology is adjusted to allow filling of micro features under normal injection pressures. This can be achieved by a reduction in the powder loading and proper selection of binders (Yin et al., 2011).

All-thermal debinding of wax-based systems can be done successfully but requires long thermal cycles. Deformation of the body can occur due to gravitational forces, especially in thick cross-sections (Martyn and James, 1994; Zorzi et al., 2003). A powder bed of alumina and graphite may be used to assist in wicking the binders out and avoiding distortion. Others have also demonstrated all-thermal debinding of wax–polymer systems (Baojun et al., 2002; Bruhn and Terselius, 1999; Qu et al., 2005), though this may result in a decrease in dimensional precision. Thermal debinding of carbides is similar to tool steels, requiring close attention to carbon control. Several researchers have shown that hydrogen may be used, but can cause decarburization (Baojun et al. 2002; Bruhn and Terselius, 1999; Martyn and James, 1994; Qu et al., 2005; Yang and German, 1998). Debinding in nitrogen/hydrogen blends and vacuum debinding have been shown to avoid decarburization. Decarburization may also be caused by reaction of carbon with oxides in the powder, and carburization is possible if binder removal is incomplete, which is most likely to occur with inert atmospheres.

The carbon content of hardmetals must be carefully controlled to achieve optimal properties. High carbon levels will cause free carbon to precipitate out in the microstructure. Low carbon levels will result in the formation of embrittling double carbides, such as $Co_3W_3C$ or $Co_6W_6C$, called η phase. The carbon window depends on the Co content. The higher the Co content, the wider the limit on carbon levels. Hardmetals with matrix alloys based on metals other than Co or Ni usually require even tighter carbon control (Fernandes and Senos, 2011).

Many factors affect carbon levels including the chemistry of the starting powder, binder composition, component size, debinding method, furnace construction, furnace load, sintering substrate, sintering atmosphere, heating rate, sintering time, and temperature. The carbon content of sintered components is usually determined non-destructively by measuring their magnetic saturation. If the hardmetal is undercarburized, the Co matrix will dissolve more W, lowering its magnetic saturation (Fang and Eason, 1993; Roebuck, 1996; Schwenke and Sturdevant, 2007). A magnetic saturation below 78% that of the equivalent carbon-saturated alloy indicates the occurrence of η phase (Schwenke and Sturdevant, 2007). Precipitated carbon may occur at readings of 100% magnetic saturation, at which point the magnetic saturation stops increasing with carbon content.

In addition to carbon control, grain size is also critical to many hardmetal applications. The grain size can be directly related to the hardness, but can be characterized non-destructively and more precisely by measuring the magnetic coercivity, which is related to the interfacial area between the WC grains and Co matrix. The interface restricts movement of magnetic domain walls within the ferromagnetic Co matrix, so a smaller grain size requires a stronger magnetic field to restore zero magnetization to a magnetically saturated hardmetal component (Fang and Eason, 1993; Roebuck, 1999, 2002; Sundin and Haglund, 2000).

Although WC–Co components have a low tendency to distort from their own weight during sintering, they are susceptible to distortion from temperature and carbon gradients in the furnace. Consistent furnace loading is important. Dimensional precision of $+/-0.2\%$ has been reported (Baojun et al., 2002; Qu et al., 2005). Hardmetals are very sensitive to contamination and should only be sintered in dedicated furnaces.

Sintered hardmetal parts occasionally come out of the furnace coated with a thin layer of Co. This cobalt-capping phenomenon can be beneficial if the parts are to be brazed, but in most cases it must be ground off. Cobalt-capping can appear to occur randomly from part to part and from furnace run to furnace run. It is related to the carbon activity of the residual gas, which can vary from one location to another in the furnace and on different surfaces of a single product (Guo et al., 2010). Cooling in both a non-decarburizing atmosphere (Guo et al., 2010) and in a decarburizing atmosphere (Janisch et al., 2010) have been recommended to avoid cobalt-capping. Some grades utilize special sintering methods to intentionally drive Co to the surface to increase the toughness of a tool's cutting edge.

In almost all cases, the sintered parts undergo post-sintering operations. The minimum operation to a cutting tool is honing of the cutting edge, which is critical to its performance. Many geometries of cutting tools require grinding after sintering. For some tools, the top and bottom of the part will be ground. Others require the periphery to be ground with or without honing of the cutting edge. In many cases, the finished part is coated. The coating provides lubricity and increased hardness. It also provides a diffusion barrier to keep the hardmetal from oxidizing when exposed to the high temperatures encountered during machining.

## 21.7 References

Aggarwal, G, Park, S J and Smid, I (2006), 'Development of niobium powder injection molding. Part I: Feedstock and injection molding', *International Journal of Refractory Metals and Hard Materials*, 24(3), 253–262.

Aggarwal, G, Smid, I, Park, S J and German, R M (2007), 'Development of niobium powder injection molding. Part II: Debinding and sintering', *International Journal of Refractory Metals and Hard Materials*, 25(3), 226–236.

ASM (1998a), 'Production of refractory metal powders', in *Powder Metal*

*Technologies and Applications, Volume 7, ASM Handbook*, ASM International, Materials Park, OH, pp. 188–201.

ASM (1998b), 'Refractory metals', in *Powder Metal Technologies and Applications, Volume 7, ASM Handbook*, ASM International, Materials Park, OH, pp.903–913.

ASTM (2011), ASTM standard F560-08: 'Specification for unalloyed tantalum for surgical implant applications (UNS R05200, UNS R05400)', *Annual Book of ASTM Standards Volume 13.01: Medical and Surgical Materials and Devices*, ASTM International, West Conshohocken, PA, USA.

Baojun, Z, Xuanhui, Q and Ying, T (2002), 'Powder injection molding of WC–8% Co tungsten cemented carbide', *International Journal of Refractory Metals and Hard Materials*, 20(5–6), 389–394.

Bin, Y and Lenel, F V (1984), 'Activated sintering of molybdenum powder electroless plated with a nickel–phosphorus alloy', *International Journal of Powder Metallurgy and Powder Technology*, 20(1), 15–21.

Blaine, D C, Gurosik, J D, Park, S J, Heaney, D F and German, R M (2006a), 'Master sintering curve concepts as applied to the sintering of molybdenum', *Metallurgical and Materials Transactions A*, 37A, 715–720.

Blaine, D C, Park, S J, Suri, P and German, R M (2006b), 'Application of work-of-sintering concepts in powder metals', *Metallurgical and Materials Transactions A*, 37A, 2827–2835.

Bose A (2011), 'A perspective on the earliest commercial PM metal–ceramic composite: Cemented tungsten carbide', *International Journal of Powder Metallurgy*, 47(2), 31–50.

Bose, A, Coque, H R A and Langford, Jr, J (1992), 'Development and properties of new tungsten-based composites for penetrators', *International Journal of Powder Metallurgy*, 28(4), 383–394.

Bose, A and German, R M (1988a), 'Microstructural refinement of W–Ni–Fe heavy alloys by alloying additions', *Metallurgical Transactions A*, 19A, 3100–3103.

Bose, A and German, R M (1988b), 'Sintering atmosphere effects on tensile properties of heavy alloys', *Metallurgical Transactions A*, 19A, 2467–2476.

Bose, A and German, R M (1990), 'Matrix composition effects on the tensile properties of tungsten–molybdenum heavy alloys', *Metallurgical Transactions A*, 21A, 1325–1327.

Bruhn, J and Terselius, B (1999), 'MIM offers increased application for submicron WC–10Co', *Metal Powder Report*, 54(1), 30–33.

Bryskin, B D and Danek, F C (1991), 'Powder processing and the fabrication of rhenium', *JOM*, 43(7), 24–26.

Buckman, R W (1997), 'Rhenium as an alloy addition to the group VA metals', in *Rhenium and Rhenium Alloys*, Bryskin, B D (ed.), The Mineral, Metals, and Materials Society, Warrendale, PA, pp. 629–638.

Caldwell, S G (1998), 'Tungsten heavy alloys', in *Powder Metallurgy Technologies, Volume 7, ASM Handbook*, ASM International, Materials Park, OH, pp. 914–921.

Cho, K C, Kellogg, F, Klotz, B R and Dowding, R J (2006), 'Plasma pressure compaction ($P^2C$) of submicron size tungsten powder', in *Proceedings of the 2006 International Conference on Tungsten, Refractory, and Hardmetals VI*,

Bose, A and Dowding, R J (eds), Metal Powder Industries Federation, Princeton, NJ, pp. 161–170.

Dropman, M C, Stover, D, Buchkremer, H P and German, R M (1992), 'Properties and processing of niobium superalloys by injection molding', in *Advances in Powder Metallurgy and Particulate Materials*, Capus, J M and German, R M, (eds), Metal Powder Industries Federation, Princeton, NJ, pp. 213–24.

Dushina, O V and Nevskaya, L V (1969), 'Activated sintering of rhenium with palladium additions', *Soviet Powder Metallurgy and Metal Ceramics*, 8, 642–644.

Fan, J, Lua, M, Chenga, H, Tiana, J and Huang, B (2009), 'Effect of alloying elements Ti, Zr on the property and microstructure of molybdenum', *International Journal of Refractory Metals and Hard Materials*, 27, 78–82.

Fang, Z Z and Eason, J W (1993), 'Nondestructive evalatuion of WC–Co composites using magnetic properties', *International Journal of Powder Metallurgy*, 29, 259–265.

Fang, Z Z and Eason, J W (1995), 'Study of nanostructured WC–Co composites', *International Journal of Refractory Metals and Hard Materials*, 13, 297–303.

Fernandes, C M and Senos, A M R (2011), 'Cemented carbide phase diagrams: A review', *International Journal of Refractory Metals and Hard Materials*, 29, 405–418.

Gaur, R P S and Wolfe, T A (2006), 'Sub-micron and low-micron size Mo metal powders made by a new chemical precursor', in *Proceedings of the 2006 International Conference on Tungsten, Refractory, and Hardmetals VI*, Bose, A and Dowding, R J (eds), Metal Powder Industries Federation, Princeton, NJ, pp. 122–131.

German, R M (1981), 'How to get more from a sintering cycle', *Progress in Powder Metallurgy*, 37, 195–211.

German, R M (1983), 'A quantitative theory of diffusional activated sintering', *Science of Sintering*, 15, 27–42.

German, R M (1985), *Liquid Phase Sintering*, Plenum Press, New York.

German, R M (2001), 'Unique opportunities in powder injection molding of refractory and hard metals', in *Proceedings of the 15th International Plansee Seminar*, Kneringer, G, Roedhammer, P and Wilder, H (eds), Metallwerk Plansee, Reutte, Austria, vol. 4, pp. 175–186.

German, R M and Bose, A (1997), *Injection Molding of Metals and Ceramics*, Metal Powder Industries Federation, Princeton, NJ.

German, R M, Bose, A and Mani, S S (1992), 'Sintering time and atmosphere influences on the microstructure and mechanical properties of tungsten heavy alloys', *Metallurgical Transactions A*, 23A, 211–219.

German, R M and Ham, V (1976), 'The effect of nickel and palladium additions on the activated sintering of tungsten', *International Journal of Powder Metallurgy*, 12, 115–125.

German, R M and Labombard, C A (1982), 'Sintering molybdenum treated with Ni, Pd, and Pt', *International Journal of Powder Metallurgy and Powder Technology*, 18(2), 147–156.

German, R M and Munir, Z A (1976a), 'Enhanced low-temperature sintering of tungsten', *Metallurgical Transactions A*, 7A, 1873–1877.

German, R M and Munir, Z A (1976b), 'Systematic trends in the chemically activated sintering of tungsten', *High Temperature Science*, 8, 267–280.

German, R M and Munir, Z A (1977), 'Rhenium activated sintering', *Journal of Less Common Metals*, 53, 141–146.

German, R M and Munir, Z A (1978), 'Heterodiffusion model for the activated sintering of molybdenum', *Journal of Less Common Metals*, 58, 61–74.

German, R M and Rabin, B H (1985), 'Enhanced sintering through second phase additions', *Powder Metallurgy*, 28, 7–12.

Gessinger, G H and Fischmeister, H F (1972), 'A modified model for the sintering of tungsten with nickel additions', *Journal of Less Common Metals*, 27, 129–141.

Green, E C, Jones, D J and Pitkin, W R (1956), 'Developments in high-density alloys', *Symposium on Powder Metallurgy*, Special Report #58, Iron and Steel Institute, London, pp. 253–256.

Guo, J, Fan, P, Wang, X and Fang, Z Z (2010), 'Formation of Co-capping during sintering of straight WC–10 wt% Co', *International Journal of Refractory Metals and Hard Materials*, 28, 317–323.

Hens, K F, Johnson, J L and German, R M (1994), 'Pilot production of advanced electronic packages via powder injection molding', in *Advances in Powder Metallurgy*, Lall, C and Neupaver, A (eds), Metal Powder Industries Federation, Princeton, NJ, 4, 217–229.

Huang, H S and Hwang, K S (2002), 'Deoxidation of molybdenum during vacuum sintering', *Metallurgical and Materials Transactions A*, 33A, 657–664.

Huppmann, W J (1979), 'The elementary mechanisms of liquid phase sintering, Part II: Solution–reprecipitation', in *Sintering and Catalysis*, Kuczynski, G C (ed.), Plenum Press, pp. 359–378.

Hwang, K S and Huang, H S (2003), 'Identification of the segregation layer and its effect on the activated sintering and ductility of Ni-doped molybdenum', *Acta Materialia*, 51, 3915–3926.

Hwang, K S and Huang, H S (2004), 'Ductility improvement of Ni-added molybdenum compacts through the addition of Cu and Fe powders', *International Journal of Refractory Metals and Hard Materials*, 22, 185–191.

Ivanov, E Y and Bryskin, B D (1997), 'The solid-state synthesis of the W–25wt.% Re using a mechanical alloying approach', in *Proceedings of the 14th International Plansee Seminar*, Kneringer, G, Rödhammer, P and Wilhartitz, P (eds), Metallwerk Plansee, Reutte, Austria, vol. 1, pp. 631–640.

Janisch, D S, Lengauer, W, Roediger, K and van den Berg, H (2010), 'Cobalt capping: Why is sintered hardmetal sometimes covered with binder', *International Journal of Refractory Metals and Hard Materials*, 28, 466–471.

Johnson, J L (2008), 'Progress in processing nanoscale refractory and hardmetal powders', in *2008 International Conference on Tungsten, Refractory, and Hardmaterials VII*, Bose, A, Dowding, R J and Shields, J A (eds), Metal Powder Industries Federation, Princeton, NJ, vol. 5, 57–71.

Johnson, J L, Campbell, L G, Park, S J and German, R M (2009), 'Grain growth in dilute sintered tungsten heavy alloys during liquid-phase sintering under microgravity conditions', *Metallurgical and Materials Transactions A*, 40A, 426–437.

Johnson, J L and German, R M (2006), 'Liquid phase sintering of W–Co–Mn heavy alloys', in *2006 International Conference on Tungsten, Refractory Metals, and*

*Hardmetals VI*, Bose, A and Dowding, R J (eds), Metal Powder Industries Federation, Princeton, NJ, pp. 257–265.

Johnson, J L, Upadhyaya, A and German, R M (1998), 'Microstructural effects on distortion and solid-liquid segregation during liquid phase sintering under microgravity conditions', *Metallurgical and Materials Transactions B*, 29B, 857–866.

Kaysser, W A and Petzow, G (1985), 'Present state of liquid phase sintering', *Powder Metallurgy*, 28, 145–150.

Kipphut, C M, Bose, A, Farooq, S and German, R M (1988), 'Gravity and configurational energy induced microstructural changes in liquid phase sintering', *Metallurgical Transactions A*, 19A, 1905–1913.

Kwon, Y-S, Wu, Y, Suri, P and German, R M (2004), 'Simulation of the sintering densification and shrinkage behavior of powder injection molded 17-4 PH stainless steel', *Metallurgical and Materials Transactions A*, 35A, 257–263.

Larsen, E I and Murphy, P C (1965), 'Characteristics and applications of high-density tungsten-based composites', *The Canadian Mining and Metallurgical Bulletin*, April, 413–420.

Lassner, E and Schubert, W-D (1999), *Tungsten: Properties, Chemistry, Technology of the Element, Alloys, and Chemical Compounds*, Kluwer Academic, New York.

Lejbrandt, M M and Rutkowski, W (1978), 'Effect of nickel additions on sintering molybdenum', *International Journal of Powder Metallurgy and Powder Technology*, 12(1), 17–30.

Leonhardt, T, Moore, N, Downs, J and Hamister, M (2001), 'Advances in powder metallurgy rhenium', in *Advances in Powder Metallurgy and Particulate Materials*, Eisen, W B and Kassam, S (eds), Metal Powder Industries Federation, Princeton, NJ, vol. 8, pp. 193–203.

Li, C J and German, R M (1983), 'The properties of tungsten processed by chemically activated sintering', *Metallurgical Transactions A*, 14A, 2031–2041.

Li, C J and German, R M (1984), 'Enhanced sintering of tungsten – phase equilibria effects on properties', *International Journal of Powder Metallurgy and Powder Technology*, 20, 149–162.

Lipetzky, P (2002), 'Refractory metals: a primer', *JOM*, 54, 4749.

Lisovksy, A F (2000), 'Cemented carbides alloyed with ruthenium, osmium, and rhenium', *Powder Metallurgy and Metal Ceramics*, 39(9–10), 428–433.

Lisovksy, A F, Tkachenko, N V and Kebko, V (1991), 'Structure of a binding phase in Re-alloyed WC–Co cemented carbides', *Refractory Metals and Hard Materials*, 10, 33–36.

Manneson, K, Borgh, I, Borgenstam, A and Agren, J (2011a), 'Abnormal grain growth in cemented carbides: Experiments and simulations', *International Journal of Refractory Metals and Hard Materials*, 29, 488–494.

Manneson, K, Jeppsson, J, Borgenstam, A and Agren, J (2011b), 'Carbide grain growth in cemented carbides', *Acta Materialia*, 59(5), 1912–1923.

Martyn, M T and James, P J (1994), 'The process of hardmetal components by powder injection moulding', *International Journal of Refractory Metals and Hard Materials*, 12, 61–69.

Massalski, T B (1986), *Binary Alloy Phase Diagrams*, ASM, Metals Park, OH.

Meredith, B and Milner, D R (1976), 'Densification mechanisms in the tungsten carbide–cobalt system', *Powder Metallurgy*, 19(1), 38–45.

Milman, Y V and Kurdyumova, G G (1997), 'Rhenium effect on the improving of mechanical properties in Mo, W, Cr, and their alloys', in *Rhenium and Rhenium Alloys*, Bryskin, B D (ed.), The Mineral, Metals, and Materials Society, Warrendale, PA, pp. 717–728.

Morton, C W, Wills, D J and Stjernberg, K (2005), 'The temperature ranges for maximum effectiveness of grain growth inhibitors in WC–Co alloys', *International Journal of Refractory Metals and Hard Materials*, 23, 287–293.

Muddle, B C (1984), 'Interphase boundary precipitation in liquid phase sintered W–Ni–Fe and W–Ni–Cu', *Metallurgical Transactions A*, 15A, 1090–1098.

Munir, Z A and German, R M (1977), 'A generalized model for the prediction of periodic trends in the activation of sintering of refractory metals', *High Temperature Science*, 9, 275–283.

Paramore, J D, Zhang, H, Wang, X, Fang, Z Z, Siddle, D and Cho, K C (2007), 'Production of nanocrystalline tungsten using ultra-high-pressure rapid hot consolidation (UPRC)', in *Proceedings of the 2007 International Conference on Powder Metallurgy and Particulate Materials*, Engquist, J and Murphy, T F (eds), Metal Powder Industries Federation, Princeton, NJ, vol. 8, pp. 1–9.

Park, J J and Jacobson, D L (1997), 'Steady-state creep rates of W-4Re-0.32HfC', in *Rhenium and Rhenium Alloys*, Bryskin, B D (ed.), The Mineral, Metals, and Materials Society, Warrendale, PA, pp. 327–340.

Park, S J, Chung, S H, Johnson, J L and German, R M (2006a), 'Finite element simulation of liquid phase sintering with tungsten heavy alloys', *Materials Transactions*, 47(11), 2745–2752.

Park, S J, Chung, S H, Martin, J M, Johnson, J L and German, R M (2008), 'Master sintering curve for densification derived from a constitutive equation with consideration of grain growth: Application to tungsten heavy alloys', *Metallurgical and Materials Transactions A*, 39A, 2941–2948.

Park, S J, Martin, J M, Guo, J F, Johnson, J L and German, R M (2006b), 'Densification behavior of tungsten heavy alloy based on master sintering curve concept', *Metallurgical and Materials Transactions A*, 37A, 2837–2848.

Park, S J, Martin, J M, Guo, J F, Johnson, J L and German, R M (2006c), 'Grain growth behavior of tungsten heavy alloys based on master sintering curve concept', *Metallurgical and Materials Transactions A*, 37A, 3337–3346.

Petzow, G, Kaysser, W A and Amtenbrink, M (1982), 'Liquid phase and activated sintering', in *Sintering – Theory and Practice*, Kolar, D, Pejovnik, S and Ristic, M M (eds), Elsevier, Amsterdam, The Netherlands, pp. 27–36.

Povarova, K B, Bannykh, O A and Zavarzina, E K (1997), 'Low- and high-rhenium tungsten alloys: Properties, production, and treatment', in *Rhenium and Rhenium Alloys*, Bryskin B D (ed.), The Mineral, Metals, and Materials Society, Warrendale, PA, pp. 691–705.

Prabhu, G, Chakraborty, A and Sarma, B (2009), 'Microwave sintering of tungsten', *International Journal of Refractory Metals and Hard Materials*, 27(3), 545–548.

Puzz, TE, Antonyraj, A, German, RM and Oakes, J J (2007), 'Binder optimization for the production of tungsten feedstocks for PIM', in *Advances in Powder Metallurgy and Particulate Materials – 2007*, Engquist, J and Murphy, T F (eds), Metal Powder Industries Federation, Princeton, NJ, USA, pp. 4.21–4.27.

Qu, X, Gao, J, Qin, M and Lei, C (2005), 'Application of a wax-based binder in PIM of WC–TiC–Co cemented carbides', *International Journal of Refractory Metals and Hard Materials*, 23(4–6), 273–277.

Roebuck, B (1996), 'Magnetic moment (saturation) measurements on hardmetals', *International Journal of Refractory Metals and Hard Materials*, 14, 419–424.

Roebuck, B (1999), 'Magnetic coercivity measurements for WC-Co hardmaterials', *NPL CMMT(MN)042*, National Physical Laboratory, UK.

Roebuck, B (2002), 'Hardmetals: Hardness and coercivity property maps', *NPL MATC(MN)14*, National Physical Laboratory, UK.

Roediger, K, Van den Berg, H, Dreyer K, Kassel, D and Orths, S (2000), 'Near-net-shaping in the hardmetal industry', *International Journal of Refractory Metals and Hard Materials*, 18, 111–120.

Samsonov, G V and Yakovlev, V I (1969), 'Activation of the sintering of tungsten by the iron-group metals', *Soviet Powder Metallurgy and Metal Ceramics*, 8, 804–880.

Sandim, H R Z, Padilha, A F and Randle, V (2005), 'Grain growth during sintering of pure niobium', in *Proceedings of the 16th International Plansee Seminar*, Kneringer, G, Rödhammer, P and Wildner, H (eds), Metallwerk Plansee, Reutte, Austria, pp. 684–695.

Santhanam, A T, Tierney, P and Hunt, J L (1998), 'Cemented carbides', *Powder Metallurgy Technologies, Volume 7, ASM Handbook*, ASM International, Materials Park, OH, pp. 950–977.

Schwenke, G K and Sturdevant, J V (2007), 'Magnetic saturation and coercivity measurements on chromium-doped cemented carbides', *International Journal of Powder Metallurgy*, 43(2), 21–31.

Sethi, G, Park S J, Johnson, J L and German, R M (2009), 'Linking homogenization and densification in W–Ni–Cu alloys through master sintering curve (MSC) concepts', *International Journal of Refractory Metals and Hard Materials*, 27, 688–695.

Snowball, R F and Milner, D R (1968), 'Densification processes in the tungsten carbide–cobalt system', *Powder Metallurgy*, 11(21), 23–40.

Solonin, S M and Kivalo, L I (1982), 'Sintering of mixtures of tungsten and rhenium powders', *Soviet Powder Metallurgy and Metal Ceramics*, 21, 451–453.

Srikanth, V and Upadhyaya, G S (1986), 'Sintered heavy alloys – a review', *International Journal of Refractory and Hard Metals*, 5, 49–54.

Stjernberg, K and Johnsson, J (1998), 'Recycling of cemented carbides', in *Advances in Powder Metallurgy and Particulate Materials*, Oakes, J J and Reinshagen, J H (eds), Metal Powder Industries Federation, Princeton, NJ, vol.1, pp.173–179.

Su, H and Johnson, D L (1996), 'Master sintering curve: A practical approach to sintering', *Journal of the American Ceramic Society*, 79(12), 3211–3217.

Sundin, S and Haglund, S (2000), 'A comparison between magnetic properties and grain size for WC-Co hard materials containing additives of Cr and V', *International Journal of Refractory Metals and Hard Materials*, 18, 297–300.

Suri, P, Atre, S V, German, R M and de Souza, J P (2003), 'Effect of mixing on the rheology and particle characteristics of tungsten-based powder injection molding feedstock', *Materials Science and Engineering A*, 356(1–2), 337–344.

Suri, P, German, R M and De Souza, J P (2009), 'Influence of mixing and effect of agglomerates on the green and sintered properties of 97W–2.1Ni–0.9Fe heavy alloys', *International Journal of Refractory Metals and Hard Materials*, 27(4), 683–687.

Trybus, C L, Wang, C, Pandheeradi, M and Meglio, C A (2002), 'Powder metallurgical processing of rhenium', *Advanced Materials and Processes*, December, 23–26.

Uskokovic, D, Petkovic, J and Ristic, M M (1976), 'Kinetics and mechanism of sintering under constant heating rates', *Science of Sintering*, 8, 129–148.

Uskokovic, D, Zivkovic, M, Zivanovic, B and Ristic, M M (1971), 'Study of the sintering of molybdenum powder', *High Temperature–High Pressures*, 3, 461–466.

Wang, C M, Cardarella, J J, Miller, K R and Trybus, C L (2001), 'Powder injection molding to fabricate tungsten and rhenium components', in *Advances in Powder Metallurgy and Particulate Materials*, Eisen, W B and Kassam, S (eds), Metal Powder Industries Federation, Princeton, NJ, pp. 8.180–8.192.

Wei, T–S and German, R M (1988), 'Injection molded tungsten heavy alloy', *International Journal of Powder Metallurgy*, 24, 327–335.

Wojcik, C G (1991), 'High temperature niobium alloys', in *High Temperature Niobium Alloys*, Stephens, J J and Ahmad, I (eds), The Mineral, Metals, and Materials Society, Warrendale, PA, pp. 1–13.

Yang, M-J and German, R M (1998), 'Nanophase and superfine cemented carbides processed by powder injection molding', *International Journal of Refractory Metals and Hard Materials*, 16, 107–117.

Yin, H, Tong, J, Qu, X and Zheng, J (2011), 'Powder injection molding for micro cemented carbide parts', in *Proceedings of 2011 International Conference on Tungsten, Refractory and Hardmaterials VIII*, Bose, A, Dowding, R J and Johnson, J L (eds), Metal Powder Industries Federation, Princeton, NJ.

Youseffi, M and Menzies, I A (1997), 'Injection molding of WC-6Co powder using two new binder systems based on montanester waxes and water-soluble gelling polymers', *Powder Metallurgy*, 40(1), 62–65.

Yunn, H S (2011), 'Critical solid loading and rheological study of WC–10%Co', *Applied Mechanics and Materials*, 42–43, 97–102.

Zhang, G-J, Sun, Y-J, Sun, J, Wie, J-F, Zhao, B-H and Yang, L-X (2005a), 'Microstructure and mechanical properties of ceria dispersion strengthened molybdenum alloy', in *Proceedings of the 16th International Plansee Seminar*, Kneringer, G, Rödhammer, P and Wildner, H (eds), Metallwerk Plansee, Reutte, Austria, pp. 1089–1095.

Zhang, X, Zhang, T, Hu, Z, Li, Q, Tan, S and Yin, W (2005b), 'Effect of hot isostatic pressing and high temperature sintering on the performance of PM Ta-W-Hf alloys', in *Proceedings of the 16th International Plansee Seminar*, Kneringer, G, Rödhammer, P and Wildner, H (eds), Metallwerk Plansee, Reutte, Austria, pp. 776–784.

Zorzi, J, Perottoni, C and Jornada, J (2003), 'A new partially isostatic method for fast debinding of low-pressure injection molded ceramic parts', *Materials Letters*, 57(24–35), 3784–3788.

Zovas, P E and German, R M (1984), 'Retarded grain boundary mobility in activated sintered molybdenum', *Metallurgical Transactions A*, 15A, 1103–1110.

Zovas, P E, German, R M, Hwang, K S and Li, C J (1983), 'Activated and liquid phase sintering – progress and problems', *Journal of Metals*, 35(1), 28–33.

# Index

activated sintering, 549–50
activation energy
   calculation, 219–21
      characteristics of binder components used in the binder system, 219
      TGA results and Kissinger method for polypropylene, 220
active heat sink design, 447
adjuncts, 5
advanced mold calculation software, 107
air vent, 214
Allied Signal feedstock, 136
ambient atmosphere, 157–8
anisotropic shrinkage, 127–9
   flow direction terminology, 128
   labels used to describe features measured for analysis, 127
   measured experimental green, 128
   mechanical coupling, 129
anodic oxidation, 425
apparent activation energy, 217
Apparent Co-Sintering Index, 342
Armstrong process, 433
ASTM B 243-09a, 147
ASTM B 417, 56
ASTM B 527, 56
ASTM B 703, 56
ASTM B 822 - 10, 56–7
ASTM B348-02, 416, 426
ASTM B817, 417
ASTM D 2638, 55–6
ASTM D 4892, 55–6
ASTM D5930, 193
ASTM E1269, 190
ASTM F2885-11, 426
ASTM F560, 529

atmosphere, 275–6
atomic force microscopy (AFM), 321
austenitic steel, 285, 394
auxiliary equipment, 115–16
   granulators, 115–16
   material drying, 115
   mold temperature controllers, 115
   part removal, 116

Bakelite, 2
batch furnaces, 174–5, 178
   graphite furnace with a graphite hot zone and elements, 173
   refractory metal MIM furnace, 174
   *vs.* continuous furnaces, 175–6
bi-viscosity approach, 329
binder, 420–1
   chemistry, 66–9
      melting temperature and tensile strength variation polymers molecular weight, 68
      temperature in relation to relative molecular mass plots, 67
   constituents, 69
   formulation and compound manufacture in MIM, 64–89
   lab scale and commercial formulation, 88
      binder composition used in injection molding, 88–9
   mixing technology, 84–9
   properties and effects on feedstock, 70–84
      rheology, 70–4
      solubility and thermal degradation, 75–84
      thermal conductivity and heat capacity, 74–5

role, 64–6
  characteristics of an ideal system for metal injection molding, 65
binder residue, 248–50
binder/powder separation, 239–40
  delamination observed after drilling, 240
  high shear rate which forms binder-rich gate marks, 240
blistering, 543
bosses, 48
boundary element method (BEM), 209
brittle intergranular failure, 550
Brunauer–Emmett–Teller (BET) particle, 533

capillary flow porosimeter, 369, 371
capillary rheometer, 70, 187
carbon contamination, 82–3
carbon control
  importance, 265–7
  material properties, 297–9
    examples affected by carbon content, 298–9
  metal injection molding (MIM), 265–99
  methods and binder elimination, 267–76
    additives, 274–5
    carbon source, 267–9
    correlation between the debinding cycle and TGA, 270
    debinding mechanism, 271–4
    measuring tools, 269–71
    process parameters, 275–6
    thermogravimetrical analysis (TGA) of a two-components binder, 270
  particular materials, 276–97
    cemented carbides, 289–90
    high-speed steels (HSS), 277–83
    magnets, 290–4
    stainless steels, 283–8
    titanium alloys, 294–7
carbon source, 267–9
  degradation mechanism of polymers, 268
carbon steels, 162
carnauba wax, 274
catalytic debinding, 139–41
  main computer screen of an Elnik oven, 141
  oven, 140

Catamold, 134–5, 341, 453
catalytic debinding, 139–41
cemented carbides, 289–90, 528
ceramic-filled feedstocks, 313
ceramics, 307
chain depolymerisation, 272–3
chemical reduction, 60–1
  SEM image of tungsten powder, 61
chromium nitride precipitation, 400
clamp tonnage, 112
clamping unit, 94–5, 111–12
closed foams, 349
co-injection molding, 339, 387
co-sintering, 379
cobalt, 554
cobalt-capping, 560
component mass, 261–2
component production, 5
computer simulation, 107
computer-aided engineering (CAE), 183, 185, 196
condensation, 149
consumable suppliers, 5
continuity equation, 204, 206
continuous furnace, 171–4, 178
  pusher furnace, 171–2
  vs. batch furnaces, 175–6
  walking beam furnace, 172
continuum modelling, 225
control computer, 95
conventional injection molding machine, 110–13
  injection unit configuration, 111
  screw tip configuration, 111
cool time, 122
cooling stage, 207–8
  analysis, 209–10
  coupled analysis, 210–11
  material properties, 212
coordinate measuring machine, 321
copper, 451–61
  debinding and sintering, 455–7
    oxide-reduced copper powder micrographs, 458
    sintering temperature effect on Cu powders density, 456
    sintering temperature effect on Cu powders oxygen content, 457
    water-atomized powder micrographs, 458
  example applications, 460–1
    MIM Cu heat sink photo, 460
  feedstock preparation, 453–4

heat pipe and outer wall and wick interface micrograph, 461
MIM Cu examples, 454
molding, 454–5
powder, 452–3
  powder characteristics, 452
  representative samples SEM, 453
thermal properties, 457–60
  iron impurities effect on thermal conductivity, 459
  porosity effect on thermal conductivity, 459
copper based heat sinks, 344
corrosion resistant low carbon steels, 163–5
  fracture surface gas atomized 17-4 PH, 164
  fracture surface of a 17-4 PH material made from carbonyl powder, 163
coupling agents, 274
Cross-WLF model, 198
Cu 'bleed-out,' 467

deagglomeration, 474
debinding, 133–47, 243–50, 271–4
  binder residue, 248–50
  binder systems, 134–5
  chemical structure of carbonaceous residue, 273
  furnaces, 169–76
  metal injection molding (MIM) components, 133–78
  MIM materials, 161–6
    powder availability, 166
    reactivity effect, 161–6
  MIM systems, 135–6
  primary, 136–44
  secondary, 144–7
  settering, 167–9
  solvent, 244–5
  thermal, 245–8
  weight loss, 262
decarburisation, 559
decorative features, 48–9
  lettering produced by MIM processing, 49
defect formation, 80–2
  observation during thermal debinding process, 81
densification, 227
designers, 4
differential scanning calorimetry, 183, 191

differential thermal analysis (DTA), 216
dimensional accuracy
  experimental results, 376–9
    impeller dimensional deviation and variance coefficient, 377
    MIM impeller vs extrusion plate, 378
    specimens dimensional deviation and variance coefficients, 378
  measurement, 374
  micro-porous MIM parts, 374–9
  MIM specimen and manufacturing method, 374–6
    constituent materials and compositions, 375
    debinding and sintering temperature control condition, 375
    impeller with micro-porous structures SEM, 376
    test specimens and locations for measurement, 375
dimensional variability, 35
direct sintering, 545
dispersants, 69
distortion, 83–4, 228, 541–2
  polymer burnout from admixed powders of gas atomized 316L stainless, 84
  shape loss during polymer burnout, 85
draft, 41–4
  inside and outside angle for easy component removal from tool, 43
  inside drafter *vs*. undrafter diameter, 43
duplex stainless steel, 284, 394

effective bulk viscosity, 225
effective shear viscosity, 225
ejector pin marks, 37–8
  blemish shown on a MIM component, 37
electrical discharge machining (EDM), 35
electron backscattering diffraction (EBSD), 431
electron beam welding, 346
electron probe microanalyser (EPMA), 510
elemental analyser, 269
elemental method, 61–2
elemental powder metallurgy, 434

Ellingham diagrams, 158
energy equation, 204–5, 206, 208
equipment suppliers, 4–5
ethylene vinyl acetate, 421
evaporation, 149

Fe–50Ni alloy, 31–2, 506–13
 experimental procedure, 506
  sintering temperature effect on sintered density, 507
 magnetic properties, 507–9
  MIM Fe–50 Ni microstructures, 508
  sintering temperature effect on MIM Fe–50Ni compacts, 508
 magnetic properties improvement, 509–13
  carbonyl Fe powder chemical composition, 512
  EPMA line analysis, 510
  heating condition effect on Fe–50Ni oxygen content, 511
  MIM Fe–50Ni microstructures and magnetic properties, 512
  powder type effect on impurity contents, 512
  rapid cooling after annealing effect on compacts, 513
  sintering temperature effect on MIM Fe–50Ni C and O contents, 509
Fe–6.5Si alloy, 489–97
 characteristics, 489
 experimental procedure
  debinding and sintering programs, 489
 magnetic properties, 493–7
  coercive force changes vs relative density, 495
  eddy current loss in each specimen, 496
  Fe–9.5Si–5.5Al chemical compositions, 497
  hysteresis loss in each specimen, 495
  iron loss at various frequencies, 496
  magnetic induction and relative density relationship, 494
  relative density influence on maximum permeability, 494
 relative density, hardness, mean grain size, and chemical composition, 490–2
  carbon content and relative density relationship, 493
  hardness change vs density, 491
  mean grain size and relative density relationship, 492
  oxygen content and relative density relationship, 493
  relative density and sintered temperature relationship, 491
  sintered compacts micrographs, 492
  sintering conditions, 490
Fe–9.5Si–5.5Al, 497–504
 gas- vs water-atomized compacts magnetic properties, 502–6
  MIM Sendust alloy compacts' microstructures, 505
  MIM Sendust alloy sintering temperature effects on different parameters, 503
  powdered and sintered carbon and oxygen contents, 505
  Sendust alloys magnetic properties, 506
  sintering temperature effects on MIM Sendust alloy magnetic properties, 504
 gas-atomized powder compacts magnetic properties, 497–502
  continuous process effect on MIM Sendust alloy magnetic properties, 500
  conventional and continuous process heating conditions, 500
  MIM Sendust alloy compacts' microstructure, 498
  MIM Sendust alloy grain size and oxygen content, 503
  MIM Sendust alloy sintering temperature effects on different parameters, 503
  process condition effects on MIM Sendust alloys' density, 501
  sintering temperature and atmosphere effect on magnetic properties, 499
  sintering temperature and atmosphere effect on relative density, 498
  sintering temperature and atmosphere effects on grain size, 499
feedstock production firms, 4
ferritic steels, 394

Index    573

filling stage, 203–5
  analysis, 209
  coupled analysis, 210–11
  material properties, 211
filling time, 214–15
  optimisation by using a CAE tool for PIM, 215
finite difference method (FDM), 209
finite element method (FEM), 198–9, 209
finite element modeling (FEM), 541
flash, 214
  high molding pressures to force feedstock into clearances between mold, 238
flash control, 557
flats, 40–1
Forschungszentrum Karlsruhe *see* Karlsruhe Institute of Technology
Fraunhofer Institut IFAM, 324–5
Fray–Farthing–Chen (FCC)-Cambridge process, 433
furnaces, 166, 169–76
  batch, 174–5
  batch *vs.* continuous, 175–6
  continuous, 171–4
  evolution, 170–1
  profiles, 176
fusion brazing, 346

gas atomization, 58
gas-atomized powder, 291, 396, 455
gate locations, 38–40
  center gate blemish, 40
  recessed tab gate blemish along parting line, 39
  tab gate blemish along parting line on MIM component, 38
  tunnel (sub-) gate blemish, 39
gating, 99–101
  tab gate, sub gate and three-plate tool, 100
glass transition temperature, 67–8
grain boundary diffusion, 150
grain growth, 226–7
granulators, 115–16
gravitational distorting, 228–30
  final distorted shape by sintering, 229

hardmetals, 554–60
  feedstock formulation, 529–34
    feedstock examples, 529–33, 535
    powder characteristics, 529–33
    powder preparation, 529–33
    solids loading, 533–4
  liquid-phase sintering, 555–6
    sintering temperature effect on WC grain size, 556
  metal injection molding (MIM), 526–60
  practical process concerns, 557–60
  sintered microstructures, 558
  typical process cycles, 556–7
heat capacity, 74–5
  variation of iron–wax feedstock, 75
heat sink design, 447–9
  machined vs MIM heat sink, 448
  MIM electronic package geometries, 449
  possible MIM geometries, 448
  typical material properties, 450
heavy alloys, 534–44
  feedstock formulation, 529–34
    feedstocks examples, 535
    powder preparation, 529–33
    solids loading, 533–4
  liquid-phase sintering, 536–43
    88W–8.4Ni–3.6Fe grain size vs sintering temperature, 538
    distortion effects, 541
    practical process concerns, 542–3
    properly and overly sintered 90W–8Ni–2Fe MIM, 541–2
    sintered 95W–4Ni–1Fe MIM heavy alloy 10 µm grain size, 540
    sintered 95W–4Ni–1Fe MIM heavy alloy 35 µm grain size, 540
    tungsten content effect on W–Ni–Fe alloy grain-growth-rate constant, 539
    tungsten heavy alloys liquid-filled region, 543
    typical process cycles, 538–40
    W–Ni–Fe and W–Ni–Cu densification, 537
  metal injection molding (MIM), 526–60
    tungsten disk drive counterweight, 528
  properties, 543
    typical properties, 544
    vacuum anneal effect on 95wt% W alloy, 544
Hele–Shaw model, 203
high alloyed carbon steels, 163
high-speed steels (HSS), 277–83

carbon content during debinding and sintering, 283
debinding and sintering cycles, 279
M2 HSS component microstructure comparison, 280
microstructure of sintered M2 HSS, 282
pseudobinary M2 phase diagram, 278
sintering curves, 281
high-speed tool steels
  mechanical properties, 523–4
    ASTM G65-94 wear results for MIM tool steels, 524
    average MIM tool steels average hardness, 524
    heat treatment conditions, 524
  metal injection molding, 516–24
  MIM processing, 517–23
    debinding, 519
    feedstock, 517–19
    heat treatment, 522–3
    M2 tool steel dilatometer plot, 520
    M2 tool steel pseudo phase diagram, 521
    M2 tool steel structures, 521
    M4, T15, and M42 microstructures after heat treatment, 523
    published sintering temperatures in various atmospheres, 522
    sintering, 519–22
high-temperature alloys, 166
hold pressure, 121
hold time, 121
Honeywell feedstock, 136
hot isostatic pressing, 31, 425, 517, 551–3
hot pressing
  multilayered porous structure formation, 379–82
    graded porous structures production by co-sintering, 380
    multilayered metals production, 380
    specimens with various PMMA particle fractions bending properties, 381
    specimens with various PMMA particle fractions fracture aspects, 382
hydraulic pressure, 94
hydride–dehydride (HDH) process, 419
hydroxyapatite, 437

IKV model, 206
incomplete binder removal, 146–7
inert gas atomization, 419
ingredient suppliers, 4
injection molding machine, 110–14
  conventional, 110–13
  design and function, 94–6
    profile and cross-sections of a feed screw, 96
  micro, 113–14
  mold, 114
injection speed, 120
injection unit, 94–5, 110–11
intermolecular lubricants, 69
interstitial elements
  titanium MIM challenges, 416–18
    ASTM B348-02 max oxygen level and tensile properties, 417
    processing parameters oxygen contents contribution, 418
ISO 13485, 22
ISO 14000, 22
ISO 3923-1, 56
ISO 3953, 56
ISO 9000/9001/9002, 21–2
iso-electric point (IEP), 275
ISO13320-1, 56–7
ISO178, 381

Karlsruhe Institute of Technology, 311, 320, 325
Kenics mixer, 200
  working principles based on flow characteristics
    two grey tints representing fluid-particles mixtures, 201
Kissinger method, 217
Kovar, 449
Kozeny–Carman's equation, 369
Kroll process, 419, 432

lab scale binder system, 88
laser brazing, 346
laser flash method, 450–1
LIGA method, 311–13
liquid infiltration test, 369
liquid phase sintering, 277, 401, 536–43, 555–6
loop tracer, 497
lost-core technology, 331

$M_{23}C_6$ carbides, 516–17
macro-sized porous structures, 351
magnets, 290–4

Index    575

densification of Fe–2Ni, 292
magnetic properties of Sm(Co, Fe, Cu, Zr)z, 294
market
  MIM by application, 14–15
    global market attention based on primary focus for MIM firms, 15
    partition by region and application, 14
    percent of global MIM sales each segment, 16
  MIM by region, 13–14
    sales partition based on major geographical region, 13
    summary of regional sales, 13
  opportunities, 15–21
    copper MIM heat transfer device, 18
    Hermetic Kovar microelectronic package, 20
    sales distribution chart for global MIM firms, 21
    stainless steel MIM medical implant device, 19
    titanium dental implant formed by MIM, 20
  statistics, 12–13
    global MIM summary, 12
martensitic steel, 399
mass transport, 148–50
  different stages of progression of sintering, 149
master alloy method, 62
master decomposition curve (MDC)
  multi-reaction step, 217–19
  multi-reaction step decomposition, 222
    Nb feedstock, 222
  single reaction step, 216–17
  single reaction step decomposition polypropylene showing all TGA, 221
    synthesis of overall decomposition behaviour for the binder system, 221
material drying, 115
melt temperature, 118–20
melt thermal conductivity, 193–4
metal injection molding (MIM), 1–23, 133–78, 197–231, 350, 352–3
  auxiliary equipment, 115–16
  carbon control, 265–99
    importance, 265–7
  material properties, 297–9
  methods and binder elimination, 267–76
  particular materials, 276–97
components, 109–31
control parameters, 260–3
  debinding, 262
  feedstock behaviour, 261
  injection molding, 261–2
  powder characteristics, 261
  sintering, 262–3
debinding, 133–47
defects, 129–31
design, 29–49
  attributes produced by the process, 30
  attributes $vs.$ other fabrication techniques, 30
design consideration, 40–9
dimensional capability, 35
equipment, 110
feedstock characterisation, 183–96
  characteristics used in the study, 186
  comparison of typical thermal properties of some materials, 184
  pressure–volume–temperature (PVT), 194–5
  rheology, 186–9
  thermal analysis, 190–2
  thermal conductivity, 193–4
furnaces, 169–76
heavy alloys, refractory metals, and hardmetals, 526–60
  applications, 527–9
  feedstock formulation, 529–34
  hardmetals, 554–60
  heavy alloys, 534–44
  refractory metals, 544–54
high-speed tool steels, 516–24
  cutting bits photo, 518
  gas-atomized steel powder characteristics, 518
  mechanical properties, 523–4
  processing, 517–23
  typical stoichiometry, 518
industry shifts, 9–10
industry structure, 4–6
injection molding process modelling and simulation, 203–15
  applications, 213–15
  material properties and verification, 211–13

numerical simulation, 208–11
theoretical background and
 governing equations, 203–8
machine, 110–14
market by application, 14–15
market by region, 13–14
market opportunities, 15–21
market statistics, 12–13
materials, 161–6
materials and properties, 31–5
 bioimplantable alloys, 34
 controlled-expansion alloys, 34
 copper property comparison, 33
 heavy alloys, 34
 overview of applications and
  features, 32
 soft magnetic alloy properties, 33
 structural properties, 33
metal powders, 50–62
 alloying methods, 61–2
 characterisation, 55–7
 characteristics, 51–5
 fabrication techniques, 57–61
microelectronics thermal management
 materials, 446–82
 copper, 451–61
 heat dissipation in microelectronics,
  447–51
 molybdenum–copper, 474–82
 tungsten–copper, 461–74
mixing process modelling and
 simulation, 197–202
molding defect, 235–52
 binder/powder separation, 239–40
 causes and remedies, 242–3
 debinding, 243–50
 feedstock, 236–8
 flash, 238
 flow marks, 241
 incomplete filling, 241
 residual stress, 239
 sintering, 250–1
 weld lines, 242
overview, 1–2
powder binders formulation and
 compound manufacture, 64–89
 chemistry and constituents, 66–9
 lab scale and commercial
  formulation, 88
 mixing technology, 84–9
 properties and effects on feedstock,
  70–84
 role, 64–6

process, 116–29, 255
 anisotropic shrinkage, 127–9
 input and output products for
  comprehension and control, 256
 molding parameters, 118–22
 overview, 116–18, 119
 PVT effect, 123–6
 shrinkage, 122
 typical gate freeze study, 118
process control, 258–60
 auditing comparison for minimum
  and precision process, 260
 parameters, 259
product qualification methods, 255–7
 logic diagram to go from concept to
  production, 256
production sophistication, 21–3
 plot showing the relative
  sophistication of the MIM, 22
prototype methodology, 257–8
qualification, 254–63
sales situation, 10–12
settering, 167–9
sintering, 147–61, 156
sintering process modelling and
 simulation, 224–30
soft magnetic metals, 487–514
 Fe–6.5Si, 489–97
 Fe–50Ni, 506–13
 Fe–9.5Si–5.5Al, 497–504
stainless steel, 393–409
 applications, 403–9
 mechanical properties orientative
  values, 395
statistical highlights, 6–9
 summary of PIM global sales, 8
 summary statistics on PIM, 8
 typical productivity ratios, 9
 typical unit manufacturing cell in
  metal PIM, 9
success history, 2–3
surface finish, 35
thermal debinding process modelling
 and simulation, 215–24
titanium and its alloys, 415–41
 cost reduction, 432–4
 future trends, 440–1
 mechanical properties, 425–32
 MIM challenges, 416–22
 processing basics, 422–5
 special applications, 435–40
tooling artifacts, 35–40
tools, 93–107

machine design and function, 94–6
  software and economic factors, 106–7
  special features and instrumentation, 104–6
  tool design options, 98–104
  tool set elements, 96–8
metal oxides, 162
metal powder
  alloying methods, 61–2
    elemental method, 61–2
    master alloy method, 62
    prealloy method, 62
  binders formulation and compound manufacture in MIM, 64–89
    chemistry and constituents, 66–9
    lab scale and commercial formulation, 88
    mixing technology, 84–9
    properties and effects on feedstock, 70–84
    role, 64–6
  characterisation, 55–7
    particle size distribution, 56–7
    pycnometer density, 55–6
    tap density, 56
  characteristics, 51–5
    chemistry, 261
    shape, 54–5
    size, 52–3
    size and size distribution, 261
    size distribution, 53–4
  effects
    catalytic effect on decomposition behaviour, 223
  fabrication techniques, 57–61
    chemical reduction, 60–1
    gas atomization, 58
    manufacturing methods and attributes, 57
    thermal decomposition, 59–60
    water atomization, 59
  metal injection molding (MIM), 50–62
metal-filled feedstocks, 313
micro in-mold labellling see sinter joining
micro injection molding machine, 113–14
  two-stage unit configuration, 114
Micro metal injection molding (Micro MIM), 307–31
  future trends, 330–1

micro-components powder injection molding, 313–25
  debinding and sintering, 320–1
  feedstocks, 313–16
  leading and innovative trends, 322–5
  metrology and handling, 321–2
  molding procedure, 316–20
multi-component micro powder injection molding, 325–8
  2C-MicroPIM, 325–6
  fixed and movable structures, 326
  micro in-mold labeling with PIM feedstocks, 327
  sinter joining, 327–8
powder injection potential for microtechnology, 308–9
simulation, 328–9
tool making micro-manufacturing methods, 309–13
  LIGA method, 311–13
  micro-tools manufacture, 309
  microstructured mold inserts manufacturing options, 309–11
micro-porous metals
  functionally graded structures, 379–87
    bending strength vs PMMA particle, 388
    multilayered porous structure formation by hot pressing, 379–82
    sequential injection molding, 383–7
    steamed bread-like porous structures by co-injection molding, 387
  liquid infiltration properties, 369–74
    acid cleaning effects, 373
    evaluation apparatus schematic, 370
    immersion length and pore size effects on infiltration rate and percentage, 373
    immersion length effect on weight change, 372
    measurement, 369
    pore size effect on weight change, 373
    porous specimens properties, 371
    principle and evaluation method, 369–71
    saturation time vs immersion length, 372

specimen and experimental results, 371–4
surface treatment effect on weight change, 373
weight change behaviour during immersion, 371
MIM parts dimensional accuracy, 374–9
  experimental results, 376–9
  measurement, 374
  specimen and manufacturing methods, 374
powder space holder metal injection molding (PSH-MIM), 349–88
  micro-porous structures formation, 354–9
  porous structure control, 360–8
  production methods, 351–3
    debinding in metal injection molding, 352–3
    porosity vs cell size, 352
    porous metals and manufacturing types, 351–2
    structured MIM parts, 353
Micro-systems technology, 308
microelectronics
  copper, 451–61
    debinding and sintering, 455–7
    example applications, 460–1
    feedstock preparation, 453–4
    molding, 454–5
    powders, 452–3
    thermal properties, 457–60
  heat dissipation, 447–51
    heat sink design, 447–9
    material selection, 449–50
    thermal property measurement, 450–1
  molybdenum–copper, 474–82
    debinding and sintering, 477–9
    example applications, 481–2
    feedstock preparation, 475–6
    infiltration, 479–80
    injection molding, 476–7
    powders, 474–5
    thermal properties, 480–1
  thermal management materials metal injection moulding (MIM), 446–82
  tungsten–copper, 461–74
    debinding and sintering, 466–9
    example applications, 473–4
    feedstock preparation, 465
    infiltration, 469–70
    injection molding, 466
    powders, 462–4
    thermal properties, 470–3
microstructural homogenisation, 477
microstructured parts, 316, 320
MIM feedstock, 236–8
  behaviour
    density, 261
    viscosity, 261
  characterisation in metal injection molding (MIM), 183–96
    pressure–volume–temperature (PVT), 194–5
  recycled, 237–8
    surface condition of the specimen after solvent debinding, 238
  rheology, 186–9
  thermal analysis, 190–2
  thermal conductivity, 193–4
    solid thermal conductivity, 194
  uniformity, 236–7
    Mo agglomerates and large voids in Mo-rich region, 236
MIM patents, 3
MIM prototype
  methodology, 257–8
    material selection, 257–8
    production, 258
mixing process, 84–9, 197–202
  applications, 200–2
    Kenics static mixer, 200
    progress characterised by the normalised entropy, 202
  modelling, 198
  numerical methods, 198–200
  variation of torque with time, 87
μMetal injection molding, 402
molybdenum, 528–9
molybdenum–copper, 474–82
  debinding and sintering, 477–9
    injection molded and liquid phase sintered Mo–15Cu microstructure, 479
    relative sintered density vs sintering temperature, 478
    relative sintered density vs sintering time, 478
  example applications, 481–2
    infiltrated Mo–15Cu photo, 481
  feedstock preparation, 475–6
    mixing torques vs solids loading, 476
  infiltration, 479–80

injection molded and infiltrated Mo–15Cu microstructure, 480
relative sintered density vs Mo particle size, 480
injection molding, 476–7
powders, 474–5
characteristics, 475
Mo powders SEM, 475
thermal properties, 480–1
monochromatic synchrotron radiation, 322
mold, 98, 114
mold cavity, 101
mold parting line, 101
mold temperature, 118–20
controllers, 115
molding defect, 129–31
binder/powder separation, 239–40
causes and remedies, 242–3
debinding, 243–50
feedstock, 236–8
flash, 238
flow marks, 241
incomplete filling, 241
metal injection molding (MIM), 235–52
residual stress, 239
sintering, 250–1
solution for MIM molding, 130–1
weld lines, 242
molding firms, 4
MP35N alloy, 34
MPIF 28, 56
MPIF 46, 56
MPIF 488, 56
MPIF 63, 55–6

Nanyang Technological University, 322–4
naphthalene, 420
Navier–Stokes equation, 204
niobium, 529
no flow temperature, 191
Nordheim's Rule, 458
numerical simulation, 208–11, 226
cooling analysis, 209–10
coupled analysis between filling, packing and cooling, 210–11
geometry modelling and mesh generation, 210
filling and packing analysis, 209

one-channel system, 339

open foams, 349
Osaka Prefectural College of Technology, 322–3
over-molding, 338–9
oxidative mechanism, 272
oxygen equivalent, 417

P.A.N.A.C.E.A. steel, 402
packing stage, 205–7
analysis, 209
coupled analysis, 210–11
material properties, 211–12
pressure–volume–temperature (PVT) data of the feedstock, 212
paraffin wax, 420
part removal, 116
particle shape, 54–5, 73–4
effects on packing density on monosized particles, 54
particle size, 52–3, 73–4
effect on surface finish for a 17-4PHSS alloy, 52
particle size distribution, 53–4, 56–7
D10, D50 and D90 MIM characterisation, 53
particle tracking method, 199–200, 201
parting line, 35–7
blemish on MIM component, 36
passive heat sink design, 447
permalloy see Fe–50Ni
persistent liquid phase sintering, 154–5
phase-hardened stainless steel, 284
PIM-specific simulation routine, 329
plasma rotating electrode processing, 419
plastic flow, 150
plasticisers, 69
plastification, 117
polyacetal binder systems, 398
polyethylene, 421
polymethylmethacrylate (PMMA), 355, 384
polypropylene, 421
porous materials, 349
post-sinter honing, 557
powder, 419–20
powder injection molding (PIM), 1, 2–3, 110, 197–8, 517
leading and innovative trends, 322–5
316L stainless steel MicroMIM samples, 324
MicroMIM Peltier elements, 323
micro-components, 313–25

debinding and sintering, 320–1
metrology and handling, 321–2
micro-components feedstocks, 313–16
  binder systems, 313–15
  metal powders, 315–16
  MicroMIM metal materials, 315
molding procedure, 316–20
  micro-injection molding state, 320
  microtechnical PIM components, 318
  Ni-free steel alloy tooth brackets, 319
  watch gears MIM components, 319
multi-component microPIM, 325–8
  2C MicroPIM, 325–6
  2C MicroPIM demonstrator, 326
  fixed and movable structures, 326
  micro in-mold labelling with PIM feedstocks, 327
  sinter joining, 327–8
powder metallurgy, 277, 278, 350, 448, 488
powder size, 157
powder space holder (PSH) method, 354–5
  debinding mechanisms, 355–6
    debinding–sintering conditions and TG curves, 356
    process schematic, 356
  micro-porous metals examples, 356–9
    experimental materials and constituents fraction, 360
    metal parts from 316L powder and 50 μm PMMA particle, 358
    Ni porous specimens surface structures, 357
    particle size and molding methods effects pore formation, 359
    SEM images and pore sizes, 357
  MIM process for micro-porous metals manufacturing, 354
powder space holder metal injection molding (PSH-MIM)
  micro-porous metals, 349–88
    functionally graded structures, 379–87
    liquid infiltration properties, 369–74
    MIM parts dimensional accuracy, 374–9
    production methods, 351–3
  micro-porous structures formation, 354–9

debinding mechanisms, 355–6
micro-porous metals by PSH method, 356–9
powder space holder (PSH) method, 354–5
porous structures control, 360–8
  geometrical analysis, 367–8
  materials and manufacturing conditions, 360
  PMMA particle size ratio vs max fraction, 368
  sintered specimen surface SEM, 361
  sintering shrinkage and porosity, 361–4
  sintering temperature effect, 364–7
  surface structures, 360–1
powder surface area, 157
prealloy method, 62
precious metals, 161
precipitate hardened stainless steels, 394, 399
pressure–volume–temperature (PVT), 117, 123–6, 194–5
  comparison of sintering shrinkage, 126
  injection molding cycle traces, 125
  melt density of feedstock, 195
  MIM volumetric shrinkage prediction behaviour, 125
  molding cycle trace, 124
  relationship of sample C, 195
primary debinding, 136–44
  catalytic debinding of Catamold feedstock, 139–41
  guidelines, 142–4
    solvents and their boiling points, 143
    temperatures and rates for different binder systems, 144
  solvent debinding of wax-based system, 137–8
  supercritical solvent debinding, 138–9
  water-soluble system, 141–2
production techniques
  titanium cost reduction, 432–3
    Armstrong process, 433
    FCC-Cambridge process, 433
    metal hydride reduction, 432–2
    plasma-quench process, 432
profile injection, 339
programmable logic controllers (PLC), 137
prosthetic knee joint, 403

Index    581

pseudo-binary phase diagram, 520
pycnometer density, 55-6

radii, 47-8
  design consideration for MIM
    processing, 47
  stress concentration factor as function
    to thickness ratio, 47
random scission, 273
rapid omni-directional compaction
  (ROC), 553
REACH regulation, 394-5, 401
refractory metals, 544-54
  activated sintering, 549-50
  feedstock formulation, 529-34
    BET particle size effect on optimal
      solids loading, 534
    feedstock examples, 535
    powder characteristics, 531
    powder preparation, 529-33
    powder SEMs, 532
    solids loading, 533-4
    solids loading effect on mixing
      torque, 533
  metal injection molding (MIM),
    526-60
    general properties, 528
  practical process concerns, 552-74
  solid-state sintering, 546-9
    Nb density and grain size vs
      sintering temperature, 552
    W particle size vs relative sintered
      density, 546-7
    W particle size vs relative sintered
      density and grain size, 548
    W solids loading vs relative sintered
      density, 547-8
  typical process cycles, 550-2
    MIM Mo mechanical properties,
      551
    MIM Mo microstructure, 551
    resintered and HIPed MIM Mo
      microstructure, 552
relative viscosity, 70-2
residual stress, 239
rhenium, 529
rheology, 186-9
  Mooney slip velocity data for samples
    C and D, 189
  sample A viscosity showing shear
    thinning behaviour, 188
  sample B viscosity showing a broad
    Newtonian region, 188

sample C viscosity, 189
sample C vs. D viscosity, 189
shear rate, temperature and particle
  attributes, 70-4
  particle shape effect on relative
    viscosity, 74
  relative viscosity variations with
    solid loading, 72
  variation of relative viscosity with
    % small particle content, 74
ribs, 45-6
  design rules in MIM tooling, 46
  good vs. poor practice in design
    thickness for MIM component, 46
sales, 10-12
  annual MIM in millions of US
    dollars, 11
  global MIM by main material
    category, 11
sandwich structure, 379
screw, 95, 112-13
secondary debinding, 144-7, 177
  incomplete binder removal, 146-7
  temperatures of common secondary
    binders, 145
  TGA curve for a solvent debound
    wax-based 4140 MIM feedstock,
    145
Sendust, 488
sensitisation process, 393-4
sequential injection molding
  multi core-in-sheath porous Ti-MIM
    parts formation, 383-7
  multi-layered Ti pipes surface
    structures, 385
  sandwich structures cross-sectional
    views, 387
  single- and multi-layered Ti pipes
    compressive stress vs strain
    curves, 386
  single-layered Ti pipes surface
    structures, 384
  sintered Ti porosity vs PMMA
    particle fraction, 385
settering, 167-9
  appendage added to the part in front
    to obtain a flat surface, 168
  high end spoon concept, 168
  special part concept, 169
shear rate, 72
short shot, 214
shrinkage

## Index

typical relationships with various parameters, 122
side-group elimination, 273
sinter joining, 327–8, 331
sintered materials technology, 2
sinterHIPs furnace, 556
sintering, 40–1, 147–61, 156, 177, 224–30, 250–1, 262–3, 422, 424
  applications, 228–30
    gravitational distorting, 228–30
    minimum grain size for a given final sinter density, 231
    optimisation, 230
  atmospheres, 157–8
  chemistry analysis, 262
  component density, 262
  component dimensions, 262
  component mass, 262
  crack detection, 263
  definitions, 147
  dimensional control and distortion, 251
    parts with long overhangs, steps and slots, 252
  furnaces, 169–76
  mass transport mechanism, 148–50
  material properties and simulation verification, 226–8
    dilatometry showing *in situ* shrinkage during constant heating, 227
    *in situ* bending test and shear viscosity property as function of time, 228
    liquid phase sintered tungsten and grain size model results, 227
    material properties for the sintering, 229
  mechanical testing, 263
  metal injection molding (MIM) components, 133–78
  microstructure, 263
  MIM materials, 161–6
    powder availability, 166
    reactivity effect, 161–6
  powder size and surface area effect, 157
  practices, 152, 154
  presence of liquid phase, 154–6
    90W–7Ni–3Fe alloy showing rounded grains of tungsten in a Ni–Fe–W alloy matrix, 155
    structure of an as-sintered M2 tool steel, 156
  results, 158–61
    sintered density in relation to time, 159
    structure of an as-sintered M2 tool steel, 159
  settering, 167–9
  shrinkage and porosity, 361–4
    porous specimens minimal pore size and surface area, 363
    sintered specimens minimal pore size distribution, 363
    vs space-holding particles fraction, 362
  sintering temperature, 364–7
    minimal pore size distributions, 365
    specimens surfaces SEM, 366
    vs shrinkage and porosity, 364
    vs specific surface, 366
    vs specific water flow, 367
  stages, 150–2
    cobalt–chrome alloy sintered at 1200°C and 1300°C, 153
    sintered iron–cobalt–vanadium Permandur pre-alloyed material, 151
  theoretical background and governing equations, 224–6
    constitutive relation, 225–6
    numerical simulation, 226
  theories, 147–8
  X-ray, 263
sintering stress, 225
sintering window, 520
small chain polymers, 68
smoothed particle hydrodynamics (SPH) method, 329
soft magnetic metals
  Fe–6.5Si, 489–97
    experimental procedure, 489
    magnetic properties, 493–7
    relative density, hardness, mean grain size, and chemical composition, 490–2
  Fe–50Ni, 506–13
    experimental procedure, 506
    magnetic properties, 507–9
    magnetic properties improvement, 509–13
  Fe–9.5Si–5.5Al, 497–504
    gas- vs water-atomized compacts magnetic properties, 502–6

gas-atomized powder compacts
  magnetic properties, 497–502
  metal injection molding (MIM), 487–514
solid thermal conductivity, 194
solid-state sintering, 546–9
solids loading, 518–19
solubility, 75–84
solvent debinding, 137–8, 244–5, 420–1
  illustration, 138
  swelling of molded specimens with increasing temperature, 244
  thickness effect on swelling, 245
  typical system marketed by Elnik Systems, 139
solvent extraction, 76–7
spark plasma sintering (SPS), 553
spherical powders, 54–5
stabilisers, 69
stainless steel, 284–8, 527
  Archimedes densities of PIM 17-4PH, 287
  binders, feedstocks, and debinding, 396–8
    catalytic debinding, 398
    commonly used stainless steel grades in MIM, 397
    solvent debinding, 398
    thermal debinding, 398
  emerging technologies, 402–3
    μMetal injection molding (μMIM), 402
    two-colour MIM, 403
  medical device made of 17-4PH, 288
  metal injection molding (MIM), 393–409
    performance, 401
    sintering aspects, 398–400
  microstructure appearance of 316L component, 286
  MIM applications, 403–9
    17-4 PH and 3R93 photo, 407
    angle plug photo, 406
    breakdown in world market, 404
    door lock casing photo, 407
    endo tips photo, 408
    main applications, 405
    quartz watch mechanical parts, 409
    swivel hinge photo, 404
    wound forceps photo, 406
  special grades and products, 401–2
    boron-based sintering improvements, 401

nickel-free stainless steels, 401–2
stainless steel metal matrix composites, 402
volume fraction of austenite and carbon content for sintered compacts, 286
static mixer *see* Kenics mixer
stearic acid, 421
steel, 393
strength model, 78–80
stress shielding effect, 436
supercritical solvent debinding, 138–9
supersolidus liquid phase sintering (SLPS), 155, 277
surface active agents, 274
surface diffusion, 150
surface finish, 35
switchover method, 120–1
switchover point, 120–1

tantalum, 529
tap density, 56
temperature, 73, 275–6
thermal analysis, 190–2
  cooling mode specific heat and transition plots, 190
  DSC cooling transitions *vs.* no-flow temperature, 192
  specific heat and transitions which have the same binder system, 192
thermal conductivity, 74–5
thermal debinding, 77–8, 215–24, 245–8, 421, 424, 455, 559
  applications, 219–24
  defects frequently found in debinding, 249
  fluorescence dye penetration test, 247
  microstructure changes during process, 79
  porosity structure for modelling the polymer burnout process, 78
  theoretical background and governing equations, 216–19
thermal decomposition, 59–60
  SEM image of iron carbonyl powder, 60
thermal degradation, 75–84
  microstructural changes of compact during debinding, 76
thermal management materials
  metal injection molding (MIM) in microelectronics, 446–82
  copper, 451–61

heat dissipation in microelectronics, 447–51
molybdenum–copper, 474–82
tungsten–copper, 461–74
thermal processing firms, 4
thermogravimetric analysis (TGA), 144, 146, 216, 246, 266, 269–71, 519
thermoplastic polymers, 66–8
thermoplastic-based feedstocks, 314
thermosetting polymers, 66, 68–9
threads, 44–5
  configuration with a flat, 45
  external type produced by metal injection molding, 44
three-channel system, 340
Ti–6Al–4V alloy, 418–19, 424
titanium, 383
  cost reduction, 432–4
    novel production techniques, 432–3
    powder blending, 433–4
  mechanical properties, 425–32
    fatigue properties, 430–2
    MIM processed Ti–6Al–4V microstructures tensile properties, 427
    MIM-processed specimens microstructures, 431
    plastic elongation dependence on oxygen equivalent, 428
    tensile properties, 426–30
    Ti alloys elongation range, 430
    Ti alloys ultimate tensile strength ranges, 429
    wrought vs MIM process Ti–6Al–4V microstructures, 426
    YS and UTS dependence on oxygen equivalent, 428
  MIM challenges, 416–22
    biocompatibility, 422
    facilities, 421–2
    interstitial elements, 416–18
    porosity, 422
    powder and feedstock, 418–21
  processing basics, 422–5
    debinding, 424
    feedstock production, 423
    further processing, 425
    injection molding, 423–4
    powder handling, 422–3
    sintering, 424–5
  pure and alloys metal injection molding, 415–41
    future trends, 440–1
  special applications, 435–40
    medical application, 435–8
    MIM processed Ti–45Al–5Nb–0.2B–0.2C tensile properties, 440
    MIM Ti–6Al–4V permanent implant, 435
    MIM-processed Ti–6Al–4V elastic modulus dependence on porosity, 437
    MIM-processed Ti–6Al–4V UTS dependence on porosity, 437
    Ti–45Al–5Nb–0.2B–0.2C tensile test curves, 439
    titanium aluminides MIM, 438–40
titanium alloys, 165–6, 294–7
  maximum debinding temperature correlation, 296
  structure of MIM Ti–6Al–4V sintered under argon, 165
titanium aluminides, 438–40
titanium powders, 295
tool cavity expansion factor, 99
tool evacuation, 328
tool making
  LIGA method, 311–13
    typical inserts, 312
  micro-manufacturing methods, 309–13
    optional methods, 310
  micro-tools manufacture, 309
  microstructured mold inserts manufacturing options, 309–11
tooling artifacts, 35–40
  ejector pin marks, 37–8
  gate locations, 38–40
  parting line, 35–7
tools
  design options, 98–104
    gating and venting, 99–101
    mold materials, 98
    oversize, 98–9
    undercut, 102–4
  injection molding machine design and function, 94–6
  metal injection molding (MIM), 93–107
    general design, 94
  set elements, 96–8
    basic elements for metal injection molding, 97
  software and economic factors, 106–7

special features and instrumentation, 104–6
  hot runner injection nozzle, 105
  single hot runner mold design, 106
transition metals, 549
tungsten alloys, 165
tungsten heavy alloys, 528, 536
tungsten–copper, 461–74
  debinding and sintering, 466–9
    sintered W–15Cu micrograph, 467
    sintering temperature effect on W–15Cu density and oxygen content, 467
    sintering temperature vs relative sintered density, 468
    W particle size vs relative sintered density, 468
    W–Cu specimen Cu bleed-out at edges, 469
  example applications, 473–4
  feedstock preparation, 465
    examples, 465
  infiltration, 469–70
  injection molding, 466
  powders, 462–4
    characteristics, 463
    co-reduced W–15Cu powders SEM, 464
    rod-milled W–15Cu powders SEM, 464
  thermal properties, 470–3
    Cu content effect on W–Cu conductivity predictions, 470
    Cu content predicted effects on W–Cu thermal expansion coefficient, 472
    porosity effect on W–15Cu thermal conductivity predictions, 472
    transition metal impurities effect on W–Cu conductivity predictions, 471
    W–Cu example, 473
twin-barrel injection molding unit, 339
two-channel system, 339
two-colour MIM, 403
two-colour powder injection molding (2C-PIM), 338–47
  2C-PIM products, 344–6
    prototypes diagrams, 345
  debinding and sintering, 341–4
    components with different densification characteristics, 344
    M2 tool steel shrinkage behaviour, 343
  future trends, 346–7
  injection molding technology, 338–41
    2C-PIM process schematic, 339
    over-molding and co-injection molding diagram, 341
    runner system and molding diagram, 340
two-component micro-injection, 331
two-domain Tait model, 124, 206
two-material powder injection molding *see* two-colour powder injection molding (2C-PIM)

undercuts, 48, 102–4
  angle lifters, 103
  collapsible core in molding and ejection position, 103
  gearbox for synchronised unscrewing, 104
  inner thread and threaded cores in mold with four cavities, 104
  part and mold design, 103
  shaping by laterally moving elements, 102

vacuum debinding, 455
vacuum sintering, 520
variothermal temperature control, 317, 328
venting, 99–101
verification, 211–13
  pressure–time plot at three points, 213
viscous flow, 149–50
volume diffusion, 150

wall thickness, 41
  good and poor practice, 42
  transition recommendations, 42
water atomization
  SEM image of stainless steel powder, 59
water-atomized powders, 396
water-soluble system, 135
  debinding, 141–2
    water debind unit, 142
wax-based binders, 314
wax-based system, 135
  solvent debinding, 137–8
WC–Co hardmetals, 555
webs, 45–6
weight –temperature–time plot, 223–4

decomposition of solvent debound 316L feedstock, 224
weld line, 214
wicking, 76
Wiedemann–Franz relationship, 451, 458

X-Cooler, 325
X-ray diffraction test, 513

Zamak, 408
zinc-reclaim process, 530